Lecture Notes in Computer Science 12638

More information about this subseries at http://www.springer.com/series/7407

Alberto Leporati · Carlos Martín-Vide ·
Dana Shapira · Claudio Zandron (Eds.)

Language
and Automata Theory
and Applications

15th International Conference, LATA 2021
Milan, Italy, March 1–5, 2021
Proceedings

 Springer

Editors
Alberto Leporati (ID)
University of Milano-Bicocca
Milan, Italy

Carlos Martín-Vide (ID)
Rovira i Virgili University
Tarragona, Spain

Dana Shapira (ID)
Ariel University
Ariel, Israel

Claudio Zandron (ID)
University of Milano-Bicocca
Milan, Italy

ISSN 0302-9743   ISSN 1611-3349 (electronic)
Lecture Notes in Computer Science
ISBN 978-3-030-68194-4   ISBN 978-3-030-68195-1 (eBook)
https://doi.org/10.1007/978-3-030-68195-1

LNCS Sublibrary: SL1 – Theoretical Computer Science and General Issues

This Springer imprint is published by the registered company Springer Nature Switzerland AG
The registered company address is: Gewerbestrasse 11, 6330 Cham, Switzerland

Preface

These proceedings contain the papers that were presented at the 15th International Conference on Language and Automata Theory and Applications (LATA 2021), held in Milan, Italy, during March 1–5, 2021.

Due to the Covid-19 pandemic, LATA 2020 and LATA 2021 were merged and held on these dates together. LATA 2020 proceedings were published as LNCS 12038.

The scope of LATA is rather broad, including: algebraic language theory; algorithms for semi-structured data mining; algorithms on automata and words; automata and logic; automata for system analysis and programme verification; automata networks; automatic structures; codes; combinatorics on words; computational complexity; concurrency and Petri nets; data and image compression; descriptional complexity; foundations of finite-state technology; foundations of XML; grammars (Chomsky hierarchy, contextual, unification, categorial, etc.); grammatical inference, inductive inference and algorithmic learning; graphs and graph transformation; language varieties and semigroups; language-based cryptography; mathematical and logical foundations of programming methodologies; parallel and regulated rewriting; parsing; patterns; power series; string-processing algorithms; symbolic dynamics; term rewriting; transducers; trees, tree languages and tree automata; weighted automata.

LATA 2021 received 52 submissions. Every paper was reviewed by three Programme Committee members. Some external experts were also consulted. After a thorough and vivid discussion phase, the committee decided to accept 26 papers (which represents an acceptance rate of exactly 50%). The conference program included 6 invited talks as well.

The excellent facilities provided by the EasyChair conference management system allowed us to deal with the submissions properly and handle the preparation of these proceedings in time.

We would like to thank all invited speakers and authors for their contributions, the Program Committee and the external reviewers for their cooperation, and Springer for its very professional publishing work.

December 2020

Alberto Leporati
Carlos Martín-Vide
Dana Shapira
Claudio Zandron

Organization

Program Committee

Jorge Almeida	University of Porto, Portugal
Franz Baader	Technical University of Dresden, Germany
Alessandro Barenghi	Polytechnic University of Milan, Italy
Marie-Pierre Béal	University of Paris-Est, France
Djamal Belazzougui	CERIST, Algeria
Marcello Bonsangue	Leiden University, The Netherlands
Flavio Corradini	University of Camerino, Italy
Bruno Courcelle	University of Bordeaux, France
Laurent Doyen	ENS Paris-Saclay, France
Manfred Droste	Leipzig University, Germany
Rudolf Freund	Technical University of Vienna, Austria
Paweł Gawrychowski	University of Wrocław, Poland
Amélie Gheerbrant	University of Paris, France
Tero Harju	University of Turku, Finland
Lane A. Hemaspaandra	University of Rochester, USA
Lukáš Holik	Brno University of Technology, Czech Republic
Jarkko Kari	University of Turku, Finland
Dexter Kozen	Cornell University, USA
Markus Lohrey	University of Siegen, Germany
Sebastian Maneth	University of Bremen, Germany
Nicolas Markey	IRISA, Rennes, France
Carlos Martín-Vide (Chair)	Rovira i Virgili University, Spain
Giancarlo Mauri	University of Milano-Bicocca, Italy
Victor Mitrana	University of Bucharest, Romania
Paliath Narendran	University at Albany, USA
Gennaro Parlato	University of Molise, Italy
Madhusudan Parthasarathy	University of Illinois at Urbana-Champaign, USA
Dominique Perrin	University of Paris-Est, France
Nir Piterman	Chalmers University of Technology, Sweden
Sanguthevar Rajasekaran	University of Connecticut, USA
Antonio Restivo	University of Palermo, Italy
Wojciech Rytter	University of Warsaw, Poland
Kai Salomaa	Queen's University, Canada
Helmut Seidl	Technical University of Munich, Germany
William F. Smyth	McMaster University, Canada
Jiří Srba	Aalborg University, Denmark
Edward Stabler	University of California, Los Angeles, USA
Benjamin Steinberg	City University of New York, USA

Frank Stephan	National University of Singapore, Singapore
Jan van Leeuwen	Utrecht University, The Netherlands
Margus Veanes	Microsoft Research, USA
Mikhail Volkov	Ural Federal University, Russia

Additional Reviewers

Bønneland, Frederik M.	Maslennikova, Marina
Cacciagrano, Diletta Romana	Mhaskar, Neerja
Cameron, Peter	Modanese, Augusto
Carayol, Arnaud	Moerman, Joshua
Castiglione, Giuseppa	Okhotin, Alexander
Corradini, Andrea	Pighizzini, Giovanni
Dolce, Francesco	Popescu, Marius
Domaratzki, Mike	Prezza, Nicola
Dudek, Bartlomiej	Pulver, Andrew
Eğecioğlu, Ömer	Rao, Michael
Erbatur, Serdar	Rinaldi, Simone
Erofeev, Evgeny	Ryzhikov, Andrew
Fernau, Henning	Sangnier, Arnaud
Gauwin, Olivier	Schou, Morten Konggaard
Giammarresi, Dora	Sciortino, Marinella
Havlena, Vojtěch	Suchy, Ashley
Inenaga, Shunsuke	Sznajder, Nathalie
Itsykson, Dmitry	Síč, Juraj
Jecker, Ismaël	Teh, Wen Chean
Jurvanen, Eija	Tesei, Luca
Klíma, Ondřej	Tesson, Pascal
Lorber, Florian	Tiezzi, Francesco
Loreti, Michele	Trivedi, Ashutosh
Maletti, Andreas	van Heerdt, Gerco
Marzouk, Reda	Zhang, Yu

Contents

Algebraic Structures

On Language Varieties Without Boolean Operations

Fabian Birkmann, Stefan Milius⬤, and Henning Urbat$^{(\boxtimes)}$

Friedrich-Alexander-Universität Erlangen-Nürnberg, Erlangen, Germany
{fabian.birkmann,stefan.milius,henning.urbat}@fau.de

Abstract. Eilenberg's variety theorem marked a milestone in the algebraic theory of regular languages by establishing a formal correspondence between properties of regular languages and properties of finite monoids recognizing them. Motivated by classes of languages accepted by quantum finite automata, we introduce *basic varieties of regular languages*, a weakening of Eilenberg's original concept that does not require closure under any boolean operations, and prove a variety theorem for them. To do so, we investigate the algebraic recognition of languages by *lattice bimodules*, generalizing Klíma and Polák's lattice algebras, and we utilize the duality between algebraic completely distributive lattices and posets.

1 Introduction

The introduction of algebraic methods into the study of regular languages provides a convenient classification system that allows to study finite automata and their languages in terms of associated finite algebraic structures. A celebrated example is Schützenberger's theorem [20] stating that a language is star-free iff its syntactic monoid is aperiodic, thus proving the decidability of star-freeness. Eilenberg's *variety theorem* [9] formalizes this type of correspondence as a bijection between *varieties of regular languages* (i.e. classes of regular languages closed under the set-theoretic boolean operations, word derivatives and preimages of monoid homomorphisms) and *pseudovarieties of monoids* (i.e. classes of finite monoids closed under finite products, submonoids and quotient monoids).

Numerous extensions and generalizations of Eilenberg's theorem have been discovered over the past four decades, differing from the original one by either changing the type of languages under consideration, e.g. from regular languages to ω-regular languages [22], or by considering notions of varieties with relaxed closure properties. On the algebraic side, such a relaxation requires to replace monoids by more complex algebraic structures. For instance, Pin [16] studied *positive varieties of regular languages*, where the closure under complement is

---

* Supported by Deutsche Forschungsgemeinschaft under projects MI 717/5-2 and MI 717/7-1

** Supported by Deutsche Forschungsgemeinschaft under project SCHR 1118/8-2.

A. Leporati et al. (Eds.): LATA 2021, LNCS 12638, pp. 3–15, 2021.
https://doi.org/10.1007/978-3-030-68195-1_1

dropped, and proved them to biject with pseudovarieties of *ordered* monoids. Subsequently, Polák [17] introduced *disjunctive varieties of regular languages*, where in addition to closure under complement also the closure under intersection is dropped, and related them to pseudovarieties of idempotent semirings.

One item is conspicuously missing from this list: a variety theorem for classes of languages that need not be closed under any boolean operations, i.e. in which only closure under word derivatives and preimages of monoid homomorphisms is required. Such *basic varieties of regular languages* subsume all the above notions of varieties and naturally arise in several areas of automata theory, most notably in the study of languages accepted by reversible finite automata [11] or quantum finite automata [13]. In the present paper, we close this gap by developing the theory of basic varieties. As the corresponding algebraic structure we introduce *lattice bimodules*, a two-sorted generalization of the *lattice algebras* recently studied by Klíma and Polák [12], as algebraic recognizers for regular languages. The two-sorted approach allows for a clearer and more conceptual view of the underlying categorical and universal algebraic concepts. As our main result, we establish the following algebraic classification of basic varieties:

Basic Variety Theorem. *Basic varieties of regular languages correspond bijectively to pseudovarieties of lattice bimodules.*

This answers the open problem of Klíma and Polák [12] about an Eilenberg-type correspondence. Our presentation of the theorem and its proof is inspired by the recently developed duality-theoretic perspective on algebraic language theory [1,10,19,21], which provides the insight that correspondences between language varieties and pseudovarieties of algebraic structures can be understood in terms of an underlying dual equivalence of categories. In our setting, we shall demonstrate that pseudovarieties of lattice bimodules can be interpreted as *theories* of lattice bimodules in the category **AlgCDL** of algebraic completely distributive lattices, while basic varieties give rise to (basic) *cotheories* of regular languages in the category **Pos** of posets. Our Eilenberg correspondence for basic varieties then boils down to an application of the duality **AlgCDL** $\simeq^{\mathrm{op}}$ **Pos**.

Let us note that our main result is not an instance of previous category-theoretic generalizations of Eilenberg's theorem [1,5,19,21] since the two-sorted nature of lattice bimodules requires to introduce the novel concept of *reduced* structures, which makes the ensuing notion of pseudovariety more intricate than the ones studied in *op. cit.* However, much of the general methodology developed there turns out to apply smoothly, which can be seen as further evidence of its scope and flexibility. In order to make the present paper accessible to readers not familiar with the previous work, we opted to give a self-contained presentation of our results, merely assuming some familiarity with basic category theory.

2   Lattice Bimodules

In this section we introduce a new algebraic structure whose aim it is to capture languages varieties that are not necessarily closed under boolean operations.

Our notion is a two-sorted generalization of Klíma and Polák's *lattice alge-bras* [12]. Intuitively, for our intended purpose the following structure should be present:

(1) a *monoid action* that corresponds to word derivation on the language side;
(2) *lattice-like operations* to compensate for the missing closure under union and intersection on the language side;
(3) *equational axioms* specifying the interaction of (1) and (2).

From a categorical perspective, the last point means that our algebras can be modeled by a *monad*. This allows us to use previous work on languages recog-nizable by monad algebras [5,19,21] as a guide towards our results.

While Klíma and Polák considered distributive lattices with an embedded monoid acting on them, we upgrade the lattice to a *completely distributive lattice* (shortly, *CDL*), i.e. a complete lattice satisfying the infinite distributive law $\bigvee_{i \in I} \bigwedge_{j \in J_i} x_{i,j} = \bigwedge_{f \in F} \bigvee_{i \in I} x_{i,f(i)}$ for every family $\{x_{i,j} : i \in I, j \in J_i\}$ of elements, where F is the set of all choice functions f mapping each $i \in I$ to some $f(i) \in J_i$. *Morphisms of CDLs* are maps preserving all joins and meets. We let **CDL** denote the category of CDLs and their morphisms. Even though completeness makes no difference for finite structures, completely distributive lattices admit a more convenient duality theory than general distributive lattices.

In addition, in lieu of an embedded monoid we use a two-sorted structure with a monoid in the first sort. This avoids partial operations, which are somewhat awkward from the perspective of (categorical) universal algebra.

Definition 2.1. (1) A *lattice bimodule* $(M, D, \iota, \triangleright, \triangleleft)$, abbreviated as (M, D), is given by a monoid $(M, \cdot, 1)$, a CDL $(D, \vee, \wedge)$, and three operations

$$\triangleright: M \times D \to D, \qquad \triangleleft: D \times M \to D, \qquad \iota: M \to D,$$

such that $\triangleright$ and $\triangleleft$ form a monoid biaction of M on D that distributes over the lattice operations, and ι translates the multiplication of M to $\triangleleft$ and $\triangleright$; that is, for all $m, n \in M$, $d \in D$ and $\{d_i\}_{i \in I} \subseteq D$, the following equational laws hold:

$$(m \cdot n) \triangleright d = m \triangleright (n \triangleright d), \qquad d \triangleleft (m \cdot n) = (d \triangleleft m) \triangleleft n,$$
$$1 \triangleright d = d, \qquad d \triangleleft 1 = d,$$
$$(m \triangleright d) \triangleleft n = m \triangleright (d \triangleleft n),$$
$$m \triangleright (\bigvee_{i \in I} d_i) = \bigvee_{i \in I} (m \triangleright d_i), \qquad (\bigvee_{i \in I} d_i) \triangleleft m = \bigvee_{i \in I} (d_i \triangleleft m),$$
$$m \triangleright (\bigwedge_{i \in I} d_i) = \bigwedge_{i \in I} (m \triangleright d_i), \qquad (\bigwedge_{i \in I} d_i) \triangleleft m = \bigwedge_{i \in I} (d_i \triangleleft m),$$
$$m \triangleright \iota(n) = \iota(m \cdot n), \qquad \iota(m) \triangleleft n = \iota(m \cdot n).$$

Note that since the least and the greatest element of D are given by $\bot = \bigvee \emptyset$ and $\top = \bigwedge \emptyset$, resp., we also have $m \triangleright \bot = \bot = \bot \triangleleft m$ and $m \triangleright \top = \top = \top \triangleleft m$.

(2) A *homomorphism* from a lattice bimodule $(M, D, \iota, \triangleright, \triangleleft)$ to a lattice bimod-ule $(M', D', \iota', \triangleright', \triangleleft')$ is given by a two-sorted map $h = (h^\star, h^\diamond): (M, D) \to (M', D')$ such that $h^\star$ is a monoid homomorphism, $h^\diamond$ is a morphism of completely distributive lattices and the following diagrams commute:

$$
\begin{array}{ccc}
M \times D \xrightarrow{\;\rhd\;} D & \qquad D \times M \xrightarrow{\;\lhd\;} D & \qquad M \xrightarrow{\;\iota\;} D \\
{\scriptstyle h^{\star} \times h^{\circ}} \downarrow \qquad\quad \downarrow {\scriptstyle h^{\circ}} & {\scriptstyle h^{\circ} \times h^{\star}} \downarrow \qquad\quad \downarrow {\scriptstyle h^{\circ}} & {\scriptstyle h^{\star}} \downarrow \qquad \downarrow {\scriptstyle h^{\circ}} \\
M' \times D' \xrightarrow{\;\rhd'\;} D' & \qquad D' \times M' \xrightarrow{\;\lhd'\;} D' & \qquad M' \xrightarrow{\;\iota'\;} D'
\end{array}
$$

Subbimodules and *quotient bimodules* of lattice bimodules are represented by sortwise injective and surjective homomorphisms, respectively.

We let **LBM** denote the category of lattice bimodules and their homomorphisms. A *free* lattice bimodule over a pair (Σ, Γ) of sets is given by a lattice bimodule $(\hat{\Sigma}, \hat{\Gamma})$ together with a sorted map $\eta = (\eta^{\star}, \eta^{\circ}) : (\Sigma, \Gamma) \to (\hat{\Sigma}, \hat{\Gamma})$ satisfying the universal mapping property: for every sorted map $h_0 : (\Sigma, \Gamma) \to (M, D)$ to a lattice bimodule (M, D) there exists a unique lattice bimodule homomorphism $h : (\hat{\Sigma}, \hat{\Gamma}) \to (M, D)$ such that $h \cdot \eta = h_0$. In the following, we denote by $\Sigma^{\star}$ the free monoid on the set Σ with neutral element $\varepsilon \in \Sigma^{\star}$ and by $\mathsf{FCDL}(\Gamma)$ the free completely distributive lattice [14] on the set Γ. The latter can be described as the lattice of downwards closed subsets of the power set $\mathcal{P}(\Gamma)$, or equivalently as the lattice of all formal expressions $\bigvee_{i \in I} \bigwedge_{j \in J_i} x_{i,j}$, where $x_{i,j} \in \Gamma$, modulo the equational laws of CDLs. We view Γ as a subset of $\mathsf{FCDL}(\Gamma)$.

Proposition 2.2. *The free lattice bimodule over (Σ, Γ) is given by $\eta : (\Sigma, \Gamma) \to (\Sigma^{\star}, \mathsf{FCDL}(\Sigma^{\star} + \Sigma^{\star} \times \Gamma \times \Sigma^{\star}))$ with $\eta^{\star}(a) = a$, $\eta^{\circ}(b) = (\varepsilon, b, \varepsilon)$, and operations uniquely determined by the following identities for $u, v, w \in \Sigma^{\star}$ and $z \in \Gamma$:*

$$\iota(u) = u, \quad u \rhd v = uv, \quad u \rhd (v, z, w) = (uv, z, w), \quad u \lhd v = uv, \quad (v, z, w) \lhd u = (v, z, wu).$$

Notation 2.3. We write $(\Sigma^{\star}, \Sigma^{\circ}) = (\Sigma^{\star}, \mathsf{FCDL}(\Sigma^{\star}))$ for the free lattice bimodule on $(\Sigma, \emptyset)$. Note that a homomorphism $h : (\Sigma^{\star}, \Sigma^{\circ}) \to (M, D)$ is completely determined by its first component $h^{\star} : \Sigma^{\star} \to M$. In fact, its second component $h^{\circ} : \Sigma^{\circ} \to D$ is the unique **CDL**-morphism extending the map $\iota \cdot h^{\star} : \Sigma^{\star} \to D$.

We now define three properties of lattice bimodules needed subsequently.

Definition 2.4. A lattice bimodule (M, D) is called

(1) *$\star$-generated* if the complete lattice D is generated by the image $\iota[M] \subseteq D$: For all $d \in D$ there exist elements $m_{i,j} \in M$ such that $d = \bigvee_{i \in I} \bigwedge_{j \in J_i} \iota(m_{i,j})$;

(2) *$\star$-embedded* if the operation $\iota : M \to D$ is injective;

(3) *reduced* if for every quotient bimodule $h : (M, D) \twoheadrightarrow (M', D')$ such that $h^{\circ} : D \twoheadrightarrow D'$ is a **CDL**-isomorphism, h is an **LBM**-isomorphism.

We note that finite $\star$-embedded lattice bimodules are precisely the finite lattice algebras of Klíma and Polák [12]. The following lemma links the above concepts:

Lemma 2.5. *(1) A lattice bimodule (M, D) is $\star$-generated if and only if there exists a surjective homomorphism from $(\Sigma^\star, \Sigma^\diamond)$ to (M, D) for some set Σ.*
(2) Every $\star$-embedded lattice bimodule is reduced.
(3) Every $\star$-generated reduced lattice bimodule is $\star$-embedded.

In Section 3 we will study lattice bimodules that are $\star$-generated and reduced (equivalently, $\star$-generated and $\star$-embedded). Intuitively, these properties capture lattice bimodules whose monoid component generates the lattice component and is "minimal" with that property. This allows us to relate the two-sorted notion of lattice bimodules to the single-sorted notion of languages recognized by them.

In the categorical approach to variety theorems [21] it was shown that the key to understanding language derivatives lies in the concept of a *unary presentation* of an algebraic structure. Informally, such a presentation expresses the structure of an algebra in terms of suitable unary operations in the underlying category, which then dualize to the derivative operations on the set of languages recognized by that algebra. The heterogeneous nature of our present setting, which regards lattice bimodules as algebraic structures over the product category **Set** $\times$ **CDL**, requires a slight adaptation of the concepts from *op. cit.*

Definition 2.6. Let (M, D) be a lattice bimodule. A *unary operation* on (M, D) is either a map of type $M \to M$ or $M \to D$, or a **CDL**-morphism $D \to D$. A set $\mathbb{U}$ of unary operations forms a *unary presentation* of (M, D) if for every pair $e = (e^\star, e^\diamond)$ of a surjective map $e^\star \colon M \twoheadrightarrow M'$ and a surjective **CDL**-morphism $e^\diamond \colon D \twoheadrightarrow D'$, the following statements are equivalent:

(1) There exists a lattice bimodule structure on (M', D') making e a homomorphism of lattice bimodules.
(2) For every $u \in \mathbb{U}$, there exists $\bar{u}$ making the respective square below commute:

$$
\begin{array}{ccc}
M \xrightarrow{u} M \\
e^\star \downarrow \qquad \downarrow e^\star \\
M' \dashrightarrow{\bar{u}} M'
\end{array}
\qquad
\begin{array}{ccc}
M \xrightarrow{u} D \\
e^\star \downarrow \qquad \downarrow e^\diamond \\
M' \dashrightarrow{\bar{u}} D'
\end{array}
\qquad
\begin{array}{ccc}
D \xrightarrow{u} D \\
e^\diamond \downarrow \qquad \downarrow e^\diamond \\
D' \dashrightarrow{\bar{u}} D'
\end{array}
$$

Lemma 2.7. *Every lattice bimodule (M, D) admits a unary presenation composed of the following unary operations ranging over $m \in M$ and $d \in D$:*

$$(m\cdot), (\cdot m) \colon M \to M, \quad \iota \colon M \to D, \quad (m \triangleright), (\triangleleft m) \colon D \to D, \quad (\triangleright d), (d \triangleleft) \colon M \to D.$$

Note that the maps $(m \triangleright)$ and $(\triangleleft m)$ are indeed **CDL**-morphisms, as required.

3 Pseudovarieties of Reduced Lattice Bimodules

In this section, we introduce *pseudovarieties* and *theories* of (reduced) lattice bimodules and show them to be in one-to-one correspondence. The concept of a pseudovariety originates in Eilenberg's classical variety theorem [9] where a *pseudovariety of monoids* is a class of finite monoids closed under finite products, submonoids, and quotient monoids. In our setting of lattice bimodules, we shall consider pseudovarieties of $\star$-generated reduced lattice bimodules. Their definition is slightly more involved than in the case of monoids because subbimodules of $\star$-generated lattice bimodules are not necessarily $\star$-generated and quotient bimodules of reduced lattice bimodules are not necessarily reduced.

Definition 3.1. A *pseudovariety of lattice bimodules* is a class $\mathcal{V}$ of $\star$-generated reduced finite lattice bimodules such that

(1) $\mathcal{V}$ is closed under reduced quotients: for every surjective homomorphism $e \colon (M, D) \twoheadrightarrow (M', D')$ of lattice bimodules, if $(M, D) \in \mathcal{V}$ and (M', D') is reduced then $(M', D') \in \mathcal{V}$.
(2) $\mathcal{V}$ is closed under $\star$-generated subbimodules of finite products: for every injective homomorphism $(M, D) \rightarrowtail \prod_{i=1}^{n}(M_i, D_i)$ of lattice bimodules, if $(M_i, D_i) \in \mathcal{V}$ for $i = 1, \ldots, n$ and (M, D) is $\star$-generated then $(M, D) \in \mathcal{V}$.

We shall also consider the notion of a *local* pseudovariety. It is local in the sense that it involves only quotient bimodules of a fixed free lattice bimodule $(\Sigma^\star, \Sigma^\diamond)$. The set of all such quotients carries a partial order: given $e_i \colon (\Sigma^\star, \Sigma^\diamond) \twoheadrightarrow (M_i, D_i)$, $i = 0, 1$, put $e_0 \leq e_1$ iff $e_0 = h \cdot e_1$ for some **LBM**-morphism h.

Definition 3.2. A *local pseudovariety of lattice bimodules* over the finite set Σ is a set $\mathcal{T}_\Sigma$ of quotient bimodules of $(\Sigma^\star, \Sigma^\diamond)$ such that

(1) The codomain of every $e \in \mathcal{T}_\Sigma$ is finite and reduced. (Note that it is also $\star$-generated by Lemma 2.5(1).)
(2) $\mathcal{T}_\Sigma$ is downwards closed: if $e \in \mathcal{T}_\Sigma$ and $e' \colon (\Sigma^\star, \Sigma^\diamond) \twoheadrightarrow (M, D)$ is a quotient bimodule with reduced codomain, then $e' \leq e$ implies $e' \in \mathcal{T}_\Sigma$.
(3) $\mathcal{T}_\Sigma$ is directed: if $e_0, e_1 \in \mathcal{T}_\Sigma$, then there exists $e \in \mathcal{T}_\Sigma$ with $e_0, e_1 \leq e$.

In order-theoretic terminology, a local pseudovariety is thus precisely an ideal in the poset of finite reduced quotient bimodules of $(\Sigma^\star, \Sigma^\diamond)$.

Definition 3.3. A *theory of lattice bimodules* is a family $\mathcal{T} = (\mathcal{T}_\Sigma)_{\Sigma \in \mathbf{Set_f}}$ of local pseudovarieties, with Σ ranging over the class $\mathbf{Set_f}$ of finite sets, such that for each homomorphism $h \colon (\Delta^\star, \Delta^\diamond) \to (\Sigma^\star, \Sigma^\diamond)$ and $e_\Sigma \in \mathcal{T}_\Sigma$ their composite $e_\Sigma \cdot h$ *lifts through* $\mathcal{T}_\Delta$, that is, there exist $e_\Delta \in \mathcal{T}_\Delta$ and $\bar{h}$ such that $e_\Sigma \cdot h = \bar{h} \cdot e_\Delta$.

$$
\begin{array}{ccc}
(\Delta^\star, \Delta^\diamond) & \xrightarrow{\ h\ } & (\Sigma^\star, \Sigma^\diamond) \\
{\scriptstyle e_\Delta}\big\downarrow & & \big\downarrow{\scriptstyle e_\Sigma} \\
(M', D') & \dashrightarrow[\bar{h}] & (M, D)
\end{array}
\tag{3.1}
$$

This notion resembles the concept of an equational theory from universal algebra: equations can be identified with quotients of free algebras, and the closure of a theory under substitution is expressed by a commutative diagram like (3.1).

Notation 3.4. (1) Given a theory $\mathcal{T}$, let $\mathcal{V}^{\mathcal{T}}$ be the class of all lattice bimodules (M, D) such that some $\mathcal{T}_\Sigma$ contains a quotient with codomain (M, D).

(2) Given a pseudovariety $\mathcal{V}$, form the family $\mathcal{T}^{\mathcal{V}} = (\mathcal{T}_\Sigma^{\mathcal{V}})_{\Sigma \in \mathbf{Set_f}}$ where $\mathcal{T}_\Sigma^{\mathcal{V}}$ consists of all quotient bimodules of $(\Sigma^\star, \Sigma^\diamond)$ with codomain in $\mathcal{V}$.

The class of all pseudovarieties of lattice bimodules forms a lattice ordered by inclusion. Similarly, the class of all theories of lattice bimodules forms a lattice ordered by pointwise inclusion: $\mathcal{T} \leq \mathcal{T}'$ iff $\mathcal{T}_\Sigma \subseteq \mathcal{T}_\Sigma'$ for each Σ.

Theorem 3.5. *The maps $\mathcal{V} \mapsto \mathcal{T}^{\mathcal{V}}$ and $\mathcal{T} \mapsto \mathcal{V}^{\mathcal{T}}$ give rise to an isomorphism between the lattice of pseudovarieties of lattice bimodules and the lattice of theories of lattice bimodules.*

We conclude this section with another characterization of theories, linking them to the concept of a unary presentation. For any set Σ, let $\mathbb{U}_\Sigma$ be the canonical unary presentation of the free lattice bimodule $(\Sigma^\star, \Sigma^\diamond)$ given by Lemma 2.7, and denote by $\overline{\mathbb{U}}_\Sigma$ its closure under composition. Then $\overline{\mathbb{U}}_\Sigma$ also forms a unary presentation of $(\Sigma^\star, \Sigma^\diamond)$. We write $\overline{\mathbb{U}}_\Sigma(S, T) \subseteq \overline{\mathbb{U}}_\Sigma$ for the set of unary operations in $\overline{\mathbb{U}}_\Sigma$ with domain S and codomain T, where $S, T \in \{\Sigma^\star, \Sigma^\diamond\}$. In particular, $\overline{\mathbb{U}}_\Sigma(\Sigma^\diamond, \Sigma^\diamond) = \{x \mapsto vxw \mid v, w \in \Sigma^\star\}$.

Definition 3.6. (1) A quotient $e \colon \Sigma^\diamond \twoheadrightarrow D$ in **CDL** is called a $\mathbb{U}$ *-quotient* if for every unary operation $u \in \overline{\mathbb{U}}_\Sigma(\Sigma^\diamond, \Sigma^\diamond)$ there exists a **CDL**-morphism $\bar{u} \colon D \to D$ such that $e \cdot u = \bar{u} \cdot e$. We call such a $\bar{u}$ a *lifting of u along e*.

(2) A *local pseudovariety of $\mathbb{U}$-quotients* over the finite set Σ is an ideal in the poset of finite $\mathbb{U}$-quotients of $\Sigma^\diamond$.

(3) A *theory* of $\mathbb{U}$-quotients is a family $\mathcal{T} = (\mathcal{T}_\Sigma)_{\Sigma \in \mathbf{Set_f}}$ of local pseudovarieties of $\mathbb{U}$-quotients such that for each lattice bimodule homomorphism $h \colon (\Delta^\star, \Delta^\diamond) \to (\Sigma^\star, \Sigma^\diamond)$ and $e_\Sigma \in \mathcal{T}_\Sigma$ their composite $e_\Sigma \cdot h^\diamond$ *lifts through $\mathcal{T}_\Delta$:* there exist morphisms $e_\Delta \in \mathcal{T}_\Delta$ and $\bar{h}$ such that $e_\Sigma \cdot h^\diamond = \bar{h} \cdot e_\Delta$.

$$\begin{array}{ccc} \Sigma^\diamond \xrightarrow{\ u\ } \Sigma^\diamond & \qquad & \Delta^\diamond \xrightarrow{\ h^\diamond\ } \Sigma^\diamond \\ {\scriptstyle e}\downarrow \qquad \downarrow{\scriptstyle e} & \qquad & {\scriptstyle e_\Delta}\downarrow \qquad \downarrow{\scriptstyle e_\Sigma} \\ D \dashrightarrow_{\bar{u}} D & \qquad & D' \dashrightarrow_{\bar{h}} D \end{array} \qquad (3.2)$$

Proposition 3.7. *The lattice of theories of lattice bimodules is isomorphic to the lattice of theories of $\mathbb{U}$-quotients. The isomorphism is given by $\mathcal{T} \mapsto \mathcal{T}^\diamond$, where $\mathcal{T}^\diamond$ consists of all quotients in $\mathcal{T}$ restricted to their $\diamond$-component.*

The advantage in using theories of $\mathbb{U}$-quotients is that they are easier to dualize but still carry as much information as theories of lattice bimodules.

4 Basic Varieties of Regular Languages

In this section, we study lattice bimodules as recognizers for regular languages. Their purpose is to capture classes of regular languages with no boolean closure at all, which we thus call *basic varieties*. Observe that since the set $2 = \{0, 1\}$ with $0 \leq 1$ forms a CDL and the set $\Sigma^\star$ generates the free completely distributive lattice Σ°, we get the correspondence $\mathcal{P}(\Sigma^\star) \cong \mathbf{Set}(\Sigma^\star, 2) \cong \mathbf{CDL}(\Sigma^\circ, 2)$. We use the term "language" for elements of any of these sets, identifying elements that correspond to each other via the bijections. Thus, we use the same symbol for a subset $L \subseteq \Sigma^\star$ and for its characteristic function. We denote the extension of $L \colon \Sigma^\star \to 2$ to a lattice morphism by $L^\circ \colon \Sigma^\circ \to 2$, and in turn denote the restriction of a lattice morphism $L \colon \Sigma^\circ \to 2$ to $\Sigma^\star$ by $L^\star = L \cdot \iota \colon \Sigma^\star \to \Sigma^\circ \to 2$.

Definition 4.1. A language $L \colon \Sigma^\star \to 2$ is *recognized* by a finite lattice bimodule (M, D) if there exists a lattice bimodule homomorphism $h \colon (\Sigma^\star, \Sigma^\circ) \to (M, D)$ and a **CDL**-morphism $p \colon D \to 2$ with $L^\circ = p \cdot h^\circ$.

Lemma 4.2. *The languages recognizable by finite lattice bimodules are precisely the regular languages.*

We now introduce our concept of a language variety that we will show to correspond to pseudovarieties of lattice bimodules. It subsumes Eilenberg's original concept [9], as well as its variants due to Pin [16] and Polák [17], by dropping the requirement of being closed under any set-theoretic boolean operations. Recall that the *derivatives* of a language $L \subseteq \Sigma^\star$ are the languages $v^{-1}Lw^{-1} = \{u \in \Sigma^\star \mid vuw \in L\}$ for $v, w \in \Sigma^\star$. The *preimage* of L w.r.t. a monoid homomorphism $g \colon \Delta^\star \to \Sigma^\star$ is given by $g^{-1}L = \{w \in \Delta^\star \mid g(w) \in L\}$. In the following we write $\mathbf{Reg}_\Sigma$ for the set of all regular languages over Σ.

Definition 4.3. (1) A *basic local variety of languages over* Σ is a set $V_\Sigma \subseteq \mathbf{Reg}_\Sigma$ closed under derivatives: If $L \in V_\Sigma$ then $v^{-1}Lw^{-1} \in V_\Sigma$ for all $v, w \in \Sigma^\star$.

(2) A *basic variety of languages* is a family $(V_\Sigma \subseteq \mathbf{Reg}_\Sigma)_{\Sigma \in \mathbf{Set}_f}$ of local varieties closed under preimages of monoid homomorphisms: If $L \in V_\Sigma$ then $g^{-1}L \in V_\Delta$ for each monoid homomorphism $g \colon \Delta^\star \to \Sigma^\star$.

Just as pseudovarieties of reduced lattice bimodules can be presented as theories, basic varieties of languages correspond uniquely to *cotheories*. As suggested by the terminology, theories and cotheories form dual concepts, see Sect. 5. In the following, $\mathcal{P}(X)$ denotes the poset of subsets of a set X. Recall that an ideal of $\mathcal{P}(X)$ is a subset $I \subseteq \mathcal{P}(X)$ that is downwards closed and upwards directed.

Definition 4.4. A *basic cotheory of regular languages* is a family

$$T = (I_\Sigma \subseteq \mathcal{P}(\mathbf{Reg}_\Sigma))_{\Sigma \in \mathbf{Set}_f}$$

of ideals with the following properties:

(1) Every element $F_\Sigma \in I_\Sigma$ is a finite basic local variety.
(2) T is closed under preimages of monoid homomorphisms: If $F_\Sigma \in I_\Sigma$, then $g^{-1}[F_\Sigma] = \{g^{-1}L \mid L \in F_\Sigma\} \in I_\Delta$ for each monoid homomorphism $g \colon \Delta^\star \to \Sigma^\star$.

In diagrammatic terms, (1) means that for every $u \in \mathbb{U}_\Sigma(\Sigma^\diamond, \Sigma^\diamond)$, viewed as a map $u \colon \Sigma^\star \to \Sigma^\star$ by restricting its domain and codomain, the preimage map $u^{-1} \colon \mathcal{P}(\Sigma^\star) \to \mathcal{P}(\Sigma^\star)$ restricts to F_Σ. Indeed, since $\mathbb{U}_\Sigma(\Sigma^\diamond, \Sigma^\diamond)$ consists of all unary operations u of the form $x \mapsto vxw$ for $v, w \in \Sigma^\star$, the map u^{-1} is given by $L \mapsto v^{-1}Lw^{-1}$. Similarly, (2) means that for every $F_\Sigma \in I_\Sigma$ and $g \colon \Delta^\star \to \Sigma^\star$, the map $g^{-1} \colon \mathcal{P}(\Sigma^\star) \to \mathcal{P}(\Delta^\star)$ restricts to one between F_Σ and some $F_\Delta \in I_\Delta$.

$$
\begin{array}{ccc}
\mathcal{P}(\Sigma^\star) \xrightarrow{\;u^{-1}\;} \mathcal{P}(\Sigma^\star) & \qquad & \mathcal{P}(\Sigma^\star) \xrightarrow{\;g^{-1}\;} \mathcal{P}(\Delta^\star) \\[4pt]
\subseteq\uparrow \qquad\qquad \uparrow\subseteq & & \subseteq\uparrow \qquad\qquad \uparrow\subseteq \\[4pt]
F_\Sigma \dashrightarrow F_\Sigma & & F_\Sigma \dashrightarrow F_\Delta
\end{array}
\tag{4.1}
$$

Theorem 4.5. *The lattice of basic varieties of regular languages (ordered by inclusion) is isomorphic to the lattice of basic cotheories of regular languages. The isomorphism and its inverse are given pointwise for $\Sigma \in \mathbf{Set_f}$ by the maps $I_\Sigma \mapsto \bigcup I_\Sigma$ and $V_\Sigma \mapsto \{\, F \subseteq V_\Sigma \mid F$ is a finite basic local subvariety of $V_\Sigma \,\}$.*

5   Duality and the Basic Variety Theorem

The glue between the algebraic concepts of Sect. 3 and the language-theoretic ones of Sect. 4 is provided by *duality*, more precisely, the dual equivalence **AlgCDL** $\simeq^{\mathrm{op}}$ **Pos** between the full subcategory **AlgCDL** of **CDL** given by algebraic completely distributive lattices and the category **Pos** of posets and monotone maps [8]. Recall that a complete lattice D is *algebraic* if every element is a join of compact elements, where $c \in D$ is *compact* if for every subset $S \subseteq D$ with $c \leq \bigvee S$ one has $c \leq \bigvee F$ for some finite $F \subseteq S$. Since all free CDLs and finite CDLs are algebraic, a theory of $\mathbb{U}$-quotients (Definition 3.6) lives in the category **AlgCDL**. Similarly, a basic cotheory of regular languages (Definition 4.4) lives in **Pos**, viewing the set $\mathcal{P}(\Sigma^\star)$ of languages as a poset ordered by inclusion. Let us now make the key observation that, under the above duality, theories of $\mathbb{U}$-quotients *dualize* to basic cotheories of regular languages: One can show that, up to isomorphism, the duals of the commutative squares (3.2) in **AlgCDL** are precisely the commutative squares (4.1) in **Pos** where $g = h^\star$ and F_Σ and F_Δ are the posets of languages recognized by e_Σ and e_Δ, respectively. We can therefore bring the results of the previous sections together to establish our main result:

Theorem 5.1 (Basic Variety Theorem). *The lattice of basic varieties of regular languages is isomorphic to the lattice of pseudovarieties of lattice bimodules.*

Proof. We simply compose all the previously established lattice isomorphisms:

Pseudovarieties of lattice bimodules

$\cong$ Theories of lattice bimodules   (Theorem 3.5)

$\cong$ Theories of $\mathbb{U}$-quotients   (Proposition 3.7)

$\cong$ Basic cotheories of regular languages   (Duality)

$\cong$ Basic varieties of regular languages   (Theorem 4.5)   □

Spelling out the four isomorphisms in the proof, from top to bottom we transform between the following collections:

$$
\left.
\begin{array}{c}
\mathcal{V} \\
\cong \\
\mathcal{T}^{\mathcal{V}} = (\{(\Sigma^\star, \Sigma^\diamond) \xrightarrow{e} (M, D) \mid (M, D) \in \mathcal{V}\})_{\Sigma \in \mathbf{Set_f}}
\end{array}
\right\} \quad \mathbf{LBM}
$$

$$
\begin{array}{c}
\cong \\
(\{\Sigma^\diamond \xrightarrow{e^\circ} D \mid e \in \mathcal{T}^{\mathcal{V}}\})_{\Sigma \in \mathbf{Set_f}} \qquad \mathbf{AlgCDL} \\
\cong^{\mathrm{op}}
\end{array}
$$

$$
\left.
\begin{array}{c}
(I_\Sigma \hookrightarrow \mathcal{P}(\mathbf{Reg}_\Sigma))_{\Sigma \in \mathbf{Set_f}} \\
\cong \\
(V_\Sigma \hookrightarrow \mathbf{Reg}_\Sigma)_{\Sigma \in \mathbf{Set_f}}
\end{array}
\right\} \quad \mathbf{Pos}
$$

Thus, starting from the top, a pseudovariety $\mathcal{V}$ of lattice bimodules is sent to the basic variety of all regular languages recognized by some lattice bimodule in $\mathcal{V}$. Conversely, starting from the bottom, a basic variety $(V_\Sigma)_{\Sigma \in \mathbf{Set_f}}$ of languages is sent to the pseudovariety of all $\star$-generated reduced finite lattice bimodules (M, D) such that every language $L \subseteq \Sigma^\star$ recognized by (M, D) lies in V_Σ.

6   Quantum Finite Automata

In this section we present a natural example of a basic variety of regular languages that is not closed under union and intersection and therefore not captured by any previously known Eilenberg-type correspondence. It is concerned with languages accepted by *quantum finite automata (QFA)*. Several different notions of QFA have been proposed and studied, varying in their expressive power; see e.g. the recent survey paper by Ambainis and Yakaryılmaz [2]. Here, we focus on the model of *Kondacs-Watrous quantum finite automata (KWQFA)* [13], also known in the literature as *measure-many quantum finite automata*.

A KWQFA $M = (Q, \Sigma, T, q_0, Q_{\mathsf{acc}}, Q_{\mathsf{rej}}, Q_{\mathsf{non}})$ is given by a finite set Q of *basis states*, an input alphabet Σ not containing the end markers κ and $\$$, an initial state $q_0 \in Q$ and a partition $Q_{\mathsf{acc}} \mathbin{\dot{\cup}} Q_{\mathsf{rej}} \mathbin{\dot{\cup}} Q_{\mathsf{non}}$ of Q into accepting, rejecting and non-halting states. The transitions are specified by a family of unitary linear maps $T_\sigma \colon \mathcal{H}_Q \to \mathcal{H}_Q$ ($\sigma \in \Sigma \cup \{\kappa, \$\}$) on the complex Hilbert space $\mathcal{H}_Q$ with orthonormal basis Q. Thus, denoting the basis vectors by $|q\rangle$ ($q \in Q$), every element $|\psi\rangle$ of $\mathcal{H}_Q$ can be uniquely expressed as a linear combination

$|\psi\rangle = \sum_{q \in Q} \alpha_q |q\rangle$ with $\alpha_q \in \mathbb{C}$. The *states* of M are those $|\psi\rangle \in \mathcal{H}_Q$ with norm $\sum_{q \in Q} |\alpha_q|^2 = 1$. Note that a unitary transformation T_σ maps states to states. A *measurement* collapses the state $|\psi\rangle$ to the basis state $|q\rangle$ with probability $|\alpha_q|^2$.

Initially, the automaton is in the basis state $|q_0\rangle$. An input $w \in \Sigma^\star$ is processed by first adding the left (κ) and right ($\$$) end markers. Then, for every successive symbol σ in $\tilde{w} = \kappa w \$$ the corresponding transformation T_σ is applied and a measurement is performed. The automaton halts and accepts if the resulting basis state lies in Q_{acc}, halts and rejects if it lies in Q_{rej}, and continues with processing the next input letter if it lies in Q_{non}. Thus, if the QFA is in the state $|\psi\rangle = \sum_{q \in Q_{\text{acc}}} \alpha_q |q\rangle + \sum_{q \in Q_{\text{rej}}} \beta_q |q\rangle + \sum_{q \in Q_{\text{non}}} \gamma_q |q\rangle$ after reading the current input symbol but before making the measurement, it accepts with probability $\sum_{q \in Q_{\text{acc}}} |\alpha_q|^2$, rejects with probability $\sum_{q \in Q_{\text{rej}}} |\beta_q|^2$ and continues processing the input with probability $\sum_{q \in Q_{\text{non}}} |\gamma_q|^2$. This yields an overall probability $p \in [0, 1]$ that the input word w is accepted, i.e. that at any stage of the computation the automaton reaches a state in Q_{acc}.

We say that M *accepts* the language $L \subseteq \Sigma^\star$ (with bounded error) if there exists a real number $p > 1/2$ such that M accepts every word in L with probability $\geq p$ and rejects every word not in L with probability $\geq p$. The class of languages accepted by KWQFA is denoted by **RMM**. It is known to be a proper subclass of the class of all regular languages; for instance, $\{a, b\}^\star a \notin \mathbf{RMM}$ [13, Proposition 7]. Subsequent work has identified certain "forbidden configurations" in the minimal deterministic finite automaton of a regular language making it unrecognizable by a KWQFA [4,6]. In this way, it was shown that **RMM** is not closed under union and intersection [4, Corollary 3.2]. However, **RMM** is closed under preimages of monoid homomorphisms and derivatives [6, Theorem 4.1] and thus forms a basic variety of regular languages.

The questions whether **RMM** is decidable and whether it has an algebraic characterization remain open problems in the theory of quantum automata [3]. Our Basic Variety Theorem provides strong evidence that such a characterization must exist: it asserts that **RMM** corresponds to a pseudovariety of reduced lattice bimodules, which by Theorem 3.5 admits an (abstract form of) equational presentation. We expect that the latter can be turned into a more concrete form using *profinite equations* over free lattice bimodules $(\Sigma^\star, \Sigma^\circ)$, analogous to Reiterman's [18] description of pseudovarieties of finite monoids in terms of profinite equations over free monoids $\Sigma^\star$. A concrete profinite axiomatization of the pseudovariety induced by **RMM** might pave the way towards the decidability of that class: deciding whether a given regular language lies in **RMM** reduces to checking whether its syntactic lattice bimodule satisfies the equational axioms.

7   Conclusion and Future Work

We have introduced a new two-sorted algebraic structure, lattice bimodules, for the recognition of regular languages. Our main result is a new Eilenberg-type correspondence between basic varieties of regular languages, which need not be closed under set-theoretic boolean operations, and pseudovarieties of reduced lattice bimodules. The proof is guided by the recent category-theoretic approach to algebraic language theory and makes use of the duality between algebraic completely distributive lattices and posets.

An immediate next step to unleash the full power of our new variety theorem is to establish a Reiterman-type theorem for lattice bimodules leading to a description of pseudovarieties of lattice bimodules in terms of profinite equations. The recent categorical account of (profinite) equational theories [7,15] should provide inspiration in this direction. This may lead to new results on the decidability of basic varieties of regular languages, e.g. language classes recognized by different models of reversible automata [11] or quantum automata (cf. Sect. 6).

Furthermore, several generalizations of our work are conceivable. The most obvious one is to replace the duality **AlgCDL** $\simeq^{op}$ **Pos** by an abstract dual equivalence $\mathscr{A} \simeq^{op} \mathscr{B}$ between suitable categories $\mathscr{A}$ and $\mathscr{B}$, and to consider the recognition of languages by $\mathscr{A}$-bimodules. We anticipate that this minor generalization already recovers results closely related to the original Eilenberg theorem for $\mathscr{A}$ being the category of sets, and to Polák's variety theorem for idempotent semirings for $\mathscr{A}$ being the category of complete semilattices. In an orthogonal direction, the monoid action on the algebra may be generalized to the action of a monad **T** on the category of sets, but the dependence between the monad **T** and the category $\mathscr{A}$ is not obvious and remains to be investigated.

References

1. Adámek, J., Milius, S., Myers, R., Urbat, H.: Generalized Eilenberg theorem: varieties of languages in a category. ACM Trans. Comput. Log. **20**(1), 3:1–3:47 (2019)
2. Ambainis, A., Yakaryılmaz, A.: Automata and quantum computing (2018). https://arxiv.org/abs/1507.01988
3. Ambainis, A., Beaudry, M., Golovkins, M., Kikusts, A., Mercer, M., Thérien, D.: Algebraic results on quantum automata. In: Diekert, V., Habib, M. (eds.) STACS 2004. LNCS, vol. 2996, pp. 93–104. Springer, Heidelberg (2004). https://doi.org/10.1007/978-3-540-24749-4_9
4. Ambainis1, A., Kikusts, A., Valdats, M.: On the class of languages recognizable by 1-way quantum finite automata. In: Ferreira, A. Reichel, H. (eds.) STACS 2001. LNCS, vol. 2010, pp. 75–86. Springer, Heidelberg (2001). https://doi.org/10.1007/3-540-44693-1_7
5. Bojańczyk, M.: Recognisable languages over monads. In: Potapov, I. (ed.) DLT 2015. LNCS, vol. 9168, pp. 1–13. Springer, Cham (2015). https://doi.org/10.1007/978-3-319-21500-6_1
6. Brodsky, A., Pippenger, N.: Characterizations of 1-way quantum finite automata. SIAM J. Comput. **31**, 73–91 (1999)
7. Chen, L.-T., Adámek, J., Milius, S., Urbat, H.: Profinite monads, profinite equations, and Reiterman's theorem. In: Jacobs, B., Löding, C. (eds.) FoSSaCS 2016. LNCS, vol. 9634, pp. 531–547. Springer, Heidelberg (2016). https://doi.org/10.1007/978-3-662-49630-5_31
8. Davey, B.A., Priestley, H.A.: Introduction to Lattices and Order, 2 edn. Cambridge University Press, Cambridge (2002)

9. Eilenberg, S.: Automata, Languages, and Machines. Academic Press, Cambridge (1974)
10. Gehrke, M., Grigorieff, S., Pin, J.É.: Duality and equational theory of regular languages. In: Aceto, L., Damgård, I., Goldberg, L.A., Halldórsson, M.M., Ingólfsdóttir, A., Walukiewicz, I. (eds.) ICALP 2008. LNCS, vol. 5126, pp. 246–257. Springer, Heidelberg (2008). https://doi.org/10.1007/978-3-540-70583-3_21
11. Golovkins, M., Pin, J.-E.: Varieties generated by certain models of reversible finite automata. In: Chen, D.Z., Lee, D.T. (eds.) COCOON 2006. LNCS, vol. 4112, pp. 83–93. Springer, Heidelberg (2006). https://doi.org/10.1007/11809678_11
12. Klíma, O., Polák, L.: Syntactic structures of regular languages. Theoret. Comput. Sci. **800**, 125–141 (2019)
13. Kondacs, A., Watrous, J.: On the power of quantum finite state automata. In: Proceedings of the FOCS, pp. 66–75. IEEE (1997)
14. Markowsky, G.: Free completely distributive lattices. Proc. Amer. Math. Soc. **74**, 227–228 (1979)
15. Milius, S., Urbat, H.: Equational axiomatization of algebras with structure. In: Bojańczyk, M., Simpson, A. (eds.) FoSSaCS 2019. LNCS, vol. 11425, pp. 400–417. Springer, Cham (2019). https://doi.org/10.1007/978-3-030-17127-8_23
16. Pin, J.E.: A variety theorem without complementation. Russ. Math. **39**, 80–90 (1995)
17. Polák, L.: Syntactic semiring of a language. In: Sgall, J., Pultr, A., Kolman, P. (eds.) MFCS 2001. LNCS, vol. 2136, pp. 611–620. Springer, Heidelberg (2001). https://doi.org/10.1007/3-540-44683-4_53
18. Reiterman, J.: The Birkhoff theorem for finite algebras. Algebra Universalis **14**(1), 1–10 (1982). https://doi.org/10.1007/BF02483902
19. Salamanca, J.: Unveiling Eilenberg-type correspondences: Birkhoff's theorem for (finite) algebras + duality (2017). https://arxiv.org/abs/1702.02822
20. Schützenberger, M.P.: On finite monoids having only trivial subgroups. Inform. Control **8**(2), 190–194 (1965)
21. Urbat, H., Adámek, J., Chen, L.T., Milius, S.: Eilenberg theorems for free. In: Proceedings of the MFCS. LIPIcs, vol. 83, pp. 43:1–43:14 (2017)
22. Wilke, T.: An Eilenberg theorem for infinity-languages. In: Proceedings of the ICALP. LNCS, vol. 510, pp. 588–599. Springer, Heidelberg (1991)

Partially Directed Animals
with a Bounded Number of Holes

Valentina Dorigatti[✉] and Paolo Massazza

Department of Theoretical and Applied Sciences, University of Insubria, Varese, Italy
{vdorigatti,paolo.massazza}@uninsubria.it

Abstract. We address the problem of the exhaustive generation of a particular class of polyominoes, corresponding to partially directed animals with a bounded number of holes. We apply an approach based on discrete dynamical systems to develop an algorithm that generates each polyomino in constant amortized time and space $O(n)$. By implementing the algorithm in C++ we have obtained new sequences that do not appear in the On-Line Encyclopedia of Integer Sequences.

Keywords: Exhaustive generation · Polyominoes · CAT algorithms

1 Introduction

A *polyomino* is a finite and connected union of unitary squares, called cells, in the plane $\mathbb{Z} \times \mathbb{Z}$, considered up to translations [12]. The number of cells of a polyomino is its *area*. The problem of determining the number of polyominoes of area n is still open. Indeed, no closed formula for the number of polyominoes of area n is known, and the most used algorithm to compute this number has exponential complexity (both in time and in space) [13]. So, it is worthwhile to design efficient generating algorithms with strict constraints on the space. Several subclasses of polyominoes have been introduced in literature, with the aim of classifying them as well as tackling some difficult questions (*e.g.* counting and exhaustive generation [3,6–9]). A good survey for (solved and unsolved) problems on polyominoes is [2].

Previous works dealt with the problem of generating (and enumerating) the class of polyominoes corresponding to partially directed animals [10], also in the case where holes are forbidden [15,17]. We also recall that a different class of combinatorial objects, namely, directed animals on the square (resp., triangular) lattice, has been enumerated in [1] by means of the so-called ECO method.

Here we are interested in CAT algorithms, that is, algorithms which generate each polyomino in Constant Amortized Time. CAT algorithms for the exhaustive generation of certain classes of polyominoes and other combinatorial structures have been already developed by exploiting discrete dynamical systems [4,5,10, 11,14,16]. In this work we apply discrete dynamical systems to develop a CAT algorithm for the class HPDA_k of polyominoes that are duals of Partially Directed

© Springer Nature Switzerland AG 2021
A. Leporati et al. (Eds.): LATA 2021, LNCS 12638, pp. 16–28, 2021.
https://doi.org/10.1007/978-3-030-68195-1_2

Animals with at most k holes. We show that the set $\mathsf{HPDA}_k(n)$ can be generated by a CAT algorithm that uses space $O(n)$ (n represents the area). We provide also the counting sequences for HPDA_k, with $1 \leq k \leq 3$, compared to the sequences associated with partially directed animals, with or without holes, up to $n = 22$. All proofs appear in the appendix which is omitted due to bounds on the number of pages for conference papers.

2 Notation and Preliminaries

The area $A(P)$ of a polyomino P is the number of its cells. A cell is identified by integer coordinates (the center of the unitary square). We say that two squares (i, j) and (i', j') are *adjacent* if $|i - i'| + |j - j'| = 1$. A *path* from a square a to a square b is a sequence $q_1, q_2, \ldots, q_k$ of squares such that $q_1 = a$, $q_k = b$ and q_i, q_{i+1} are adjacent for all i with $1 \leq i < k$. A path in P is a path whose squares are cells of P. A polyomino can be seen as a finite sequence of *columns*. A column of a polyomino consists of a sequence of *vertical segments* separated by empty unitary squares. A *vertical segment* is a finite sequence of adjacent cells in the same column. The *position* of a cell is its y-coordinate. The position of the top (resp., bottom) cell of a segment s is denoted by $\mathrm{Top}(s)$ (resp., $\mathrm{Bot}(s)$). We represent a segment s of a column by means of the pair $(A(s), \mathrm{Top}(s))$. Segments belonging to the same column are numbered from top to bottom, so a column with p segments is simply a sequence of disjoint segments $\mathbf{c} = (s_1, \ldots, s_p)$, with $\mathrm{Top}(s_{i+1}) < \mathrm{Top}(s_i) - A(s_i)$ for $1 \leq i < p$. Furthermore, the position of $\mathbf{c}$ is the position of the top cell of the first segment, $\mathrm{Top}(\mathbf{c}) = \mathrm{Top}(s_1)$. Similarly, we set $\mathrm{Bot}(\mathbf{c}) = \mathrm{Bot}(s_p)$. Given a segment s and an integer j such that $\mathrm{Top}(s) > j \geq \mathrm{Bot}(s)$, we denote by $s_{>j}$ (resp., $s_{\leq j}$) the part of s consisting of the cells with position greater than j (resp., smaller than or equal to j). The part of a column $\mathbf{c}$ that is above the position j is $\mathbf{c}_{>j}$ ($\mathbf{c}_{\geq j}$, $\mathbf{c}_{<j}$ and $\mathbf{c}_{\leq j}$ are defined similarly). Given two segments s and t with $\mathrm{Bot}(s) > \mathrm{Top}(t)$ or $\mathrm{Bot}(t) > \mathrm{Top}(s)$, their *vertical distance* is $\mathrm{d}(s, t) = \min(|\mathrm{Bot}(s) - \mathrm{Top}(t)|, |\mathrm{Bot}(t) - \mathrm{Top}(s)|) - 1$. We set $\mathrm{d}(s, t) = \infty$ if either s or t is the null segment (denoted by ϵ). Segments can be ordered with respect to their position and their area.

Definition 1. *(< on segments) Let u and v be two segments. Then, $u < v$ iff $\mathrm{Top}(u) > \mathrm{Top}(v)$ or $\mathrm{Top}(u) = \mathrm{Top}(v) \wedge A(u) > A(v)$.*

We can extend $<$ in order to obtain a total order on columns, denoted by $\prec$.

Definition 2. *($\prec$ on columns) Let $\mathbf{b} = (s_1, \ldots, s_p)$ and $\mathbf{c} = (t_1, \ldots, t_q)$ be two columns. Then, $\mathbf{b} \prec \mathbf{c}$ iff $A(\mathbf{b}) > A(\mathbf{c})$, or $A(\mathbf{b}) = A(\mathbf{c})$ and there exists m, with $1 \leq m \leq \min(p, q)$, such that $s_m < t_m$ and $s_j = t_j\ \forall j < m$.*

We assume that the position of the bottom cell of the last segment of the first column of a polyomino is 0. We denote by $\mathsf{Pol}(n)$ the set of polyominoes of area n. Given $P \in \mathsf{Pol}(n)$, we denote by $P_{\leq i}$ (resp., P_i) the i-prefix of P (resp., the i-th column of P), that is, the sequence of the first i columns of P. Notice that $P_{\leq i}$ is not necessarily a polyomino. The *width* $w(P)$ of P is the number of its

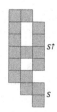

Fig. 1. The gender of $s^\uparrow$ is 2, ($h_1 = h_2 = 1$ in Definition 5), while s has gender 0.

columns. A segment s of P_i is *left-adjacent* (l-adjacent, for short) to a segment t of P_{i-1} if there exists a cell of s that is adjacent to a cell of t. A segment that is not l-adjacent is called *left-detached* (l-detached, for short).

From here on, given a segment s of P_i, the segment of P_i immediately above (resp., below) s is denoted by $s^\uparrow$ (resp., $s_\downarrow$). The second segment below s is $s_{\downarrow\downarrow}$, and the last (resp., first) segment to which s is l-adjacent is $_\downarrow s$ (resp., $^\uparrow s$).

We consider a class of polyominoes (called prefix-closed in [10]) that correspond to partially directed animals on the square grid [18,19].

Definition 3. *(The class* PDA*)* *The class* PDA(n) *is the class containing all* $P \in$ Pol(n) *such that* $P_{\leq i}$ *is a polyomino for all* $i \in \mathbb{N}$ *with* $1 \leq i \leq w(P)$.

Definition 4. *(Hole)* *A hole of* $P \in$ Pol *is a finite maximal set* S *of empty unit squares such that:*

- *for any two squares* a *and* b *in* S*, there exists a path in* S *connecting* a *to* b*;*
- *for any two empty unit squares* $c \notin S$ *and* $a \in S$*, there is not a path connecting* c *to* a *that crosses only empty unit squares.*

In this work we deal with prefix-closed polyominoes with at most k holes. They are duals of Partially Directed Animals with at most k holes, hence we are going to call HPDA$_k$ such a class.

Since we want to limit the number of holes, we define the concept of *gender of a segment* s, denoted by Gen(s), as the number of holes the segment creates when considering the columns to its left.

Definition 5. *(Gender of a segment)* *Given* $P \in$ PDA *and* $i \in \mathbb{N}$*, with* $1 < i \leq w(P)$*, let* s *be a segment of* P_i*. Then, Gen(s) is defined as the number*

$$Gen(s) = \sharp\{t \in P_{i-1} : s \text{ is l-adjacent to } t\} - 1 + h_1 + h_2$$

where $h_1 = 1$ *if* $d(({}^\uparrow s)^\uparrow, s)) = 0$ *(*$h_1 = 0$ *if* $d(({}^\uparrow s)^\uparrow, s)) > 0$*), and* $h_2 = 1$ *if* $d(s, (_\downarrow s)_\downarrow) = 0$ *(*$h_2 = 0$ *if* $d(s, (_\downarrow s)_\downarrow) > 0$*).*

An example of gender of a segment is shown in Fig. 1.

The *number of holes* in $P \in$ PDA is $H(P)$. For any $k \in \mathbb{N}$ we define a class of polyominoes called HPDA$_k$.

Definition 6. *(The class* HPDA$_k$*)* *Let* $k \in \mathbb{N}$*. A polyomino* P *belongs to* HPDA$_k$ *if* $P \in$ PDA *and* $H(P) \leq k$.

In the sequel, we generate polyominoes column-by-column by exploiting a relation between polyominoes in HPDA_k and columns, called k-*compatibility*.

Definition 7. *(k-compatibility) Fix $k \in \mathbb{N}$. Given $P \in \mathsf{HPDA}_k$ and a column* **b**, *we say that* **b** *is k-compatible with P, denoted by $P \,\square_k\, \mathbf{b}$, if every segment s of* **b** *is l-adjacent to at least one segment of $P_{w(P)}$ and*

$$\sum_{s \in \mathbf{b}} Gen(s) + H(P) \leq k.$$

The (right) *column concatenation* | is the operation which takes a polyomino $P \in \mathsf{HPDA}_k$ and a column **c** such that $P \,\square_k\, \mathbf{c}$, and produces a polyomino $P' = P|\mathbf{c}$, with $w(P) + 1$ columns, which is still in HPDA_k. Notice that the left hand side of the inequality in Definition 7 is precisely the number of holes of $P|\mathbf{b}$.

Remark 1. For any $n > 0$, the one-column polyomino $(((n), n - 1))$ (consisting of one segment of area n) is in $\mathsf{HPDA}_k(n)$. So, a polyomino P is in HPDA_k if and only if for all i, with $1 < i \leq w(P)$, one has $P_{<i} \in \mathsf{HPDA}_k$ and $P_{<i} \,\square_k\, P_i$.

The set of all columns that are k-compatible with $P \in \mathsf{HPDA}_k$ and have area r is indicated by $\mathrm{Comp}_k(P, r) = \{\mathbf{b} : P \,\square_k\, \mathbf{b} \text{ and } \mathrm{A}(\mathbf{b}) = r\}$.

   The main idea of the paper is to generate all polyominoes in $\mathsf{HPDA}_k(n)$ according to the following total order.

Definition 8. *(Order on HPDA_k) Given $P, Q \in \mathsf{HPDA}_k(n)$, set $P < Q$ if there exists i such that $P_{<i} = Q_{<i}$ and $P_i \prec Q_i$.*

3   A Dynamical System for HPDA_k

We are going to define a specific family $f_{P,k,r}$ of discrete dynamical systems depending on three parameters $k, r \in \mathbb{N}$ and $P \in \mathsf{HPDA}_k$. For fixed values of the parameters, $f_{P,k,r}$ takes in input a column $\mathbf{b} \in \mathrm{Comp}_k(P, r)$ and outputs a column $\mathbf{c} \in \mathrm{Comp}_k(P, r)$ such that $\mathbf{b} \prec \mathbf{c}$.

   The initial state of the system is the column $\mathbf{c}_{\min} = \min_{\prec}(\mathrm{Comp}_k(P, r))$. This is a column consisting of one segment of area r whose bottom cell has position $\mathrm{Top}(P_{w(P)})$, $\mathbf{c}_{\min} = ((r, \mathrm{Top}(P_{w(P)}) + r - 1))$. The evolution rule $f_{P,k,r}$ rearranges the cells of the column in input according to four operations called *moves*. A move in a column $\mathbf{b} \in \mathrm{Comp}_k(P, r)$ may occur only in a position j of a cell of a segment s of $\mathbf{b}$. We have four different moves called **shift**, **split**, **shift-and-split** and **split-and-shift**. The behaviour of these moves depends on $H(P)$, the number of holes in P.

   A *shift* move corresponds to sliding s downward (this implies $j = \mathrm{Top}(s)$). More precisely, the position of s is decreased by x, where x is the smallest integer greater than zero such that the resulting column is in $\mathrm{Comp}_k(P, r)$, see Fig. 2.

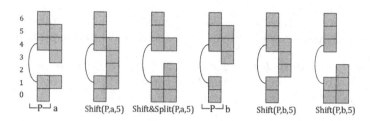

Fig. 2. Shift$(P, \mathbf{a}, 5)$ is defined only if $H(P) < k$, whereas Shift&Split$(P, \mathbf{a}, 5)$ is defined only if $H(P) = k$. The move Shift$(P, \mathbf{b}, 5)$ is always defined ($x = 1$ if $H(P) < k$, $x = 3$ if $H(P) = k$).

A *split* move is to split s into two and slide the lower portion downward (this implies $\text{Bot}(s) \leq j < \text{Top}(s)$). It suffices to consider only splits somewhere strictly below $\text{Top}(^\uparrow s)$, because the upper portion has to be l-adjacent to $^\uparrow s$. So, $s_{>j}$ maintains the position, whereas the portion $s_{\leq j}$ slides x positions downward, where x is the smallest integer greater than zero such that the resulting column belongs to $\text{Comp}_k(P, r)$, see Fig. 3.

We point out that the number of holes before a move differs from the number of holes after a move by at most 1.

The two moves *shift-and-split* and *split-and-shift* are needed to guarantee that the phase state of $f_{P,k,r}$ is equal to $\text{Comp}_k(P, r)$. They are defined only for the last or the second to last segment s of a column, and require that $H(P|\mathbf{b}_{\geq \text{Bot}(s)}) = k$. Both moves operate in two steps (each one consisting of sliding a suitable portion of a segment one position downward).

A *shift-and-split* move can occur at j only if both the following conditions holds: $j = \text{Top}(s)$, $d((^\uparrow s)^\uparrow, s) > 0$ and either $d(s, s_\downarrow) = 1 \wedge {}_\downarrow s \neq {}^\uparrow(s_\downarrow)$ or $d(s, s_\downarrow) > 1 \wedge d(s, ({}_\downarrow s)_\downarrow) = 1 \wedge A(s) > 1$. Furthermore, if $A(s) > 1$ the relations $\text{Bot}(s) = \text{Top}({}_\downarrow s)$ and $^\uparrow s = {}_\downarrow s$ must hold. In the first step, s slides one position downward and creates the $(k + 1)$-th hole. Indeed, no existing hole disappears since $d((^\uparrow s)^\uparrow, s) > 0$, whereas a new hole arises since $d(s, s_\downarrow) = 1$ or $d(s, ({}_\downarrow s)_\downarrow) = 1$. Then, a split move (or a shift move if $A(s) = 1$) is made at $\text{Bot}({}_\downarrow s) - 1$, and the segment slides exactly one position downward. The shift-and-split move is not defined if the resulting column is not in $\text{Comp}_k(P, r)$. Figure 2 shows an example of shift-and-split.

A *split-and-shift* move may occur only at the position $\text{Bot}(s)$ of a segment s with $A(s) > 1$. The move is enabled only if $d(s, s_\downarrow) = 1$, $\text{Bot}(s) = \text{Bot}({}_\downarrow s)$ and $\text{Top}(^\uparrow s) > \text{Bot}(s)$. Firstly, s is split at $\text{Bot}(s)$ and the bottom cell of s slides one position downward joining $s_\downarrow$ (this step creates the $(k + 1)$-th hole). The new segment then slides one position downward to reduce the number of holes (see Fig. 3). The split-and-shift move is not defined if the resulting column is not in $\text{Comp}_k(P, r)$.

Given $j \in \mathbb{N}$, $P \in \text{HPDA}_k$ and a column $\mathbf{b} \in \text{Comp}_k(P, r)$, we denote by Split$(P, \mathbf{b}, j)$ (resp., Shift$(P, \mathbf{b}, j)$, Shift&Split$(P, \mathbf{b}, j)$ and Split&Shift$(P, \mathbf{b}, j)$) the function that returns the column obtained by applying the corresponding

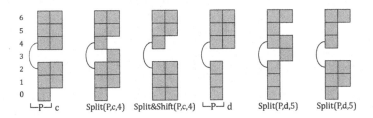

Fig. 3. Split$(P, \mathbf{c}, 4)$ is defined if $H(P) < k$, whereas Split&Shift$(P, \mathbf{c}, 4)$ is defined if $H(P) = k$. Split$(P, \mathbf{d}, 5)$ is defined in all cases ($x = 1$ if $H(P) < k$, $x = 3$ if $H(P) = k$).

operation to the segment of $\mathbf{b}$ with a cell in position j. Notice that there is at most one move defined at j.

We simply write $\mathbf{b} \xrightarrow{j} \mathbf{b}'$ if $\mathbf{b}'$ is the column obtained by a move at j in $\mathbf{b}$. We introduce a partial order $\lhd$ on the set $\mathrm{Comp}_k(P, r)$, which corresponds to the notion of derivation by means of a sequence of moves.

Definition 9. *Let $P \in \mathsf{HPDA}_k$ and $b, c \in \mathrm{Comp}_k(P, r)$. We set $\mathbf{b} \lhd \mathbf{c}$ if $\mathbf{b} = \mathbf{c}$ or there exists a sequence of integers $i_1, ..., i_k$ such that:*

$$\mathbf{b} \xrightarrow{i_1} \mathbf{b}^{(1)} \xrightarrow{i_2} \mathbf{b}^{(2)} ... \mathbf{b}^{(k-1)} \xrightarrow{i_k} \mathbf{b}^{(k)} = \mathbf{c}.$$

The four moves on columns defined above allow us to obtain any k-compatible column starting from the smallest column $\mathbf{c}_{\min}$ and applying a suitable sequence of moves. Indeed, one has the following theorem.

Theorem 1. *Let $P \in \mathsf{HPDA}_k$. Then, one has*

$$\mathrm{Comp}_k(P, r) = \{\mathbf{b} : \mathbf{c}_{\min} \lhd \mathbf{b}\}.$$

We indicate by $\mathrm{Mc}(P, \mathbf{b})$ the set of positions of moves in $\mathbf{b} \in \mathrm{Comp}_k(P, r)$, that is, $\mathrm{Mc}(P, \mathbf{b}) = \{j \in \mathbb{N} : \mathbf{b} \xrightarrow{j} \mathbf{b}'\}$. It is immediate that $\mathrm{Mc}(P, \mathbf{b}) = \emptyset$ if and only if $\mathbf{b}$ is equal to $\mathbf{c}_{\max} = \max_{\lhd}(\mathrm{Comp}_k(P, r))$, that is, $\mathbf{b} = ((r, \mathrm{Bot}(P_{w(P)})))$. In particular, we are interested in the move occurring at $\min(\mathrm{Mc}(P, \mathbf{b}))$. This move occurs in the last segment s of $\mathbf{b}$ if and only if $\mathrm{Top}(s) > \mathrm{Bot}(t)$, where t is the last segment of $P_{w(P)}$. Otherwise, one has $\mathrm{Top}(s) = \mathrm{Bot}(t)$ and the move occurs in $s^{\uparrow}$. More precisely, the following lemma characterizes the position of the lowest move in a column.

Lemma 1. *Given $P \in \mathsf{HPDA}_k$ and $\mathbf{b} \in \mathrm{Comp}_k(P, r)$, let s (resp., t) be the last segment of $\mathbf{b}$ (resp., $P_{w(P)}$). Suppose that we have $\mathrm{Mc}(P, \mathbf{b}) \neq \emptyset$ and let $j = \min(\mathrm{Mc}(P, \mathbf{b}))$. If $\mathrm{Top}(s) > \mathrm{Bot}(t)$ then*

$$
j = \begin{cases}
\mathrm{Top}(s) & \text{if } A(s) = 1 \lor \mathrm{Bot}(s) = \mathrm{Top}(^{\uparrow}s) \lor \,_{\downarrow}s = \,^{\uparrow}s = t \land A(t) \le 2, \\
\mathrm{Bot}(s) & \text{if } A(s) > 1 \land \mathrm{Bot}(t) < \mathrm{Bot}(s) < \mathrm{Top}(^{\uparrow}s), \\
\mathrm{Bot}(t) + 1 & \text{if } \mathrm{Bot}(s) \le \mathrm{Bot}(t) \land A(t) > 2.
\end{cases}
$$

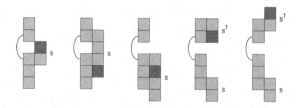

Fig. 4. The shaded cell is where the lowest move occurs.

Otherwise, one has $Top(s) = Bot(t)$ and

$$j = \begin{cases} Bot(s^\uparrow) & \text{if } Bot(s^\uparrow) < Top(^\uparrow(s^\uparrow)), \\ Top(s^\uparrow) & \text{if } Bot(s^\uparrow) = Top(^\uparrow(s^\uparrow)). \end{cases}$$

Figure 4 illustrates the cases of Lemma 1.

The lowest move in **b** allows us to determine a particular column, called the *grand ancestor* of **b**, denoted by $GA_k(P, \mathbf{b})$. The column $GA_k(P, \mathbf{b})$ is defined so that, by making a move at $j = \min(Mc(P, \mathbf{b}))$ in $GA_k(P, \mathbf{b})$ one gets the column that follows **b** in the ordered sequence of columns belonging to $Comp_k(P, A(\mathbf{b}))$.

Intuitively, the column $GA_k(P, \mathbf{b})$ differs from **b** only if j refers to the second-last segment of **b**. In this case, $GA_k(P, \mathbf{b})$ is obtained from **b** by deleting the last segment s and increasing the area of $s^\uparrow$ by the largest integer e (with $e \leq A(s)$) such that the resulting column, say $\mathbf{b}'$, is k-compatible with P and the move at j in $\mathbf{b}'$ won't generate a $(k+1)$-th hole. The remaining $A(s) - e$ cells are placed below $s^\uparrow$, in the highest possible positions such that the column remains in $Comp_k(P, A(\mathbf{b}))$, while ensuring a move at j. More formally, one has:

Definition 10. (Grand Ancestor) *Fix $k \in \mathbb{N}$ and let $P \in \mathsf{HPDA}_k$. Given a column $\mathbf{b} \in Comp_k(P, r)$, with $\mathbf{b} \neq \mathbf{c}_{\max}$, let s be the last segment of $\mathbf{b}$, $j = \min(Mc(P, \mathbf{b}))$, $v = {}_\downarrow(s^\uparrow)$ and $w = {}^\uparrow(s^\uparrow)$. If $j \leq Top(s)$ then $GA_k(P, \mathbf{b}) = \mathbf{b}$.*

Otherwise, let $\mathbf{b}'$ be the column obtained from $\mathbf{b}$ by replacing s and $s^\uparrow$ with a segment t with $Top(t) = Top(s^\uparrow)$ and $A(t) = A(s^\uparrow) + A(s)$. We distinguish three main cases:

$(H(P|\mathbf{b}') < k)$
   in this case one has $GA_k(P, \mathbf{b}) = \mathbf{b}'$;   (i)
$(H(P|\mathbf{b}') = k)$
   there are two cases:
   $(d(t, ({}_\downarrow t)_\downarrow) \neq 1)$ *one has $GA_k(P, \mathbf{b}) = \mathbf{b}'$;*   (ii)
   $(d(t, ({}_\downarrow t)_\downarrow) = 1)$ *if $j = Top(t) \wedge d(w^\uparrow, t) = 0$ or $j = Bot(s^\uparrow) \wedge (j < Bot(v) \vee j = Bot(v) \wedge v = {}_\downarrow t)$ then $GA_k(P, \mathbf{b}) = \mathbf{b}'$.*   (iii)
   Otherwise, one has $GA_k(P, \mathbf{b}) = \mathbf{b}''$, where $\mathbf{b}''$ is obtained from $\mathbf{b}'$ by deleting the bottom cell of t and placing a one-cell segment with position $Top(({}_\downarrow t)_\downarrow)$;   (iv)

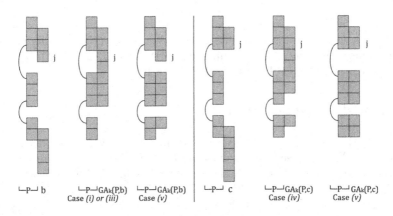

Fig. 5. The grand ancestor.

$(H(P|\mathbf{b}') > k)$ (v)

Let m be the smallest integer such that by replacing t with a segment $\hat{t}$ with $A(\hat{t}) = A(t) - m$ and $Top(\hat{t}) = Top(t)$, one gets a column $\mathbf{b}'' \in Comp_k(P, r - m)$ with $d(\hat{t}, ({}_\downarrow \hat{t})_\downarrow) = p$, where $p = 1$ (if $j < Bot(v) \vee j = Bot(v) \wedge H(P|\mathbf{b}) < k$) or $p = 2$ (if $j > Bot(v) \vee j = Bot(v) \wedge H(P|\mathbf{b}) = k$).

Then, $GA_k(P, \mathbf{b}) = \mathbf{b}'''$, where

$$\mathbf{b}''' = \min{}_{\prec}\{\mathbf{c} \in Comp_k(P, r) : \mathbf{c}_{\geq Bot(\mathbf{b}'')-p} = \mathbf{b}'', j \in Mc(P, \mathbf{c})\}.$$

The construction of column $\mathbf{b}'''$ in Definition 10 is straightforward. Indeed, let $P_{w(P)} = (s_1, \ldots, s_e, \ldots, s_{e+q})$, where $s_e = {}_\downarrow(\hat{t})$. We simply add below the segment $\hat{t}$ a sequence of segments $(s_1', \ldots, s_h')$, such that $A\left(\mathbf{b}_{> Top(\hat{t})}\right) + A(\hat{t}) + \sum_{i=1}^{h} A(s_i') = r$ and $Top(s_i') = Top(s_{e+i})$, $A(s_i') = A(s_{e+i}) + d(s_{e+i}, s_{e+i+1}) - 1$, for $1 \leq i < h$, with $Top(s_h') = Top(s_{e+h})$. Figure 5 illustrates the construction of the grand ancestor in some of the cases of Definition 10. We define the discrete dynamical system $f_{P,k,r} : Comp_k(P, r) \to Comp_k(P, r)$ over columns as follows.

Definition 11. *($f_{P,k,r}$) Let* $P \in \mathsf{HPDA}_k$ *and* $\mathbf{b} \in Comp_k(P, r)$. *Then,*

$$f_{P,k,r}(\mathbf{b}) = \begin{cases} \mathbf{c} & \text{if } GA_k(P, \mathbf{b}) \xrightarrow{j} \mathbf{c}, \ j = \min(Mc(P, \mathbf{b})) \\ (r, Bot(P_{w(P)})) & \text{if } Mc(P, \mathbf{b}) = \emptyset \end{cases}$$

We denote by $f_{P,k,r}^n$ the n-fold composition of with itself. The main properties of the dynamical system are stated in the following lemma.

Lemma 2. *Given* $r, k \in \mathbb{N}$ *and* $P \in \mathsf{HPDA}_k$, *for all* $\mathbf{b} \in Comp_k(P, r)$:

1. $f_{P,k,r}^n(\mathbf{b}) \prec f_{P,k,r}^{n+1}(\mathbf{b})$ *if* $f_{P,k,r}^n(\mathbf{b}) \neq \mathbf{c}_{\max}$;
2. $\bigcup_{n \in \mathbb{N}} f_{P,k,r}^n(\mathbf{c}_{\min}) = Comp_k(P, r)$.

Given two columns $\mathbf{b}, \mathbf{c} \in \mathrm{Comp}_k(P, r)$ such that $\mathbf{b} \neq \mathbf{c}$, one possibly has $\mathrm{GA}_k(P, \mathbf{b}) = \mathrm{GA}_k(P, \mathbf{c})$. Nevertheless, the column $f_{P,k,r}(\mathbf{b})$ is different from $f_{P,k,r}(\mathbf{c})$, since $\min(\mathrm{Mc}(P, \mathbf{b})) \neq \min(\mathrm{Mc}(P, \mathbf{c}))$. Indeed, if $\min(\mathrm{Mc}(P, \mathbf{b})) = \min(\mathrm{Mc}(P, \mathbf{c}))$ and $\mathrm{GA}_k(P, \mathbf{b}) = \mathrm{GA}_k(P, \mathbf{c})$ then $\mathbf{b} = \mathbf{c}$.

4   Exhaustive Generation

The algorithm for generating $\mathsf{HPDA}_k(n)$ exploits Lemma 2 and adopts an inductive approach. We suppose that a polyomino $P \in \mathsf{HPDA}_k(n-r)$ has been already generated. Thus, the aim is to add all columns in $\mathrm{Comp}_k(P, j)$ (one at a time, by exploiting Lemma 2), for all j with $1 \leq j \leq r$. Then, for each $\mathbf{c} \in \mathrm{Comp}_k(P, j)$ we recursively add all columns in $\mathrm{Comp}_k(P|\mathbf{c}, h)$, with $1 \leq h \leq n - r - j$, and so on until we obtain a polyomino in $\mathsf{HPDA}_k(n)$.

The computation starts by calling $\mathrm{HPDAGEN}(n, k)$ (see Algorithm 1), which sets the first column of the polyomino (a segment of area r, with $1 \leq r \leq n$) and then calls COLGEN to (recursively) add all subsequent columns, until a polyomino of area n is obtained. Procedure $\mathrm{COLGEN}(i, r, k)$ (see Algorithm 2) assumes that a polyomino $P \in \mathsf{HPDA}_k(n-r)$ with $i-1$ columns has been already generated and adds all i-th columns of area at most r that are k-compatible with P. Notice that the order of generation derives from Definition 2. As a matter of fact, procedure COLGEN is an application of Lemma 2 and consists of a while-loop, where at each iteration a move is executed. The polyomino is given in output if the area is n; otherwise a recursive call is made. Procedure $\mathrm{GRAN}(P_i, k)$ restores the grand ancestor of the current column P_i and returns $\min(\mathrm{Mc}(P, P_i))$, the position used to make a move in the next step. Lastly, $\mathrm{MOVE}(i, j, k)$ makes a move at j in the grand ancestor (either a Shift or a Split or a Shift&Split or Split&Shift), producing the next column (with respect to $\prec$).

Algorithm 1. Generation of $\mathsf{HPDA}_k(n)$

---

procedure $\mathrm{HPDAGEN}(n, k)$
  $P = ((n, n-1))$; Output(P);
  for $r = n - 1$ downto 1 **do**
    $P_1 = ((r, r-1))$;
    $\mathrm{COLGEN}(2, n - r, k)$;
  end for
end procedure

4.1   Data Structure

A polyomino P is simply a sequence of columns, where a column $\mathbf{b}$ is represented by a doubly-linked list L_b associated with the sequence of segments in $\mathbf{b}$ (as many nodes as segments). So, a node of the list corresponds to a segment s and contains seven entries $(\mathrm{A}(s), \mathrm{Top}(s), e, l_1, l_2, l_3, l_4)$ where:

Algorithm 2. Generation of columns

---

  procedure COLGEN(i, r, k)
    for $d = r$ **downto** 1 **do**
      $P_i = ((d, \text{Top}(P_{i-1}) + d - 1))$; #the smallest column w.r.t. $\prec$
      if $d < r$ **then** COLGEN$(i + 1, r - d, k)$;
      else OUTPUT(P);
      end if
      while !(ISFIXEDPOINT(P_i)) **do**
        $j = \text{GRAN}(P_i, k)$; # set P_i to the grand ancestor of the current column
        MOVE(i, j, k); #P_i is changed by a move at j
        if $d < n$ **then** COLGEN$(i + 1, r - d, k)$;
        else OUTPUT(P);
        end if
      end while
    end for
  end procedure

- e is the number of holes in the polyomino consisting of s, the columns to the left of s and the segments above s;
- l_1 is the link to the preceding node in the list L_b (associated with $s^\uparrow$);
- l_2 is the link to the next node in the list L_b (associated with $s_\downarrow$);
- l_3 is a link to the node representing $^\uparrow s$ (in the previous list);
- l_4 is a link to the node representing $_\downarrow s$ (in the previous list).

Lemma 1 states that the move at $\min(\text{Mc}(P, b))$ regards either the last or the second-last segment of **b**. Thus, P is represented by an array of records, where the i-th record has three fields, namely, the area of P_i, and two links to the nodes associated with the first and the last segment of P_i, respectively.

4.2 Complexity

The data structure used to represent a polyomino of area n requires space $O(n)$. In order to determine the time complexity, notice that the execution of HPDA-GEN(n, k) is described by a tree with the following properties:

- the root corresponds to the procedure call HPDAGEN(n, k);
- an internal node v at level i, with $i > 0$, corresponds to a procedure call COLGEN$(i + 1, r, k)$ for a suitable $r > 0$. This call adds all k-compatible $(i + 1)$-th columns of area at most r (associated with the children of v) to a particular polyomino with i columns and area $n - r$, uniquely identified by the path from the root to v;
- there is a one-to-one mapping between leaves and polyominoes in HPDA$_k(n)$;
- each internal node has at least two children or the only child is a leaf.

The complexity of COLGEN depends on the complexity of ISFIXEDPOINT, GRAN and MOVE. Firstly, by using the data structure described in Sect. 4.1, one can easily develop a function ISFIXEDPOINT that runs in constant time. Indeed, a

Table 1. Enumeration of some classes of polyominoes up to area 22.

Area	PDA	HFPDA	HPDA$_1$	HPDA$_2$	HPDA$_3$
1	1	1	1	1	1
2	2	2	2	2	2
3	6	6	6	6	6
4	19	19	19	19	19
5	62	62	62	62	62
6	206	206	206	206	206
7	694	692	694	694	694
8	2362	2341	2362	2362	2362
9	8101	7961	8101	8101	8101
10	27951	27186	27951	27951	27951
11	96905	93157	96900	96905	96905
12	337298	320149	337221	337298	337298
13	1177935	1103017	1177242	1177935	1177935
14	4125287	3808621	4120447	4125281	4125287
15	14482481	13176367	14453400	14482370	14482481
16	50950871	45663745	50792682	50949682	50950871
17	179584544	158494947	178782399	179574832	179584536
18	634020055	550882849	630157968	633952924	634019865
19	2241695805	1917098855	2223822306	2241281792	2241693374
20	7936423123	6679141228	7856231219	7934072801	7936400300
21	28131425351	23293978379	27780457688	28118894420	28131248036
22	99822752304	81315858059	98317497644	99759138360	99821541631

column P_i of area d is a fixed point if and only if $P_i = ((d, \mathrm{Bot}(P_{i-1})))$. Procedure $\mathrm{GRAN}(P_i, k)$ is used to restore the grand ancestor of the current column P_i. Lemma 1 allows us to compute the value $j = \min(\mathrm{Mc}(P, P_i))$ (corresponding to a cell in the last or in the second-last segment of P_i) in time $O(1)$. When j is the position of a cell in the last segment, the construction of the grand ancestor runs in time $O(1)$, since $\mathrm{GA}_k(P, P_i) = P_i$. Unfortunately, this is not true in general, and the construction runs in time $O(\min(q, \mathrm{A}(s)))$, where q is the number of segments in P_{i-1} with position smaller than j, and s is the last segment of P_i. Nevertheless, an amortized analysis leads to the following result.

Lemma 3. *Let $P \in \mathrm{HPDA}_k$. Consider two columns $\mathbf{b}, \mathbf{c}$ in $\mathrm{Comp}_k(P, r)$ such that $\mathrm{GRAN}(\mathbf{b}, k)$ and $\mathrm{GRAN}(\mathbf{c}, k)$ with cost $\Theta(p_1)$ and $\Theta(p_2)$, respectively. Then, there exist two sets $T_1, T_2 \subseteq \mathrm{Comp}_k(P, r)$, with $|T_1| = \Omega(p_1)$, $|T_2| = \Omega(p_2)$ and $T_1 \cap T_2 = \emptyset$, such that $\mathrm{GRAN}(\mathbf{d}, k)$ has cost $O(1)$ for any column $\mathbf{d} \in T_1 \cup T_2$.*

From Lemma 3 one easily obtains the following lemma and the main result.

Lemma 4. *Let $P \in \mathsf{HPDA}_k$ and $i = w(P)$. Then, $\mathrm{COLGEN}(i+1,r,k)$ runs in time $O(t)$, where $t = \sum_{e=1}^{r} |Comp_k(P,e)|$ is the number of all columns $\mathbf{c}$ (of area at most r) that can be added to P so that $P|\mathbf{c} \in \mathsf{HPDA}_k$.*

Theorem 2. $\mathrm{HPDAGEN}(n,k)$ *runs in constant amortized time.*

5   Conclusion and Future Works

In this paper we consider the approach to polyominoes generation based on discrete dynamical systems (previously used in [10,15]) and, for any fixed integer $k > 0$, we define a dynamical system that is used for the exhaustive generation of the class HPDA_k of polyominoes corresponding to partially directed animals with at most k holes. As in previous works, the properties of the dynamical system ensure that the generation is as efficient as possible, that is, each polyomino is generated in constant amortized time. We have implemented the algorithm in C++, obtaining the counting sequences of $\mathsf{HPDA}_k(n)$ up to $n = 22$ and $1 \leq k \leq 3$. None of these sequences appears in the OEIS database. A comparison of these sequences with the sequences associated with PDA and HFPDA (Hole-Free Partially Directed Animals) is given in Table 1.

We plan to develop a parallel version of the algorithm to compute more items of the counting sequences of $\mathsf{HPDA}_k(n)$.

References

1. Barcucci, E., Lungo, A.D., Pergola, E., Pinzani, R.: Directed animals, forests and permutations. Discrete Math. **204**(1–3), 41–71 (1999)
2. Barequet, G., Golomb, S.W., Klarner, D.A.: Polyominoes. In: Handbook of Discrete and Computational Geometry, 3rd edn., pp. 359–380. Chapman and Hall/CRC Press (2017)
3. Bousquet-Mélou, M.: A method for the enumeration of various classes of column-convex polygons. Discrete Math. **154**(1–3), 1–25 (1996)
4. Brocchi, S., Castiglione, G., Massazza, P.: On the exhaustive generation of k-convex polyominoes. Theor. Comput. Sci. **664**, 54–66 (2017)
5. Castiglione, G., Massazza, P.: An efficient algorithm for the generation of Z-convex polyominoes. In: Barneva, R.P., Brimkov, V.E., Šlapal, J. (eds.) IWCIA 2014. LNCS, vol. 8466, pp. 51–61. Springer, Cham (2014). https://doi.org/10.1007/978-3-319-07148-0_6
6. Castiglione, G., Restivo, A.: Reconstruction of L-convex polyominoes. Electron. Notes Discrete Math. **12**, 290–301 (2003)
7. Del Lungo, A., Duchi, E., Frosini, A., Rinaldi, S.: On the generation and enumeration of some classes of convex polyominoes. Electron. J. Comb. **11**(1) (2004)
8. Delest, M.P., Viennot, G.: Algebraic languages and polyominoes enumeration. Theor. Comput. Sci. **34**(1–2), 169–206 (1984)
9. Duchi, E., Rinaldi, S., Schaeffer, G.: The number of Z-convex polyominoes. Adv. Appl. Math. **40**(1), 54–72 (2008)
10. Formenti, E., Massazza, P.: From tetris to polyominoes generation. Electron. Notes Discrete Math. **59**, 79–98 (2017)

11. Formenti, E., Massazza, P.: On the Generation of 2-Polyominoes. In: Konstantinidis, S., Pighizzini, G. (eds.) DCFS 2018. LNCS, vol. 10952, pp. 101–113. Springer, Cham (2018). https://doi.org/10.1007/978-3-319-94631-3_9

12. Golomb, S.W.: Checker boards and polyominoes. Amer. Math. Monthly **61**, 675–682 (1954)

13. Jensen, I.: Enumerations of lattice animals and trees. J. Stat. Phys. **102**(3), 865–881 (2001)

14. Mantaci, R., Massazza, P.: From linear partitions to parallelogram polyominoes. In: Mauri, G., Leporati, A. (eds.) DLT 2011. LNCS, vol. 6795, pp. 350–361. Springer, Heidelberg (2011). https://doi.org/10.1007/978-3-642-22321-1_30

15. Massazza, P.: A dynamical system approach to polyominoes generation (2020 submitted)

16. Massazza, P.: On the generation of convex polyominoes. Discrete Appl. Math. **183**, 78–89 (2015)

17. Massazza, P.: Hole-free partially directed animals. In: Hofman, P., Skrzypczak, M. (eds.) DLT 2019. LNCS, vol. 11647, pp. 221–233. Springer, Cham (2019). https://doi.org/10.1007/978-3-030-24886-4_16

18. Privman, V., Barma, M.: Radii of gyration of fully and partially directed lattice animals. Zeitschrift für Physik B Condensed Matter **57**(1), 59–63 (1984)

19. Redner, S., Yang, Z.R.: Size and shape of directed lattice animals. J. Phys. A: Math. Gen. **15**(4), L177–L187 (1982)

On the Computational Power
of Programs over $\mathbf{BA_2}$ Monoid

Manasi S. Kulkarni, Jayalal Sarma$^{(\boxtimes)}$, and Janani Sundaresan

Indian Institute of Technology Madras, Chennai, India
`jayalal@cse.iitm.ac.in`

Abstract. The PLP conjecture for monoids states that for every monoid M, either M is universal (that is, *for every language $L \subseteq \Sigma^*$ there is a program over M which accepts the language L*) or it has the polynomial length property (that is, *every program over the monoid M has an equivalent program of length* poly(n)). The conjecture has been confirmed (Tesson-Therien (2001)) for the case of groups and several subclasses of aperiodic monoids such as the variety **DA** and the monoids divided by the monoid U. However, the case of the set of monoids divided by the monoid BA_2 is still open, which if resolved, confirms the conjecture for all aperiodic monoids.

In this paper, we make progress towards confirming the conjecture for the case when the monoid is BA_2. It is known (Tesson-Therien 2001) already that the monoid BA_2 is not universal.

Towards proving that the monoid BA_2 has polynomial length property, we show the following results: we define a program over a monoid M to be a *non-nullable program* if there is no input for which the yield of the program is the zero of the monoid. We prove the following:

- If a program over BA_2 is non-nullable, then there is an equivalent program with length at most poly(n).
- If a program over BA_2 is nullable, then it should be *exponentially non-nullable* - that is there should be at least $2^{\Omega(n)}$ many inputs which send the output of the program to 0 of BA_2. We show that for any program P over BA_2, if the zeroes of the program have a *witness subprogram* of polynomial length, then there is a program of length poly(n) equivalent to program P.

On the universality front, Tesson and Therien(2001) have already shown that PARITY cannot be computed by programs over BA_2. We strengthen this in two ways. Firstly, we show that programs over BA_2 cannot accept any subset of PARITY or $\overline{\text{PARITY}}$ of size $n^{\omega(1)}$. Secondly, we generalize the model of programs to allow parity queries to the input instead of variables. We show that BA_2 cannot compute parity of n input bits even when parity queries are allowed of size $k < \frac{n}{3}$. In contrast, we show that there are polynomial length programs over BA_2 to compute parity when queries are allowed as parity of $\frac{n}{3}$ bits or higher.

M. S. Kulkarni—Supported by postdoctoral fellowship from National Board of Higher Mathematics, Department of Atomic Energy (Government of India).

© Springer Nature Switzerland AG 2021
A. Leporati et al. (Eds.): LATA 2021, LNCS 12638, pp. 29–40, 2021.
https://doi.org/10.1007/978-3-030-68195-1_3

Keywords: Combinatorics on words · Polynomial length program (PLP) conjecture · Programs over monoids · Monoids and computation

1   Introduction

A program $P_k \in \{P_n\}_{n \geq 0}$ over a monoid M is a tuple $(\mathcal{I}, A)$ where $A \subset M$ and $\mathcal{I}$ is sequence $\mathcal{I} = I_1, I_2, \ldots I_\ell$ of triplets of the form $\langle i, g_0, g_1 \rangle$ where i is any position in the input $(1 \leq i \leq k)$, and $g_0, g_1 \in M$. Each triplet is called an *instruction* and ℓ is the length of the program. For a given input $x \in \{0,1\}^n$, the acceptance of the program P_n is defined as follows : For $0 \leq t \leq \ell$, for each instruction of the form $I_t = \langle i, g_0, g_1 \rangle$, define : $\mathsf{Yield}(I_t, x) = g_0$ if $x_i = 0$ and g_1 otherwise. The input x is said to be accepted by the program if $\mathsf{Yield}(P_n, x) = \prod_{i=1}^{\ell} \mathsf{Yield}(I_i, x) \in A$ where the product operation is the operation in the monoid.

The above model of computation caught a lot of attention due the surprising result due to Barrington [2] (partly observed earlier in [9]) which showed that the circuit complexity class of NC^1, exactly coincides with the languages that can be accepted by polynomial length programs over the group S_5 (and in fact, any monoid M which has a non-solvable group inside in it). This led to lot of research efforts [1,3,8,10,12,13], using the extensive tools from algebraic automata theory [5,7,11] in understanding the power of programs over different varieties of monoids.

Despite the set of deep tools, a particularly striking question that remains elusive is a characterization of efficient computation and universal computation by programs over monoids (See [14] and references therein). A monoid M is said to have the *polynomial length property*, if for any program P over the monoid M, accepting the set of strings $L_n \subseteq \{0,1\}^n$, there is a program P' equivalent to P (the set of accepted strings remains same) and length of P' is bounded by n^c (where n is the input length and c is a constant). A monoid M is said to be *universal*, if for any language $L = \{L_n\}_{n \geq 0}$ there is a family of programs $\{P_n\}_{n \geq 0}$ accepting L. Tesson and Therien [14] formulated the following conjecture:

PLP Conjecture: For any monoid M, M is universal if and only if M does not have the polynomial length property.

Known Results: Forward direction of the conjecture follows from a standard counting argument [14]. Indeed, we can argue that there must be a language $L \subseteq \Sigma^*$ that requires exponential length for any program that computes it. Hence, the challenge in the conjecture is to prove that if a monoid M is not universal then it must necessarily have polynomial length property. The PLP conjecture has been confirmed [1,14] for the case when M is a group. This is achieved by showing that if a group is nilpotent then it has polynomial length property and if it is not nilpotent then it is universal and hence cannot have polynomial length property.

For monoids, notable research efforts were in the case of aperiodic monoids. A variety of monoids contained in the set of aperiodic monoids is that of **DA**.

A celebrated classification of aperiodic monoids [4]: every aperiodic monoid is either in the variety **DA** or it is divided[1] by one of the two special monoids named U or BA_2 (See Sect. 2 for a precise definition).

Tesson and Therien [14] established the polynomial length property (and hence the PLP conjecture) for the variety[2] DA. In addition, Grosshands et al. [6] also showed explicit languages that cannot be computed by programs (with *any* length) over monoids in DA.

Hence, it is interesting to study the conjecture for the monoids U and BA_2. It is known that U [14] is universal and hence cannot have polynomial length property by counting arguments. Hence, the conjecture is confirmed for U and any monoid that it divides. However, the monoid BA_2 behaves differently and is less understood [13] - it is known [14] that no program (of any length) over BA_2 can compute parity function on n bits. Hence, to confirm the conjecture, a starting point[3] is to prove that BA_2 itself has polynomial length property.

Our Results: In this work, we make progress towards confirming the conjecture for the case when the monoid is BA_2. Our results are on both frontiers of universality and polynomial length property.

On the Polynomial Length Property: Towards proving that the monoid BA_2 has polynomial length property, we show the following results: we define a program over a monoid M to be a *non-nullable program* if there is no input which takes the output of the program to the zero of the monoid.

PLP for Non-nullable Programs: If a program over BA_2 is non-nullable, then we prove that there is an equivalent program with length at most poly(n) (Theorem 1).

Zeroes of Nullable Programs: If a program over BA_2 is nullable, then we first show a dichotomy that it should be *exponentially non-nullable* - that is there should be at least $2^{\Omega(n)}$ many inputs which sends the output of the program to 0 of BA_2 (Proposition 1).

PLP for (Restricted) Nullable Programs: We show that for any program P over BA_2, if the zeros of the program has a *witness subprogram* of polynomial length, then there is program of length poly(n) equivalent to P (Theorem 2).

Universality Property: Tesson and Therien [14] proved that the language PARITY $= \{w \in \{0,1\}^* \mid w$ has odd number of 1s$\}$ cannot be computed by programs over BA_2. We strengthen this in two ways.

---

[1] A monoid M is said to be divided by another monoid N if there is a homomorphism from a submonoid of M onto N.

[2] Languages recognized by polynomial length programs over monoids in DA are contained in AC^0 (with depth 3). And more generally, polynomial length programs over aperiodic monoids are known [3] to capture exactly the complexity class AC^0.

[3] Note, however, that even if BA_2 is proven to be having the polynomial length property, it does not imply the conjecture for all aperiodic monoids since monoids such as $BA_2 \times U$ which is divided by BA_2 does not have the polynomial length property.

Programs over BA_2 and subsets of PARITY: We show that programs over BA_2 cannot accept any subset of PARITY or $\overline{\text{PARITY}}$ of size $n^{\omega(1)}$. (Theorem 3).

Computing PARITY with smaller parity queries: We generalize the model of programs to allow parity queries to the input instead of variables. We show that BA_2 cannot compute parity of n input bits even when parity queries are allowed of size $k < \frac{n}{3}$ (Sect. 4). In contrast, we show that there are polynomial length programs over BA_2 to compute parity when queries are allowed as parity of size $\frac{n}{3}$ or higher.

2   Preliminaries

A monoid M is a set with an associative binary operation where the set is closed under the operation and has an identity with respect to it. That is, there is an $e \in M$ $ae = ea = a$ for all $a \in M$. In addition, if for every $a \in M$, there is a unique $b \in M$ such that $ab = ba = e$, then the set is a group.

Monoids are associated with regular languages through their syntactic congruence. U is the syntactic monoid of the regular language $(aa^*b)^*$, that is, one or more as separated by exactly one b. For the purpose of this paper, we only need the multiplicative structure of this monoid. The monoid has six elements $a, b, ab, ba, e, 0$ with e as the identity and 0 as the sink state. Two bs occurring consecutively gives 0. Similarly, BA_2 is the syntactic monoid of the regular language $(ab)^*$. The monoid has six elements $a, b, ab, ba, e, 0$ with e as the identity and 0 as the sink state with following relations between the elements of the monoid: $aa = 0, bb = 0, aba = a, bab = b$.

Normal Form for Programs over BA_2: Recall the definition of programs over monoids and acceptance of languages by programs from the introduction. By using a standard transformation we force each instruction to have one of $\{a, b, 0\}$ as a possible output when the literal queried is 1 and identity in the other case, or vice-versa. We do this by observing that the following instruction sequences are equivalent for $1 \leq i \leq n$. (1) $(i, x, y) \equiv (i, x, e)(i, e, y)$ for $x, y \in BA_2$ and $x \neq e$ and $y \neq e$ (2) $(i, ab, e) \equiv (i, a, e)(i, b, e)$. (3) $(i, e, ab) \equiv (i, e, a)(i, e, b)$. (4) $(i, ba, e) \equiv (i, b, e)(i, a, e)$. (5) $(i, e, ba) \equiv (i, e, b)(i, e, a)$.

The position of the instructions with 0 and e do not matter, as operations with 0 and e are commutative. We push them towards the beginning of the program. In a program P, we call a segment P_{ij}, as an x-sequence if it is a largest sequence of contiguous instructions such that the yield of each instruction is either x or e. We call these instructions of x-type (x-instructions for short).

The program, after the above changes, can be viewed as a 0-sequence, followed by alternating a-sequences and b-sequences. Here onward, we will consider every program over BA_2 to be of this form. The length of our program will be at most 4 times the length of the original program. Ignoring 0-instructions, we have only $4n$ types of instructions. (n choices exist for the input bit queried, whether it is an a-instruction or a b-instruction, and on which value of the input literal queried (1 or 0) it becomes identity.) We will use this fact repeatedly.

As a warm up for the arguments in this paper, consider a special monoid $U_k = \{a_1, a_2, \ldots, a_k\} \cup \{e\}$, defined by the operation $a_i a_j = a_j$ for each $i, j \in \{1, \ldots, k\}$. While it is known that U_k is a monoid in the variety **DA** (hence it is known to have polynomial length property already [14]), this serves as a simpler case to work with for the arguments in this paper. However, due to space limitations, we refer the reader to the full version.

3 PLP for (Restricted) Programs over **BA$_2$**

In this section, we prove the that programs over BA$_2$, satisfying some constraints can be reduced to polynomial size programs accepting the same language.

Let our program be P and let it have ℓ instructions. Instruction at position k will be called I_k, querying index i_k. P_{i-j} denotes the part of program P from the instruction at position i to that in j, both inclusive. We will split the analysis into two cases, in the next two sections. A program P is said to be non-nullable if $\forall w\ P(w) \neq 0$. The program is said to be nullable if $\exists w$ such that $P(w) = 0$.

3.1 PLP for Non-nullable Programs

In this subsection, we show that for any non-nullable program there is an equivalent program with length at most $\mathsf{poly}(n)$. Recall that for such programs, $\forall w P(w) \neq 0$. For some word $w \in \{0,1\}^n$ and program P, we call $|P(w)|$ as the total number of instructions in P which are not identity on w. w_i denotes the value at position i in w. $I(w)$ denotes the output of a particular instruction on w. We observe that any program P with $P(w) \neq 0$ and $|P(w)| > 2$ on BA$_2$, removing two consecutive *abs* or *bas* does not change the output.

Theorem 1. *For any non-nullable program P, there is a program P' with $|P'| \leq n^c$ for some constant $c > 0$ such that P and P' are equivalent. That is, $\forall w \in \{0,1\}^n, P(w) = P'(w)$.*

Proof. The idea of the proof is as follows. To reduce the length of the program, we need to remove instructions. We will first identify which instructions, if removed will necessarily change the output of the program. An instruction I in P is said to be non-removable if $\exists w$ such that $I(w) \neq e$ and $|P(w)| \leq 2$. If we remove such an instruction we will change the output on w. Instructions which do not obey this condition are called removable. We show that if all the instructions in a program are non-removable, then the program cannot be too long. Hence, it suffices to remove the actual removable instructions without changing the output.

We start with the following lemma, which bounds the length of a program that contains only non-removable instructions. Details appear in the full version.

Lemma 1. *If all instructions in P are non-removable, then $|P| \leq 8n$.*

We proceed to the proof of the theorem. We define a *good pair* of instructions as two instructions I_i and I_j of the same type which query different indices, a_i and a_j respectively. We handle the rest of the proof in two different cases:

Case 1: A good pair exists: Consider a good pair such that $|i - j|$ is minimum. We will call it the *closest good pair*. Without loss of generality, assume I_i and I_j are a-type instructions. By definition, I_i and I_j cannot have any other a-type instruction between them. Let P' be the sequence that lies in between them. Since every instruction in P' is a b-type instruction, they can be re-ordered without changing the output of the program. We have the following two exhaustive cases.

Case 1(a): $[\forall w \in \{0, 1\}^n : P'(w) = b]$. Since the program P is non-nullable, and P' can contain only b-type instructions, P' is equivalent to (a_i, b, e) and (a_i, e, b) next to each other (or in the reverse order, but notice that the order can be interchanged). We can combine one of these with I_i so that the output of I_i is ab or identity. We remove this as we know the output is never zero, and leave the other b-type instruction behind. For example, if I_i was (a_i, a, e), the segment $I_i b I_j$ would be replaced by $(a_i, e, b) I_j$. The pair of instructions we want to eliminate must be removable because when $a_i = 1$, for some w, $|P(w)| \geq 3$.

Case 1(b) - $[\exists w \in \{0, 1\}^n : P'(w) = e]$. If no instruction in the sequence P' queries a_i or a_j, we show that P should be nullable by constructing a $w' \in \{0, 1\}^n$ with $P(w') = 0$ as follows: we fix all indices which P' queries such that $P'(w) = e$ set a_i, a_j so that both I_i and I_j yields a. If P' does not query a_i or a_j, we are done because we can ensure that $P'(w) = e$ and $P(w) = 0$. Otherwise, P' must query at least one of a_i or a_j. Now we reorder the instructions in P' so that either the b-type instruction querying a_i is adjacent to I_i (or respectively a_j with I_j). We can combine them to get new instructions I_i' and I_j'. At least one of them can be eliminated if they are of the form (a_i, ab, e) or (a_i, e, ab).

Case 2: No good pair exists: This implies that the program queries only two input indices - one for the a-type instructions, and another for b-type instructions. Let the two indices queried be i and j, i.e., $(i, a, e), (i, e, a), (j, e, b), (j, b, e)$ are the only four types of instructions produced in the program. On setting $w_i = 1$, each instance of the instruction (i, a, e) must be separated by an instruction producing b irrespective of the values of w_j. If we have a single instruction producing b, we proceed according to the above first case to reduce the length. If not, j must be equal to i which is a contradiction.

Thus, the length of the program can always be reduced till only non- removable instructions, and a constant number of removable remain (this may be left by the end points of the last good pair). This proves the theorem. □

The next section talks of the other case, the nullable programs and gives a lower bound on the number of zeroes.

3.2 PLP for (Restricted) Nullable Programs

We now handle the programs where there are inputs such that the program outputs zero on it. A natural approach is to bound the number of inputs which makes it zero. However, we show the following strong dichotomy using the structure of the monoid BA$_2$. More precisely, we show that if the set of inputs which make the program go to zero output in the monoid (which we denote by Z) is non-empty then it has to be exponential large in size.

Proposition 1. *If P is nullable (that is, $|Z| > 0$), then $|Z| \geq 2^{n-3}$.*

The main idea of the proof is that, if program P is nullable, only the following reasons are possible: (a) One instruction I_k which has 0 as the output. (b) Two instructions I_p and I_{p+m} of the same type, with $x = I_p(w) = I_{p+m}(w)$ and x is either a or b, along with the condition that $I_{p+i}(w) = e$ for $1 \leq i < m$. The first case is easier to handle. If i is the index queried by I_k, and if we set w_i for some input such that $I_k(w) = 0$, we have 2^{n-1} choices for the remaining indices. There are at least 2^{n-1} inputs for which the output of the program is zero. The second one requires a more careful argument (see full version for details) keeping track of the pair of instructions which influences the output. Now we prove the polynomial length property for some restricted nullable programs.

Definition 1 (Witness for Zeroes). *For a program P, we call a set of instructions $\mathcal{I}$ of the program P, a witness for the zeroes of P if for all w with $P(w) = 0$ where either $\exists I \in \mathcal{I}, I(w) = 0$ or $\exists i < j$ with $I_i, I_j \in \mathcal{I}$ such that $I_i(w) = I_j(w) \neq e$ while $I_k(w) = e, \forall k \in (i, j)$.*

We now show that nullable programs which have polynomial size witness $\mathcal{I}$ can be reduced to equivalent programs of polynomial length.

Theorem 2. *For any P such that $\exists \mathcal{I}$, a witness for the zeroes of P and $|\mathcal{I}| \leq q(n)$ where q is some polynomial in n, then there exists a program P' with $|P'| \leq n^c$, for constant c such that $P'(w) = P(w)$ for all $w \in \{0,1\}^n$.*

Given a witness $\mathcal{I}$ for P, the sequence of instructions from I_{p+1} to I_{q-1} is called a *witness-free interval* if $I_r \notin \mathcal{I}$ for all $p < r < q$ and I_p, I_q if they exist, belong to $\mathcal{I}$. We will show that the size of any witness-free interval can be reduced to poly(n) while preserving output of the program, by removing certain subsets of instructions.

Lemma 2 (Witness-free Interval Shortening Lemma). *Given a sequence of instructions Q as a witness-free interval, there exists Q' such that $\forall w \in \{0,1\}^n$ if $Q(w) \neq 0$ then $Q(w) = Q'(w)$ and $|Q'| \leq q'(n)$ for some polynomial q'.*

We first give the proof of Theorem 2, followed by the proof of witness-free interval shortening lemma.

Proof of Theorem 2: For a program P, let P' be the program obtained from Lemma 2. If $P(w) \neq 0$, then $P'(w) = P'_0(w)I_1(w)P'_1(w)I_2(w)\ldots I_k(w)P'_k(w)$ and which after rewriting gives $P_0(w)I_1(w)P_1(w)I_2(w)\ldots I_k(w)P_k(w)$ which gives $P(w)$, where each P_i is a witness-free interval, and P'_i is the corresponding reduced equivalent interval given by the above lemma. The output is preserved by the lemma. Indeed, in the case when $P(w) = 0$ as all instructions from the witness $\mathcal{I}$ are retained, the output is unchanged.

To analyze the length of the program, let the total number of elements in $\mathcal{I}$ be bounded by the polynomial $q(n)$, and the length of any witness-free interval be bounded by the polynomial $q'(n)$. The total number of witness-free intervals will be at most $q(n) + 1$, at most one between any two elements of $\mathcal{I}$ if they were ordered by their position in P. The size of the entire program will be less than $q'(n)(q(n) + 1)$ which is indeed $\mathsf{poly}(n)$. $\qquad\qquad\square$

The Witness-boundary Graph Representation: We now introduce the main technical tool which is the graph representation of such programs. For each witness-free interval, we will define a graph which captures the program contained in the interval. Let NZ be the set of inputs for which the program yields a non-zero element, i.e., $NZ = \{w | P(w) \neq 0\}$.

For the witness-free interval in consideration, we first remove all instructions I_j such that $\forall w \in NZ$, $I_j(w) = e$. Let the resulting sequence of instructions be $I_1 I_2 \ldots I_\ell$ which we denote by Q for the rest of the argument.

Definition 2 (Witness-boundary Graph). *The witness-boundary (directed) graph $G : (V, E)$ for Q is as follows: $V = \{1, 2, \ldots, \ell\}$. We associate each vertex with the corresponding instruction I_j for $1 \leq j \leq \ell$. We define the edges $g = (I_i, I_j) \in E$ if $\exists w \in NZ$ such that both $I_i(w) \neq e$, $I_j(w) \neq e$ and $I_k(w) = e$, $\forall k \in (i, j)$ in the program Q.*

In addition, for each vertex $j \in V$, associate the non-empty subset $N_j = \{w \in NZ \mid I_j(w) \neq e\}$, and for each edge $g = (i, j) \in E$ associate $D_g = \{w \in NZ \mid I_i(w) \neq e, I_j(w) \neq e$, and $\forall j, \in (i, j), I_j(w) = e$ in the program $Q\}$. Note that, by definition, both N_j and D_g are non-empty subsets of NZ for all vertices j and edges g.

Each witness-free interval gives a graph, and these graphs in order by the position of the witness-free interval capture the program. We will prove that a witness-free interval Q' can be constructed from the graph of Q alone which has the same output as Q on all inputs $w \in NZ$. Q' will have as many instructions as the vertices of the graph of Q. This will prove that the graph model is equivalent to the corresponding witness-free interval, and that we do not lose any information about the program in the graph model even though we do not retain positional information.

Proposition 2. *From the witness-boundary graph G for the witness-free interval Q, we can construct a program Q' back such that $\forall w \in NZ, Q'(w) = Q(w)$.*

We will identify how to reduce the size of the graph without affecting the output. Towards this, we will introduce a set of reduction rules which enables us to remove sets of vertices.

A Reduction Rule for Witness-boundary Graphs: As mentioned before, if we remove ab's or ba's which occur consecutively in the word problem $P(w)$ when $P(w) \neq 0$, we will not change the output. We will use this to design a way to remove vertices repeatedly from the graph.

Consider any induced subgraph of G with vertex set V'. A path in the subgraph is said to be a w-maximal path if it is a path $\chi = (u_1, u_2, \ldots u_k)$ where $\forall 1 \le i < k$, if $g = (u_i, u_{i+1})$, $w \in D_g$ and if $v = u_j$, $\forall 1 \le j \le k$, $w \in N_v$, and the path cannot be extended, obeying this condition.[4].

Definition 3 (Removable Vertices). *A subset of instructions (vertices) C is removable if the following is true: let Q' be the program obtained after removing the instructions in the set C. (1) $\forall w \in NZ$, $Q'(w) = e$ implies $Q(w) = e$. (2) $\forall w \in S = \bigcup_{j \in C} N_j$, all w-maximal paths have odd length.*

We will refer to the above as a reduction rule. We justify the removal by showing that $\forall w \in NZ$, $Q'(w) = Q(w)$. For any $w \notin S$, the output is not changed, as all instructions in C are anyway yielding e on w. If $w \in S$, each w-maximal path corresponds to a sequence of instructions that has non-identity output on w and occur continuously in $P(w)$. Note that since $Q(w)$ is not zero, it must be in one of the four forms $(ab)^*a$, $(ab)^*$, $(ba)^*b$ or $(ba)^*$.

Let $\chi = \{a_1, a_2, \ldots a_k\}$ be the path we remove. If $k = 2m + 1$, the sequence in $Q(w)$ corresponding to χ would be either $(ab)^*a$ or $(ba)^*b$. On removing such a sequence, $Q(w)$ will become zero. Otherwise, if $k = 2m$ for some m and the path p was of odd length, the sequence in $Q(w)$ corresponding to χ would be of the form $(ab)^*$ or $(ba)^*$ which can be removed. We also do not remove the entire sequence to make $Q(w) = e$ from the first condition.

To reinterpret the reduction rules, we define a chain for any $w \in NZ$ as follows. A chain for w is a sequence of instructions I'_j from $j = 1$ to k such that $I'_j(w) \ne e$ $\forall 1 \le j \le k$ and I_j occurs before I_{j+1} in Q. It will be a path in the graph, as edges (I_j, I_{j+1}) will be present. In subset C of V, if for some w $\exists v \in C$ which appears in the chain for w, then the intersections of the chain for w with the subgraph induced by C will be w-maximal paths. It is sufficient if these intersections between chains and the subgraph are of odd length.

It remains to prove that on successively using the reduction rule, the size of the graph reduces to polynomial in input size. To this end, we prove a bound on the number nodes where a chain can start or end. See full version for a proof.

Lemma 3. *The number of vertices at which chains start or end is at most $O(n)$.*

Proof of the Witness-free Interval Shortening Lemma: We are now ready to prove the witness-free interval shortening lemma from the following:

Lemma 4. *Any given graph $G : (V, E)$ for witness-free interval Q, can be reduced to $G' : (V', E)$ corresponding to the polynomial length program Q' such that $Q(w) = Q'(w)$ for all $w \in NZ$.*

Proof. The algorithm considers each subset which appears consecutively in the topological ordering without a vertex where a chain starts or ends. The size of each subset is reduced to $\mathsf{poly}(n)$ from the reduction rules, and the total number of such sets is also $\mathsf{poly}(n)$.

---

[4] Such paths can also have zero length, when it is just a vertex v with input $w \in N_v$.

Algorithm 1: Reducing Size of G with Reduction Rules.

---

1 Let $[v_1v_2 \ldots v_\ell]$ be the topological ordering of graph G.
2 $V_1 \leftarrow \{i \mid \exists w, \text{ the chain for } w \text{ starts or ends at } v_i\}$.
3 Let $V_1 = \{a_1, a_2, \ldots a_k\}$ in ascending order.
4 $a_0 = 0$, $a_{k+1} = \ell + 1$.
5 $R \leftarrow V_1$
6 **for** $i = 1$ *to* $k + 1$ **do**
7 $\quad\quad C \leftarrow \{v_p \mid a_{i-1} < p < a_i\}$.
8 $\quad\quad \Gamma \leftarrow C$.
9 $\quad\quad$ **while** $\Gamma \neq \phi$ **do**
10 $\quad\quad\quad\quad \Gamma_1 \leftarrow \{v \in \Gamma \mid \exists v' \in V \setminus \Gamma, (v', v) \in E\}$
11 $\quad\quad\quad\quad \Gamma_2 \leftarrow \{v \in \Gamma \mid \exists v' \in V \setminus \Gamma, (v, v') \in E\}$
12 $\quad\quad\quad\quad \Gamma \leftarrow \Gamma \setminus (\Gamma_1 \cap \Gamma_2)$
13 $\quad\quad\quad\quad R \leftarrow R \cup (\Gamma_1 \cap \Gamma_2)$
14 $\quad\quad\quad\quad$ **if** $\forall w$ *w-maximal path in subgraph induced by* $\Gamma > 2$ **then**
15 $\quad\quad\quad\quad\quad\quad \Gamma_1 \leftarrow \{v \in \Gamma \mid \exists v' \in V \setminus \Gamma, (v', v) \in E\}$
16 $\quad\quad\quad\quad\quad\quad \Gamma_2 \leftarrow \{v \in \Gamma \mid \exists v' \in V \setminus \Gamma, (v, v') \in E\}$
17 $\quad\quad\quad\quad\quad\quad \Gamma' \leftarrow \Gamma \setminus (\{v : v \in \Gamma_1, v \text{ is } b\text{-type}\} \cup \{v : v \in \Gamma_2, v \text{ is } a\text{-type}\})$
18 $\quad\quad\quad\quad\quad\quad \Gamma \leftarrow \Gamma \setminus \Gamma'$
19 $\quad\quad\quad\quad$ **end**
20 $\quad\quad\quad\quad$ **else**
$\quad\quad\quad\quad\quad\quad$ // $\exists w$ **w-maximal path in subgraph induced by** Γ **is edge**
$\quad\quad\quad\quad\quad\quad$ (v_i, v_j).
21 $\quad\quad\quad\quad\quad\quad R \leftarrow R \cup \{v_i, v_j\}$
22 $\quad\quad\quad\quad\quad\quad \Gamma \leftarrow \Gamma \setminus \{v_i, v_j\}$
23 $\quad\quad\quad\quad$ **end**
24 $\quad\quad$ **end**
25 **end**
26 **return** $G : (R, E)$.

Correctness: The set R is the set of vertices which will be retained in the graph. The only step where vertices are removed from the set Γ, but not added to R is line 18. We will prove that Γ' is a removable set. From lines 15 to 18, we know that all w-maximal paths in Γ are of length at least 3. From Γ, each vertex of a-type having an incoming edge from $V \setminus \Gamma$, and each vertex of b-type having an outgoing edge to $V \setminus \Gamma$ is removed. Notice that, Γ' will be non-empty, as we remove at most two vertices from each w-maximal path and all such paths were of length at least three. Now, $\forall w$ each w-maximal path starts with an a and ends with a b in C', forming a sequence in $Q(w)$ of the form $(ab)^*$ and an odd length path. Γ' is removable. Therefore, the program Q' constructed after running the algorithm has the same output as Q on all $w \in NZ$.

We still have to prove the truth of the comment before line 21 of the algorithm. $\Gamma_1 \cap \Gamma_2$ is the set of all vertices from Γ which have both an incoming edge and outgoing edge from $V \setminus \Gamma$. Γ' is removed from Γ and added to R. For each w, any w-maximal path in Γ, if present, must be of length at least two now, as all w-maximal paths of length one in the subgraph induced by Γ must be present as vertices in $\Gamma_1 \cap \Gamma_2$.

Bounding the Size of the Retained Set R: We will look at the final size of R, and show that it is polynomial. $|R| \leq k + (k + 1)(2 + k_1)(k_2)$ where $k = |V_1|$, $(k + 1)$ is the number of iterations of the for loop starting at line 6, k_1 is the maximum size of $\Gamma_1 \cap \Gamma_2$ during the course of the algorithm and k_2 is the maximum number of iterations of the while loop starting at line 10.

V_1 is the set of all vertices where a chain starts or ends, and all these vertices are added to R initially, which accounts for the first term. We know $|V_1| = k = O(n)$ from Lemma 3. The total number of intervals in the topological ordering between vertices in V_1 will be $k+1$, the number of iterations of the loop at line 6.

The size of $(\Gamma_1 \cap \Gamma_2)$ is $O(n)$ for all values of Γ, as each instruction I of the $4n$ types can be associated with only one vertex in $(\Gamma_1 \cap \Gamma_2)$. If not, if two vertices have the same instruction I associated with them, then they must have a path between them and cannot belong to both Γ_1 and Γ_2. Thus, $k_1 \leq O(n)$.

Let the instructions associated with a w-maximal path (v_i, v_j) be I_1 and I_2. No vertex in $\Gamma \setminus \{v_i, v_j\}$ can have I_1 or I_2 as the instruction associated with them as otherwise the length of w-maximal path (v_i, v_j) would be higher. Every time we remove a w-maximal path of length two from Γ, we pick two instructions which cannot occur in any subset of Γ from the $4n$ types of instructions. The total number of iterations of the while loop starting at line 10 can be at most $2n$. $k_2 = O(n)$. This shows that $|R| \leq k + (k + 1)(2 + k_1)(k_2) \leq O(n^3)$ which concludes the proof.   □

4   Limitations of Programs over **BA₂** Monoid

The results in this section establish structure and limitations of programs over BA₂. Both of them extend the previous known result about how the parity language cannot be computed.

Computing Subsets of PARITY: The first one demonstrates that programs over BA₂ are very much more limited than just not being able to compute PARITY. They cannot accept any subset of PARITY which is more than poly(n) size.

Theorem 3. *Programs over* BA₂ *cannot compute any* $L \subseteq$ PARITY *where* $|L \cap \{0,1\}^n| \geq n^{\omega(1)}$.

We actually prove a stronger theorem (defer the details to the full version), that a language accepted by a program over BA₂ which accepts a set of strings either all with even parity or all with odd parity has poly(n) strings in it.

Lemma 5. *If* $L \subseteq$ PARITY *or* $L \subseteq \overline{\text{PARITY}}$ *is accepted by a program family* $\{P_n\}$ *over* BA₂, *with accepting set, say* $A \subset$ BA₂, *then* $|L \cap \{0,1\}^n| \leq O(n^3)$.

Computing PARITY with smaller PARITY Queries over BA₂: We study a generalization of the programs over BA₂ that computes parity using parity queries on smaller number of variables. We first define the model. A k-parity-program $\{P_n\}_{n \geq 0}$ over a monoid M is a tuple $(\mathcal{I}, A)$ where $A \subset M$ and $\mathcal{I}$ is sequence $\mathcal{I} = I_1, I_2, \ldots I_\ell$ of triplets of the form $\langle A, g_0, g_1 \rangle$ where $A \subseteq [n]$ with $|A| \leq k$, and $g_0, g_1 \in M$. Each triplet is called an *instruction*, the query is said to be k-*parity query* and ℓ is the length of the program. For a given input $x \in \{0, 1\}^n$, the acceptance of the program P_n is defined as follows : For $0 \leq t \leq \ell$, for each instruction of the form $I_t = \langle i, g_0, g_1 \rangle$, define : $\mathsf{Yield}(I_t, x) = g_0$ if $\oplus_{i \in A} x_i = 0$ and g_1 otherwise. The input x is said to be accepted by the program if $\mathsf{Yield}(P_n, x) = \prod_{i=1}^{\ell} \mathsf{Yield}(I_i, x) \in A$ where the product operation is the operation in the monoid. We show that even k-parity-programs (with $k < \frac{n}{3}$) over BA_2 still cannot still accept PARITY language. We state this as the theorem.

In contrast, we show that k-parity-programs can actually compute PARITY for $k \geq \frac{n}{3}$, which is based on a program without k-parity queries computing PARITY when $n = 3$. We also show that programs over BA_2 cannot compute PARITY even when $n = 4$. We defer the proof details to the full version.

References

1. Barrington, D., Straubing, H., Thérien, D.: Non-uniform automata over groups. Inf. Comput. **89**(2), 109–132 (1990). Preliminary version appeared in ICALP 1987
2. Barrington, D.A.: Bounded-width polynomial-size branching programs recognize exactly those languages in NC^1. J. Comput. Syst. Sci. **38**(1), 150–164 (1989)
3. Barrington, D.A.M., Thérien, D.: Finite monoids and the fine structure of NC^1. J. ACM **35**(4), 941–952 (1988)
4. Beaudry, M., McKenzie, P., Thérien, D.: The membership problem in aperiodic transformation monoids. J. ACM **39**(3), 599–616 (1992)
5. Eilenberg, S.: Automata, Languages, and Machines Vol A and B. Academic press (1974)
6. Grosshans, N., McKenzie, P., Segoufin, L.: The power of programs over monoids in DA. In: 42nd International Symposium on Mathematical Foundations of Computer Science (MFCS 2017), vol. 83, pp. 2:1–2:20 (2017)
7. Krohn, K., Rhodes, J.: Algebraic theory of machines. I. Prime decomposition theorem for finite semigroups and machines. Trans. Am. Math. Soc. **116**, 450–464 (1965)
8. Maciel, A., Péladeau, P., Thérien, D.: Programs over semigroups of dot-depth one. Theor. Comput. Sci. **245**(1), 135–148 (2000)
9. Maurer, W.D., Rhodes, J.L.: A property of finite simple non-Abelian groups. Proc. Am. Math. Soc. **16**(3), 552–554 (1965)
10. McKenzie, P., Péladeau, P., Thérien, D.: NC^1: the automata-theoretic viewpoint. Comput. Complex. **1**, 330–359 (1991)
11. Pin, J.E., Miller, R.E.: Varieties of Formal Languages. Plenum Publishing Co. (1986)
12. Péladeau, P., Straubing, H., Therien, D.: Finite semigroup varieties defined by programs. Theor. Comput. Sci. **180**(1), 325–339 (1997)
13. Tesson, P.: Computational Complexity Questions Related to Finite Monoids and Semigroups. Ph.D. thesis, McGill University (2003)
14. Tesson, P., Thérien, D.: The computing power of programs over finite monoids. J. Autom. Lang. Comb. **7**(2), 247–258 (2001)

Automata

Location Based Automata for Expressions with Shuffle

Sabine Broda[ID], António Machiavelo[ID], Nelma Moreira[(✉)][ID],
and Rogério Reis[ID]

CMUP & DM-DCC, Faculdade de Ciências da Universidade do Porto,
Rua do Campo Alegre, 4169-007 Porto, Portugal
{sabine.broda,antonio.machiavelo,nelma.moreira,rogerio.reis}@fc.up.pt

Abstract. We define the notion of location for regular expressions with shuffle by extending the notion of position in standard regular expressions. Locations allow for the definition of the sets Follow, First and Last with their usual semantics. From these, we construct an automaton for regular expressions with shuffle ($\mathcal{A}_{\mathrm{POS}}$), which generalises the standard position/Glushkov automaton. The sets mentioned above are also the foundation for other constructions, such as the Follow automaton, and automata based on pointed expressions. As a consequence, all these constructions can now be directly generalised to regular expressions with shuffle, as well as their known relationships. We also show that the partial derivative automaton ($\mathcal{A}_{\mathrm{PD}}$) is a (right) quotient of the new position automaton, $\mathcal{A}_{\mathrm{POS}}$. In previous work an automaton construction based on positions was studied ($\mathcal{A}_{\partial pos}$), and here we relate $\mathcal{A}_{\mathrm{POS}}$ and $\mathcal{A}_{\partial pos}$. Finally, we extend the construction of the prefix automaton $\mathcal{A}_{\mathrm{Pre}}$ to the shuffle operator and show that it is not a quotient of $\mathcal{A}_{\mathrm{POS}}$.

Keywords: Regular expressions · Position automaton · Shuffle operator

1 Introduction

Regular expressions with shuffle provide succinct representations for modelling concurrent systems [9,13]. Recently, several automata constructions for expressions with shuffle operators were considered [4,8,16]. For the standard interleaving shuffle operator ($\sqcup\!\sqcup$), Broda et al. [4] defined the partial derivative automaton ($\mathcal{A}_{\mathrm{PD}}$) and a position automaton ($\mathcal{A}_{\partial pos}$), showing that $\mathcal{A}_{\mathrm{PD}}$ is a right-quotient of $\mathcal{A}_{\partial pos}$. For standard regular expressions there is a one-to-one correspondence between (non initial) states in the position/Glushkov automaton [10] and occurrences of letters in the expression. This is no longer true for $\mathcal{A}_{\partial pos}$. Moreover, unlike most constructions of position automata, the definition of $\mathcal{A}_{\partial pos}$ did not

This work was partially supported by CMUP, which is financed by national funds through FCT – Fundação para a Ciência e a Tecnologia, I.P., under the project with reference UIDB/00144/2020.

© Springer Nature Switzerland AG 2021
A. Leporati et al. (Eds.): LATA 2021, LNCS 12638, pp. 43–54, 2021.
https://doi.org/10.1007/978-3-030-68195-1_4

rely on the sets First, Last, and Follow [4]. The former two sets characterise the positions of letters that can, respectively, begin or end words of the language; while the latter contains, for each letter position, the positions of letters that can follow that letter in words of the language. In order to define these sets for expressions containing the shuffle operator, we introduce novel and more complex structures, which we call locations. Locations are defined in such a way that, given an expression with nested shuffles, it allows to specify how far a word has advanced in each of the components (shuffles) of the expression. Each location in First corresponds to a position of a letter that can begin a word in the language. The positions that appear in a location in Last are the ones that can end a word. In the same way, the members of Follow represent pairs of positions of letters such that the second can follow the first in a word. From these sets, using locations, the definition of the position automaton $\mathcal{A}_{POS}$ is the usual one. Each location is the label of a state, and the incoming transitions of a state are labelled with letters corresponding to positions in that location.

This new construction is presented in Sect. 3, where we give an upper bound for the number of states of $\mathcal{A}_{POS}$ in the worst case. In Sect. 4 we show that the partial derivative automaton $\mathcal{A}_{PD}$ is a right-quotient of $\mathcal{A}_{POS}$. A comparison of $\mathcal{A}_{POS}$ and $\mathcal{A}_{\partial pos}$ is considered in Sect. 5, where their average number of states is discussed. Restricted to expressions without shuffle both constructions coincide with the standard position automaton. The same holds for $\mathcal{A}_{PD}$ [1]. Thus, the proofs in Sect. 4 are alternatives to show that, for standard regular expressions, $\mathcal{A}_{PD}$ is a quotient of $\mathcal{A}_{POS}$.

The sets First, Last and Follow are also the base for other constructions, such as the Follow automaton [11], as well as (deterministic) automata based on pointed expressions [2,3,15]. As a consequence, it is now straightforward to extend those constructions to expressions with shuffle, solving a problem stated by Asperti et al. [2]. Moreover, the known relationships between those constructions [3] extend to expressions with shuffle. Finally, we generalize the construction of the prefix automaton $\mathcal{A}_{Pre}$ [12,17] to the shuffle operator and show that $\mathcal{A}_{Pre}$ is not a quotient of $\mathcal{A}_{POS}$. The resulting taxonomy is presented in Sect. 6.

2   Preliminaries

Let RE denote the set of standard regular expressions over an alphabet Σ. The *language* associated with an expression $\alpha \in$ RE is denoted by $\mathcal{L}(\alpha)$. The empty word is denoted by ε. We define $\varepsilon(\alpha)$ by $\varepsilon(\alpha) = \varepsilon$ if $\varepsilon \in \mathcal{L}(\alpha)$, and $\varepsilon(\alpha) = \emptyset$ otherwise. Given a set of expressions S, the *language* associated with S is $\mathcal{L}(S) = \bigcup_{\alpha \in S} \mathcal{L}(\alpha)$. Moreover, we consider $\varepsilon S = S\varepsilon = S$ and $\emptyset S = S\emptyset = \emptyset$, for any set S of expressions. The *alphabetic size* $|\alpha|_\Sigma$ is its number of letters. We denote the subset of Σ containing the symbols that occur in α by Σ_α.

A *nondeterministic finite automaton* (NFA) is a quintuple $A = \langle Q, \Sigma, \delta, I, F \rangle$ where Q is a finite set of states, Σ is a finite alphabet, $I \subseteq Q$ is the set of initial states, $F \subseteq Q$ is the set of final states, and $\delta : Q \times \Sigma \to 2^Q$ is the transition function. If $|I| = 1$ and $|\delta(q, \sigma)| \leq 1$, for all $q \in Q, \sigma \in \Sigma$, A is *deterministic* (DFA). The *language* of A is denoted by $\mathcal{L}(A)$ and two automata are *equivalent* if they have the same language. Given an automaton $A = \langle Q, \Sigma, \delta, I, F \rangle$ its *reversal* is $A^R = \langle Q, \Sigma, \delta^R, F, I \rangle$, where $\delta^R(q, \sigma) = \{ p \mid q \in \delta(p, \sigma) \}$, and $\mathcal{L}(A^R) = \mathcal{L}(A)^R$, which is the language obtained by reversing the words in $\mathcal{L}(A)$. Two automata $A_1 = \langle Q_1, \Sigma, \delta_1, I_1, F_1 \rangle$ and $A_2 = \langle Q_2, \Sigma, \delta_2, I_2, F_2 \rangle$ are *isomorphic*, $A_1 \simeq A_2$, if there is a bijection $\varphi : Q_1 \longrightarrow Q_2$ such that $\varphi(I_1) = I_2$, $\varphi(F_1) = F_2$, and $\varphi(\delta_1(q_1, \sigma)) = \delta_2(\varphi(q_1), \sigma)$, for all $q_1 \in Q_1$, $\sigma \in \Sigma$. An equivalence relation $\equiv$ defined on the set of states Q is *right-invariant* w.r.t. A if and only if $\equiv \, \subseteq (Q - F)^2 \cup F^2$ and if $p \equiv q$, then $\forall \sigma \in \Sigma$, $p' \in \delta(p, \sigma)$, $\exists q' \in \delta(q, \sigma)$ such that $p' \equiv q'$, for all $p, q \in Q$. If $\equiv$ is a right-invariant relation on Q, the *right-quotient automaton* $A/\equiv$ is given by $A/\equiv \, = \langle Q/\equiv, \Sigma, \delta/\equiv, I/\equiv, F/\equiv \rangle$, where $\delta/\equiv ([p], \sigma) = \{ [q] \mid q \in \delta(p, \sigma) \}$. Then, $\mathcal{L}(A/\equiv) = \mathcal{L}(A)$. An equivalence relation on Q is *left-invariant* w.r.t. A if it is right-invariant w.r.t. A^R.

The Position Automaton. Given $\alpha \in \mathrm{RE}$, one can mark each occurrence of a letter σ with its position in α, reading it from left to right. The resulting regular expression is a *marked* regular expression $\overline{\alpha}$ with all letters occurring only once (linear) and belonging to $\Sigma_{\overline{\alpha}}$. Thus, a *position* $i \in [1, |\alpha|_\Sigma]$ corresponds to the symbol σ_i in $\overline{\alpha}$, and consequently to exactly one occurrence of σ in α. For instance, if $\alpha = a(bb + aba)^\star b$, then $\overline{\alpha} = a_1(b_2 b_3 + a_4 b_5 a_6)^\star b_7$. The same notation is used for unmarking, $\overline{\overline{\alpha}} = \alpha$. Let $\mathsf{Pos}(\alpha) = \{1, 2, \ldots, |\alpha|_\Sigma\}$, and $\mathsf{Pos}_0(\alpha) = \mathsf{Pos}(\alpha) \cup \{0\}$. Positions were used by Glushkov [10] to define an NFA equivalent to α, usually called the *position* or *Glushkov automaton*, $\mathcal{A}_{\mathrm{POS}}(\alpha)$. Each state of the automaton, except for the initial one, corresponds to a position, and there exists a transition from i to j by σ such that $\overline{\sigma_j} = \sigma$, if σ_i can be followed by σ_j in some word represented by $\overline{\alpha}$. The sets that are used to define the position automaton are $\mathsf{First}(\overline{\alpha}) = \{ i \mid \exists w \in \Sigma_{\overline{\alpha}}^\star . \sigma_i w \in \mathcal{L}(\overline{\alpha}) \}$, $\mathsf{Last}(\overline{\alpha}) = \{ i \mid \exists w \in \Sigma_{\overline{\alpha}}^\star . w \sigma_i \in \mathcal{L}(\overline{\alpha}) \}$ and, given $i \in \mathsf{Pos}(\alpha)$, $\mathsf{Follow}(\overline{\alpha}, i) = \{ j \mid \exists u, v \in \Sigma_{\overline{\alpha}}^\star . u \sigma_i \sigma_j v \in \mathcal{L}(\overline{\alpha}) \}$. For the sake of readability, whenever an expression α is not marked, we take $f(\alpha) = f(\overline{\alpha})$, for any function f that have marked expressions as arguments. We define the position automaton using the approach in Broda et al [3], where the transition function is expressed as the composition of functions Select and Follow. Given a letter σ and a set of positions S, the function Select selects the subset of positions in S that correspond to letter σ. Formally, given $S \subseteq \mathsf{Pos}(\alpha)$ and $\sigma \in \Sigma$, let $\mathsf{Select}(S, \sigma) = \{ i \mid i \in S \wedge \overline{\sigma_i} = \sigma \}$. Then, the *position automaton* for α is $\mathcal{A}_{\mathrm{POS}}(\alpha) = \langle \mathsf{Pos}_0(\alpha), \Sigma, \delta_{\mathrm{POS}}, 0, \mathsf{Last}_0(\alpha) \rangle$, where $\delta_{\mathrm{POS}}(i, \sigma) = \mathsf{Select}(\mathsf{Follow}(\alpha, i), \sigma)$.

Regular Expressions with Shuffle. Given an alphabet Σ, the shuffle of two words in $\Sigma^\star$ is the finite set of words defined inductively as follows: $x \amalg \varepsilon = \varepsilon \amalg x = \{x\}$ and $ax \amalg by = \{ az \mid z \in x \amalg by \} \cup \{ bz \mid z \in ax \amalg y \}$, for $x, y \in \Sigma^\star$, and $a, b \in \Sigma$. This definition is extended to languages in the natural way by $L_1 \amalg L_2 = \bigcup_{x \in L_1, y \in L_2} x \amalg y$. It is well known that $\amalg$ is a regular operator. One can, hence, extend regular expressions to include the $\amalg$ operator. The set of regular expressions with shuffle over Σ, $\mathcal{R}(\amalg)$, contains $\emptyset$ plus all terms generated by the grammar

$$\alpha \rightarrow \varepsilon \mid a \mid (\alpha + \alpha) \mid (\alpha \cdot \alpha) \mid (\alpha \amalg \alpha) \mid \alpha^{\star} \quad (a \in \Sigma).$$

The language represented by an expression $\alpha \amalg \beta$ is $\mathcal{L}(\alpha \amalg \beta) = \mathcal{L}(\alpha) \amalg \mathcal{L}(\beta)$.

3   A Location Based Position Automaton

In this section we define a new construction for a position automaton for expressions with shuffle, which is based on the sets First, Last, and Follow. In order to define those sets for expressions containing the shuffle operator, we need to consider more complex structures, which we call locations. Locations are defined in such a way that, given an expression with nested shuffles, it allows to specify how far a word has advanced in each of the components (shuffles) of this expression. Given $\alpha \in \mathcal{R}(\amalg)$, we associate a *set of locations* $\mathsf{Loc}(\alpha) = \mathsf{Loc}(\overline{\alpha})$, which is inductively defined on the structure of the expression $\overline{\alpha}$ as follows (where the concatenation operator $\cdot$ is omitted).

$$\mathsf{Loc}(\varepsilon) = \emptyset, \ \mathsf{Loc}(\sigma_i) = \{i\}, \ \mathsf{Loc}(\alpha^{\star}) = \mathsf{Loc}(\alpha),$$
$$\mathsf{Loc}(\alpha_1 + \alpha_2) = \mathsf{Loc}(\alpha_1 \alpha_2) = \mathsf{Loc}(\alpha_1) \cup \mathsf{Loc}(\alpha_2),$$
$$\mathsf{Loc}(\alpha_1 \amalg \alpha_2) = \mathsf{Loc}(\alpha_1) \times \mathsf{Loc}(\alpha_2) \cup \mathsf{Loc}(\alpha_1) \times \{0\} \cup \{0\} \times \mathsf{Loc}(\alpha_2).$$

Note that each location p in α is either a position $i \in \mathsf{Pos}(\alpha)$, or of the form $(0, p_2), (p_1, 0)$, or (p_1, p_2), where p_1, p_2 are also locations in α. *The set of positions of a location* p, $\mathsf{ipos}(p)$, is defined inductively by $\mathsf{ipos}(i) = \{i\}$, $\mathsf{ipos}((0, p)) = \mathsf{ipos}((p, 0)) = \mathsf{ipos}(p)$, and finally $\mathsf{ipos}((p_1, p_2)) = \mathsf{ipos}(p_1) \cup \mathsf{ipos}(p_2)$.

Example 1. For $\alpha = (a^{\star}b \amalg cd)^{\star} \amalg (ac)^{\star}$ and $\overline{\alpha} = (a_1^{\star}b_2 \amalg c_3 d_4)^{\star} \amalg (a_5 c_6)^{\star}$,

$$\mathsf{Loc}(\alpha) = \{ \, ((0,3), n), ((0,4), n), ((1,0), n), ((2,0), n), (0,5), (0,6), ((1,3), n),$$
$$((2,3), n), ((1,4), n), ((2,4), n) \mid n = 0, 5, 6 \, \},$$

$\mathsf{ipos}(((2,3), 0)) = \{2, 3\}$, and $\mathsf{ipos}(((2,3), 5)) = \{2, 3, 5\}$. For instance, the location $((2,3), 5)$ corresponds to words for which the last letters read in the subexpressions $a^{\star}b$, cd and $(ac)^{\star}$, are respectively b, c, and a.

Lemma 1. *Given* $\alpha \in \mathcal{R}(\amalg)$ *and* $i \in \mathsf{Pos}(\alpha)$, *there is* $p \in \mathsf{Loc}(\alpha)$ *with* $i \in \mathsf{ipos}(p)$ *if and only if there are words* $w, w' \in \Sigma_{\overline{\alpha}}^{\star}$, *such that* $w\sigma_i w' \in \mathcal{L}(\overline{\alpha})$.

Given $\alpha \in \mathcal{R}(\amalg)$, the states in the position automaton will be labelled by the elements in $\mathsf{Loc}(\alpha)$, except for the initial state labelled as 0. The following proposition gives an upper bound on the size of $\mathsf{Loc}(\alpha)$, and the next example exhibits an expression for which this upper bound is reached.

Proposition 1. *Given* $\alpha \in \mathcal{R}(\amalg)$, *one has* $|\mathsf{Loc}(\alpha)| \leq 2^{|\alpha|_{\Sigma}} - 1$.

Example 2. Consider $\alpha_n = a_1 \amalg \cdots \amalg a_n$, where $n \geq 1$, $a_i \neq a_j$ for $1 \leq i \neq j \leq n$. Then, $\mathsf{ipos}(\mathsf{Loc}(\alpha_n)) = 2^{\mathsf{Pos}(\alpha_n)} \setminus \{\emptyset\}$, which is of size $2^n - 1$.

The sets First, Last and Follow are defined extending the usual definitions, [5, 11], to the shuffle operator. The set $\mathsf{First}(\alpha)$ for $\alpha = \alpha_1 \sqcup\!\sqcup \alpha_2$ is defined as follows,

$$\mathsf{First}(\alpha_1 \sqcup\!\sqcup \alpha_2) = \mathsf{First}(\alpha_1) \times \{0\} \cup \{0\} \times \mathsf{First}(\alpha_2).$$

Fact 1. *One has* $\mathsf{First}(\alpha) \subseteq \mathsf{Loc}(\alpha)$. *Furthermore, every location* $p \in \mathsf{First}(\alpha)$ *contains exactly one non-null component* $i \in \mathsf{Pos}(\alpha)$. *Thus,* $\mathsf{ipos}(p) = \{i\}$.

Lemma 2. *Given* $\alpha \in \mathcal{R}(\sqcup\!\sqcup)$, *there is a location* $p \in \mathsf{First}(\alpha)$ *with* $\mathsf{ipos}(p) = \{i\}$ *if and only if there is some* $w \in \Sigma_{\overline{\alpha}}^\star$, *such that* $\sigma_i w \in \mathcal{L}(\overline{\alpha})$.

Proof. The proof is by structural induction on the marked expression $\overline{\alpha}$. For ε and marked singletons the result is obvious. For union, concatenation and Kleene star the proof is similar to the one for standard expressions. Consider the case of an expression $\alpha_1 \sqcup\!\sqcup \alpha_2$. Let $(p, 0)$, with $p \in \mathsf{First}(\alpha_1)$ and $\mathsf{ipos}((p, 0)) = \mathsf{ipos}(p) = \{i\}$. By the induction hypothesis, there is some $w \in \Sigma_{\overline{\alpha}}^\star$, such that $\sigma_i w \in \mathcal{L}(\alpha_1)$. Consider any word $w' \in \mathcal{L}(\alpha_2) \neq \emptyset$. Then, $\sigma_i w w' \in \mathcal{L}(\alpha_1) \sqcup\!\sqcup \mathcal{L}(\alpha_2) = \mathcal{L}(\alpha_1 \sqcup\!\sqcup \alpha_2)$. The case of $(0, p)$, with $p \in \mathsf{First}(\alpha_2)$ is analogous. For the other direction, consider a word $\sigma_i w \in \mathcal{L}(\alpha_1 \sqcup\!\sqcup \alpha_2)$. By definition, either there is some $\sigma_i w_1 \in \mathcal{L}(\alpha_1)$ and some $w_2 \in \mathcal{L}(\alpha_2)$ such that $w \in w_1 \sqcup\!\sqcup w_2$, or vice-versa. By the induction hypothesis, there exists a location $p \in \mathsf{First}(\alpha_1)$ with $\mathsf{ipos}(p) = \{i\}$. Consequently $(p, 0) \in \mathsf{First}(\alpha_1 \sqcup\!\sqcup \alpha_2)$. □

The set $\mathsf{Last}(\alpha)$ for $\alpha = \alpha_1 \sqcup\!\sqcup \alpha_2$ is defined by

$$\mathsf{Last}(\alpha_1 \sqcup\!\sqcup \alpha_2) = \mathsf{Last}(\alpha_1) \times \mathsf{Last}(\alpha_2)$$
$$\bigcup \varepsilon(\alpha_1)\Big(\{0\} \times \mathsf{Last}(\alpha_2)\Big) \bigcup \varepsilon(\alpha_2)\Big(\mathsf{Last}(\alpha_1) \times \{0\}\Big).$$

Lemma 3. *Given* $\alpha \in \mathcal{R}(\sqcup\!\sqcup)$, *there is a location* $p \in \mathsf{Last}(\alpha)$ *and* $i \in \mathsf{ipos}(p)$ *if and only if there is some* $w \in \Sigma_{\overline{\alpha}}^\star$, *such that* $w\sigma_i \in \mathcal{L}(\overline{\alpha})$.

Proof. The proof is by structural induction on $\overline{\alpha}$. We need only to consider the case of an expression $\alpha_1 \sqcup\!\sqcup \alpha_2$. Let $(p_1, p_2) \in \mathsf{Last}(\alpha_1) \times \mathsf{Last}(\alpha_2)$ and $i \in \mathsf{ipos}(p_1)$. By the induction hypothesis, there is some $w_1 \in \Sigma_{\alpha_1}^\star$, such that $w_1 \sigma_i \in \mathcal{L}(\alpha_1)$. For any $w_2 \in \mathcal{L}(\alpha_2) \neq \emptyset$, $w_2 w_1 \sigma_i \in \mathcal{L}(\alpha_1 \sqcup\!\sqcup \alpha_2)$. Next, consider a location $(0, p) \in \varepsilon(\alpha_1)(\{0\} \times \mathsf{Last}(\alpha_2))$ and $i \in \mathsf{ipos}((0, p)) = \mathsf{ipos}(p)$. By the induction hypothesis, there is some $w_2 \in \Sigma_{\alpha_2}^\star$, such that $w_2 \sigma_i \in \mathcal{L}(\alpha_2)$. On the other hand $\varepsilon \in \mathcal{L}(\alpha_1)$. Thus, $w_2 \sigma_i \in \mathcal{L}(\alpha_1 \sqcup\!\sqcup \alpha_2)$. The remaining cases are analogous.

For the other direction, consider $w\sigma_i \in \mathcal{L}(\alpha_1 \sqcup\!\sqcup \alpha_2)$. By definition, there is some $w_1 \sigma_i \in \mathcal{L}(\alpha_1)$ and some $w_2 \in \mathcal{L}(\alpha_2)$ such that $w \in w_1 \sqcup\!\sqcup w_2$ (or vice-versa). By the induction hypothesis, there exists a location $p_1 \in \mathsf{Last}(\alpha_1)$ with $i \in \mathsf{ipos}(p_1)$. If $w_2 = w_2' \sigma_j$, by induction there is some $p_2 \in \mathsf{Last}(\alpha_2)$ with $j \in \mathsf{ipos}(p_2)$. Thus, $(p_1, p_2) \in \mathsf{Last}(\alpha_1) \times \mathsf{Last}(\alpha_2)$ and $i \in \mathsf{ipos}(p_1) \subseteq \mathsf{ipos}((p_1, p_2))$. If $w_2 = \varepsilon$, then $(p_1, 0) \in \varepsilon(\alpha_2)(\mathsf{Last}(\alpha_1) \times \{0\})$ and $i \in \mathsf{ipos}(p_1) = \mathsf{ipos}((p_1, 0))$. □

For expressions without shuffle, each position i corresponds exactly to one marked letter σ_i and, consequently, all incoming transitions of state i are labelled by σ. This is no longer true for expressions with shuffle. In this case a location p labelling a state can have incoming transitions labelled by different letters (corresponding to the positions in $\mathsf{ipos}(p)$), depending on the source state. For this reason we will include letters in the definition of the Follow set.

Let $\mathsf{FirstL}(\alpha) = \{ (\sigma_i, p) \mid p \in \mathsf{First}(\alpha) \wedge \mathsf{ipos}(p) = \{i\} \}$. We also define $\mathsf{Loc}_0(\alpha) = \mathsf{Loc}(\alpha) \cup \{0\}$ and $\mathsf{Last}_0(\alpha) = \mathsf{Last}(\alpha) \cup \varepsilon(\alpha)\{0\}$. Finally, we define $\mathsf{Follow} : \mathcal{R}(\sqcup\!\sqcup) \times \mathsf{Loc}_0(\alpha) \to 2^{\Sigma \times \mathsf{Loc}(\alpha)}$ by setting $\mathsf{Follow}(\alpha, 0) = \mathsf{FirstL}(\alpha)$, and for $p \in \mathsf{Loc}(\alpha)$,

$$\mathsf{Follow}(\varepsilon, p) = \mathsf{Follow}(\sigma_i, p) = \emptyset,$$

$$\mathsf{Follow}(\alpha_1 + \alpha_2, p) = \begin{cases} \mathsf{Follow}(\alpha_1, p), & \text{if } p \in \mathsf{Loc}(\alpha_1), \\ \mathsf{Follow}(\alpha_2, p), & \text{if } p \in \mathsf{Loc}(\alpha_2). \end{cases}$$

$$\mathsf{Follow}(\alpha_1 \alpha_2, p) = \begin{cases} \mathsf{Follow}(\alpha_1, p), & \text{if } p \in \mathsf{Loc}(\alpha_1) \setminus \mathsf{Last}(\alpha_1), \\ \mathsf{Follow}(\alpha_1, p) \cup \mathsf{FirstL}(\alpha_2), & \text{if } p \in \mathsf{Last}(\alpha_1), \\ \mathsf{Follow}(\alpha_2, p), & \text{if } p \in \mathsf{Loc}(\alpha_2). \end{cases}$$

$$\mathsf{Follow}(\alpha_1^\star, p) = \begin{cases} \mathsf{Follow}(\alpha_1, p), & \text{if } p \notin \mathsf{Last}(\alpha_1), \\ \mathsf{Follow}(\alpha_1, p) \cup \mathsf{FirstL}(\alpha_1), & \text{otherwise}. \end{cases}$$

$$\mathsf{Follow}(\alpha_1 \sqcup\!\sqcup \alpha_2, p) = \{ (\sigma_i, (p_1', p_2)) \mid (\sigma_i, p_1') \in \mathsf{Follow}(\alpha_1, p_1) \}$$
$$\cup \{ (\sigma_i, (p_1, p_2')) \mid (\sigma_i, p_2') \in \mathsf{Follow}(\alpha_2, p_2) \}$$
$$\text{if } p = (p_1, p_2) \wedge p \in \mathsf{Loc}(\alpha_1 \sqcup\!\sqcup \alpha_2).$$

Furthermore, given $S \in 2^{\mathsf{Loc}_0(\alpha)}$ set $\mathsf{Follow}(\alpha, S) = \bigcup_{p \in S} \mathsf{Follow}(\alpha, p)$.

Example 3. For $\alpha = a^\star \sqcup\!\sqcup b^\star$ and $\overline{\alpha} = a_1^\star \sqcup\!\sqcup b_2^\star$, $\mathsf{Last}(\alpha) = \{(1,0),(0,2),(1,2)\}$, $\mathsf{Follow}(\alpha, 0) = \mathsf{FirstL}(\alpha) = \{(a_1, (1,0)), (b_2, (0,2))\}$, and

$$\mathsf{Follow}(\alpha, (1,0)) = \{(a_1, (1,0)), (b_2, (1,2))\},$$
$$\mathsf{Follow}(\alpha, (0,2)) = \{(a_1, (1,2)), (b_2, (0,2))\},$$
$$\mathsf{Follow}(\alpha, (1,2)) = \{(a_1, (1,2)), (b_2, (1,2))\}.$$

Lemma 4. *For an expression $\alpha \in \mathcal{R}(\sqcup\!\sqcup)$ and $i, j \in \mathsf{Pos}(\alpha)$, there are locations $p, q \in \mathsf{Loc}(\alpha)$ with $(\sigma_j, q) \in \mathsf{Follow}(\alpha, p)$ and $i \in \mathsf{ipos}(p)$ if and only if there are $w, w' \in \Sigma_{\overline{\alpha}}^\star$, such that $w \sigma_i \sigma_j w' \in \mathcal{L}(\overline{\alpha})$.*

For a set $S \subseteq \Sigma_{\overline{\alpha}} \times \mathsf{Loc}(\alpha)$ and $\sigma \in \Sigma$, let

$$\mathsf{Select}(S, \sigma) = \{ p \mid (\sigma_i, p) \in S \wedge \overline{\sigma_i} = \sigma \}.$$

The *position automaton* for α is $\mathcal{A}_{\mathrm{POS}}(\alpha) = \langle \mathsf{Loc}_0(\alpha), \Sigma, \delta_{\mathrm{POS}}, 0, \mathsf{Last}_0(\alpha) \rangle$, where $\delta_{\mathrm{POS}}(p, \sigma) = \mathsf{Select}(\mathsf{Follow}(\alpha, p), \sigma)$, for $p \in \mathsf{Loc}_0(\alpha), \sigma \in \Sigma$.

Example 4. For $\alpha = (ab)^\star \sqcup\!\sqcup (bc)^\star$ with $\overline{\alpha} = (a_1 b_2)^\star \sqcup\!\sqcup (b_3 c_4)^\star$, $\mathsf{Loc}_0(\alpha) = \{0, (0,3), (0,4), (1,0), (2,0), (1,3), (1,4), (2,3), (2,4)\}$, $\mathsf{First}(\alpha) = \{(1,0), (0,3)\}$, and $\mathsf{Last}_0(\alpha) = \{0, (0,4), (2,0), (2,4)\}$. The position automaton $\mathcal{A}_{\mathrm{POS}}(\alpha)$ is depicted below.

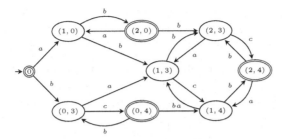

The correctness of this construction follows from the previous four lemmas.

Proposition 2. $\mathcal{L}(\mathcal{A}_{\mathrm{POS}}(\alpha)) = \mathcal{L}(\alpha)$.

Note that, for an expression α without shuffle we have $\mathsf{Loc}(\alpha) = \mathsf{Pos}(\alpha)$ and $\mathcal{A}_{\mathrm{POS}}(\alpha)$ is exactly the standard position automaton. In fact, the usual *Follow* set for a position j is equal to $\{\, i \mid (\sigma_i, i) \in \mathsf{Follow}(\alpha, j)\,\}$.

4 $\mathcal{A}_{\mathrm{PD}}(\alpha)$ as a Quotient of $\mathcal{A}_{\mathrm{POS}}(\alpha)$

In this section we show that the partial derivative automaton $\mathcal{A}_{\mathrm{PD}}(\alpha)$ for expressions $\alpha \in \mathcal{R}(\sqcup\!\sqcup)$ [4] is a quotient of the position automaton as defined in the previous section. We recall the definition of the set of partial derivatives of $\alpha \in \mathcal{R}(\sqcup\!\sqcup)$ w.r.t. a letter $\sigma \in \Sigma$, for $\alpha = \alpha_1 \sqcup\!\sqcup \alpha_2$:

$$\partial_\sigma(\alpha_1 \sqcup\!\sqcup \alpha_2) = \partial_\sigma(\alpha_1) \sqcup\!\sqcup \{\alpha_2\} \cup \{\alpha_1\} \sqcup\!\sqcup \partial_\sigma(\alpha_2).$$

As usual, the set of partial derivatives of $\alpha \in \mathcal{R}(\sqcup\!\sqcup)$ w.r.t. a word $w \in \Sigma^\star$ is inductively defined by $\partial_\varepsilon(\alpha) = \{\alpha\}$ and $\partial_{w\sigma}(\alpha) = \partial_\sigma(\partial_w(\alpha))$. Let $\partial(\alpha) = \bigcup_{w \in \Sigma^\star} \partial_w(\alpha)$, and $\partial^+(\alpha) = \partial(\alpha) \setminus \partial_\varepsilon(\alpha)$. The partial derivative automaton of $\alpha \in \mathcal{R}(\sqcup\!\sqcup)$ is $\mathcal{A}_{\mathrm{PD}}(\alpha) = \langle \partial(\alpha), \Sigma, \{\alpha\}, \delta_{\mathrm{PD}}, F_{\mathrm{PD}}\rangle$, where $F_{\mathrm{PD}} = \{\beta \in \partial(\alpha) \mid \varepsilon(\beta) = \varepsilon\}$ and $\delta_{\mathrm{PD}}(\beta, \sigma) = \partial_\sigma(\beta)$, for $\beta \in \mathcal{R}(\sqcup\!\sqcup), \sigma \in \Sigma$.

The partial derivative automaton for $\alpha_1 \sqcup\!\sqcup \alpha_2$, where $\alpha_1 = (ab)^\star$ and $\alpha_2 = (bc)^\star$, from Example 4, is depicted on the right.

Note that both $\partial^+(\alpha)$ and $\mathsf{Loc}_0(\alpha)$ are at most of size $2^{|\alpha|_\Sigma}$, cf. [4]. Champarnaud and Ziadi [7] proved that, for standard regular expressions, $\mathcal{A}_{\mathrm{PD}}$ is a quotient of the position automaton $\mathcal{A}_{\mathrm{POS}}$. It was shown that, given a position i, there exists some expression, $c_i(\overline{\alpha})$, such that for all $w \in \Sigma^\star_{\overline{\alpha}}$, either $\partial_{w\sigma_i}(\overline{\alpha})$ is empty or is $\{c_i(\overline{\alpha})\}$. This naturally induces a right-invariant relation on the set of posi-

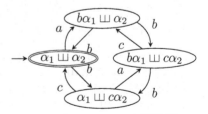

tions, where $i \equiv_o j$ if $c_i(\overline{\alpha}) = c_j(\overline{\alpha})$, and such that $\mathcal{A}_{\mathrm{POS}}(\alpha)/\!\equiv_o \simeq \overline{\mathcal{A}_{\mathrm{PD}}(\overline{\alpha})}$. For expressions in $\mathcal{R}(\sqcup\!\sqcup)$ it is no longer true that given a position i there exists a unique expression $c_i(\overline{\alpha})$ satisfying the conditions described above. The following is an example of this.

Example 5. Consider $\alpha = (a^\star + b)^\star \shuffle (c^\star + d)^\star$ and $\overline{\alpha} = \alpha_1 \shuffle \alpha_2$, where $\alpha_1 = (a_1^\star + b_2)^\star$ and $\alpha_2 = (c_3^\star + d_4)^\star$. For the letter a_1 we have $\partial_{a_1}(\overline{\alpha}) = \{a_1^\star \alpha_1 \shuffle \alpha_2\}$, while $\partial_{c_3 a_1}(\overline{\alpha}) = \{a_1^\star \alpha_1 \shuffle c_3^\star \alpha_2\}$.

However, for expressions with shuffle we can associate a unique expression $c(\overline{\alpha}, p)$ to each location p, which will be used to show that, also in this case, $\mathcal{A}_{PD}$ is a quotient of $\mathcal{A}_{POS}$. The c-*continuation* $c(\overline{\alpha}, p)$ of $\overline{\alpha}$ w.r.t. p is defined as:

$$c(\sigma_i, i) = \varepsilon, \quad c(\alpha^\star, p) = c(\alpha, p)\alpha^\star,$$

$$c(\alpha_1 + \alpha_2, p) = \begin{cases} c(\alpha_1, p), & \text{if } p \in \mathsf{Loc}(\alpha_1), \\ c(\alpha_2, p), & \text{if } p \in \mathsf{Loc}(\alpha_2), \end{cases}$$

$$c(\alpha_1 \alpha_2, p) = \begin{cases} c(\alpha_1, p)\alpha_2, & \text{if } p \in \mathsf{Loc}(\alpha_1), \\ c(\alpha_2, p), & \text{if } p \in \mathsf{Loc}(\alpha_2), \end{cases}$$

$$c(\alpha_1 \shuffle \alpha_2, (p, 0)) = c(\alpha_1, p) \shuffle \alpha_2, \quad c(\alpha_1 \shuffle \alpha_2, (0, p)) = \alpha_1 \shuffle c(\alpha_2, p),$$

$$c(\alpha_1 \shuffle \alpha_2, (p_1, p_2)) = c(\alpha_1, p_1) \shuffle c(\alpha_2, p_2), \quad \text{if } p_1, p_2 \neq 0.$$

Example 6. Consider again $\alpha = (ab)^\star \shuffle (bc)^\star$ from Example 4. For the elements in $\mathsf{Loc}_0(\alpha)$ we have $c(\overline{\alpha}, 0) = c(\overline{\alpha}, (2, 0)) = c(\overline{\alpha}, (2, 4)) = c(\overline{\alpha}, (0, 4)) = \overline{\alpha}$, $c(\overline{\alpha}, (0, 3)) = c(\overline{\alpha}, (2, 3)) = (a_1 b_2)^\star \shuffle c_4(b_3 c_4)^\star$, $c(\overline{\alpha}, (1, 0)) = c(\overline{\alpha}, (1, 4)) = b_2(a_1 b_2)^\star \shuffle (b_3 c_4)^\star$, and $c(\overline{\alpha}, (1, 3)) = b_2(a_1 b_2)^\star \shuffle c_4(b_3 c_4)^\star$. The partial derivative automaton of the expression given above is obtained by merging the states in the $\mathcal{A}_{POS}(\alpha)$ labelled with locations that have the same c-continuation.

To show that $\mathcal{A}_{PD}(\overline{\alpha})$ is a quotient of $\mathcal{A}_{POS}(\overline{\alpha})$, we first prove that the set of all c-continuations is precisely $\partial^+(\overline{\alpha})$. Furthermore, p is a final state in $\mathcal{A}_{POS}(\overline{\alpha})$ if and only if $c(\overline{\alpha}, p)$ is a final state in $\mathcal{A}_{PD}(\overline{\alpha})$. Finally, in Proposition 3 we relate $\partial_{\sigma_i}(c(\overline{\alpha}, p))$ with $\mathsf{Follow}(\overline{\alpha}, p)$.

Lemma 5. $\partial^+(\overline{\alpha}) = \{ c(\overline{\alpha}, p) \mid p \in \mathsf{Loc}(\alpha) \}$.

Lemma 6. *For* $\alpha \in \mathcal{R}(\shuffle)$ *and* $p \in \mathsf{Loc}(\alpha)$, $\varepsilon(c(\overline{\alpha}, p)) = \varepsilon \iff p \in \mathsf{Last}(\alpha)$.

Proposition 3. *For* $\alpha \in \mathcal{R}(\shuffle)$, $p \in \mathsf{Loc}_0(\alpha)$, *and* $\sigma_i \in \Sigma_{\overline{\alpha}}$, *one has*
$$\beta \in \partial_{\sigma_i}(c(\overline{\alpha}, p)) \iff \exists q \in \mathsf{Loc}(\alpha) \ (\beta = c(\overline{\alpha}, q) \ \wedge \ (\sigma_i, q) \in \mathsf{Follow}(\overline{\alpha}, p)).$$

Now, the equivalence relation $\equiv_o$ on $\mathsf{Loc}_0(\alpha)$, that defines $\mathcal{A}_{PD}(\overline{\alpha})$ as a quotient of $\mathcal{A}_{POS}(\overline{\alpha})$, is defined by $p \equiv_o q$ if $c(\overline{\alpha}, p) = c(\overline{\alpha}, q)$.

Lemma 7. *The relation* $\equiv_o$ *is right-invariant w.r.t.* $\mathcal{A}_{POS}(\alpha)$.

Example 7. Consider again $\alpha = (ab)^\star \shuffle (bc)^\star$ from Example 4. Recall that $c(\overline{\alpha}, (2, 0)) = c(\overline{\alpha}, (0, 4)) = \overline{\alpha}$, i.e. $(2, 0) \equiv_o (0, 4)$. Furthermore, $(a_1 b_2)^\star \shuffle c_4(b_3 c_4)^\star \in \partial_{b_3}(\overline{\alpha})$, $(b_3, (2, 3)) \in \mathsf{Follow}(\overline{\alpha}, (2, 0))$, $(b_3, (0, 3)) \in \mathsf{Follow}(\overline{\alpha}, (0, 4))$, and $(2, 3) \equiv_o (0, 3)$.

Given an expression α, one can naturally apply any automaton construction $\mathcal{A}$ to the marked expression $\overline{\alpha}$, where transitions are labelled with marked letters. We denote by $\overline{\mathcal{A}(\overline{\alpha})}$ the automaton obtained from $\mathcal{A}(\overline{\alpha})$ by unmarking the labels of transitions, but without changing the labels of the states.

Proposition 4. $\mathcal{A}_{\mathrm{POS}}(\alpha)/\equiv_o \, \simeq \, \overline{\mathcal{A}_{\mathrm{PD}}(\overline{\alpha})}$.

Proof. We show that the function $\varphi_c : \mathrm{Loc}_0(\alpha)/\equiv_o \, \longrightarrow \, \partial^+(\alpha)$, defined by $\varphi_c([p]) = c(\overline{\alpha}, p)$, is an isomorphism. Injectivity follows from Lemma 7 and surjectivity from Lemma 5. For the initial state we have $\varphi_c([0]) = c(\overline{\alpha}, 0) = \overline{\alpha}$. Furthermore, by Lemma 6, $[p]$ is a final state in $\mathcal{A}_{\mathrm{POS}}(\alpha)/\equiv_o$ if and only if $\varphi_c([p])$ is a final state in $\partial^+(\alpha)$. Finally, $\varphi_c(\delta_{\mathrm{POS}}/\equiv_o([p], \sigma)) = \varphi_c(\{\,[q] \mid (\sigma_i, q) \in \mathsf{Follow}(\overline{\alpha}, p) \wedge \overline{\sigma_i} = \sigma \,\}) = \{\, c(\overline{\alpha}, q) \mid (\sigma_i, q) \in \mathsf{Follow}(\overline{\alpha}, p) \wedge \overline{\sigma_i} = \sigma \,\} = \bigcup_{\overline{\sigma_i} = \sigma} \partial_{\sigma_i}(c(\overline{\alpha}, p)) = \delta_{\mathrm{PD}}(\varphi_c([p]), \sigma)$. $\qquad\square$

Broda et al. [4] showed that $\mathcal{A}_{\mathrm{PD}}(\alpha)$ is a quotient of $\overline{\mathcal{A}_{\mathrm{PD}}(\overline{\alpha})}$ by the right-invariant equivalence relation $\equiv_2$ defined on the states of $\overline{\mathcal{A}_{\mathrm{PD}}(\overline{\alpha})}$ by $\beta_1 \equiv_2 \beta_2$ if $\overline{\beta_1} = \overline{\beta_2}$. Let $\equiv_c$ be the relation $\equiv_2 \circ \equiv_o$. Thus, we have the following result.

Proposition 5. $\mathcal{A}_{\mathrm{POS}}(\alpha)/\equiv_c \, \simeq \, (\overline{\mathcal{A}_{\mathrm{PD}}(\overline{\alpha})})/\equiv_2 \, \simeq \, \mathcal{A}_{\mathrm{PD}}(\alpha)$.

Example 8. It follows from the c-continuations computed in Example 6 for $\alpha = (ab)^\star \sqcup\!\!\sqcup (bc)^\star$, that there are no $\beta_1 \neq \beta_2 \in \partial^+(\overline{\alpha})$ such that $\beta_1 \equiv_2 \beta_2$. Consequently, in this particular case, $\overline{\mathcal{A}_{\mathrm{PD}}(\overline{\alpha})} \simeq \mathcal{A}_{\mathrm{PD}}(\alpha)$.

5 $\mathcal{A}_{\mathrm{POS}}(\alpha)$ Vs. $\mathcal{A}_{\partial pos}(\alpha)$

In this section, we relate the position automaton defined in this paper with the one presented by Broda et al. [4]. The states of $\mathcal{A}_{\partial pos}$ are labelled by pairs (γ, i), where i is a position of a letter in the original expression, and $\gamma \in \mathcal{R}(\sqcup\!\!\sqcup)$ describes the right-language of the state. Given $\alpha \in \mathcal{R}(\sqcup\!\!\sqcup)$, the automaton obtained by that construction will be denoted by $\mathcal{A}_{\partial pos}(\alpha)$, and is defined by $\mathcal{A}_{\partial pos}(\alpha) = \langle S^0_{\partial pos}(\alpha), \Sigma, \{(\overline{\alpha}, 0)\}, \delta_{\partial pos}, F_{\partial pos}\rangle$, where $S^0_{\partial pos}(\alpha) = \{(\overline{\alpha}, 0)\} \cup \{(\gamma, i) \mid \gamma \in \partial_{\sigma_i}(\partial(\overline{\alpha})), \sigma_i \in \Sigma_{\overline{\alpha}}\}$, $F_{\partial pos} = \{(\gamma, i) \in S^0_{\partial pos}(\alpha) \mid \varepsilon(\gamma) = \varepsilon\}$ and $\delta_{\partial pos}((\gamma, i), \sigma) = \{(\beta, j) \mid \beta \in \partial_{\sigma_j}(\gamma), \sigma = \overline{\sigma_j}\}$.

Consider the expression $\alpha = (ab)^\star \sqcup\!\!\sqcup (bc)^\star$ from Example 4, with $\overline{\alpha} = \alpha_1 \sqcup\!\!\sqcup \alpha_2$, where $\alpha_1 = (a_1 b_2)^\star$ and $\alpha_2 = (b_3 c_4)^\star$. $\mathcal{A}_{\partial pos}(\alpha)$ is depicted on the right. Merging the states whose labels contain the same expression, we obtain $\overline{\mathcal{A}_{\mathrm{PD}}(\overline{\alpha})}$, which in this case coincides with $\mathcal{A}_{\mathrm{PD}}(\alpha)$, displayed in page 7.

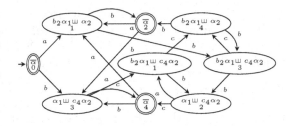

However, neither $\mathcal{A}_{\mathrm{POS}}(\alpha)$ is a quotient of $\mathcal{A}_{\partial pos}(\alpha)$, nor vice-versa. It was shown [4] that $\overline{\mathcal{A}_{\mathrm{PD}}(\overline{\alpha})}$ is a quotient of $\mathcal{A}_{\partial pos}(\alpha)$ by the right-invariant equivalence relation $\equiv_1$ defined on the set of states in $\mathcal{A}_{\partial pos}(\alpha)$ by $(\beta_1, i) \equiv_1 (\beta_2, j)$ if $\beta_1 = \beta_2$. Thus, we obtain the following relation between $\mathcal{A}_{\mathrm{POS}}(\alpha)$ and $\mathcal{A}_{\partial pos}(\alpha)$.

Corollary 1. $\mathcal{A}_{\partial pos}(\alpha)/\equiv_1 \, \simeq \, \mathcal{A}_{\mathrm{POS}}(\alpha)/\equiv_o \, \simeq \, \overline{\mathcal{A}_{\mathrm{PD}}(\overline{\alpha})}$.

We note that, for standard regular expressions, $\mathcal{A}_{\partial pos}$ coincides (up to isomorphism) with the standard position automaton. This is due to the fact that, whenever $\gamma, \gamma' \in \partial_{\sigma_i}(\partial(\overline{\tau}))$ then $\gamma = \gamma' = c(\overline{\tau}, i)$. The average number of states of $\mathcal{A}_{\mathrm{PD}}$, which is $(\frac{4}{3} + o\,(1))^{|\alpha|_\Sigma}$, was estimated using an upper bound $p(\alpha)$ for the number of elements in $\partial(\alpha)$ (see [4]). The value of $p(\alpha)$ is precisely $|\mathsf{Loc}(\alpha)|$, obtained by the definition of locations. Thus, we conclude that the average number of states of $\mathcal{A}_{\mathrm{POS}}$ is the same. However, an analogous analysis gives an upper bound for the average number of states for $\mathcal{A}_{\partial pos}$ of $(\frac{5}{3} + o\,(1))^{|\alpha|_\Sigma}$ (see [4]).

6   Relation with Other Constructions

A conversion from regular expressions to automata that has been recently studied is the prefix-automaton $\mathcal{A}_{\mathrm{Pre}}$ [3,12,17]. For standard regular expressions it is a left-quotient of the $\mathcal{A}_{\mathrm{POS}}$, and for linear expressions $\overline{\alpha}$ one has $\mathcal{A}_{\mathrm{POS}}(\overline{\alpha}) \simeq \mathcal{A}_{\mathrm{Pre}}(\overline{\alpha})$. Being a left-quotient also implies that the determinisation of $\mathcal{A}_{\mathrm{Pre}}$ coincides with the determinisation of $\mathcal{A}_{\mathrm{POS}}$ [3]. Below we define an extension of the $\mathcal{A}_{\mathrm{Pre}}$-construction for expressions in $\mathcal{R}(\sqcup\!\sqcup)$. However, the relationship with the position automaton doesn't hold any more neither for $\mathcal{A}_{\mathrm{POS}}$, nor for $\mathcal{A}_{\partial pos}$. Every state in $\mathcal{A}_{\mathrm{Pre}}$ is labelled either with ε or with an expression of the form $\alpha\sigma$, which describes the left-language of that state. To obtain those expressions, one uses a function R that, given an expression α, computes a set of normalised expressions of the form $\alpha'\sigma$. For $\alpha = \alpha_1 \sqcup\!\sqcup \alpha_2$, $\mathsf{R}(\alpha)$ is given by

$$\mathsf{R}(\alpha_1 \sqcup\!\sqcup \alpha_2) = \{\, (\alpha_1' \sqcup\!\sqcup \alpha_2)\sigma \mid \alpha_1'\sigma \in \mathsf{R}(\alpha_1) \,\} \cup \{\, (\alpha_1 \sqcup\!\sqcup \alpha_2')\sigma \mid \alpha_2'\sigma \in \mathsf{R}(\alpha_2) \,\}.$$

One can see that $\mathsf{R}_\varepsilon(\alpha) = \mathsf{R}(\alpha) \cup \varepsilon(\alpha)$ is such that $\mathcal{L}(\mathsf{R}_\varepsilon(\alpha)) = \mathcal{L}(\alpha)$. Thus, this is the set of final states of $\mathcal{A}_{\mathrm{Pre}}(\alpha)$. Then, the remaining construction of the automaton is done backwards. For each state of the form $\alpha'\sigma$ the set $\mathsf{R}_\varepsilon(\alpha')$ is computed and a transition by σ is added from each element $\alpha'' \in \mathsf{R}_\varepsilon(\alpha')$ to $\alpha'\sigma$. The state labelled by ε is the initial state of $\mathcal{A}_{\mathrm{Pre}}(\alpha)$. Formally, consider the function $\mathsf{p}_w(\alpha)$ for words $w \in \Sigma^*$ defined as follows: $\mathsf{p}_\varepsilon(\alpha) = \mathsf{R}_\varepsilon(\alpha)$, and $\mathsf{p}_{\sigma w}(\alpha) = \bigcup_{\alpha'\sigma \,\in\, \mathsf{p}_w(\alpha)} \mathsf{R}_\varepsilon(\alpha')$. We have that $\mathcal{L}(\mathsf{p}_w(\alpha)) = \{\, x \mid xw \in \mathcal{L}(\alpha) \,\}$. The *prefix* automaton of α is $\mathcal{A}_{\mathrm{Pre}}(\alpha) = \langle \mathsf{Pre}(\alpha), \Sigma, \delta_{\mathrm{Pre}}, \varepsilon, \mathsf{R}_\varepsilon(\alpha) \rangle$, where $\mathsf{Pre}(\alpha) = \bigcup_{w \in \Sigma^*} \mathsf{p}_w(\alpha)$, $\delta_{\mathrm{Pre}} = \{\, (\alpha'', \sigma, \alpha'\sigma) \mid \alpha'\sigma \in \mathsf{Pre}(\alpha),\ \alpha'' \in \mathsf{R}_\varepsilon(\alpha'),\ \sigma \in \Sigma \,\}$, that is, for all $\alpha'\sigma \in \mathsf{Pre}(\alpha)$, $\delta_{\mathrm{Pre}}^{\mathrm{R}}(\alpha'\sigma, \sigma) = \mathsf{R}_\varepsilon(\alpha')$.

Proposition 6. $\mathcal{L}(\mathcal{A}_{\mathrm{Pre}}(\alpha)) = \mathcal{L}(\alpha)$.

For expressions with shuffle $\mathcal{A}_{\mathrm{Pre}}$ is neither a quotient of $\mathcal{A}_{\mathrm{POS}}$, nor of $\mathcal{A}_{\partial pos}$. Considering the expression $\alpha_1 = (ab)^\star \sqcup\!\sqcup (bc)^\star$ of Example 4, $\overline{\mathcal{A}_{\mathrm{Pre}}(\overline{\alpha}_1)}$ does not coincide with $\mathcal{A}_{\mathrm{POS}}(\alpha_1)$. The automaton $\mathcal{A}_{\mathrm{Pre}}(\alpha_1)$ is obtained from $\overline{\mathcal{A}_{\mathrm{Pre}}(\overline{\alpha}_1)}$ by merging states that after unmarking are labelled with identical expressions. One can verify that $\mathcal{A}_{\mathrm{Pre}}(\alpha_1)$ is not a quotient of $\mathcal{A}_{\mathrm{POS}}(\alpha_1)$. Also, the determinisation of $\mathcal{A}_{\mathrm{Pre}}(\alpha_1)$ does not coincide with the determinisation of $\mathcal{A}_{\mathrm{POS}}(\alpha_1)$. Moreover, for the linear expression $\alpha_2 = (a^* + b^*) \sqcup\!\sqcup c$, the automata $\mathcal{A}_{\partial pos}(\alpha_2)$ and $\mathcal{A}_{\mathrm{Pre}}(\alpha_2)$ do not coincide. Nevertheless, the relationship between $\mathcal{A}_{\mathrm{Pre}}$ and

$\mathcal{A}_{\mathrm{PD}}$ established in Broda et al. [3] still holds for the set $\mathcal{R}(\sqcup\!\sqcup)$. To show that, it is enough to consider the dual reversal of $\mathcal{A}_{\mathrm{Pre}}$, i.e. $\mathcal{A}_{\overleftarrow{\mathrm{Pre}}}(\alpha) \simeq \mathcal{A}_{\mathrm{Pre}}(\alpha^{\mathrm{R}})^{\mathrm{R}}$. Defining $\mathsf{L}(\alpha) = \mathsf{R}(\alpha^{\mathrm{R}})^{\mathrm{R}}$ and $\overleftarrow{\mathsf{p}}_w(\alpha)$ as $\mathsf{p}_w(\alpha)$, but using L instead of R, we have $\mathcal{A}_{\overleftarrow{\mathrm{Pre}}}(\alpha) = \langle \overleftarrow{\mathrm{Pre}}(\alpha), \Sigma, \delta_{\overleftarrow{\mathrm{Pre}}}, \mathsf{L}_\varepsilon(\alpha), \varepsilon \rangle$, where $\overleftarrow{\mathrm{Pre}}(\alpha) = \bigcup_{w \in \Sigma^\star} \overleftarrow{\mathsf{p}}_w(\alpha)$ and $\delta_{\overleftarrow{\mathrm{Pre}}}(\alpha', \sigma) = \mathsf{L}_\varepsilon(\alpha'')$ if $\alpha' = \sigma\alpha''$, and $\delta_{\overleftarrow{\mathrm{Pre}}}(\alpha', \sigma) = \emptyset$ otherwise, for $\sigma \in \Sigma$. The following lemma establishes the relationship of L with partial derivatives for $\alpha \in \mathcal{R}(\sqcup\!\sqcup)$.

Lemma 8. $\mathsf{L}_\varepsilon(\alpha) = \bigcup_{\sigma \in \Sigma} \sigma \partial_\sigma(\alpha) \cup \varepsilon(\alpha)$.

Then, one can prove that the determinisation of $\mathcal{A}_{\overleftarrow{\mathrm{Pre}}}$ is isomorphic to a quotient of the determinisation of $\mathcal{A}_{\mathrm{PD}}$ by a right-invariant relation $(\equiv_{\mathsf{L}_\varepsilon})$ [3]. The same holds if one considers Brzozowski derivatives [6] extended with shuffle and the correspondent deterministic automaton (B in Fig. 1).

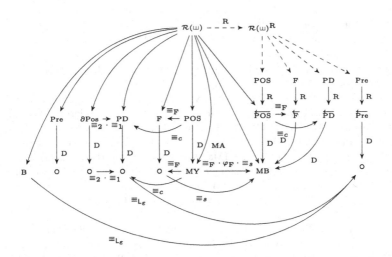

Fig. 1. Taxonomy of conversions for regular expressions with shuffle to finite automata. Edges correspond to operations/conversions between models (which are the nodes). The edges labelled by R correspond to the reversal operation, and the ones labelled by D to determinisation. The remaining labelled edges correspond to quotients where the labels identify the defining relation (see [3] for details).

Broda et al. [3] established relations between different conversions from regular expressions to equivalent finite automata, using the notion of position, the sets Follow and Select, and operations such as quotients, determinisation and reversal. These constructions are the Follow automaton ($\mathcal{A}_{\mathrm{F}}$) [11], the *Au Point* automaton ($\mathcal{A}_{\mathrm{MB}}$) [2,15], the McNaughton-Yamada automaton ($\mathcal{A}_{\mathrm{MY}}$, $\mathcal{A}_{\mathrm{MA}}$) [14,15], as well as some dual constructions using a double reversal. Considering locations instead of positions and the definitions of Follow and Select given in this paper, these constructions are now automatically defined for expressions in $\mathcal{R}(\sqcup\!\sqcup)$. Moreover, all the relationships established between them extend to expressions with shuffle. Those relationships are depicted in Fig. 1. In contrast to the situation for RE, for $\mathcal{R}(\sqcup\!\sqcup)$ we cannot ensure that $D(\mathcal{A}_{\overleftarrow{\mathrm{Pre}}})$ is always the smallest DFA among the ones represented, as it is incomparable (for instance) with $\mathcal{A}_{\mathrm{MB}}$: for $\alpha_1 = (a + b^\star)^\star \sqcup\!\sqcup (bc)^\star$, $\mathcal{A}_{\mathrm{MB}}(\alpha_1)$ has three states and $D(\mathcal{A}_{\overleftarrow{\mathrm{Pre}}}(\alpha_1))$ has eight states; while for $\alpha_2 = b \sqcup\!\sqcup ab$, $\mathcal{A}_{\mathrm{MB}}(\alpha_2)$ has seven states and $D(\mathcal{A}_{\overleftarrow{\mathrm{Pre}}}(\alpha_2))$ has six states. Exploring the applicability of the notion of locations to other shuffle operators or even intersection is left for future work.

References

1. Antimirov, V.M.: Partial derivatives of regular expressions and finite automaton constructions. Theoret. Comput. Sci. **155**(2), 291–319 (1996)
2. Asperti, A., Coen, C.S., Tassi, E.: Regular expressions, au point. CoRR abs/1010.2604 (2010)
3. Broda, S., Holzer, M., Maia, E., Moreira, N., Reis, R.: A mesh of automata. Inf. Comput. **265**, 94–111 (2019)
4. Broda, S., Machiavelo, A., Moreira, N., Reis, R.: Automata for regular expressions with shuffle. Inf. Comput. **259**(2), 162–173 (2018)
5. Brüggemann-Klein, A.: Regular expressions into finite automata. Theoret. Comput. Sci. **48**, 197–213 (1993)
6. Brzozowski, J.: Derivatives of regular expressions. J. ACM **11**(4), 481–494 (1964)
7. Champarnaud, J.M., Ziadi, D.: Canonical derivatives, partial derivatives and finite automaton constructions. Theoret. Comput. Sci. **289**, 137–163 (2002)
8. Estrade, B.D., Perkins, A.L., Harris, J.M.: Explicitly parallel regular expressions. In: Ni, J., Dongarra, J. (eds.) 1st IMSCCS, pp. 402–409. IEEE (2006)
9. Garg, V., Ragunath, M.: Concurrent regular expressions and their relationship to Petri nets. Theoret. Comput. Sci. **96**(2), 285–304 (1992)
10. Glushkov, V.M.: The abstract theory of automata. Russ. Math. Surv. **16**, 1–53 (1961)
11. Ilie, L., Yu, S.: Follow automata. Inf. Comput. **186**(1), 140–162 (2003)
12. Maia, E., Moreira, N., Reis, R.: Prefix and right-partial derivative automata. In: Beckmann, A., Mitrana, V., Soskova, M. (eds.) CiE 2015. LNCS, vol. 9136, pp. 258–267. Springer, Cham (2015). https://doi.org/10.1007/978-3-319-20028-6_26
13. Mayer, A.J., Stockmeyer, L.J.: Word problems-this time with interleaving. Inf. Comput. **115**(2), 293–311 (1994)
14. McNaughton, R., Yamada, H.: Regular expressions and state graphs for automata. IEEE Trans. Comput. **9**, 39–47 (1960)
15. Nipkow, T., Traytel, D.: Unified decision procedures for regular expression equivalence. In: Klein, G., Gamboa, R. (eds.) ITP 2014. LNCS, vol. 8558, pp. 450–466. Springer, Cham (2014). https://doi.org/10.1007/978-3-319-08970-6_29
16. Sulzmann, M., Thiemann, P.: Derivatives for regular shuffle expressions. In: Dediu, A.-H., Formenti, E., Martín-Vide, C., Truthe, B. (eds.) LATA 2015. LNCS, vol. 8977, pp. 275–286. Springer, Cham (2015). https://doi.org/10.1007/978-3-319-15579-1_21
17. Yamamoto, H.: A new finite automaton construction for regular expressions. In: Bensch, S., Freund, R., Otto, F. (eds.) 6th NCMA. books@ocg.at, vol. 304, pp. 249–264. Österreichische Computer Gesellschaft (2014)

Succinct Representations for (Non)Deterministic Finite Automata

Sankardeep Chakraborty[1], Roberto Grossi[2], Kunihiko Sadakane[3],
and Srinivasa Rao Satti[4(✉)]

[1] National Institute of Informatics, Tokyo, Japan
sankar@nii.ac.jp
[2] Università di Pisa, Pisa, Italy
grossi@di.unipi.it
[3] The University of Tokyo, Tokyo, Japan
sada@mist.i.u-tokyo.ac.jp
[4] Seoul National University, Seoul, South Korea
ssrao@cse.snu.ac.kr

Abstract. Deterministic finite automata are one of the simplest and most practical models of computation studied in automata theory. Their extension is the non-deterministic finite automata which also have plenty of applications. In this article, we study these models through the lens of succinct data structures where our ultimate goal is to encode these mathematical objects using information theoretically optimal number of bits along with supporting queries on them efficiently. Towards this goal, we first design a succinct data structure for representing any deterministic finite automaton $\mathcal{D}$ having n states over a σ-letter alphabet Σ using $(\sigma-1)n \log n(1+o(1))$ bits, which can determine, given an input string x over Σ, whether $\mathcal{D}$ accepts x in optimal $O(|x|)$ time. We also consider the case when there are $N < \sigma n$ non-failure transitions, and obtain various time-space trade-offs in both the cases. When the input deterministic finite automaton $\mathcal{A}$ is acyclic, not only we can improve the above space bound significantly to $(\sigma - 1)(n - 1) \log n + O(n + \log^2 \sigma)$ bits, we can also check if an input string x over Σ can be accepted by $\mathcal{A}$ optimally in $O(|x|)$ time. We also exhibit a succinct data structure for representing a non-deterministic finite automaton $\mathcal{N}$ having n states over a σ-letter alphabet Σ using $\sigma n^2 + n$ bits of space, such that given an input string x, we can decide whether $\mathcal{N}$ accepts x efficiently in $O(n^2|x|)$ time. Finally, we also provide time and space efficient algorithms for performing several standard operations such as union, intersection and complement on the languages accepted by deterministic finite automata.

Keywords: Data and image compression · Computational complexity

---

The full version of this paper appears as [4]. The work of the first author is supported by JSPS KAKENHI Grants Number 18H05291.

© Springer Nature Switzerland AG 2021
A. Leporati et al. (Eds.): LATA 2021, LNCS 12638, pp. 55–67, 2021.
https://doi.org/10.1007/978-3-030-68195-1_5

1   Introduction

Automata theory is a branch of theoretical computer science that deals exclusively with the definitions, properties and applications of different mathematical models of computation. These models play a major role in multiple applied areas of computer science. One of the most basic and fundamental models that is studied in automata theory since a long time ago is called the *finite automata*. They primarily come in two different types, *deterministic finite automata* (henceforth DFA) and *non-deterministic finite automata* (henceforth NFA) among others. There exists more complex and sophisticated models as well, for example, *Context-free grammars, Turing machines* etc. In what follows, let us formally define DFA and NFA in a nutshell as these are our primary subjects of study in this article. A DFA $\mathcal{D}$ is a quintuple $\mathcal{D} = (\Sigma, Q, q_0, \delta, F)$ where: Σ is an *alphabet* i.e., a finite set of letters, Q is the finite set of *states*, $q_0 \in Q$ is the *initial state*, $\delta : Q \times \Sigma \to Q$ is the *transition function* and finally, $F \subseteq Q$ is the *set of final states*. We often extend the transition function to $\delta : Q \times \Sigma^* \to Q$ which is defined recursively as follows: $\delta(q, \epsilon) = q$ for all $q \in Q$, where ϵ is the empty string; and $\delta(q, aw) = \delta(\delta(q, a), w)$ for all $q \in Q$, $a \in \Sigma$, and $w \in \Sigma^*$. Given the above definition, we say that the DFA accepts a string x over the alphabet Σ if and only if $\delta(q, x) \in F$. The *language* $\mathcal{L}$ accepted by a DFA $\mathcal{D}$ is defined as the set of all strings accepted by the DFA $\mathcal{D}$, and is denoted by $\mathcal{L}(\mathcal{D})$. In the rest of this paper, we assume that the alphabet Σ is $\{1, 2, \ldots, \sigma\}$, and the state set Q is $\{q_0, q_1, \ldots, q_{n-1}\}$. A deterministic automaton $\mathcal{A}$ is called *acyclic* [18] if it has a unique recurrent state where a state q is defined as *recurrent* if there exists a non-empty string x over Σ such that $\delta(q, x) = q$. Non-recurrent states are typically called *transient*, and the unique recurrent state (denoted by $q'' \in Q$) is classically called the *dead state* as $\delta(q'', \sigma) = q''$ for all $\sigma \in \Sigma$. An NFA is a conceptual extension of DFAs where the definition of the transition function is mainly extended. More specifically, for DFA, the transition function is defined as $\delta: Q \times \Sigma \to Q$ whereas for NFA, the same is defined as $\delta: Q \times \Sigma \to \mathcal{P}(Q)$ where $\mathcal{P}(Q)$ denotes the power set of Q. Another extension, which is sometimes used in the literature, is to simply allow more than one initial state in an NFA, and in this case, the third item in the tuple becomes I denoting the set of initial states, instead of singleton $\{q_0\}$. The rest of above quintuple definition remains as it is for NFA. Thus, in the case of NFA $\mathcal{N}$, the language $\mathcal{L}(\mathcal{N})$ is defined as $\{x \mid \exists_{q \in I} \exists_{q' \in F} [q' \in \delta(q, x)]\}$. We refer the readers to the classic texts of [16,22] for a thorough discussions on these mathematical models and automata theory in general.

Even if a DFA is defined as an abstract mathematical concept, still it has a myriad of practical applications. More specifically, it is used in text processing, compilers, and hardware design [22]. Quite often it is implemented in small hardware and software tools for solving various specific tasks. For example, a DFA can model a software that can figure out whether or not online user input such as email addresses are valid. DFAs/NFAs are also used for network packet filtering. In some of these applications, the alphabet is large and there is a failure/exit state so that only a subset of transitions go to non-failure states; so we call the latter ones *non-failure* transitions.

Despite having so many applications in practically motivated problems, we are aware of only a few studies of encoding arbitrary DFAs and NFAs from the point of view of *succinct data structures* (refer to the related work section) where the goal is to store an arbitrary element from a set Z of objects using the information theoretic minimum $\log(|Z|) + o(\log(|Z|))$ bits of space while being able to support the relevant set of queries efficiently, which is what we focus on in this paper. We also assume the usual model of computation, namely a $\Theta(\log n)$-bit word RAM model where n is the size of the input.

Related Work: The field of *succinct data structures* originally started with the work of Jacobson [17], and by now it is a relatively mature field in terms of breadth of problems considered. To illustrate this further, there already exists a large body of work on representing various combinatorial objects succinctly, such as trees [19, 20], interval graphs [1] etc. Regarding encoding DFA/NFA, the work on Wheeler NFA [15] can be considered as an attempt to encode particular classes of NFA succinctly, and this work has recently been generalized to arbitrary NFA in [7]. In the later part of our paper, we will compare these results with ours.

For DFA/NFA, other than the basic structure that is mentioned in the introduction, there exists many extensions/variations in the literature, for example, two-way finite automata, Büchi automata and many more. Researchers generally study the properties, limitations and applications of these mathematical structures. One such line of study that is particularly relevant to us for this paper is the research on counting DFAs/NFAs. Since the fifties there are plenty of attempts in exactly counting the number of DFAs/NFAs with n states over the alphabet Σ, and the state-of-the-art result is due to [3] for DFAs and [11] for NFAs. We refer the readers to the survey of Domaratzki [10] for more details. Basically, from these results, we can deduce the information theoretic lower bounds on the number of bits required to represent any DFA or NFA. Then we augment these lower bounds by designing succinct data structures capable of executing algorithms efficiently using this representation, and this is our main contribution.

DFA and NFA Enumeration: After a number of efforts by several authors, finally Bassino and Nicaud [3] found a matching upper and lower bound on the number of non-isomorphic initially-connected (i.e., all the states are reachable from the initial state) DFA's. They showed that the number of DFAs with n states over an alphabet of size σ is $\Theta(n2^{2n}S_2(\sigma n, n))$ where $S_2(n, m)$ denotes the Stirling numbers of the second kind. Using the approximation of the Stirling numbers of the second kind [14], which states that $S_2(n, m) \approx \frac{m^n}{m!}$, we can obtain the information theoretic lower bound for representing any DFA having n states and σ-sized alphabet is given by $\lg(n2^{2n}S_2(\sigma n, n)) = (\sigma - 1)n\lg n + O(n)$ bits. On the other hand, Domaratzki et al. [11] showed that there are asymptotically $2^{\sigma n^2 + n}$ initially connected NFAs on n states over a σ-letter alphabet with a fixed initial state and one or more final states. Thus, information theoretically, we need at least $\sigma n^2 + n$ bits to represent any NFA. In what follows later, we show that we can represent any given DFA/NFA using asymptotically optimal number of bits as mentioned here. Throughout this paper, we assume that the input DFAs/NFAs that we want to encode succinctly are initially connected.

1.1   Our Main Results and Paper Organization

The classical representation of DFAs/NFAs consists of explicitly writing the transition function δ in a two dimensional array $J[0..n-1][1..\sigma]$ having n rows corresponding to the n states of the DFA/NFA and σ (where $|\Sigma| = \sigma$) columns corresponding to the alphabet Σ such that $J[i][j] = \delta(q_i, j)$ where $q_i \in Q, j \in \Sigma$. For DFA, the entry in $J[i][j]$ is a singleton set whereas for NFA it could possibly contain a set having more than one state. Thus, the space requirement for representing any given DFA (NFA respectively) is given by $O(n\sigma \log n)$ ($O(n^2\sigma \log n)$ respectively) bits. These space bounds are clearly not optimal – for the DFAs, it is off by an additive $n \log n$ term from the information theoretic minimum, while for the NFAs, it is off by a multiplicative factor of $\log n$ from the optimal bound. We alleviate this discrepancy in the space bounds by designing optimal succinct data structures for these objects.

Towards this goal, in Sect. 2.1 we first discuss the relevant prior work from [3], and show that, by using suitable data structures, their work already gives a succinct encoding of DFA. But the major drawback of this encoding is that it is not capable of handling the problem of checking whether a string is accepted by the DFA extremely efficiently. In Section 2.3, we overcome this problem by designing a succinct data structure for DFA, which can also check the string acceptance almost optimally. We summarize our results below.

Theorem 1. *Given a DFA $\mathcal{D}$ having n states and working over an alphabet Σ of size σ, and a query string x for which we want to check the membership in $\mathcal{L}(\mathcal{D})$, there exists a succinct encoding for $\mathcal{D}$ taking:*

- $(\sigma - 1)n \log n + \sigma n + o(\sigma n)$ *bits, to support queries in $O(|x|)$ time;*
- $(\sigma - 1)n \log n + O(n \log \sigma)$ *bits, to support queries in $O(|x| \log \sigma)$ time; and*
- $(\sigma - 1)n \log n + n \log \sigma + O(n)$ *bits, to support queries in $O(|x| \log n)$ time.*

If $\mathcal{D}$ has $N < \sigma n$ non-failure transitions, then there exists an encoding taking:

- $(N - n) \log n + O(n\sigma)$ *bits, to support queries in $O(|x|)$ time;*
- $(N - n) \log n + O(N \log \sigma)$ *bits, to support queries in $O(|x| \log \sigma)$ time; and*
- $N(\log n + \log \sigma + 1.45)$ *bits, to support queries in $O(|x| \log n)$ time.*

The upper bounds in Theorem 1 save roughly $n \log n$ bits with respect to the immediate representation of the DFA. The former upper bound is optimal as it matches the information-theoretical lower bound in Sect. 1, up to lower order terms. As for the latter upper bound, we do not know its optimality but it is smaller than the information-theoretical lower bound of $\lceil \log \binom{n^2}{N} \rceil + \Theta(N \log \sigma)$

bits derived for edge-labeled deterministic directed graphs [13]. Indeed, DFAs can be seen as a special case of these graphs where n is the number of nodes, $N \geq n-1$ is the number of arcs, and σ is the maximum node degree.[1]

We can improve the above space bound significantly if the given DFA is acyclic along with obtaining optimal query time for string acceptance checking. More specifically, in full version of this paper, we obtain the following result in this case.

Theorem 2. *Given an acyclic DFA $\mathcal{A}$ having $n-1$ transient states, a unique dead state and working over an alphabet Σ of size σ, there exists a succinct encoding for $\mathcal{A}$ taking $(\sigma-1)(n-1)\log n + 3n + O(\log^2 \sigma) + o(n)$ bits of space, which can optimally determine, given an input string x over Σ, whether $\mathcal{A}$ accepts x in time proportional to the length of x, using constant words of working space.*

This is followed by the succinct data structure for NFA (see full version of the paper for the proof) where we prove the following result. Note that the running time for string acceptance checking in the following theorem is close to optimal, due to the lower bound of Equi et al. [12].

Theorem 3. *Given an NFA $\mathcal{N}$ having n states and working over an alphabet Σ of size σ, there exists a succinct encoding for $\mathcal{N}$ taking $\sigma n^2 + n$ bits of space, which can determine, given an input string x over Σ, whether $\mathcal{N}$ accepts x in $O(n^2|x|)$ time, using $2n$ bits of working space.*

Next we move on to discuss how one can support several standard operations such as union and intersection of two languages accepted by the deterministic finite automata. Classically it is done via the product automaton construction [16,22], and here we provide a time and space efficient algorithm for performing this construction. More specifically, we show the following theorem in the full version of our paper.

Theorem 4. *Suppose we are given the succinct representations for two DFAs $\mathcal{D}_1$ (having n states) and $\mathcal{D}_2$ (having n' states) respectively such that both are working over the same alphabet Σ. Also suppose that the product automaton (denoted by $\mathcal{P}$) has n'' states where $n'' \leq nn'$. Then, using $O(\sigma n'')$ expected time and $O(n'' \log n'')$ bits of working space, we can directly construct a succinct representation for $\mathcal{P}$. Moreover, $\mathcal{P}$ can be represented optimally using $(\sigma-1)n'' \log n'' + O(n'' \log \sigma)$ bits overall, and by suitably defining the final states of $\mathcal{P}$, we can make $\mathcal{P}$ accept either $\mathcal{L}(\mathcal{D}_1) \cup \mathcal{L}(\mathcal{D}_2)$ or $\mathcal{L}(\mathcal{D}_1) \cap \mathcal{L}(\mathcal{D}_2)$. Finally, given an input string x over Σ, we can decide whether $x \in \mathcal{L}(\mathcal{P})$ in $O(|x| \log \sigma)$ time using constant words of working space.*

---

[1] A directed graph with labels on its arcs is deterministic if no two out-neighbor arcs have the same label. Since there are $\lceil \log \binom{n^2}{N} \rceil$ directed graphs [13] with n nodes and N arcs, each deterministic graph $G = (V, E)$ can have $L = \prod_{u \in V} d_u!$ label assignments for its arcs, where d_u s the out-degree of node u and $N = \sum_{u \in V} d_u$. Note that $\log L = \Theta(N \log \sigma)$ when labels are from Σ and thus $d_u \leq \sigma$.

We conclude in Sect. 3 with some remarks on future research avenues.

Preliminaries: We will assume the knowledge of basic graph theoretic terminology (like trees, paths etc.) as given in [8] and basic graph algorithms (mostly the depth first search (henceforth DFS) traversal of a graph and its related concepts) as given in [6].

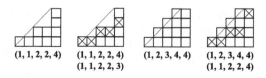

Fig. 1. A diagram of width $m = 5$ and height $n = 4$, a boxed diagram, a k-Dyck diagram and a k-Dyck boxed diagram with $k = 2$.

2 Succinct Representations for DFA and NFA

In this section, we provide all the upper bound results of our paper dealing with DFA/NFA. Throughout this section, whenever we mention DFA (NFA resp.), it should refer to an initially-connected deterministic (non-deterministic resp.) finite automata having n states and working over an alphabet Σ of size σ. With this notation in mind, we start with the succinct encoding of DFA first.

2.1 Succinct Encoding of DFA

Bassino and Nicaud [3] proved a beautiful bijection between the state transition diagram of any DFA and pairs of integer sequences which can be represented by boxed diagrams (will be defined shortly) along with providing an efficient algorithm to perform this construction. We will refer the readers to [3] for complete details regarding the bijection, counting and many other details that we choose to not repeat here. Later, Almeida et al. [2] also analysed the construction of Bassino and Nicaud [3] from which one can obtain a succinct encoding of DFA. These encodings do not support membership queries efficiently.

In what follows, we provide another succinct encoding based on the construction of Bassino and Nicaud [3] which is then used in Sect. 2.3 for efficient membership query support as well. Following [3], a *diagram* of width m and height n is defined as a sequence $(x_1, \ldots, x_m)$ of non-decreasing non-negative integers such that $x_m = n$. See Fig. 1 for better visual description and understanding. A *boxed diagram* can be defined as a pair of sequences $((x_1, \ldots, x_m), (y_1, \ldots, y_m))$ where $(x_1, \ldots, x_m)$ is a diagram and for all i (such that $1 \leq i \leq m$), the y_i-th box of the column i of the diagram is marked. Note that $1 \leq y_i \leq x_i$. Thus, a diagram can lead to $\prod_{i=1}^{m} x_i$ boxed diagrams. A *k-Dyck diagram* of size n is defined as a diagram of width $m := (k-1)n+1$ and height n such that $x_i \geq \lceil i/(k-1) \rceil$ for all $i \leq m-1$. Finally, a k-*Dyck boxed diagram* of size n is boxed diagram where the first coordinate $(x_1, \ldots, x_{(k-1)n+1})$ is a k-Dyck diagram of size n. Given these definitions, Bassino and Nicaud [3] proved the following theorem.

Theorem 5. *[3] The set $\mathcal{D}_n$ containing non-isomorphic initially-connected DFAs having n states and working over a σ-letter alphabet is in bijection with the set $\mathcal{B}_n$ of σ-Dyck boxed diagrams of size n. Moreover, the algorithms to construct the transition diagram of the DFA from k-Dyck boxed diagram, and vice versa run in linear time and space.*

Thus, by applying the above theorem, from any given DFA with n states and σ-letter alphabet, [3] produces a σ-Dyck boxed diagrams of size n, which can be in turn represented by two integer arrays $Max[1..m]$ and $Boxed[1..m]$ of length $m := (\sigma-1)n+1$ each. Furthermore, from these two arrays, it is possible to entirely reconstruct the DFA using the algorithm of Theorem 5. Thus, it is sufficient to store just these two arrays in order to encode any given DFA. For more details, readers are referred to [3]. For an example, see Fig. 2 which will also serve as the working example for this part of our paper.

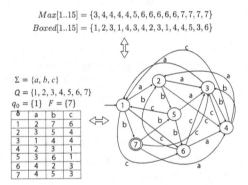

$Max[1..15] = \{3,4,4,4,4,5,6,6,6,6,6,7,7,7,7\}$
$Boxed[1..15] = \{1,2,3,1,4,3,4,2,3,1,4,4,5,3,6\}$

$\Sigma = \{a,b,c\}$
$Q = \{1,2,3,4,5,6,7\}$
$q_0 = \{1\}$ $F = \{7\}$

δ	a	b	c
1	2	7	6
2	3	5	4
3	1	4	4
4	2	3	1
5	3	6	1
6	4	2	3
7	4	5	3

Fig. 2. Three ways to define the same DFA. This DFA will serve as the working example for our discussion. By using the techniques of [3], this DFA can be entirely represented by the $Max[1..15] = \{3,4,4,4,4,5,6,6,6,6,6,7,7,7,7\}$ and $Boxed[1..15] = \{1,2,3,1,4,3,4,2,3,1,4,4,5,3,6\}$ arrays of length $(\sigma-1)n+1 = 15$ each.

First, we observe that, $1 \leq Max[1] \leq Max[2] \leq \cdots \leq Max[m] \leq n$ and $1 \leq Boxed[i] \leq Max[i]$ for each $i = 1, 2, \ldots, m$. This happens precisely because the translation is obtained by following a DFS on the DFA using the lexicographic order of words, and on each backtracking edge adding to the first vector the number of states scanned so far, and to the second vector the state reached. This also explains why each entry of these two arrays are upper bounded by n, the number of states of the given DFA. Now we consider the number of bits needed to encode the array $Max[1..m]$. It is easy to transform this array into an equivalent bit vector M of length $n + m$ with n ones in it (by storing the multiplicities of each of the n values from 1 to n in unary). Now, we encode the bit vector M using the structure of [21] using $\log \binom{(n+m)}{n} + o(n) + O(\log \log(m+n))$ bits. Since $m = (\sigma-1)n+1$, this space is $n \log \sigma + O(n)$ bits. Next we consider the number of bits required for array $Boxed[1..m]$. Because each entry of this array is an integer from 1 to n, we can use the result of [9] to represent the $Boxed[1..m]$ array using $(\sigma - 1)n \log n + O(\log^2 m)$ bits. Thus, in total, the size of the representation is $(\sigma-1)n \log n + n \log \sigma + O(n)$ bits. Because the information theoretic lower bound is $(\sigma - 1)n \log n + O(n)$ bits for the representation of DFA, this representation is succinct. In what follows, we show how to further reduce the encoding space.

2.2   Reducing the Space Further

Now consider the case when there is a failure/exit state labeled 0, and only N transitions among all the σn transitions go to non-failure states, for some $n \leq N \leq \sigma n$. Note that $Boxed$ has $N - n + 1$ non-zero values. In this case we can reduce the space for $Boxed[1..m]$ by using a new bitvector $Z[1..m]$ which has $N - n + 1$ ones. We use a new array $Boxed'[1..N - n + 1]$ which stores non-zero values of $Boxed[1..m]$. Then $Boxed[i]$ is computed as follows. If $Z[i] = 0$, $Boxed[i] = 0$ (transition to the failure state). If $Z[i] = 1$, $Boxed[i] = Boxed'[partial\_rank_1(Z, i)]$. If we use the data structure of [5], Z is represented in $\sigma n + o(\sigma n)$ bits, which is asymptotically smaller than the space lower bound of $(\sigma - 1)n \log n + O(n)$. But, by using the data structure of [21], the bitvector Z can be represented in $\log \binom{\sigma n}{N} + o(N) + O(\log \log(\sigma n)) = N \log \frac{\sigma n e}{N} + o(N) \leq N \log \sigma + N \log e + o(N)$ bits. The space for $Boxed'$ is $(N - n + 1) \log n$ bits. Therefore the total space for representing a DFA with N non-failure transitions is $(N - n) \log n + N \log \sigma + N \log e + o(N)$ bits, i.e., less than $\log n + \log \sigma + 1.45$ bits per transition.

Given a string x over Σ, it takes linear time (in the size of the DFA, i.e., $O(\sigma n)$ time) to decide whether the DFA accepts the string x, which is clearly not optimal as ideally it should be performed in time $O(|x|)$. This happens because the algorithm of Theorem 5 actually unravels the DFA from these two arrays $Max[1..m]$ and $Boxed[1..m]$, and then checks whether the input string can be accepted or not. Thus, from the point of view of string acceptance, these encodings of DFA (including the encodings of Bassino and Nicaud [3], and Almeida et al. [2]) are not optimal, whereas from the space requirement point of view, these are optimal. This motivates the need for a succinct encoding of a given DFA, where the problem of string acceptance can be performed in optimal time. In what follows, we provide such an encoding.

2.3   Succinct Data Structure for DFA

To design a succinct data structure for DFA, we need the following three bitvectors F, P and T in addition to an integer array $NewBoxed[1..m]$ (that can be obtained from the $Boxed[1..m]$ array of the previous section, as described later), which are defined as follows. P is a balanced parentheses sequence of length $2n$ obtained from the lexicographic depth-first search (DFS) tree of the given input automaton $\mathcal{D}$. More specifically, given any DFA $\mathcal{D}$, we first perform the lexicographic DFS on $\mathcal{D}$ to generate the lexicographic DFS tree R of $\mathcal{D}$, i.e., while looking for a new edge to traverse during DFS, the algorithm always searches in lexicographic order of edge labels. For example, in Fig. 2, from any vertex, lexicographic DFS first tries to traverse the edge labeled a, followed by b and finally c. The tree R is represented as a balanced parenthesis sequence P together with auxiliary structures to support the navigational queries on R, as mentioned in [19], using $2n + o(n)$ bits. The bitvector F is used to mark all the final states of the input DFA, hence it takes n bits.

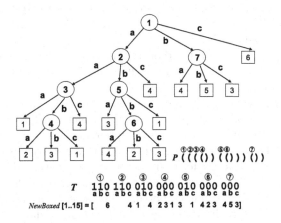

Fig. 3. The extended lex-DFS tree S of the automaton of Fig. 2 along with the corresponding bitvectors P, T, and the *NewBoxed*[1..15] array (the elements of this array are drawn exactly below the corresponding 0s with which they share one to one correspondence with). Note that, for the same automaton *Boxed*[1..15] array is given as *Boxed*[1..15] = {1, 2, 3, 1, 4, 3, 4, 2, 3, 1, 4, 4, 5, 3, 6}.

Before explaining the other bitvector, T, required for our succinct encoding, we want to explain the contents of Fig. 3. The tree depicted in the figure is what we call an *extended lexicographic DFS tree* or *extended lex-DFS tree* (denoted by S) in short. If we delete the squared nodes and their incident edges (originating from the circled nodes), we obtain the lexicographic DFS tree of the automaton $\mathcal{D}$. Actually these edges represent the *back edges/cross edges/forward edges* [6] (i.e., non-tree edges) in the DFS tree of the automaton $\mathcal{D}$. Traditionally the vertices in the square are not drawn (as in our case of Fig. 3), rather the edges point to the nodes in the circle only (hence all the nodes appear only once). We have chosen to draw and define the extended lex-DFS tree this way as it helps us to design and explain our succinct data structure well. Also note that, edges originating from a circled node and going to another circled node represents tree edges whereas edges from circled to squared nodes represent non-tree edges.

Now given the extended lex-DFS tree S, we visit the nodes of S in DFS order and append a bit string of length σ for each vertex v of S marking which of its children are attached to v via tree edges (marked with 1) and which are attached to v via non-tree edges (marked with 0) in the lexicographic order of the edge labels. The string obtained this way is referred to as T. Thus, T is a bit-vector of length σn which captures the information about the tree and non-tree edges of S. More specifically, it has exactly $n - 1$ ones, which have one-to-one correspondence with the tree edges of the lexicographic DFS tree of DFA $\mathcal{D}$, and has exactly $(\sigma - 1)n + 1$ zeros, which correspond to non-tree edges of the lexicographic DFS tree of DFA $\mathcal{D}$. See Fig. 3 for an example. We relabel all the states of $\mathcal{D}$ such that the i-th vertex (state) in R in preorder has label i, and also modify the transition function accordingly. Now it is easy to see that, for the state with label i ($1 \le i \le n$), the corresponding node in the lexicographic DFS tree has exactly σ outgoing edges, and we encode the tree edges among them using the bits in the range $T[\sigma(i - 1) + 1..\sigma i]$. More specifically, $T[\sigma(i-1)+c] = 1$ if and only if the outgoing edge labeled c is a tree edge ($1 \le c \le \sigma$). Similarly, we can also find the j-th outgoing tree edge from the state i by $select_1(T, j + rank_1(T, \sigma(i - 1)))$.

In what follows, we show three different ways to represent the T array. (1) In the first method, we simply encode T using the structure of [5] as a bitvector of length σn having exactly $n - 1$ ones while supporting constant time $rank/select$ queries in T. (2) In the second method, we compress T by observing that the positions of 1s in the T array form an increasing sequence, hence by using the data structure $D(n - 1, \sigma n, \epsilon)$ of [23], T can be encoded in $O(n \log \sigma)$ bits (by setting $\epsilon = 1/\log(\sigma - 1)$) while supporting $access$, $rank$ and $select$ operations in $O(\log \sigma)$ time. (3) Finally, we can encode T using the data structure of [21] in $n \log \sigma + O(n)$ bits (like we did for Max array in Sect. 2.1) while supporting the $rank$ query in $O(\log n)$ time and $select$ in $O(1)$ time.

Now let us define the new integer array $NewBoxed[1..m]$. First, observe that elements of the array $Boxed[1..m]$ are nothing but the leaves (i.e., node labels in the squared nodes) of the extended lex-DFS tree S in the left to right order. More specifically, they are the node labels of the destinations of the non-tree edges emanating from the nodes of the lexicographic DFS tree of the automaton $\mathcal{D}$ in their preorder. Instead of this specific ordering (followed in the $Boxed[1..m]$ array), $NewBoxed[1..m]$ lists the same node labels in the order of their appearance in the T bitvector (from left to right). Note that, as mentioned previously, these node are marked by 0s in T and they are in one-to-one correspondence with all the non-tree edges of the lexicographic DFS tree of the automaton $\mathcal{D}$. Thus, the $NewBoxed[1..m]$ array contains the same node labels as the $Boxed[1..m]$ array, but in a different order. See Fig. 3 for an example. This completes the description of our succinct data structure for DFA. Note that Max is no longer used in our data structure.

Space Complexity. We now analyze the space complexity of our data structure. The array $NewBoxed[1..m]$ takes $(\sigma - 1)n \log n + O(\log^2 m)$ bits (by similar analysis as before for the $Boxed[1..m]$ array). The bitvector F consumes n bits. The bitvector P is stored using the structure of [19], hence it occupies $2n + o(n)$ bits. Depending on the three different choices for storing T, the space requirement is (1) $\sigma n + o(\sigma n)$, (2) $O(n \log \sigma)$ or (3) $n \log \sigma + O(n)$ bits. Thus, the overall space usage is as stated in the Theorem 1.

It is easy to further reduce the size if the DFA has only $N < \sigma n$ non-failure transitions. Using the bitvector $Z[1..m]$ (as defined in Sect. 2.2) for indicating non-failure transitions, the array $NewBoxed[1..m]$ is compressed to $N - n + 1$ non-zero values, which can be stored in $(N - n) \log n + O(\log^2 n)$ bits. Since Z is a bit vector of length σn with N ones in it, we can use the three choices as above for representing T, using (1) $\sigma n + o(\sigma n)$, (2) $O(N \log \sigma)$ or (3) $N \log \sigma + 1.44N + o(N)$ bits. Thus the overall space usage is (1) $(N - n) \log n + O(n\sigma)$, (2) $(N - n) \log n + O(N \log \sigma)$, or (3) $(N - n) \log n + (N + n) \log \sigma + N \log e + o(N) \le N(\log n + \log \sigma + 1.45)$ bits, depending on the representation used for T and Z.

Query Algorithm. Suppose we are given an input string x of length y over Σ, and we need to decide if the DFA $\mathcal{D}$ accepts x or not. We start the following procedure from the initial state (stored explicitly using $O(\log n)$ bits) and repeat until the end of the input string x. At any generic step, to figure out the transition function $\delta(q, c) := q'$ where $1 \le q, q' \le n$ are the states, we first look at the bit $T[\sigma(q - 1) + c]$. If it is 1, the outgoing edge labeled c from state q is a tree edge. Let $j := rank_1(T, \sigma(q-1)+c) - rank_1(T, \sigma(q-1))$. Then the outgoing edge is the j-th tree edge of node q in the lex DFS tree. Therefore $q' = child(q, j)$ (supported using the structure of [19]). If the bit is 0, the outgoing edge labeled c from state q is a non-tree edge. Let $j := rank_0(T, \sigma(q - 1) + c)$. Then the edge is the j-th non-tree edge in the DFA, and q' is obtained by $q' := NewBoxed[j]$. Hence, when we reach the end of x, and if we are at an accepting/final states (can be figured out from the bitvector F), we say that the DFA $\mathcal{D}$ accepts x. Now, depending on the three choices for storing the T array and supporting $rank/select$ queries in it, we obtain three different query time bounds for accepting x. In option (1), the input string x can be accepted optimally in $O(|x|)$ time. In (2), as the $rank$ operations on T take $O(\log \sigma)$ time while all other operations, at each step, take $O(1)$ time, the overall run time for checking the membership of x is $O(|x| \log \sigma)$. Finally, in (3), because of the $O(\log n)$ time $rank$ operation on T, the membership of x can be checked in $O(|x| \log n)$ time. The query times remain the same, even in the case where we have $N < \sigma n$ non-failure transitions (using the data structures mentioned in the previous paragraph). This completes the proof of Theorem 1.

Note that Cotumaccio and Prezza [7] recently gave an encoding for DFAs using $\log p + \log \sigma + 2$ bits per transition (where $1 \le p \le n$ is compressibility parameter), while our data structure takes $\log n + \log \sigma + 1.45$ bits per transition with efficient support for membership checking.

3   Concluding Remarks

We considered the problem of succinctly encoding any given DFA $\mathcal{D}$, acyclic DFA $\mathcal{A}$ or NFA $\mathcal{N}$ so as to check efficiently if they accept a given input string. To this end, we successfully designed succinct data structures for them that also support the string acceptance query efficiently for DFAs, acyclic DFAs, and NFAs. We believe that our work will spur further interest in designing succinct data structures for other mathematical models from the world of automata theory in future. For example, it would be interesting to see if other variants of automata can also be succinctly encoded with efficient query support mechanism.

References

1. Acan, H., Chakraborty, S., Jo, S., Satti, S.R.: Succinct data structures for families of interval graphs. In: Friggstad, Z., Sack, J.-R., Salavatipour, M.R. (eds.) WADS 2019. LNCS, vol. 11646, pp. 1–13. Springer, Cham (2019). https://doi.org/10.1007/978-3-030-24766-9_1
2. Almeida, M., Moreira, N., Reis, R.: Enumeration and generation with a string automata representation. Theor. Comput. Sci. **387**(2), 93–102 (2007)
3. Bassino, F., Nicaud, C.: Enumeration and random generation of accessible automata. Theor. Comput. Sci. **381**(1–3), 86–104 (2007)
4. Chakraborty, S., Grossi, R., Sadakane, K., Satti, S.R.: Succinct representation for (non) deterministic finite automata. CoRR abs/1907.09271 (2019)
5. Clark, D.R.: Compact pat trees. Ph.D. thesis, University of Waterloo, Canada (1996)
6. Cormen, T.H., Leiserson, C.E., Rivest, R.L., Stein, C.: Introduction to Algorithms, 3 edn. MIT Press, Cambridge (2009)
7. Cotumaccio, N., Prezza, N.: On indexing and compressing finite automata. CoRR abs/2007.07718 (2020)
8. Diestel, R.: Graph Theory, 4th edn. Graduate texts in mathematics, vol. 173. Springer, Heidelberg (2012)
9. Dodis, Y., Patrascu, M., Thorup, M.: Changing base without losing space. In: STOC, pp. 593–602 (2010)
10. Domaratzki, M.: Enumeration of formal languages. Bull. EATCS **89**, 117–133 (2006)
11. Domaratzki, M., Kisman, D., Shallit, J.: On the number of distinct languages accepted by finite automata with n states. J. Autom. Lang. Comb. **7**(4), 469–486 (2002)
12. Equi, M., Grossi, R., Mäkinen, V., Tomescu, A.I.: On the complexity of string matching for graphs. In: 46th ICALP. LIPIcs, vol. 132, pp. 55:1–55:15 (2019)
13. Farzan, A., Munro, J.I.: Succinct encoding of arbitrary graphs. Theor. Comput. Sci. **513**, 38–52 (2013)
14. Flajolet, P., Sedgewick, R.: Analytic Combinatorics. Cambridge University Press, Cambridge (2009)
15. Gagie, T., Manzini, G., Sirén, J.: Wheeler graphs: a framework for BWT-based data structures. Theor. Comput. Sci. **698**, 67–78 (2017)
16. Hopcroft, J.E., Motwani, R., Ullman, J.D.: Introduction to Automata Theory, Languages, and Computation - International Edition, 2 edn. Addison-Wesley, Boston (2003)

17. Jacobson, G.J.: Succinct static data structures. Ph.D. thesis, Carnegie Mellon University (1998)
18. Liskovets, V.A.: Exact enumeration of acyclic deterministic automata. Discrete Appl. Math. **154**(3), 537–551 (2006)
19. Munro, J.I., Raman, V.: Succinct representation of balanced parentheses and static trees. SIAM J. Comput. **31**(3), 762–776 (2001)
20. Navarro, G., Sadakane, K.: Fully functional static and dynamic succinct trees. ACM Trans. Algorithms **10**(3), 16 (2014)
21. Raman, R., Raman, V., Satti, S.R.: Succinct indexable dictionaries with applications to encoding k-ary trees, prefix sums and multisets. ACM Trans. Algorithms **3**(4), 43 (2007)
22. Sipser, M.: Introduction to the Theory of Computation. PWS Publishing Company, Boston (1997)
23. Sumigawa, K., Sadakane, K.: An efficient representation of partitions of integers. In: Iliopoulos, C., Leong, H.W., Sung, W.-K. (eds.) IWOCA 2018. LNCS, vol. 10979, pp. 361–373. Springer, Cham (2018). https://doi.org/10.1007/978-3-319-94667-2_30

Optimising Attractor Computation in Boolean Automata Networks

Kévin Perrot[2], Pacôme Perrotin[1(✉)], and Sylvain Sené[2]

[1] Aix Marseille University, Université de Toulon, CNRS, LIS, Marseille, France
pacome.perrotin@lis-lab.fr
[2] Université Publique, Marseille, France

Abstract. This paper details a method for optimising the size of Boolean automata networks in order to compute their attractors under the parallel update schedule. This method relies on the formalism of modules introduced recently that allows for (de)composing such networks. We discuss the practicality of this method by exploring examples. We also propose results that nail the complexity of most parts of the process, while the complexity of one part of the problem is left open.

Keywords: Boolean automata networks · Modularity · Optimisation

1 Introduction

Boolean automata networks (BANs) are studied for their capacity to succicntly expose the complexity that comes with the composition of simple entities into a network. They belong to a wide family of systems which include cellular automata and neural networks, and can be described as cellular automata with arbitrary functions and on arbitrary graph structures.

Understanding and predicting the dynamics of computing with BANs has been a focus of the scientific community which studies them, in particular since their applications include the modelling of gene regulatory networks [13,15,22] [5,6]. In those applications, fixed points of a BAN are often viewed as cellular types and limit cycles as biological rhythms [13,22]. It follows that most biological studies relying on BANs require the complete computation of their dynamics to propose conclusions. The complete computation of the dynamics of BANs is an exponentially costly process. Indeed, for n the size of a BAN, the size of its dynamics is precisely 2^n. The dynamics of a BAN is usually partitioned in two sorts of configurations: the recurring ones that are parts of attractors and either belong to a limit cycle or are fixed points; the others that evolve towards these attractors and belong to their attraction basins. The questions of characterising, computing or counting those attractors from a simple description of the network have been explored [1,2,7,8,10,16], and have been shown to be difficult problems [3,4,8,17,18].

In this paper, we propose a new method for computing the attractors of a BAN under the parallel update schedule. For any input network, this method

A. Leporati et al. (Eds.): LATA 2021, LNCS 12638, pp. 68–80, 2021.
https://doi.org/10.1007/978-3-030-68195-1_6

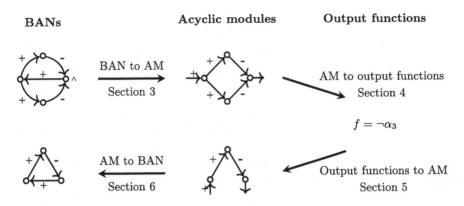

Fig. 1. Illustration of the optimisation pipeline explored in this paper. Each arrow corresponds to a part of the pipeline, and a section in this article.

generates another network which is possibly smaller and which is guaranteed to possess attractors isomorphic to those of the input network. Computing the dynamics of this smaller network therefore takes as much time as needed to compute the dynamics of the input networks, divided by some power of two.

This method uses tools and results developed in previous works by the authors [19,20]. These works involve adding inputs to BANs, in a generalisation called modules. In some cases, the entire computation of a module can be understood as functions of the states of its inputs, disregarding the network itself. In particular, a result (Theorem 16) states that two networks that have equivalent such computations share isomorphic attractors.

Section 2 starts by exposing all the definitions needed to read this paper. Section 3 explores the question of obtaining an acyclic module (AM) from a BAN. Section 4 explains how to extract so called output functions from a module. Section 5 details how to generate a minimal module from a set of output functions. Finally Sect. 6 shows the final step of the method, which implies constructing a BAN out of an AM and computing its dynamics. Each section explores complexity results of the different parts of the process, and details examples along the way. An illustrative outline of the paper can be found in Fig. 1.

2 Definitions

2.1 Boolean Functions

In this paper, we consider a Boolean function as any function $f : \mathbb{B}^A \to \mathbb{B}$, for A a finite set. An affectation x of f is a vector in $\mathbb{B}^A$. When considered as the input or output of a complexity problem, we encode Boolean functions as Boolean circuits. A *Boolean circuit* of f is an acyclic digraph in which nodes without incoming edges are labelled by an element in A, and every other node by a Boolean gate in $\{\wedge, \vee, \neg\}$, with a special node marked as the output of the circuit. The evaluation $f(x)$ is computed by mapping x to the input nodes of the

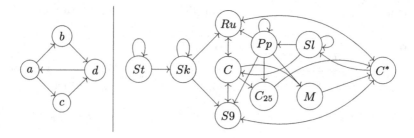

Fig. 2. On the left, the interaction digraph of F_A, as described in Example 1. On the right, the interaction digraph of F_B, as described in Example 2.

circuit, and propagating the evaluation along the circuit using the gates until the output node is reached.

2.2   Boolean Automata Networks and Acyclic Modules

Boolean Automata Networks. BANs are composed of a set S of automata. Each automaton in S, or node, is at any time in a state in $\mathbb{B}$. Gathering those isolated states into a vector of dimension $|S|$ provides us with a configuration of the network. More formally, a *configuration* of S over $\mathbb{B}$ is a vector in $\mathbb{B}^S$. The state of every automaton is bound to evolve as a function of the configuration of the entire network. Each node has a unique function, called a local function, that is predefined and does not change over time. A *local function* is thus a function f defined as $f : \mathbb{B}^S \to \mathbb{B}$. Formally, a BAN F is a set that assigns a local function f_s over S for every $s \in S$.

BANs are usually represented by the influence that automata hold on each other. As such the visual representation of a BAN is a digraph, called an *interaction digraph*, whose nodes are the automata of the network, and arcs are the influences that link the different automata. Formally, s *influences* s' if and only if there exist two configurations x, x' such that $f_{s'}(x) \neq f_{s'}(x')$ and for all r in S, $r \neq s$ if and only if $x_r = x'_r$.

Example 1. Let $S_A = \{a, b, c, d\}$. Let F_A be the BAN defined by $f_a(x) = x_d$, $f_b(x) = f_c(x) = x_a$, and $f_d(x) = \neg x_b \vee \neg x_c$. The interaction digraph of this BAN is depicted in Fig. 2 (left panel).

Example 2. Let $S_B = \{St, Sl, Sk, Pp, Ru, S9, C, C25, M, C^*\}$. Let F_B be the BAN defined by $f_{St}(x) = \neg x_{St}$, $f_{Sl}(x) = \neg x_{Sl} \vee x_{C^*}$, $f_{Sk}(x) = x_{St} \vee \neg x_{Sk}$, $f_{Pp}(x) = x_{Sl} \vee \neg x_{Pp}$, $f_{Ru}(x) = f_{S9}(x) = \neg x_{Sk} \vee x_{Pp} \vee \neg x_C \vee \neg x_{C^*}$, $f_C(x) = \neg x_{Ru} \vee \neg x_{S9} \vee \neg x_{Sl}$, $f_{C25}(x) = \neg x_{Pp} \vee x_C$, $f_M(x) = x_{Pp} \vee \neg x_C$, and $f_{C^*}(x) = \neg x_{Ru} \vee \neg x_{S9} \vee x_{C25} \vee \neg x_M$. The interaction digraph of this BAN is depicted in Fig. 2 (right panel).

In the scope of this paper, BANs (and modules) are udated according to the parallel update schedule. Formally, for F a BAN and x a configuration of F, the update of x under F is denoted by configuration $F(x)$, and defined as for all s in S, $F(x)_s = f_s(x)$.

Example 3. Consider F_A of Example 1, and $x \in \mathbb{B}^{S_A}$ such that $x = 1001$. We observe that $F_A(x) = 1111$. Configurations 1000 and 0111 are recurring and form a limit cycle of size 2, as well as configurations 0000, 0001, 1001, 1111, 1110 and 0110 that form a limit cycle of size 6.

Dynamics and Attractors. We define the *dynamics* of a BAN F as the digraph with $\mathbb{B}^S$ as its set of vertices. There exists an edge from x to y if and only if $F(x) = y$. An *attractor* of F is a strongly connected component of its dynamics. Computing the dynamics of a BAN from the description of its local function is an exponential process. See [21] for a more throughout introduction to BANs and related subjects.

Modules. Modules were first introduced in [19]. A module M is a BAN with added inputs. It is defined on two sets: S a set of automata, and I a set of inputs, with $S \cap I = \varnothing$. Similarly to standard BANs, we can define configurations as vectors in $\mathbb{B}^S$, and we define input configurations as vectors in $\mathbb{B}^I$. A local function of a module updates itself based on a configuration x and an input configuration i, concatenated into one configuration. Formally, a local function is defined from $\mathbb{B}^{S \cup I}$ to $\mathbb{B}$. The module M defines a local function for every node s in S.

Example 4. Let M_e be the module defined on $S_e = \{p, q, r\}$ and $I = \{\alpha, \beta\}$, such that $f_p(x) = x_\alpha$, $f_q(x) = \neg x_p$, and $f_r(x) = x_q \vee \neg x_\beta$.

We represent modules with an interaction digraph, in the same way as for BANs. The interaction digraph of a module has added arrows that represent the influence of the inputs over the nodes; for every node s and every input α, the node s of the interaction digraph has an ingoing arrow labelled α if and only if α influences s, that is, there exists two input configurations i, i' such that for all β in I, $\beta \neq \alpha$ if and only if $i_\beta = i'_\beta$, and x a configuration such that $f_s(x \cdot i) \neq f_s(x \cdot i')$, where $\cdot$ denotes the concatenation operator.

A module is *acyclic* if and only if its interaction digraph is cycle-free.

Recursive Wirings. A recursive wiring over a module M is defined by a partial function $\omega : I \nrightarrow S$. The result of such a wiring is denoted $\circlearrowleft_\omega M$, a module defined over sets S and $I \setminus \mathrm{dom}(\omega)$, in which the local function of node s is denoted f'_s and defined as

$$\forall x \in \mathbb{B}^{S \cup I \setminus \mathrm{dom}(\omega)}, \ f'_s(x) = f_s(x \circ \hat{\omega}), \text{ with } \hat{\omega}(i) = \begin{cases} \omega(i) & \text{if } i \in \mathrm{dom}(\omega) \\ i & \text{if } i \in I \setminus \mathrm{dom}(\omega) \end{cases}.$$

Output Functions. Output functions were first introduced in [20] and present another way of computing the evolution of an acyclic module. In the Boolean case, those functions are defined on $\mathbb{B}^{I \times \{1,\dots,D\}} \to \mathbb{B}$, for I the input set of the module, and D some integer. We interpret an input in $\mathbb{B}^{I \times \{1,\dots,D\}}$ as an evaluation over $\mathbb{B}$ of a set of variables $I \times \{1,\dots,D\}$, and for $\alpha \in I$ and $d \leq D$, we denote this variable by α_d. In the context of an acyclic module M, α_d is refering to the evaluation of the input α on the dth update of the module. A vector $j \in \mathbb{B}^{I \times \{1,\dots,D\}}$ simply describes an evaluation of all the inputs of the network over D iterations. With such a vector, and $x \in \mathbb{B}^S$, it is easy to see that the acyclic module M can be updated k times in a row, for any $k \leq D$. The result of this update is denoted by $M(x, j_{[1,\dots,k]})$. The *delay* of an output function O is the maximal value in the set of all the $d \in \mathbb{N}$ for which there exists $\alpha \in I$ such that variable α_d has an influence on the computation of O. That is, there exists a couple of vectors $x, x' \in \mathbb{B}^{I \times \{1,\dots,D\}}$ which are equal except for $x_{(\alpha,d)} \neq x'_{(\alpha,d)}$, and $O(x) \neq O(x')$. Finally, for M an acyclic module defined on the sets S and I, for D a large enough integer, for $x \in \mathbb{B}^S$ and $j \in \mathbb{B}^{I \times \{1,\dots,D\}}$ some vectors, and for s a node in S, we define the output function of s, denoted O_s, as the output function with minimal delay d such that $O_s(j) = M(x, j_{[1,\dots,k]})_s$. Such a function always exists, and since it has minimal delay it is always unique.

2.3   Promise Problems and Classes of Function Problems

In this paper, we make the hypothesis that every module that is part of an instance of a complexity problem follows the property that each of its local functions has only *essential* variables. That is, a variable is included as input of the circuit encoding the function if and only if the automaton or input represented by that variable has an influence on said function. This hypothesis will be implemented throughout this paper by the use of promise problems [9], which include a decision method which can dismiss instances of the problem without that method's complexity cost being included in the complexity of the problem.

This approach is motivated by the fact that obfuscating the relation between automatons by building redundant variables in a circuit increases the complexity of most considered problems. We justify our decision in two points: first, the approach of this paper is that of providing and studying an applicable method in a context where misleading inputs in local functions are unlikely. Second, despite the inclusion of these promises, high complexity issues arise in our pipeline. As such, we consider that they help understanding the precise issues that prevent our method from being efficient.

Additionally, we consider the FP and FNP classes as defined in [14].

3   From BANs to AMs

The first step of our process is to unfold a BAN into an AM. This simply requires the removal of any cycle in the interaction digraph of the BAN, and their replacement by inputs. In the scope of this paper, the number of inputs generated is required to be minimal. This is justified by the fact that the complexity of most of the problems addressed in the pipeline highly depends on the number of inputs of the considered AM.

▶ Acyclic Unfolding Functional Problem

Input: A Boolean automata network F, an integer k.

Promise: The encoding of the local functions of F only has essential variables.

Output: An acyclic module M with at most k inputs and a recursive wiring ω such that $\circlearrowleft_\omega M = F$.

Theorem 5. *The Acyclic Unfolding Functional Problem is in FNP.*

Proof. The promise of this problem allows us to compute the interaction digraph of F in polynomial time.

Consider the following simple non-deterministic algorithm: first guess a module M and a wiring ω; then check that the number of inputs in M is no more than k and that $\circlearrowleft_\omega M$ syntactically equals F.

This algorithm operates in polynomial non-deterministic time since the recursive wiring is a simple substitution of variables, and thanks to the fact that one only needs to compare $\circlearrowleft_\omega M$ and F at a syntactical level. Indeed, if any solution exists, then a solution exists with the same number of nodes, the same inputs, the same wirings, and such that the substitution operated by ω on M leads to a syntaxical copy of the local functions of F. □

Theorem 6. *The Acyclic Unfolding Functional Problem is NP-hard.*

Sketch of Proof. There is a straightforward reduction from the Feedback Vertex Set problem: given G, k we construct a BAN F with OR local functions whose interaction digraph is isomorphic to G. Then the inputs of a solution M to F, k correspond to a feedback vertex set (which is given by the codomain of ω). □

Example 7. Consider S_A and F_A of Example 1. Let us define $I_A = \{\alpha\}$. Let M_A be the acyclic module that defines $f'_a(x) = x_\alpha$, $f'_b(x) = f'_c(x) = x_a$, and $f'_d(x) = \neg x_b \vee \neg x_c$. The module M_A is a valid answer to the instance $F_A, k = 1$ of the Acyclic Unfolding Functional Problem. The interaction digraph of this module is represented in Fig. 3 (left panel).

Example 8. Consider S_B and F_B of Example 2. Let us define $I_B = \{\alpha_{St}, \alpha_{Sl}, \alpha_{Sk}, \alpha_{Pp}, \alpha_C, \alpha_{C*}\}$. Let M_B be the acyclic module that defines $f'_{St}(x) = \neg x_{\alpha_{St}}$, $f'_{Sl}(x) = \neg x_{\alpha_{Sl}} \vee x_{\alpha_{C*}}$, $f'_{Sk}(x) = x_{\alpha_{St}} \vee \neg x_{\alpha_{Sk}}$, $f'_{Pp}(x) = x_{\alpha_{Sl}} \vee \neg x_{\alpha_{Pp}}$, $f'_{Ru}(x) = f_{S9}(x) = \neg x_{\alpha_{Sk}} \vee x_{\alpha_{Pp}} \vee \neg x_{\alpha_C} \vee \neg x_{\alpha_{C*}}$, $f'_C(x) = \neg x_{Ru} \vee \neg x_{S9} \vee \neg x_{\alpha_{Sl}}$, $f'_{C25}(x) = \neg x_{\alpha_{Pp}} \vee x_{\alpha_C}$, $f'_M(x) = x_{\alpha_{Pp}} \vee \neg x_{\alpha_C}$, and $f'_{C*}(x) = \neg x_{Ru} \vee \neg x_{S9} \vee x_{C25} \vee \neg x_M$. The module M_B is a valid answer to the instance $F_B, k = 6$ of the Acyclic Unfolding Functional Problem. The interaction digraph of this module is represented in Fig. 3 (right panel).

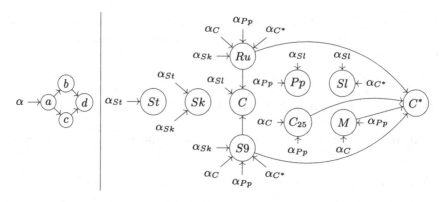

Fig. 3. On the left, the interaction digraph of M_A, as described in Example 7. On the right, the interaction digraph of M_B, as described in Example 8.

4   Output Functions

Output functions were first introduced in [20]. They are a way to characterise the asymptotic behaviour of an AM as a set of Boolean functions that are computed from the local functions of the AM. Computing the output functions of an AM is a crucial step in the pipeline proposed in this work.

▸ Output Circuit Computation Problem
  Input: An acyclic module M, and $X \subseteq S$ a set of output nodes.
  Promise: The encoding of the local functions of M only has essential variables.
  Output: An output function for each node in X, encoded as a Boolean circuit.

Theorem 9. *The Output Circuit Computation Problem is in FP.*

Sketch of Proof. To build the circuit that encodes some output function of the network, we first construct a list of every output function at different delays that are required to build it, and prove that this list can be constructed in polynomial time. We then replace every entry on that list by the circuit that encodes the corresponding local function, and merge them together to obtain the circuit encoding the result.   □

Example 10. Consider M_A of Example 7. Let $X_A = \{d\}$ be an instance of the Output Circuit Computation Problem. The circuit $O_d = \neg\alpha_3$ is a valid answer to that instance.

Example 11. Consider M_B of Example 8. Let $X_B = \{St, Sk, Sl, Pp, C, C^*\}$ be an instance of the Output Circuit Computation Problem. The circuits $O_{St} = \neg\alpha_{St,1}$, $O_{Sl} = \neg\alpha_{Sl,1} \vee \alpha_{C^*,1}$, $O_{Sk} = \alpha_{St,1} \vee \neg\alpha_{Sk,1}$, $O_{Pp} = \alpha_{Sl,1} \vee \neg\alpha_{Pp,1}$, $O_C = (\alpha_{Sk,2} \wedge \neg\alpha_{Pp,2} \wedge \alpha_{C,2} \wedge \alpha_{C^*,2}) \vee \neg\alpha_{Sl,1}$ and $O_{C^*} = \alpha_{C,2} \vee \neg\alpha_{Pp,2}$ taken altogether are a valid answer to that instance.

5   Optimal Acyclic Module Synthesis

This part of the process takes as input a set of output functions and generates a module that realizes these functions with an hopefully minimal number of nodes. In this part the actual optimisation of the pipeline, if any, can be directly observed. It is also the part of the pipeline which bears most of the computational cost.

▶ Module Synthesis Problem
   Input: A set I of input labels, a finite set of output functions O, encoded as Boolean circuits, defined on those labels, and k an integer.
   Output: An acyclic module M with at most k nodes such that every function in O is the output function of at least one node in M.

Theorem 12. *The Module Synthesis Problem is coNP-hard.*

Proof. Consider an instance f of the Tautology problem, with I the set of propositional variables contained in f. We define f' as the output function defined on the labels I such that f' is obtained from f by substituting all variables $\alpha \in I$ by their equivalent of delay 1, α_1. Let us also define f_1 as the constant output function of delay 0 which value is always 1. We compose an instance of the Module Synthesis Problem with I the set of input labels, $O = \{f', f_1\}$ and $k = 1$. This instance has a solution if and only there exists an acyclic module with only one node such that the output function of this node is equivalent to all the output functions in O. This implies that, if the problem has a solution, f' is equivalent to f_1, which proves that f' and f are tautologies. Therefore computing the output of the Module Synthesis Problem requires solving a coNP-hard decision problem.   □

Theorem 13. *The Module Synthesis Problem is in FNP^{coNP}.*

Proof. Consider the following algorithm. First, guess an acyclic module M, with size k. Compute every output function of the network, which is in FP. Then simply check that every function in O is equivalent to at least one output function in M, which requires at most $|M| \times |O|$ calls to a coNP oracle.   □

It is unclear whether the synthesis problem can be proven to be in FNP or to be NP^{coNP}-hard. An attempt has been made to prove the former by using a greedy algorithm which would fuse nodes in an acyclic module, starting from a trivially large enough module. However this method seems to require a singular fusion operation which does not seem to be computable in polynomial time. This leads us to believe that a greedy algorithm would not prove the Optimal Module Synthesis Problem to be in FNP. Similarly, it is interesting to consider the open question of whether or not the Module Synthesis Problem can be proven NP^{coNP}-hard. This implies to prove, between other things, that the problem is NP-hard. This is, to us, another open problem as the Module Synthesis Problem does not seem equiped to compute the satisfaction of a Boolean formula or circuit.

This open question bears strong ressemblance to another open problem that concerns Boolean circuits. The Circuit Minimisation Problem is known to be in NP but it is not known whether the problem is in P or NP-hard, as both possibilities have deep consequences on famous open questions in theoretical computer sciences [12]. The same problem has been found to be NP-complete in both restricted (DNFs) and generalised (unrestricted Boolean circuits) variations of the Boolean circuit model [11].

There are strong similarities between acyclic modules and Boolean circuits. Both are defined on acyclic digraphs, have inputs and outputs, and compute Boolean functions. It is important to note that this analogy is misleading when talking about the optimisation of their size. Optimising a Boolean circuit requires the optimisation of a Boolean function in terms of the number of gates that computes it. Optimising an acyclic module, however, requires the optimisation of a network of functions with respect to a notion of delay of the inputs, whereas in this case one node may contain an arbitrary Boolean function. As such these problems seem too independent to provide any reduction between them.

Example 14. Consider the output function O_d defined in Example 10. Let us define M'_A as the module defined on $S'_A = \{a, b, d\}$ and $I_A = \{\alpha\}$, such that $f''_a = x_\alpha$, $f''_b = x_a$ and $f_d = \neg x_b$. The module M'_A is a valid answer to the instance $I_A, \{O_d\}, k = 3$ of the Module Synthesis Problem. The interaction digraph of this module is depicted in Fig. 4 (left panel).

Example 15. Consider the output functions $O_B = \{O_{St}, O_{Sl}, O_{Sk}, O_{Pp}, O_C, O_{C^*}\}$ defined in Example 11. Let us define M'_B as the module defined on $S'_B = \{St, Sl, Sk, Pp, Ru, C25, C, C^*\}$ and $I_B = \{\alpha_{St}, \alpha_{Sl}, \alpha_{Sk}, \alpha_{Pp}, \alpha_C, \alpha_{C^*}\}$, such that $f''_{St}(x) = \neg x_{\alpha_{St}}$, $f''_{Sl}(x) = \neg x_{\alpha_{Sl}} \lor x_{\alpha_{C^*}}$, $f''_{Sk}(x) = x_{\alpha_{St}} \lor \neg x_{\alpha_{Sk}}$, $f''_{Pp}(x) = x_{\alpha_{Sl}} \lor \neg x_{\alpha_{Pp}}$, $f''_{Ru}(x) = \neg x_{\alpha_{Sk}} \lor x_{\alpha_{Pp}} \lor \neg x_{\alpha_C} \lor \neg x_{\alpha_{C^*}}$, $f''_C(x) = \neg x_{Ru} \lor \neg x_{\alpha_{Sl}}$, $f''_{C25}(x) = \neg x_{\alpha_{Pp}} \lor x_{\alpha_C}$, and $f'_{C^*}(x) = x_{C25}$. The module M'_B is a valid answer to the instance $I_B, O_B, k = 8$ of the Module Synthesis Problem. The interaction digraph of this module is depicted in Fig. 4 (right panel).

6   Final Wiring and Analysis

The final step in the pipeline is simply to wire the module obtained in Sect. 5 so that the obtained networks hold isomorphic attractors to the input network. This is ensured by application of the following result.

Theorem 16 [20]. *Let M and M' be two acyclic modules, with T and T' subsets of their nodes such that $|T| = |T'|$. If there exists a bijection g from I to I' and a bijection h from T to T' such that for every $s \in T$, O_s and $O'_{h(s)}$ have same delay, and for every input sequence j with length the delay of O_s,*

$$O_s(j) = O'_{h(s)}(j \circ g^{-1})$$

then for any function $\omega : I \to T$, the networks $\circlearrowleft_\omega M$ and $\circlearrowleft_{h \circ \omega \circ g^{-1}} M'$ have isomorphic attractors (up to the renaming of automata given by h).

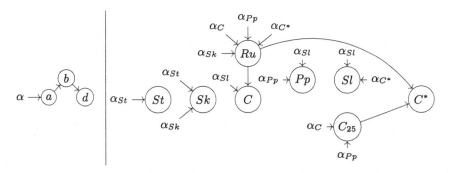

Fig. 4. On the left, the interaction digraph of M'_A, as described in Example 14. On the right, the interaction digraph of M'_B, as described in Example 15.

Applying this theorem to the current problem is simple: the module M is the module obtained in Sect. 3, and the module M' is the module obtained in Sect. 5. The set T is the set of nodes which are substituted by new inputs in the process described in Sect. 3. The set T' is the set of nodes in M' which are considered as the output of the module, for example when the module M' is obtained as the result of the application of the functional problem defined in Sect. 5.

As modules M and M' are defined over the same set of inputs, the bijection g is the identity. The bijection h is directly constructed so that for all $s \in T$, $h(s)$ in M' has an equivalent output function as s in M, which is always possible thanks to the careful structure of our pipeline. It follows quite clearly that for any $s \in T$, and for any input sequence j, $O_s(j) = O'_{h(s)}(j \circ g^{-1})$ holds, and the theorem applies.

Example 17. Consider M'_A of Example 14. Let $\omega_A(\alpha) = d$. The AN $\circlearrowleft_{\omega_A} M'_A$ is defined over $S'_A = \{a, b, d\}$ such that $f'''_a(x) = x_d$, $f'''_b(x) = x_a$, $f'''_d(x) = \neg x_b$. The interaction digraph of this module is depicted in Fig. 5 (left panel).

Example 18. Consider M'_B of Example 15. Let $\omega_B(\alpha_s) = s$, for all $s \in X_B$. The AN $\circlearrowleft_{\omega_B} M'_B$ is defined over $S'_B = \{St, Sl, Sk, Pp, Ru, C25, C^*\}$ such that $f'''_{St}(x) = \neg x_{St}$, $f'''_{Sl}(x) = \neg x_{Sl} \vee x_{C^*}$, $f'''_{Sk}(x) = x_{St} \vee \neg x_{Sk}$, $f'''_{Pp}(x) = x_{Sl} \vee \neg x_{Pp}$, $f'''_{Ru}(x) = \neg x_{Sk} \vee x_{Pp} \vee \neg x_C \vee \neg x_{C^*}$, $f'''_C(x) = \neg x_{Ru} \vee \neg x_{Sl}$, $f'''_{C25}(x) = \neg x_{Pp} \vee x_C$, and $f'_{C^*}(x) = x_{C25}$. The interaction digraph of this module is depicted in Fig. 5 (right panel).

This allows us to compute the attractors of any BAN by computing the dynamics of another BAN with possibly less nodes, thus dividing the number of computed configurations by some power of two. Examples throughout this paper showcase the application of the pipeline over two initial examples.

Examples 1, 7, 10, 14 and 17 show the optimisation of a simple four nodes network into a three nodes equivalent network. The optimisation proceeds here by 'compacting' two trivially equivalent nodes, b and c, into one. The resulting BAN has dynamics 2^1 times smaller than the initial network, with isomorphic attrac-

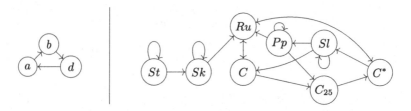

Fig. 5. On the left, the interaction digraph of F_A', as described in Example 17. On the right, the interaction digraph of F_B', as described in Example 18.

tors. Examples 2, 8, 11, 15 and 17 show the optimisation of a larger, more intricate network which is drawn from a model predicting the cell cycle sequence of fission yeast [5]. This practical example, processed through our pipeline, reduces from 10 nodes to 8. This implies a reduction in dynamics size of 2^2, while keeping isomorphic attractors. Both sets of examples are illustrated throughout the paper in Figs. 2, 3, 4 and 5.

7   Conclusion

The present paper showcases an innovative way of reducing the cost of computing the attractors of Boolean automata networks. The method provides better optimisation on networks showing structural redundancies, which are removed by the pipeline. The limitations of this method are still significant; it requires solving a problem that is at least coNP-hard, and believed to be FNP^{coNP}-complete. As it presently stands, this method is not as much a convincing practical tool as it is a good argument in favor of the powerfulness of acyclic modules, their output functions, and the approaches they allow together towards the computation of BAN dynamics.

Other future perspectives include finding better complexity bounds to the Module Synthesis Problem, finding efficient heuristical or approximate implementations of the pipeline, and generalising the formalism of output functions and the optimisation pipeline to different update schedules distinct from parallel.

Acknowledgements. The works of Kévin Perrot and Sylvain Sené were funded mainly by their salaries as French State agents, affiliated to Aix-Marseille Univ., Univ. de Toulon, CNRS, LIS, UMR 7020, Marseille, France (both) and to Univ. Côte d'Azur, CNRS, I3S, UMR 7271, Sophia Antipolis, France (KP), and secondarily by ANR-18-CE40-0002 FANs project, ECOS-Sud C19E02 project, STIC AmSud CoDANet 19-STIC-03 (Campus France 43478PD) project.

References

1. Aracena, J.: Maximum number of fixed points in regulatory Boolean networks. Bull. Math. Biol. **70**, 1398–1409 (2008). https://doi.org/10.1007/s11538-008-9304-7
2. Aracena, J., Richard, A., Salinas, L.: Number of fixed points and disjoint cycles in monotone Boolean networks. SIAM J. Discr. Math. **31**, 1702–1725 (2017)
3. Bridoux, F., Durbec, N., Perrot, K., Richard, A.: Complexity of maximum fixed point problem in Boolean networks. In: Manea, F., Martin, B., Paulusma, D., Primiero, G. (eds.) CiE 2019. LNCS, vol. 11558, pp. 132–143. Springer, Cham (2019). https://doi.org/10.1007/978-3-030-22996-2_12
4. Bridoux, F., Gaze-Maillot, C., Perrot, K., Sené, S.: Complexity of limit-cycle problems in Boolean networks (2020, submitted) arXiv:2001.07391
5. Davidich, M.I., Bornholdt, S.: Boolean network model predicts cell cycle sequence of fission yeast. PLoS One **3**, e1672 (2008)
6. Demongeot, J., Goles, E., Morvan, M., Noual, M., Sené, S.: Attraction basins as gauges of robustness against boundary conditions in biological complex systems. PLoS One **5**, e11793 (2010)
7. Demongeot, J., Noual, M., Sené, S.: Combinatorics of Boolean automata circuits dynamics. Discrete Appl. Math. **160**, 398–415 (2012)
8. Floreen, P., Orponen, P.: Counting stable states and sizes of attraction domains in Hopfield nets is hard. In: Proceedings of IJCNN 1989, pp. 395–399. IEEE (1989)
9. Goldreich, O.: On promise problems: a survey. In: Goldreich, O., Rosenberg, A.L., Selman, A.L. (eds.) Theoretical Computer Science. LNCS, vol. 3895, pp. 254–290. Springer, Heidelberg (2006). https://doi.org/10.1007/11685654_12
10. Goles, E., Salinas, L.: Comparison between parallel and serial dynamics of Boolean networks. Theor. Comput. Sci. **396**, 247–253 (2008)
11. Ilango, R., Loff, B., Oliveira, I.C.: NP-hardness of circuit minimization for multi-output functions. In: Proceedings of ECCC 2020, pp. TR20-021 (2020)
12. Kabanets, V., Cai, J.: Circuit minimization problem. In: Proceedings of STOC 2000, pp. 73–79. ACM (2000)
13. Kauffman, S.A.: Metabolic stability and epigenesis in randomly constructed genetic nets. J. Theor. Biol. **22**, 437–467 (1969)
14. Meggido, N., Papadimitriou, C.: A note on total functions, existence theorems, and computational complexity. Technical report, IBM (1989)
15. Mendoza, L., Alvarez-Buylla, E.R.: Dynamics of the genetic regulatory network for Arabidopsis thaliana flower morphogenesis. J. Theor. Biol. **193**, 307–319 (1998)
16. Noual, M.: Dynamics of circuits and intersecting circuits. In: Dediu, A.-H., Martín-Vide, C. (eds.) LATA 2012. LNCS, vol. 7183, pp. 433–444. Springer, Heidelberg (2012). https://doi.org/10.1007/978-3-642-28332-1_37
17. Noûs, C., Perrot, K., Sené, S., Venturini, L.: #P-completeness of counting update digraphs, cacti, and a series-parallel decomposition method. In: Proceedings of CiE 2020 (2020, accepted), arXiv:2004.02129
18. Orponen, P.: Neural networks and complexity theory. In: Havel, I.M., Koubek, V. (eds.) MFCS 1992. LNCS, vol. 629, pp. 50–61. Springer, Heidelberg (1992). https://doi.org/10.1007/3-540-55808-X_5
19. Perrot, K., Perrotin, P., Sené, S.: A framework for (De)composing with Boolean automata networks. In: Durand-Lose, J., Verlan, S. (eds.) MCU 2018. LNCS, vol. 10881, pp. 121–136. Springer, Cham (2018). https://doi.org/10.1007/978-3-319-92402-1_7

20. Perrot, K., Perrotin, P., Sené, S.: On the complexity of acyclic modules in automata networks. In: Chen, J., Feng, Q., Xu, J. (eds.) TAMC 2020. LNCS, vol. 12337, pp. 168–180. Springer, Cham (2020). https://doi.org/10.1007/978-3-030-59267-7_15. arXiv:1910.07299
21. Robert, F.: Discrete Iterations: A Metric Study. Springer, Heidelberg (1986). https://doi.org/10.1007/978-3-642-61607-5
22. Thomas, R.: Boolean formalization of genetic control circuits. J. Theor. Biol. **42**, 563–585 (1973)

On the Transformation of Two-Way Deterministic Finite Automata to Unambiguous Finite Automata

Semyon Petrov[ID] and Alexander Okhotin[(✉)][ID]

Department of Mathematics and Computer Science, St. Petersburg State University,
7/9 Universitetskaya nab., Saint Petersburg 199034, Russia
semenuska2010@yandex.ru, alexander.okhotin@spbu.ru

Abstract. The paper estimates the number of states in an unambiguous finite automaton (UFA) that is sufficient and in the worst case necessary to simulate an n-state two-way deterministic finite automaton (2DFA). It is proved that a 2DFA with n states can be transformed to a UFA with fewer than $2^n \cdot n!$ states. On the other hand, for every n, there is a language recognized by an n-state 2DFA that requires a UFA with at least $\Omega((4\sqrt{2})^n \cdot n^{-1/2})$ states. The latter result is proved by estimating the rank of a certain matrix.

Keywords: Descriptional complexity · Two-way finite automata · Unambiguous finite automata

1 Introduction

Many variants of finite automata are known, and although all of them define the same class of *regular languages*, they differ in terms of succinctness of description. In particular, it is well-known that every *nondeterministic finite automaton* (NFA) with n states can be transformed to a *deterministic finite automaton* (DFA) with 2^n states, and this number of states is in the worst case necessary. This kind of succinctness tradeoffs have been studied for quite a few types of finite automata.

Transformations involving *two-way finite automata*, deterministic (2DFA) and nondeterministic (2NFA), have received particular attention in the literature [1–3,8,10,11,15,17,18]. In particular, the question of whether two-way automata can be determinized using polynomially many states is one of the most important open problems of automata theory, due to its connection to the L vs. NL problem [7]. Their transformation to one-way automata was studied over the years [1,11,17], until Kapoutsis [6] presented an optimal transformation. Kapoutsis [6] showed how to transform an n-state 2DFA to an NFA with $\binom{2n}{n+1}$ states, and proved that this number of states is necessary in the worst

Research supported by Russian Science Foundation, project 18-11-00100.

A. Leporati et al. (Eds.): LATA 2021, LNCS 12638, pp. 81–93, 2021.
https://doi.org/10.1007/978-3-030-68195-1_7

case; transforming a 2DFA to a DFA takes $n(n^n - (n-1)^n)$ states in the worst case [6].

Between these two perfectly conclusive results, there is an open question involving an intermediate model between DFA and NFA: the *unambiguous finite automata* (UFA), which can use nondeterminism, yet are bound to accept each string in at most one computation (as in the unambiguous complexity classes, such as UL and UP). What is the complexity of transforming a 2DFA to a UFA?

The size of UFA has received some attention in the literature. As shown by Leung [9], transforming an n-state UFA to a DFA requires 2^n states in the worst case, whereas the NFA-to-UFA transformation incurs a blowup from n to $2^n - 1$ states. In the case of a unary alphabet, first studied by Ravikumar and Ibarra [14], it is now known that transforming a UFA to a DFA in the worst case takes $e^{\Theta(\sqrt[3]{n \log^2 n})}$ states, and the NFA-to-UFA transformation requires $e^{\Theta(\sqrt{n \log n})}$ states [12]. Jirásek Jr. et al. [5] showed that complementing a UFA requires at least $2^{0.79n}$ states, with an upper bound of 2^n states. In the unary case, the known lower bound on complementing a UFA is $n^{\Omega(\log \log \log n)}$ [13].

Turning to the complexity of the 2DFA-to-UFA transformation, it is bound to lie between the two bounds of Kapoutsis [6], $\binom{2n}{n+1}$ and $n(n^n - (n-1)^n)$, and it is natural to ask what is the exact function in this case. This question is addressed in the present paper.

The first task is to establish an upper bound that would improve over the 2DFA-to-DFA transformation. This is achieved by augmenting the NFA constructed by Kapoutsis [6] to store extra data that allows it to ensure the uniqueness of its accepting computation. The resulting UFA, presented in Sect. 3, has fewer than $2^n \cdot n!$ states.

Turning to a lower bound on the 2DFA-to-UFA transformation, a witness language is defined in Sect. 4 by constructing a 2DFA. The plan is to prove a lower bound on the size of every UFA recognizing the same language using *Schmidt's theorem* [16], which relies on the rank of a certain matrix related to the language. The rank of the matrix constructed in the paper is estimated by first applying some linear transformation, and then reducing the problem to finding the rank of another matrix $P^{(k)}$, defined entirely in terms of permutations.

A lower bound on the rank of $P^{(k)}$ is established in Sect. 5 by showing that each $P^{(k)}$ contains a submatrix $\begin{pmatrix} 0 & P^{(k-1)} \\ P^{(k-1)} & 0 \end{pmatrix}$, and hence the rank is at least 2^{k-1}. This estimation yields a lower bound of $\Omega((4\sqrt{2})^n \cdot n^{-1/2})$ states on the 2DFA-to-UFA tradeoff.

It is also shown that the matrix constructed in this paper is "optimal" in the sense that any lower bound on the 2DFA-to-UFA tradeoff obtained using Schmidt's theorem cannot exceed the rank of this matrix.

2 Definitions

The paper uses standard finite automata models: two-way deterministic automata and one-way unambiguous automata.

Definition 1. *A two-way deterministic finite automaton (2DFA) is a quintuple $\mathcal{A} = (\Sigma, Q, q_0, \delta, F)$, in which Σ is a finite alphabet; Q is a finite set of states; $q_0 \in Q$ is the initial state; $\delta \colon Q \times (\Sigma \cup \{\vdash, \dashv\}) \to Q \times \{-1, +1\}$ is the transition function, which defines a transition in a given state while observing a given tape symbol; $F \subseteq Q$ is the set of accepting states, effective at the right end-marker $\dashv$.*

Given an input string $w = a_1 \ldots a_\ell$, a 2DFA operates on a read-only tape $\vdash w \dashv$. It begins its computation in the initial state, with the head at the left end-marker ($\vdash$). At every step of the computation, the automaton is in a state $q \in Q$ and observes a symbol $a \in \Sigma \cup \{\vdash, \dashv\}$; the transition function gives a pair $\delta(q, a) = (r, d)$ representing the next state and the direction in which the head moves. The set of strings, on which the computation eventually reaches the right end-marker in an accepting state, is denoted by $L(\mathcal{A})$.

Definition 2. *A nondeterministic finite automaton (NFA) is a quintuple $\mathcal{B} = (\Sigma, Q, Q_0, \delta, F)$, in which Σ is a finite alphabet; Q is a finite set of states; $Q_0 \subseteq Q$ is the set of initial states; the transition function $\delta \colon Q \times \Sigma \to 2^Q$ defines possible next states after reading a given symbol in a given state; $F \subseteq Q$ is the set of accepting states. On an input string $w = a_1 \ldots a_\ell$, a computation is a sequence of states $p_0, p_1, \ldots, p_\ell$ satisfying $p_0 \in Q_0$ and $p_{i+1} \in \delta(p_i, a_{i+1})$ for all i. It is accepting if, furthermore, $p_\ell \in F$. The set of strings, on which there is at least one accepting computation, is denoted by $L(\mathcal{B})$.*

An NFA is said to be *unambiguous* (UFA), if there is at most one accepting computation on each string.

3   Upper Bound

The proposed new transformation of 2DFA to UFA is derived from the known 2DFA-to-NFA transformation by Kapoutsis [6].

For a 2DFA with a set of states Q, Kapoutsis [6] constructs an NFA with states of the form (P, R), with $P, R \subseteq Q$ and $|P| + 1 = |R|$. For an input string uv, after reading a prefix u, the NFA guesses a *frontier* of the 2DFA's computation on $\vdash uv \dashv$: this is a pair (P, R), where the set R consists of all states, in which the 2DFA moves to the right from the last symbol of u; states in P are those, in which the 2DFA moves to the last symbol of u from the left. The constructed NFA guesses a 2DFA computation's frontier at every step of its computation.

Theorem A (Kapoutsis [6]) *For every 2DFA with n states, there exists an NFA with $\binom{2n}{n+1}$ states that recognizes the same language.*

The NFA constructed by the method of Kapoutsis is, in general, ambiguous, because, while guessing the next frontier, it may produce a closed cycle alongside the main computation. This closed cycle shall eventually be cancelled out, without the NFA's noticing, whereas the correctly guessed computation of the 2DFA would drive the NFA to acceptance. This yields multiple accepting computations.

The above construction shall now be elaborated to ensure unambiguity. Besides a pair (P, R), the automaton shall remember a bijection $f\colon P \cup \{\text{START}\} \to R$ representing the states in R reached from each state in P, as well as from the initial configuration. Such a triple (P, R, f), illustrated in Fig. 1, shall be called a *(prefix) profile*.

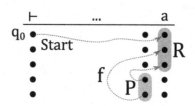

Fig. 1. A profile (P, R, f) of a computation that holds short of reading the symbol a.

How many prefix profiles are there? For every $k = |R|$, there are $\binom{n}{k-1}$ ways to choose the set P, and $\binom{n}{k}$ ways to choose the set R, and $k!$ different bijections f. Overall, there are $\sum_{k=1}^{n} \binom{n}{k-1}\binom{n}{k}k!$ profiles. With the frontiers replaced by profiles, the following theorem is obtained.

Theorem 1. *For every n-state 2DFA, there is a UFA with $\sum_{k=1}^{n} \binom{n}{k-1}\binom{n}{k}k!$ states that recognizes the same language.*

This sum is less than $2^n \cdot n!$, which is in turn asymptotically less than $n(n^n - (n-1)^n)$. This confirms that the proposed transformation to UFA is more efficient than transforming a 2DFA to a DFA, as per another construction by Kapoutsis [6] (a comparison for small values of n shall be given later on in Table 1).

4   Lower Bound

A lower bound on the state complexity of transforming a 2DFA to a UFA is based on a witness language recognized by a small 2DFA, for which every equivalent UFA would require a substantial number of states.

The witness language is defined over an alphabet $\Gamma = (\{1, \ldots, n\} \cup \{f \mid f\colon \{1, \ldots, n\} \to \{1, \ldots, n\}$ is a partial function $\}) \times \{l, r\}$, and is recognized by a 2DFA $\mathcal{D}_n$ with the set of states $Q = \{1, \ldots, n\}$, with $q_0 = 1$ and $F = \{1\}$. It uses the following transitions, defined for all $x, y \in Q$ and $f, g\colon \{1, \ldots, n\} \to$

$\{1, \ldots, n\}$.

$$\delta(q_0, \vdash) = (q_0, +1), \quad \delta(q, (x, 1)) = (x, +1),$$

$$\delta(q, (f, 1)) = \begin{cases} (f(q), +1), & \text{if } f(q) \text{ is defined} \\ (q_0, -1), & \text{otherwise} \end{cases}$$

$$\delta(q, (g, \mathtt{r})) = \begin{cases} (g(q), -1), & \text{if } g(q) \text{ is defined} \\ (q, +1), & \text{otherwise} \end{cases}$$

$$\delta(q, (y, \mathtt{r})) = \begin{cases} (q_0, +1), & \text{if } q = y \\ (q, -1), & \text{otherwise} \end{cases}$$

This is the automaton used by Kapoutsis [6] in his lower bound for the transformation of a 2NFA to an NFA. Following Kapoutsis, the subsequent proof uses four-symbol strings of the form $(x, 1)(f, 1)(g, \mathtt{r})(y, \mathtt{r})$, with $x, y \in Q$ and with partial functions f, g, where $f(x)$ is defined and $g(y)$ is not. These strings correspond to directed (n, n)-bipartite graphs, with f representing arrows from left to right, and g, from right to left. The automaton $\mathcal{D}_n$ then verifies, whether there is a path from x in the left part to y on the right.

The rest of this paper is concerned with proving a lower bound on the size of every UFA recognizing the language $L(\mathcal{D}_n)$. The only known method for proving such lower bounds is the following theorem.

Theorem B (Schmidt [16], see also Leung [9]) *Let L be a regular language, and let $(x_1, y_1), \ldots, (x_n, y_n)$ be pairs of strings. Let M be an integer matrix defined by $M_{i,j} = 1$, if $x_i y_j \in L$, and $M_{i,j} = 0$ otherwise. Then, every UFA for L has at least* rank M *states.*

In addition to the prefix profiles, describing a computation of an automaton on a prefix, a new type of profile shall be introduced.

Definition 3. *A* suffix profile *is a triple (g, P, R), where $P, R \subseteq Q$, $|P|+1 = |R|$, and $g: R \to P \cup \{\text{ACCEPT}\}$ is a bijection.*

For a given computation of automaton on the string uv, the suffix profile of v complements the prefix profile of u. The function g in the suffix profile is constructed in a similar way as in the prefix profile: for any state q, the state $g(q)$ is the state in which the automaton first crosses the border between u and v in this computation after visiting the first symbol of v in the state q—or $g(q) = \text{ACCEPT}$, if it accepts without crossing this border.

There are as many suffix profiles as prefix profiles. Indeed, for fixed P and R with $|R| = k$, there are $k!$ ways to choose a prefix profile (P, R, f) and $k!$ ways to choose a suffix profile (g, P, R).

The strings for Schmidt's theorem are chosen as follows. Let (P, R, f) be a prefix profile, where $f: P \cup \{\text{START}\} \to R$ is a bijection, and let $x \in \{1, \ldots, n\}$, with $x \notin P$; such an x exists, since $|P| < n$. Let $f': P \cup \{x\} \to R$ be a function defined by $f'(p) = f(p)$ for $p \in P$, and $f'(x) = f(\text{START})$. Then, define $x_{P,R,f} = (x, 1)(f', 1)$. The state x is added to create a bijection.

Next, for a suffix profile (g, S, T), let y be the element of T such that $g(y) = \text{ACCEPT}$. Let $g' : T \setminus \{y\} \to S$ be a function defined by $g'(r) = g(r)$. Define $y_{g,S,T} = (g', r)(y, r)$. Again, the state y is removed to create a bijection.

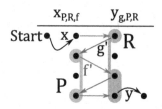

Fig. 2. An example of interaction between a prefix profile and a matching suffix profile.

For the computation of some 2DFA on a string uv, the prefix profile (P, R, f) of u and the suffix profile (g, P, R) of v define strings $x_{P,R,f}$ and $y_{g,P,R}$ that contain functions f' and g' which complement each other to a full path. Figure 2 illustrates this relation.

Note that $f'(x)$ is defined and $g'(y)$ is not, otherwise the string $x_{P,R,f} \, y_{g,S,T}$ would have no chance to be accepted by $\mathcal{D}_n$.

Let $M^{(n)}$ be the square matrix defined for these strings in Schmidt's theorem. Each row corresponds to a prefix profile (P, R, f), each column corresponds to a suffix profile (g, S, T), and the element at their intersection is denoted by $M^{(n)}_{(P,R,f),(g,S,T)}$. The order of the matrix is the total number of profiles, that is, $\sum_{k=1}^{n} \binom{n}{k-1} \binom{n}{k} k!$.

In order to estimate the rank of this matrix, it shall first be subjected to a series of rank-preserving transformations.

Definition 4. *A prefix profile (P', R', f') shall be called a subprofile of a prefix profile (P, R, f), if $P' \subseteq P$, $R' \subseteq R$ and $f'(p) = f(p)$ for all $p \in P' \cup \{\text{START}\}$. Notation: $(P', R', f') \preccurlyeq (P, R, f)$.*

Definition 5. *On an input $(x, 1)(f, 1)(g, r)(y, r)$, the automaton $\mathcal{D}_n$ is said to use the left-to-right arrow from p to q, if, at some point in its computation, $\mathcal{D}_n$ is in the state p on the symbol $(f, 1)$, and $f(p) = q$.*

Similarly, $\mathcal{D}_n$ uses the right-to-left arrow from q to p, if, at some point, $\mathcal{D}_n$ is in the state q at (g, r), and $g(q) = p$.

Definition 6. *Define a new square integer matrix $L^{(n)}$ of the same order as $M^{(n)}$, with rows and columns indexed by prefix and suffix profiles, respectively. Each element $L^{(n)}_{(P,R,f),(g,S,T)}$ is defined as 1, if $\mathcal{D}_n$ accepts the string $x_{P,R,f} \, y_{g,S,T}$ and uses all left-to-right arrows in the corresponding graph. Otherwise, let this element be 0.*

It turns out that the rows of $L^{(n)}$ are linear combinations of the rows of $M^{(n)}$ expressed by the inclusion–exclusion principle.

Lemma 1. *Let (P, R, f) be a prefix profile and let (g, S, T) be a suffix profile. Then,*

$$L^{(n)}_{(P,R,f),(g,S,T)} = \sum_{(P',R',f') \preccurlyeq (P,R,f)} (-1)^{|P|-|P'|} M^{(n)}_{(P',R',f'),(g,S,T)}$$

$$M^{(n)}_{(P,R,f),(g,S,T)} = \sum_{(P',R',f') \preccurlyeq (P,R,f)} L^{(n)}_{(P',R',f'),(g,S,T)}$$

Accordingly, $\operatorname{rank} L^{(n)} = \operatorname{rank} M^{(n)}$.

The columns of the matrix shall now be transformed by the same method.

Definition 7. *A suffix profile (g', S', T') shall be called a subprofile of a suffix profile (g, S, T), if $S' \subseteq S$, $T' \subseteq T$ and $g'(q) = g(q)$ for all $q \in T'$. Notation: $(g', S', T') \preccurlyeq (g, S, T)$.*

Definition 8. *Define yet another integer matrix $K^{(n)}$ of the same dimensions as $M^{(n)}$ and $L^{(n)}$, with its rows and columns again indexed by prefix and suffix profiles, respectively. Let $K^{(n)}_{(P,R,f),(g,S,T)}$ be 1, if $\mathcal{D}_n$ accepts $x_{P,R,f}\, y_{g,S,T}$ and uses all left-to-right and right-to-left arrows in the corresponding graph. Otherwise, let this element be 0.*

Lemma 2. *Let (P, R, f) be a prefix profile and let (g, S, T) be a suffix profile. Then,*

$$K^{(n)}_{(P,R,f),(g,S,T)} = \sum_{(g',S',T') \preccurlyeq (g,S,T)} (-1)^{|S|-|S'|} L^{(n)}_{(P,R,f),(g',S',T')}$$

$$L^{(n)}_{(P,R,f),(g,S,T)} = \sum_{(g',S',T') \preccurlyeq (g,S,T)} K^{(n)}_{(P,R,f),(g',S',T')}$$

In particular, $\operatorname{rank} K^{(n)} = \operatorname{rank} L^{(n)}$.

The above transformations of the matrix $M^{(3)}$, which is of size 39×39, are given in Fig. 3. The profiles are enumerated by ordering them first by $|P|$, and then lexicographically by P, by R and finally by the values of f.

The figure suggests that **the matrix $K^{(3)}$ is block diagonal.** This is proved as follows.

Lemma 3. *Let (P, R, f) be a prefix profile, and let (g, S, T) be a suffix profile with $(P, R) \neq (S, T)$. Then, $K^{(n)}_{(P,R,f),(g,S,T)} = 0$.*

Proof. Consider the string $x_{P,R,f}\, y_{g,S,T} = (x, 1)(f', 1)(g', \mathbf{r})(y, \mathbf{r})$. Assume the contrary, that $K^{(n)}_{(P,R,f),(g,S,T)} = 1$. Then, $\mathcal{D}_n$ uses all left-to-right arrows in the corresponding graph. In particular, the set of states in which $\mathcal{D}_n$ arrives to $(g', \mathbf{r})$

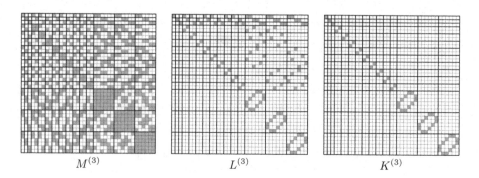

$$M^{(3)} \qquad\qquad L^{(3)} \qquad\qquad K^{(3)}$$

Fig. 3. Transformation of $M^{(3)}$. Each filled square contains 1, each empty square has 0.

from $(f', 1)$ is R, since these are the heads of all left-to-right arrows, and the set of states in which $\mathcal{D}_n$ arrives to $(f', 1)$ from $(g', \mathbf{r})$ is P: the tails of all left-to-right arrows, except the arrow from x to $f(\text{START})$.

At the same time, $\mathcal{D}_n$ uses all right-to-left arrows, and the set of states in which $\mathcal{D}_n$ arrives to $(g', \mathbf{r})$ from $(f', 1)$ is T, these are the tails of all right-to-left arrows with the addition of y, for which $g(y) = \text{ACCEPT}$; the states in which $\mathcal{D}_n$ arrives to $(f', 1)$ from $(g', \mathbf{r})$ is S, these are the heads of right-to-left arrows.

Therefore, $P = S$ and $R = T$, which contradicts the assumption.

Thus, the matrix $K^{(n)}$ is organized into blocks corresponding to different pairs (P, R).

The next important observation is that the **blocks corresponding to pairs (P_1, R_1) and (P_2, R_2), with $|P_1| = |P_2|$, are identical up to permutations of rows and columns.** With bijections $g\colon P_2 \to P_1$ and $h\colon R_1 \to R_2$ fixed, for a profile (P_1, R_1, f), set $g(\text{START}) = \text{START}$, and let the corresponding profile be $(P_2, R_2, h \circ f \circ g)$. The existence of a path in a bipartite graph is invariant to such permutations of vertices. This block is denoted by $P^{(k)}$, where $k = |R_1| = |R_2|$.

Definition 9. *Let $1 \leqslant k \leqslant n$. The matrix for permutations $P^{(k)}$ is a $k! \times k!$ submatrix of $K^{(n)}$ that consists of rows and columns corresponding to profiles (P, R, f) with $P = \{1, \ldots, k-1\}$ and $R = \{1, \ldots, k\}$. Its rows and columns are still indexed by prefix and suffix profiles prespectively, that is, the element corresponding to the functions f and g is denoted by $P^{(k)}_{(P,R,f),(g,P,R)}$.*

The form of the matrix $P^{(k)}$ for $k = 2, 3, 4$ is presented in Fig. 4. White squares represent zeroes, the rest of the squares contain 1.

Lemma 4. $\operatorname{rank} K^{(n)} = \sum_{k=1}^{n} \binom{n}{k-1} \binom{n}{k} \operatorname{rank} P^{(k)}$

Proof. By Lemma 3, the matrix $K^{(n)}$ is block diagonal, and hence $\operatorname{rank} K^{(n)}$ is a sum of ranks of independent blocks. Since the blocks with the same $|P|$ are equivalent, each block is equivalent to the matrix for permutations $P^{(k)}$ with

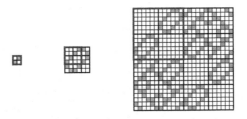

Fig. 4. The matrices $P^{(2)}$, $P^{(3)}$ and $P^{(4)}$.

$k - 1 = |P|$. There are $\binom{n}{k-1}\binom{n}{k}$ different ways to choose a pair (P, R) so that $|P| = k - 1$, hence for every k from 1 to n there are $\binom{n}{k-1}\binom{n}{k}$ blocks in $K^{(n)}$ that are equivalent to $P^{(k)}$. This gives the formula.

Thus, it is sufficient to estimate the rank of the matrix for permutations.

Definition 10. *Let (P, R, f) be a prefix profile with $P = \{1, \ldots, k - 1\}$ and $R = \{1, \ldots, k\}$. Then, the permutation corresponding to (P, R, f) is a function $g\colon \{1, \ldots, k\} \to \{1, \ldots, k\}$ defined by $g(p) = f(p)$ for $p \in P$, and $g(k) = f(\text{START})$.*

Definition 11. *Let (f, P, R) be a suffix profile with $P = \{1, \ldots, k - 1\}$ and $R = \{1, \ldots, k\}$. Then, the permutation corresponding to (f, P, R) is a function $g\colon \{1, \ldots, k\} \to \{1, \ldots, k\}$ defined by $g(p) = f(p)$ for $p \in R \setminus \{y\}$, and $g(y) = k$, where $y \in R$ is the state such that $f(y) = \text{ACCEPT}$.*

Note that, conversely, each permutation has a unique corresponding prefix profile, and an unique corresponding suffix profile of this form.

The elements of $P^{(k)}$ are characterized entirely in terms of permutations as follows. Let (P, R, f_1) be a prefix profile, and let (f_2, P, R) be a suffix profile, with $P = \{1, \ldots, k - 1\}$ and $R = \{1, \ldots, k\}$. Let g_1 and g_2 be the corresponding permutations. Denote $P^{(k)}_{g_1, g_2} = P^{(k)}_{(P, R, f_1), (f_2, P, R)}$.

Lemma 5. $P^{(k)}_{g_1, g_2} = 1$ *if and only if the permutation $g_2 \circ g_1$ is cyclic.*

Proof. Let (P, R, f_1) and (f_2, P, R) be profiles corresponding to g_1 and to g_2, respectively. Let $x_{P, R, f_1}\, y_{f_2, P, R} = (x, 1)(f_1', 1)(f_2', \mathtt{r})(y, \mathtt{r})$ be the corresponding string for this pair of profiles.

Consider the computation of $\mathcal{D}_k$ on the string $(x, 1)(f_1', 1)(f_2', \mathtt{r})(y, \mathtt{r})$. Then $x = k$, because $P = \{1, \ldots, k - 1\}$, and $y \in R$ is the state such that $f_2(y) = \text{ACCEPT}$, and $g_2(y) = k$. The automaton first moves to $(f_1', 1)$ in the state k, and then alternates between the second and the third symbols. This computation is depicted on Fig. 5.

Consider the sequence of states, in which it visits the second symbol $(f_1', 1)$. The sequence begins with k. In a state q, the automaton moves to the third symbol in the state $g_1(q)$, and then immediately returns to the second symbol in the state $g_2 \circ g_1(q)$, as long as $g_1(q)$ is not equal to $y = g_2^{-1}(k)$, which is equivalent to $g_2 \circ g_1(q) \neq k$.

⊖ If $P_{g_1,g_2}^{(k)} = 1$, then the element $P_{(P,R,f_1),(f_2,P,R)}^{(k)}$ is 1. As $P^{(k)}$ is a submatrix of $K^{(k)}$, the element $K_{(P,R,f_1),(f_2,P,R)}^{(k)}$ is 1 as well. By definition, this means that $\mathcal{D}_k$ accepts $x_{P,R,f_1} \, y_{f_2,P,R}$, using all arrows in both directions in its computation.

The above sequence of states must contain all states, since the automaton uses all left-to-right arrows in its computation. Therefore, one can reach all states by applying $g_2 \circ g_1$ starting from k, and this exactly means that this permutation is cyclic.

⊖ Assuming that the permutation $g_2 \circ g_1$ is cyclic, the above sequence must contain all states, which means that all arrows in both directions are used. The sequence is concluded with a state q satisfying $g_2 \circ g_1(q) = k$, and then $\mathcal{D}_k$ accepts. Therefore, $P_{g_1,g_2}^{(k)} = 1$.

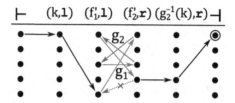

Fig. 5. A computation of $\mathcal{D}_k$ on the string $x_{P,R,f_1} \, y_{f_2,P,R}$, where profiles (P, R, f_1) and (f_2, P, R) correspond to permutations g_1 and g_2, respectively.

Theorem 2. *For every $n \geqslant 1$, there exists a language recognized by an n-state 2DFA, for which every UFA requires at least $\sum_{k=1}^{n} \binom{n}{k-1}\binom{n}{k} \operatorname{rank} P^{(k)}$ states, where $P^{(k)}$ is a $k! \times k!$ matrix, with its rows and columns corresponding to permutations, and $P_{g_1,g_2}^{(k)} = 1$ if and only if $g_2 \circ g_1$ is a cyclic permutation.*

Proof. The desired 2DFA is $\mathcal{D}_n$, defined in the beginning of this section.

The strings $x_{P,R,f}, y_{g,S,T}$ and the corresponding matrix $M^{(n)}$ are constructed, and, by Theorem B, every UFA recognizing this language has at least $\operatorname{rank} M^{(n)}$ states. Then, by Lemma 1, $\operatorname{rank} M^{(n)} = \operatorname{rank} L^{(n)}$. Next, by Lemma 2, $\operatorname{rank} L^{(n)} = \operatorname{rank} K^{(n)}$, and, by Lemma 4, $\operatorname{rank} K^{(n)} = \sum_{k=1}^{n} \binom{n}{k-1}\binom{n}{k} \operatorname{rank} P^{(k)}$. Finally, the matrix $P^{(k)}$ has the stated description in terms of permutations by Lemma 5.

There is a companion result that it is not possible to achieve a better lower bound via Schmidt's theorem by choosing another 2DFA and different pairs of strings.

Theorem 3. *Let $\mathcal{D}$ be a 2DFA over an alphabet Γ with n states that recognizes a regular language L. Let $X = \{x_1, \ldots, x_\ell\}$ and $Y = \{y_1, \ldots, y_m\}$ be sets of*

Table 1. The bounds established in this paper for small values of n, compared to the known 2DFA-to-NFA and 2DFA-to-DFA tradeoffs.

n	2DFA $\to$ NFA	2DFA $\to$ UFA (lower bound)	2DFA $\to$ UFA (upper bound)	2DFA $\to$ DFA
	$\binom{2n}{n+1}$	$\sum_{k=1}^{n} \binom{n}{k-1}\binom{n}{k} 2^{k-1}$	$\sum_{k=1}^{n} \binom{n}{k-1}\binom{n}{k} k!$	$n(n^n - (n-1)^n)$
1	1	1	1	1
2	4	6	6	6
3	15	33	39	57
4	56	180	292	700
5	210	985	2 505	10 505
6	792	5 418	24 306	186 186
7	3 003	29 953	263 431	3 805 249
8	11 440	166 344	3 154 824	88 099 230
9	43 758	927 441	41 368 977	2 278 824 849
10	167 960	5 188 590	589 410 910	65 132 155 990

strings over the alphabet Γ. Let M be an $\ell \times m$ matrix defined by $M_{i,j} = 1$ if $x_i y_j \in L$, and $M_{i,j} = 0$ otherwise. Then, $\operatorname{rank} M \leqslant \sum_{k=1}^{n} \binom{n}{k-1}\binom{n}{k} \operatorname{rank} P^{(k)}$.

Everything is thus reduced to the task of finding the rank of one particular matrix, $P^{(k)}$.

5   Estimating the Rank of the Matrix for Permutations

Theorem 4. $\operatorname{rank} P^{(k)} \geqslant 2^{k-1}$.

Proof (a sketch). The proof is based on an argument that $P^{(k)}$ has a submatrix $\begin{pmatrix} 0 & P^{(k-1)} \\ P^{(k-1)} & 0 \end{pmatrix}$. For $k = 3, 4$, this submatrix can be observed in Fig. 4.

The first claim is that the main diagonal is made of $(k-1)! \times (k-1)!$ blocks of zeroes. Formally, for every pair of permutations (g_1, g_2) satisfying $g_1(k) = g_2(k)$, the value of $P^{(k)}_{g_1, g_2^{-1}}$ is 0.

The second claim is that the matrix $P^{(k)}$ is symmetric, that is, $P^{(k)}_{g_1, g_2^{-1}} = P^{(k)}_{g_2, g_1^{-1}}$, for all permutations g_1, g_2.

The third claim is that the matrix $P^{(k)}$ contains blocks that are the same as $P^{(k-1)}$. In terms of matrices, this is expressed as follows. Let $g_1, g_2 \colon \{1, \ldots, k\} \to \{1, \ldots, k\}$ be permutations with $g_1(k) = k$ and $g_2(k) = k - 1$. Define two permutations $\tilde{g}_1, \tilde{g}_2 \colon \{1, \ldots, k-1\} \to \{1, \ldots, k-1\}$ as follows: $\tilde{g}_1(p) = g_1(p)$, thus ignoring k; $\tilde{g}_2(p) = g_2(p)$, as long as $g_2(p) \neq k$; and $\tilde{g}_2(p) = k - 1$, otherwise, thus redirecting from k to $k - 1$. Under these assumptions, it is proved that $P^{(k)}_{g_1, g_2^{-1}} = P^{(k-1)}_{\tilde{g}_1, \tilde{g}_2^{-1}}$. Since different permutations g_2 yield different $\tilde{g}_2$, this proves the claim.

Using the above three properties, one can prove that $\operatorname{rank} P^{(k)} \geqslant 2 \cdot \operatorname{rank} P^{(k-1)}$. The theorem follows from this by induction on k.

Corollary 1. *For every n, there is a language recognized by an n-state 2DFA, for which every UFA requires at least $\sum_{k=1}^{n} \binom{n}{k-1}\binom{n}{k} 2^{k-1} = \Omega\left(\frac{(4\sqrt{2})^n}{\sqrt{n}}\right)$ states.*

6   Conclusion

The bounds on the state complexity of transforming 2DFA to UFA established in this paper put it asymptotically between $\Omega\left(\frac{(2\sqrt{2})^n}{\sqrt{n}}\right)$ and $O(2^n \cdot n!)$, which shows that this is actually a new function different from the known 2DFA-to-NFA and 2DFA-to-DFA tradeoffs [6]. For small values of n, the bounds proved in this paper are compared in Table 1. All lower bounds, including the precise bounds by Kapoutsis [6], rely on using alphabet of exponential size, if the size of the alphabet is subexponential, the upper bounds are improved [4].

It would be interesting to determine the 2DFA-to-UFA tradeoff precisely. However, as shown in Theorem 3, the lower bound methods based on Schmidt's theorem have virtually been exhausted: it remains to establish the exact rank of the matrix $P^{(k)}$. New methods would be needed for any further improvements.

References

1. Birget, J.: State-complexity of finite-state devices, state compressibility and incompressibility. Math. Syst. Theory **26**, 237–269 (1993). https://doi.org/10.1007/BF01371727
2. Geffert, V., Mereghetti, C., Pighizzini, G.: Converting two-way nondeterministic unary automata into simpler automata. Theor. Comput. Sci. **295**, 189–203 (2003)
3. Geffert, V., Mereghetti, C., Pighizzini, G.: Complementing two-way finite automata. Inf. Comput. **205**, 1173–1187 (2007)
4. Geffert, V., Okhotin, A.: One-way simulation of two-way finite automata over small alphabets. In: NCMA (2013)
5. Jirásek, J., Jirásková, G., Sebej, J.: Operations on unambiguous finite automata. Int. J. Found. Comput. Sci. **29**, 861–876 (2018)
6. Kapoutsis, C.: Removing bidirectionality from nondeterministic finite automata. In: Jędrzejowicz, J., Szepietowski, A. (eds.) MFCS 2005. LNCS, vol. 3618, pp. 544–555. Springer, Heidelberg (2005). https://doi.org/10.1007/11549345_47
7. Kapoutsis, C.A.: Two-way automata versus logarithmic space. Theory Comput. Syst. **55**, 421–447 (2013). https://doi.org/10.1007/s00224-013-9465-0
8. Kunc, M., Okhotin, A.: Describing periodicity in two-way deterministic finite automata using transformation semigroups. In: Mauri, G., Leporati, A. (eds.) DLT 2011. LNCS, vol. 6795, pp. 324–336. Springer, Heidelberg (2011). https://doi.org/10.1007/978-3-642-22321-1_28
9. Leung, H.: Descriptional complexity of NFA of different ambiguity. Int. J. Found. Comput. Sci. **16**, 975–984 (2005)
10. Mereghetti, C., Pighizzini, G.: Optimal simulations between unary automata. SIAM J. Comput. **30**, 1976–1992 (2001)

11. Moore, F.R.: On the bounds for state-set size in the proofs of equivalence between deterministic, nondeterministic, and two-way finite automata. IEEE Trans. Comput. **C–20**, 1211–1214 (1971)
12. Okhotin, A.: Unambiguous finite automata over a unary alphabet. Inf. Comput. **212**, 15–36 (2012)
13. Raskin, M.: A superpolynomial lower bound for the size of non-deterministic complement of an unambiguous automaton. In: 45th International Colloquium on Automata, Languages, and Programming (ICALP 2018). Leibniz International Proceedings in Informatics (LIPIcs), vol. 107, pp. 138:1–138:11 (2018)
14. Ravikumar, B., Ibarra, O.: Relating the type of ambiguity of finite automata to the succinctness of their representation. SIAM J. Comput. **18**, 1263–1282 (1989)
15. Sakoda, W.J., Sipser, M.: Nondeterminism and the size of two way finite automata. In: STOC 1978 (1978)
16. Schmidt, E.M.: Succinctness of descriptions of context-free, regular and finite languages. Ph.D. thesis, Cornell University, Ithaca, New York (1977)
17. Shepherdson, J.: The reduction of two-way automata to one-way automata. IBM J. Res. Dev. **3**, 198–200 (1959)
18. Vardi, M.Y.: A note on the reduction of two-way automata to one-way automata. Inf. Process. Lett. **30**, 261–264 (1989)

Complexity

Deciding Non-emptiness of Hypergraph Languages Generated by Connection-preserving Fusion Grammars is NP-complete

Aaron Lye[(✉)]

Department of Computer Science, University of Bremen, P.O.Box 33 04 40,
28334 Bremen, Germany
lye@informatik.uni-bremen.de

Abstract. Fusion grammars are a novel approach to the generation of hypergraph languages. A fusion grammar is a hypergraph grammar which provides a start hypergraph of small connected components. To get large connected hypergraphs, they can be copied multiple times and can be fused by the application of fusion rules. In this paper, we analyze the non-emptiness problem for connection-preserving fusion grammars and show that this is an NP complete problem. We show this by relating language generation by connection-preserving fusion grammars to some variant of integer linear programming.

1 Introduction

Fusion grammars are a novel approach to the generation of hypergraph languages (cf. [1]). They are motivated by the fact, that fusion processes occur in various scientific fields, like DNA computing, chemistry, tiling, fractal geometry, visual modeling, etc. A fusion grammar is a hypergraph grammar which provides a start hypergraph of small connected components. To get large connected hypergraphs, they can be copied multiple times and can be fused by the application of fusion rules.

To get some insights into the power of these grammars their relation to other well known (graph) grammars and the knowledge about the decidability and complexity of the usual decision problems is of importance. We have shown that fusion grammars can simulate hyperedge replacement grammars (cf. [1, Theorem 2]). For hyperedge replacement the emptiness problem is decidable (cf. [2, Corollary 1.3]). The procedure is similar to the corresponding proof in the string case (cf. [3, Theorems 4.2, 4.3]) both are polynomial in the number of non-terminals and rules. We have also shown that fusion grammars are more powerful than hyperedge replacement grammars as they can generate hypergraphs with unbounded tree-width (cf. [1, Proposition 5]). Generalizations like splitting/fusion grammars (cf [4]) and context-dependent/sensitive fusion grammars (cf. [5,6]) turn out to be powerful enough to generate all recursively enumerable

© Springer Nature Switzerland AG 2021
A. Leporati et al. (Eds.): LATA 2021, LNCS 12638, pp. 97–108, 2021.
https://doi.org/10.1007/978-3-030-68195-1_8

string languages up to representation of strings as hypergraphs and are universal in this respect. It is an open question if fusion grammars are also universal and which decision problems are deciable. In [1, Theorem 1] it is shown that for a very restricted subclass – the so-called substantial fusion grammars – the membership problem is decidable. Substantial fusion grammars are a subclass of connection-preserving fusion grammars (defined in Sect. 3).

In this paper, we show that deciding the non-emptiness problem for connection-preserving fusion grammars, i.e., deciding if the generated hypergraph language of the grammar is not empty, is NP-complete. Compared to substantial fusion grammars this result applies to a much bigger subclass.

Any derived hypergraph can serve as a witness, that the generated language is not empty. Therefore, the non-emptiness problem of these grammars is in NP, if this derivation can be guessed and checked in non-deterministic polynomial time.

The proof for NP-completeness is done in two steps. The key idea is that addition and multiplications of vectors and fusion and multiplications of connected components are related.

1. We introduce vector fusion grammars and present a linear-time reduction of the non-emptiness problem of connection-preserving fusion grammars to the non-emptiness problem of vector fusion grammars.
2. Together with a polynomial-time equivalence of the non-emptiness problem for vector fusion grammars and solvability of an NP-complete variant of integer linear programming this implies NP-completeness of the problem.

The paper is structured as follows. Section 2 introduces basic notions and notations of hypergraphs as far as needed. Section 3 recalls the notion of fusion grammars in normal form. Section 4, presents a linear-time reduction of the non-emptiness problem of connection-preserving fusion grammars to the non-emptiness problem of vector fusion grammars. Section 5, presents a polynomial-time equivalence of the non-emptiness problem for vector fusion grammars and solvability of a variant of integer linear programming. Afterwards, in Sect. 6 we conclude that the non-emptiness problem of connection-preserving fusion grammars is NP-complete. Section 7 contains a conclusion.

2   Preliminaries

We consider hypergraphs the hyperedges of which are attached to a sequence of vertices and labeled in a given label alphabet Σ.

A *hypergraph* over Σ is a system $H = (V, E, att, lab)$ where V is a finite set of *vertices*, E is a finite set of *hyperedges*, $att\colon E \to V^*$ is a function, called *attachment*, where V^* is a sequence of vertices, and $lab\colon E \to \Sigma$ is a function, called *labeling*.

The length of the attachment $att(e)$ for $e \in E$ is called *type* of e, and e is called A-hyperedge if A is its label. The components of $H = (V, E, att, lab)$

may also be denoted by V_H, E_H, att_H, and lab_H respectively. The class of all hypergraphs over Σ is denoted by $\mathcal{H}_\Sigma$.

By $[k]$ we denote the discrete hypergraph with the vertices $1, \ldots, k$ for some $k \in \mathbb{N}_{>0}$ and by $A^\bullet$ we denote the hypergraph consisting of the vertices $1, \ldots, k$ to which a single A-hyperedge is attached.

In drawings, a hyperedge e with attachment $att(e) = v_1 \cdots v_k$ is depicted by

i.e., numbered tentacles connect the label with the corresponding attachment vertices. Moreover, an A-hyperedge of type 2 may be depicted by $\bullet \xrightarrow{A} \bullet$ instead of $\bullet \xrightarrow{1} A \xrightarrow{2} \bullet$. We assume the existence of a special label $* \in \Sigma$ that is omitted in drawings. In this way, unlabeled hyperedges are represented by hyperedges labeled with $*$.

Given $H, H' \in \mathcal{H}_\Sigma$, the *disjoint union* of H and H' is denoted by $H + H'$. Further, $k \cdot H$ denotes the disjoint union of H with itself k times.

Given $H, H' \in \mathcal{H}_\Sigma$, a *hypergraph morphism* $g \colon H \to H'$ consists of two mappings $g_V \colon V_H \to V_{H'}$ and $g_E \colon E_H \to E_{H'}$ such that $att_{H'}(g_E(e)) = g_V^*(att_H(e))$ and $lab_{H'}(g_E(e)) = lab_H(e)$ for all $e \in E_H$, where $g_V^* \colon V_H^* \to V_{H'}^*$ is the canonical extension of g_V, given by $g_V^*(v_1 \cdots v_n) = g_V(v_1) \cdots g_V(v_n)$ for all $v_1 \cdots v_n \in V_H^*$.

The *fusion* of vertices is defined as a quotient by means of an equivalence relation $\equiv$ on the set of vertices V_H as follows: $H/\equiv = (V_H/\equiv, E_H, att_{H/\equiv}, lab_H)$ with $att_{H/\equiv}(e) = [v_1] \cdots [v_k]$ for $e \in E_H$, $att_H(e) = v_1 \cdots v_k$ where $[v]$ denotes the equivalence class of $v \in V_H$ and $V_H/\equiv$ is the set of equivalence classes. It is easy to see that $f \colon H \to H/\equiv$ given by $f_V(v) = [v]$ for all $v \in V_H$ and $f_E(e) = e$ for all $e \in E_H$ defines a *quotient morphism*.

Let $H \in \mathcal{H}_\Sigma$. Then a sequence of triples $(i_1, e_1, o_1) \ldots (i_n, e_n, o_n) \in (\mathbb{N} \times E_H \times \mathbb{N})^*$ is a *path* from $v \in V_H$ to $v' \in V_H$ if $v = att_H(e_1)_{i_1}$, $v' = att_H(e_n)_{o_n}$ and $att_H(e_j)_{o_j} = att_H(e_{j+1})_{i_{j+1}}$ for $j = 1, \ldots, n-1$ where, for each $e \in E_H$, $att_H(e)_i = v_i$ for $att_H(e) = v_1 \cdots v_k$ and $i = 1, \ldots, k$. H is *connected* if each two vertices are connected by a path. A subgraph C of H, denoted by $C \subseteq H$, is a *connected component* of H if it is connected and there is no larger connected subgraph, i.e. $C \subseteq C' \subseteq H$ and C' connected implies $C = C'$. The set of connected components of H is denoted by $\mathcal{C}(H)$.

We use the *multiplication* of H defined by means of $\mathcal{C}(H)$ as follows. Let $m \colon \mathcal{C}(H) \to \mathbb{N}_{>0}$ be a mapping, called *multiplicity*, then $m \cdot H = \sum\limits_{C \in \mathcal{C}(H)} m(C) \cdot C$.

3 Fusion Grammars

Besides a start hypergraph, a fusion grammar provides a set of fusion rules. The application of a fusion rule merges certain vertices which are given by two complementary hyperedges. Complementarity is defined on a set F of fusion labels that comes together with a complementary label $\overline{A}$ for each $A \in F$. Given a

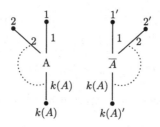

Fig. 1. The fusion rule $fr(A)$ with $type(A) = k(A)$

hypergraph, the set of all possible fusions is finite as fusion rules never create anything. To overcome this limitation, we allow arbitrary multiplications of disjoint components within derivations. The generated language does not contain derived hypergraphs, but only the terminal part of a selected connected component which is only terminal, marker and possibly connector labeled. Markers are used as a feature to distinguish between wanted and unwanted connected components. Connectors are used to preserve connectedness during the derivation process and are removed at the end. The language consists of all resulting connected components that contain no fusion symbols and at least one marker symbol, where marker and connector symbols are removed in the end.

Definition 1 (Fusion grammar and its generated language).

1. Let Σ be a finite alphabet. Then a subset $F \subseteq \Sigma$ is called *fusion alphabet* if each $A \in F$ is accompanied by a *complementary label* $\overline{A}$ such that $\overline{A} \neq \overline{B}$ for all $A, B \in F$ with $A \neq B$. Moreover, there is a *type function* $k \colon F \cup \overline{F} \to \mathbb{N}$ with $k(A) = k(\overline{A})$ for all $A \in F$ where $\overline{F} = \{\overline{A} \mid A \in F\}$. The complementary label of $\overline{A}$ is A.
2. A *fusion grammar* is a system $FG = (Z, F, T, \mu, \kappa)$ where Z is a finite *start hypergraph*, $F \subseteq \Sigma$ is a finite fusion alphabet, $T \subseteq \Sigma$ with $T \cap (F \cup \overline{F}) = \emptyset$ is a finite set of *terminal labels*, $\mu, \kappa \notin F \cup \overline{F} \cup T$ are special labels called *marker* and *connector*, respectively.
3. Each $A \in F$ induces the *fusion rule* $fr(A)$ being the hypergraph with $V_{fr(A)} = \{v_i, v_i' \mid i = 1, \ldots, k(A)\}$, $E_{fr(A)} = \{e, \overline{e}\}$, $att_{fr(A)}(e) = v_1 \cdots v_{k(A)}$, $att_{fr(A)}(\overline{e}) = v_1' \cdots v_{k(A)}'$, and $lab_{fr(A)}(e) = A$ and $lab_{fr(A)}(\overline{e}) = \overline{A}$. The rule is depicted in Fig. 1. The rule is fully specified by the label itself.
4. A *derivation step* $H \Longrightarrow H'$ for some $H, H' \in \mathcal{H}_\Sigma$ is either a rule application $H \underset{r}{\Longrightarrow} H'$ for some rule $r \in fr(F) = \{fr(A) \mid A \in F\}$ or a multiplication $H \underset{m}{\Longrightarrow} m \cdot H$ for some multiplicity m.
   The application of $fr(A)$ to a hypergraph $H \in \mathcal{H}_\Sigma$ proceeds according to the following steps: (1) Choose a *matching hypergraph morphism* $g \colon fr(A) \to H$. (2) Remove the images of the two hyperedges of $fr(A)$ yielding $X = H - (\emptyset, \{g(e), g(\overline{e})\})$. (3) Fuse the corresponding source and target vertices of the removed hyperedges yielding the hypergraph $H' = X/\equiv_R$ where the equivalence relation $\equiv_R$ is generated by the relation $R = \{(g(v_i), g(v_i')) \mid i = 1, \ldots, k(A)\}$.

5. A *derivation* $H \stackrel{n}{\Longrightarrow} H'$ of length n is a sequence $H_0 \Longrightarrow H_1 \Longrightarrow \ldots \Longrightarrow H_n$ with $H = H_0$ and $H' = H_n$ including the case $n = 0$ with $H_0 = H = H' = H_n$. One may write $H \stackrel{*}{\Longrightarrow} H'$ if the length does not matter.

6. $L(FG) = \{rem_{\{\mu,\kappa\}}(Y) \mid Z \stackrel{*}{\Longrightarrow} H, Y \in \mathcal{C}(H) \cap (\mathcal{H}_{T \cup \{\kappa,\mu\}} - \mathcal{H}_{T \cup \{\kappa\}})\}$ is the *generated language* of FG where $rem_{\{\mu,\kappa\}}(Y)$ is the hypergraph obtained when removing all hyperedges with labels in $\{\kappa, \mu\}$ from Y.

Remark 1. 1. As each hyperedge belongs to a single connected component, the fusion of two hyperedges fuses either one connected component or two connected components. In particular, this can produce three different effects.

(a) Two connected components may be fused into a single one.

(b) Fusion may be a kind of folding transforming a single connected component. , e.g., $B\!-\!\bullet\!-\!\bullet\!-\!\overline{B} \underset{fr(B)}{\Longrightarrow} \bullet\!\!\circ$.

(c) It can result in a disconnection, e.g., $\bullet\!\stackrel{1}{-}\!C\!\stackrel{2}{-}\!\bullet \quad \bullet\!\stackrel{1}{-}\!\overline{C}\!\stackrel{2}{-}\!\bullet \underset{fr(C)}{\Longrightarrow} \bullet \quad \bullet$

with respect to two connected components or $D \overset{1}{\underset{2}{\diamond}} \overline{D} \underset{fr(D)}{\Longrightarrow} \bullet$ with respect

to one connected component.

In this paper, we restrict derivations in such a way that the third case does not occur, i.e., no result of fusion is a disconnection. We call these grammar *connection-preserving fusion grammars*.

2. Without loss of generality we assume that every fusion grammar is in normal form where the start hypergraph contains exactly one marked connected component and every derivation starts with all the multiplications, but the marked component is never multiplied, then applies all fusions afterwards, and derives a single connected component.

Example 1. Consider the fusion grammar $FG = (\sum_{j=1}^{5} z_i, \{A, B\}, \{*\}, \mu, \kappa)$ with $type(A) = type(B) = 2$ and z_i for $i = 1, \ldots, 5$ as follows.

where a box is a graph with four vertices and four edges (hyperedges of lenght 2). The labels of the edges are written next to the edges (*-labels are omitted). The numbering of the attachment vertices is omitted to clarify the drawings. For A- and $\overline{A}$-hyperedges we assume the vertex on the left to be the first and the vertex to the right to be the second attachment vertex. For B- and $\overline{B}$-hyperedges we assume the vertex on the top to be the first and the vertex to the bottom to be the second attachment vertex. It is easy to see that the grammar generates graphs like the following.

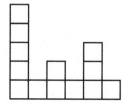

4   A Linear-Time Reduction of the Non-emptiness Problem of Connection-preserving Fusion Grammars to the Non-emptiness Problem of Vector Fusion Grammars

In this section, we introduce a special subclass of fusion grammars, which we call vector fusion grammars, and show that every connection-preserving fusion grammar can be transformed in linear-time into a vector fusion grammar such that the generated language of the fusion grammar is empty if and only if this is also the case for the generated language of the vector fusion grammar.

Recall that connected components of the start hypergraph of a fusion grammar may be multiplied and fused. A fusion consumes exactly one pair of complementary labeled fusion-hyperedges. This means that the multiplied connected components need equally many f- and $\overline{f}$-hyperedges to obtain members of the generated language.

The basic idea is that each connected component can be associated to as an integer vector and vice versa. The scalars in such a vector are the difference between the number of fusion-labeled hyperedges and the respective complementary ones within the respective connected component. Note that this applies the equivalence relation $(m, n) \sim (m', n') \iff m + n' = m' + n$ defined in $\mathbb{N} \times \mathbb{N}$ yielding $\mathbb{Z}$ as the set of equivalence classes, where in our case m, m' are the number of f-labeled hyperedges and n, n' are the number of $\overline{f}$-labeled hyperedges in some connected component z, z', respectively.

We apply a quantitative argument with respect to fusion labels only. Vertices, attachments and terminal labels are irrelevant. Therefore, very different connected components yield the same vector as long as the difference between fusion-labeled hyperedges and their corresponding complementary labeled hyperedges are the same. If fusion never yields disconnection, then this representation is meaningful as fusion of connected components and multiplication of connected components is reflected in addition and multiplication of vectors.

Definition 2. 1. Let $F = \{f_1, \ldots, f_r\}$ be a fusion alphabet and let z be a connected hypergraph with fusion labels in $F \cup \overline{F}$. Then

$$v(z) = (p_1 - n_1, \ldots, p_r - n_r)$$

is the integer vector associated to z, where p_i is the number of f_i-labeled hyperedges and n_i is the number of $\overline{f}_i$-labeled hyperedges in z.

2. Let $v = (v_1, \ldots, v_r) \in \mathbb{Z}^r$ and let $\{f_1, \ldots, f_r\}$ be a fusion alphabet, where all fusion labels of type 1. Then the connected hypergraph associated to v is

$$h(v) = ([1], \{(i,j) \mid 1 \le i \le r, 1 \le j \le |v_i|\}, att_{h(v)}, lab_{h(v)})$$

where $att_{h(v)}((i,j)) = 1$ for each i, j, and $lab_{h(v)}((i,j)) = f_i$ if $v_i > 0$ and $\overline{f_i}$ otherwise. $|v_i|$ denotes the absolute value of v_i.

By $h(v)_x$ we denote $h(v)$ with an additional designating x-flag (type-1 hyper-edge) attached. The label of these flags may be of a terminal alphabet or μ.

The designation flag enables to have and distinguish multiple hypergraph representations of the same vector.

Example 2. 1. Let $\{A, B\}$ be a fusion alphabet with $type(A) = type(B) = 2$

and let $z_1 = \mu \begin{array}{c} A \\ \boxed{} \end{array} B$, $z_2 = \begin{array}{c} A \\ \boxed{} \\ \overline{A} \end{array}$, $z_3 = \begin{array}{c} \boxed{} \\ \overline{A} \end{array}$, $z_4 = \overline{B} \begin{array}{c} A \\ \boxed{} \end{array} B$, $z_5 = \overline{B} \boxed{}$.

The respective vectors $v(z_1), \ldots, v(z_5) \in \mathbb{Z}^2$ are the following: $v(z_1) = (1, 1)$, $v(z_2) = (0, 0)$, $v(z_3) = (-1, 0)$, $v(z_4) = (1, 0)$, $v(z_5) = (0, -1)$.

2. Let $v = (-2, 1, 3)$ and let μ be its designation. Then $h(v)_\mu = $ μ. The

numbering of the attachments is omitted as it is clear from context.

This leads to the transformation of connection-preserving fusion grammars into vector fusion grammars.

Construction 1. *Let $FG = (Z, F, T, \mu, \kappa)$ be a connection-preserving fusion grammar with one marked connected component (z_μ) in its start hypergraph. Let $z_1, \ldots, z_c$ be the remaining connected components in the start hypergraph. Then $VFG(FG) = (Z(V(z_1, \ldots, z_c), v(z_\mu)), \{f_1, \ldots, f_{|F|}\}, \{1, \ldots, c\}, \mu, \kappa)$, where $Z(V(z_1, \ldots, z_c), v(z_\mu)) = h(v(z_1))_1 + \ldots + h(v(z_c))_c + h(-v(z_\mu))_\mu$, is the corresponding vector fusion grammar.*

The vector fusion grammar can be seen as a corresponding fusion grammar where the structure of the connected components is significantly reduced. For connection-preserving fusions grammars the transformation yields a meaningful vector fusion grammar due to the following reasoning. A conneced component is a largest connected subhypergraph. Because terminal-, μ- or κ-labeled hyper-edges are never modified during some derivation and fusion is by assumption connection-preserving, all vertices of a connected component can be identified. Then all terminal and κ-labeled hyperedges can be removed and the attach-ment length of fusion and μ-labeled hyperedges is set to 1. Fusion within each connected component as long a possible and adding a terminal flag with the components identifier to all except the μ-connected component yields the start hypergraph of the vector fusion grammar.

Proposition 1. *Let FG be a connection-preserving fusion grammar with one marked connected component in its start hypergraph and VFG(FG) be the corresponding vector fusion grammar. Then $L(FG) \neq \emptyset$ if and only if $L(VFG(FG)) \neq \emptyset$.*

Proof. Let $FG = (Z, F, T, \mu, \kappa)$, where $F = \{f_1, \ldots, f_r\}$, and let $H \in L(FG)$. Then w.l.o.g. $Z \overset{*}{\Longrightarrow} H_\mu$, where $H = rem_{\{\mu,\kappa\}}(H_\mu)$ and $H_\mu \in \mathcal{H}_{T\cup\{\kappa,\mu\}} - \mathcal{H}_{T\cup\{\kappa\}}$. Let the derivation be $Z \underset{m}{\Longrightarrow} m \cdot Z = z_\mu + \sum_{i=1}^{c} m_i \cdot z_i \overset{k_1}{\underset{fr(f_1)}{\Longrightarrow}} \ldots \overset{k_c}{\underset{fr(f_c)}{\Longrightarrow}} H_\mu$.

Because $H \in L(FG)$, and as H_μ is the fusion of $z_\mu + \sum_{i=1}^{c} m_i \cdot z_i$ and because H' does not have any hyperedges labeled in $F \cup \overline{F}$, all those $F \cup \overline{F}$-hyperedges disappear in the fusions. As a single fusion consumes exactly one pair $f_i, \overline{f}_i$ for some i, the number of f_i-hyperedges and the number of number of fusion hyperedges and their complements in the involved connected components must be equal, i.e., using notation of Definition 2 and denoting the number of f_i-labeled hyperedges and $\overline{f}_i$-labeled hyperedges in z_μ by $p_{i\mu}$ and $n_{i\mu}$, respectively, $k_i = p_{i\mu} + \sum_{j=1}^{c} p_{ij} \cdot m_j = n_{i\mu} + \sum_{j=1}^{c} n_{ij} \cdot m_j$ for $i = 1, \ldots, r$. Applying these multiplicies to the corresponding connected components $h(v(z_i))_i$ in the start hypergraph of $VFG(FG)$ yield the derivation $Z(V(z_1, \ldots, z_c), v(z_\mu)) \underset{m}{\Longrightarrow} h(-b)_\mu + \sum_{i=1}^{c} m_i \cdot h(v(z_i))_i$. Moreover, because $\sum_{j=1}^{c} (p_{ij} - n_{ij}) \cdot m_j = p_{i\mu} - n_{i\mu}$ for $i = 1, \ldots, r$ there exists a derivation $h(-b)_\mu + \sum_{i=1}^{c} m_i \cdot h(v(z_i))_i \overset{k}{\Longrightarrow} H'_\mu$, where $k = \sum_{i=1}^{r} k_i$ and $H'_\mu \in \mathcal{H}_{[c]\cup\{\kappa,\mu\}} - \mathcal{H}_{[c]\cup\kappa}$. Consequently, $L(VFG(FG)) \neq \emptyset$.

Conversely, let $H \in L(VFG(FG))$. Then there exists a derivation $Z(V(z_1, \ldots, z_c), v(z_\mu)) \overset{*}{\Longrightarrow} H_\mu$, where $H = rem_\mu(H)$. H_μ contains no fusion or complementary fusion hyperedges. Using similar arguments as before we can conclude that there is also a derivation $Z \overset{*}{\Longrightarrow} H'_\mu$ in FG. Note that the connected components z_i corresponding to $h(v(z_i))_i$ may have more fusion and complementary fusion hyperedges than $h(v(z_i))_i$. However, these can be removed by internal fusion within the respective connected component. Consequently, H'_μ is only terminal, κ- and μ-labeled and, therefore, contributes to the generated language after removing the marker, i.e., $H' = rem_\mu(H'_\mu) \in L(FG)$. Hence, $L(FG) \neq \emptyset$. □

5   A Polynomial-Time Equivalence of the Non-emptiness Problem for Vector Fusion Grammars and Solvability of a Variant of Integer Linear Programming

In this section, we present a polynomial-time equivalence of the non-emptiness problem for vector fusion grammars and solvability of a variant of integer linear programming.

The integer linear programming variant used in our reduction is the following. Let $(A, b) \in \mathbb{Z}^{r \times c} \times \mathbb{Z}^r$. The problem is to decide if there exists an $x \in \mathbb{N}^c$ such that $Ax = b$. The problem can be seen as a submonoid membership problem. Let $(\mathbb{Z}^r, +, 0)$ be the free commutative monoid of r-dimensional vectors over $\mathbb{Z}$, where the operation is defined componentwise and 0 is the zero-vector. Let $(S, +, 0)$ be a submonoid, where the set S is generated by the column vectors of A, i.e., $A_1, \ldots, A_c$, by adding 0 and closing S under the operation. Asking if $b \in S$ is by definition equivalent to asking for a linear combination $x_1 \cdot A_1 + \ldots + x_c \cdot A_c$ of the generating vectors $A_1, \ldots, A_c$ yielding b, where $x = (x_1, \ldots, x_c) \in \mathbb{N}^c$.

The basic idea of our reduction is to represent these vectors $(A_1, \ldots, A_c)$ by connected components in the start hypergraph of a vector fusion grammar. The submonoid membership problem is then reduced to the membership problem in vector fusion grammars (which is an instance of the membership problem for connection-preserving fusion grammars). It turns out that this reduction is in fact a p-isomorphism. The construction is crafted in such a way that the language generated by the vector fusion grammar is not empty if and only if there is a nonnegative solution for $Ax = b$.

Besides the representation of vectors in $\mathbb{Z}^r$, we need a hypergraph representation for solution vector in $\mathbb{N}^c$. Furthermore, the hypergraph representations of vectors in $\mathbb{Z}^r$ and $\mathbb{N}^c$ must be distinguishable. We represent such a vector by a connected hypergraph consisting of a single vertex and as many terminal-labeled flags labeled i as the ith scalar in the vector.

Definition 3. Let $x = (x_1, \ldots, x_c) \in \mathbb{N}^c$ and let $\{1, \ldots, c\}$ be a set of terminal labels. Then the to x associated connected hypergraph is

$$\hat{h}(x) = ([1], \{(i,j) \mid 1 \le i \le c, 1 \le j \le x_i\}, att_{\hat{h}(x)}, lab_{\hat{h}(x)}),$$

where $att_{\hat{h}(x)}((i,j)) = 1$ for all i, j and $lab_{\hat{h}(x)}((i,j)) = i$.

By $\hat{h}(x)_\mu$ we denote $\hat{h}(x)$ with an additional μ-hyperedge attached.

Example 3. Let $x = (0, 3, 5)$. Then $\hat{h}(x) =$. The hypergraph contains as many flags labeled with the index i as the ith entry of the vector holds. There are no flags labeled 1, because the first entry is 0. But there are three flags labeled 2 and five flags labeled 3 as the 2nd and 3rd entry hold the scalar 3 and 5, respectively.

Remark 2. x and $\hat{h}(x)$ are in one-to-one correspondence to each other. Therefore, we have a one-to-one mapping between the two sets $\{\hat{h}(x) \mid Ax = b, x \in \mathbb{N}^c\}$ and $\{x \in \mathbb{N}^c \mid Ax = b\}$.

Construction 2. *Let $(A, b) \in \mathbb{Z}^{r \times c} \times \mathbb{Z}^r$. Let $A_1, \ldots, A_c$ be the columns of A. Let $F = \{f_1, \ldots, f_r\}$ be a fusion alphabet with $type(f) = 1$ for all $f \in F$. Then*

$$VFG(A, b) = (Z(A, b), F, \{1, \ldots, c\}, \mu, \kappa),$$
$$where \quad Z(A, b) = h(A_1)_1 + \ldots + h(A_c)_c + h(-b)_\mu,$$

is the corresponding vector fusion grammar.

Remark 3. This reduction takes polynomial time because for each column vector exactly one connected component is created in linear time. Furthermore, this reduction is a p-isomorphism as the mapping is bijective and also the inverse mapping can be computed analogously in polynomial time.

Proposition 2. $L(VFG(A, b)) = \{\hat{h}(x) \mid Ax = b, x \in \mathbb{N}^c\}$.

Proof. Let $L(VFG(A, b)))$ be not empty. Then there exists some hypergraph $H \in L(VFG(A, b))$. Let $Z \overset{*}{\Longrightarrow} H_\mu$ be a derivation generating a hypergraph containing a connected component H_μ such that $H = rem_{\{\kappa,\mu\}}(H_\mu)$ and $H_\mu \in \mathcal{H}_{[c] \cup \{\kappa,\mu\}} - \mathcal{H}_{[c] \cup \kappa}$. Let m_j be the number of copies of $h(A_j)_j$ needed to construct H' for $j = 1, \ldots, c$. Then, using the same argument as in the proof of Proposition 1, we conclude that the number of number of fusion hyperedges and their complements in the involved connected components must be equal. Formally, using the notation as in the proof of Proposition 1, $p_{i\mu} + \sum\limits_{j=1}^{c} p_{ij} \cdot m_j = n_{i\mu} + \sum\limits_{j=1}^{c} n_{ij} \cdot m_j$ for $i = 1, \ldots, r$. This equation is equivalent to $\sum\limits_{j=1}^{c} (p_{ij} - n_{ij}) \cdot m_j = -(p_{i\mu} - n_{i\mu}) = -b_i$ for $i = 1, \ldots, r$, where b_i is the ith entry in b. This means that the vector $x = (m_1, \ldots, m_c)$ is a solution of the linear diophantine equation system $Ax = b$ because by definition $A = (a_{ij})_{i=1,\ldots,r,j=1,\ldots,c}$ with $a_{ij} = p_{ij} - n_{ij}$. Furthermore, each connected component involved in this derivation contributes with one flag to the derived hypergraph. More specifically, $h(A_i)_i$ adds one flag labeled i and $h(-b)_\mu$ adds one flag labeled μ. Consequently, H is of the form $\hat{h}(x)_\mu$ where $x = (m_1, \ldots, m_c)$. Hence, $H \in \{\hat{h}(x) \mid Ax = b\}$.

Conversely, let $\hat{h}(\overline{x}) \in \{\hat{h}(x) \mid Ax = b, x \in \mathbb{N}^c\}$. Then $\overline{x} = (\overline{x}_1, \ldots, \overline{x}_c) \in \mathbb{N}^c$ is a solution for the linear diophantine equation system $Ax = b$. Further, this vector gives the multiplicity of the respective connected components of $Z(A, b)$ such that the number of f_i-hyperedges and the number of $\overline{f}_i$-hyperedges for $i = 1, \ldots, r$ are equal. After respective multiplications the connected components in the hypergraph $h(-b)_\mu + \sum\limits_{j=1}^{c} m_j \cdot h(A_j)_j$ can be fused, i.e., $Z(A, b) \underset{m(\overline{x})}{\Longrightarrow} m(\overline{x}) \cdot$ $Z(A, b) = h(-b)_\mu + \sum\limits_{j=1}^{c} \overline{x}_i \cdot h(A_j)_j \overset{k_1}{\underset{fr(f_1)}{\Longrightarrow}} \ldots \overset{k_r}{\underset{fr(f_r)}{\Longrightarrow}} \hat{h}(\overline{x})_\mu$, where $k_i = p_{i\mu} + \sum\limits_{j=1}^{c} p_{ij} \cdot$ $\overline{x}_j = n_{i\mu} + \sum\limits_{j=1}^{c} n_{ij} \cdot \overline{x}_j$ for $i = 1, \ldots, r$. Because this connected component is only marker and terminal labeled, this connected components contributes to the language $L(VFG(A, b))$ after removing the marker. $\square$

Corollary 1. $L(VFG(A, b)) \neq \emptyset$ if and only if there exists a derivation $Z(A, b) \overset{*}{\Longrightarrow} \hat{h}(x)_\mu$ for some $x \in \mathbb{N}^c$.

6   Complexity-Theoretic Implications of the Reductions

In this section, we conclude the following theorem.

Theorem 1. *The non-emptiness problem for connection-preserving fusion grammars is NP-complete.*

Proof. The integer linear programming variant used in our reduction in strongly NP-complete (cf. [7–9]).

The equivalences

$$L(VFG(A,b)) \neq \emptyset \iff \exists Z(A,b) \overset{*}{\Longrightarrow} \hat{h}(x)_\mu \text{ for some } x \in \mathbb{N}^c$$
$$\iff \exists x = (x_1, \dots, x_c) \in \mathbb{N}^c : x_1 \cdot A_1 + \dots + x_c \cdot A_c = b$$
$$\iff \exists x \in \mathbb{N}^c : Ax = b$$

and p-isomorphism presented in Sect. 5 imply that the non-emptiness problem for vector fusion grammars is in NP. Together with the reduction presented in Sect. 4 (Proposition 1 and the following equivalence of the two notions of vector fusion grammars) this implies that the non-emptiness problem for connection-preserving fusion grammars is in NP.

Using the notation of Construction 1 and 2, we have $VFG(FG) = VFG(A,b)$,

$$\text{where } A = V(z_1, \dots, z_c) = \begin{pmatrix} d_{11} & d_{12} & \dots & d_{1c} \\ d_{21} & d_{22} & \dots & d_{2c} \\ \vdots & \vdots & & \vdots \\ d_{r1} & d_{r2} & \dots & d_{rc} \end{pmatrix}, \text{ where } r = |F|, d_{ij} = p_{ij} - n_{ij},$$

where p_{ij} is the number of f_i-labeled hyperedges and n_{ij} is the number of $\overline{f}_i$-labeled hyperedges in z_j, and $b = v(z_\mu)$.

The NP hardness of the integer linear programming variant and the reduction presented in Sect. 5 imply that the membership problem for vector fusion grammars is NP hard. Using Corollary 1 this implies that deciding non-emptiness of the language generated by a vector fusion grammars is NP hard. As vector fusion grammars are a special case of connection-preserving fusion grammars, we get that the non-emptiness problem for connection-preserving fusion grammars is NP hard. □

7   Conclusion

In this paper, we have analyzed the non-emptiness problem for connection-preserving fusion grammars and have showed that this is an NP-complete problem. The proof relates language generation by connection-preserving fusion grammars to some NP-complete variant of integer linear programming using vector fusion grammar. In particular, we have presented a transformation from connection-preserving fusion grammars into vector fusion grammars such that non-emptiness of the generated languages is preserved. Together with the presented polynomial-time equivalence of the non-emptiness problem for vector fusion grammars and solvability of the variant of integer linear programming this implies NP-completeness of the problem.

It is very interesting that the proof for deciding non-emptiness relies only on a quantitative argument, i.e., it is irrelevant how the connected components are fused. As a consequence, it may be interesting to analyze this closure property of permutations further.

Connection-preserving fusion grammars are a greater subclass then substantial fusion grammars. It would be interesting to analyze and relate their generative power. It would also be interesting to know if the non-emptiness problem is decidable for fusion grammars, where disconnection may happen during the derivation. Moreover, it would be interesting to discover further subclasses for which the non-emptiness problem is efficiently decidable. Further studies may be other decision problems like finiteness or equivalence of languages generated by fusion grammars. This will provide us with better understanding of the computational power of fusion grammars.

Acknowledgment. We are grateful to Hans-Jörg Kreowski and Sabine Kuske for valuable discussions and remarks. We also thank the anonymous reviewers for their valuable comments.

References

1. Kreowski, Hans-Jörg., Kuske, Sabine, Lye, Aaron: Fusion Grammars: a novel approach to the generation of graph languages. In: de Lara, Juan, Plump, Detlef (eds.) ICGT 2017. LNCS, vol. 10373, pp. 90–105. Springer, Cham (2017). https://doi.org/10.1007/978-3-319-61470-0_6
2. Habel, A.: Hyperedge Replacement: Grammars and Languages. LNCS, vol. 643. Springer, Heidelberg (1992). https://doi.org/10.1007/BFb0013875
3. Hopcroft, J.E., Ullman, J.D.: Formal Languages and Their Relation to Automata. Addison-Wesley Series in Computer Science and Information Processing. Addison-Wesley, Boston (1969)
4. Kreowski, H.-J., Kuske, S., Lye, A.: Splicing/Fusion grammars and their relation to hypergraph grammars. In: Lambers, L., Weber, J. (eds.) ICGT 2018. LNCS, vol. 10887, pp. 3–19. Springer, Cham (2018). https://doi.org/10.1007/978-3-319-92991-0_1
5. Lye, A.: Transformation of turing machines into context-dependent fusion grammars. In: Post-Proceedings of the 10th International Workshop on Graph Computation Models, (GCM 2019). Electronic Proceedings in Theoretical Computer Science (EPTCS) (2019). https://doi.org/10.4204/EPTCS.309.3
6. Lye, A.: Context-sensitive fusion grammars are universal. In: Leporati, A., Martín-Vide, C., Shapira, D., Zandron, C. (eds.) LATA 2020. LNCS, vol. 12038, pp. 275–286. Springer, Cham (2020). https://doi.org/10.1007/978-3-030-40608-0_19
7. Sahni, S.: Computationally related problems. SIAM J. Comput. **3**(4), 262–279 (1974)
8. Garey, M.R., Johnson, D.S.: "Strong" NP-completeness results: motivation, examples, and implications. J. ACM **25**(3), 499–508 (1978)
9. Garey, M.R., Johnson, D.S.: Conputers and Intractability: A Guide to the Theory of NP-Completeness. W. H. Freeman, New York (1979)

On the Power of Nondeterministic Circuits and Co-Nondeterministic Circuits

Hiroki Morizumi[✉]

Shimane University, Matsue, Japan
morizumi@cis.shimane-u.ac.jp

Abstract. Revealing the power of nondeterministic computation and co-nondeterministic computation is one of the central problems in computational complexity. In this paper, we consider the two computation and deterministic computation in Boolean circuits. We give the first separations on the power of deterministic circuits, nondeterministic circuits, and co-nondeterministic circuits in general circuits. More precisely, we prove the following facts.

- There is an explicit Boolean function f such that the nondeterministic U_2-circuit complexity of f is at most $2n + o(n)$ and the deterministic and co-nondeterministic U_2-circuit complexity of f is $3n - o(n)$.
- There is an explicit Boolean function f such that the co-nondeterministic U_2-circuit complexity of f is at most $2n + o(n)$ and the deterministic and nondeterministic U_2-circuit complexity of f is $3n - o(n)$.

Keywords: Circuit complexity · Nondeterministic circuit · Co-nondeterministic circuit · U_2-circuit

1 Introduction

In this paper, we give the first separations on the power of deterministic circuits, nondeterministic circuits, and co-nondeterministic circuits in general circuits. The results of this paper are the following two theorems. We denote by $size^d(f)$ the size of the smallest deterministic U_2-circuit computing a Boolean function f, and denote by $size^{nd}(f)$ and $size^{cnd}(f)$ the size of the smallest nondeterministic and co-nondeterministic U_2-circuit computing a Boolean function f, respectively.

Theorem 1. *There is an explicit Boolean function f such that $size^{nd}(f) \leq 2n + o(n)$ and $size^d(f) = size^{cnd}(f) = 3n - o(n)$.*

Theorem 2. *There is an explicit Boolean function f such that $size^{cnd}(f) \leq 2n + o(n)$ and $size^d(f) = size^{nd}(f) = 3n - o(n)$.*

© Springer Nature Switzerland AG 2021
A. Leporati et al. (Eds.): LATA 2021, LNCS 12638, pp. 109–117, 2021.
https://doi.org/10.1007/978-3-030-68195-1_9

Boolean circuits are one of computation models in computation theory, and the size of the smallest circuit computing a Boolean function f, which is called *circuit complexity* of f, means the hardness to compute f by the circuits. In this paper, we consider nondeterministic circuits and co-nondeterministic circuits, which are Boolean circuits with the power of nondeterministic computation and co-nondeterministic computation. Thus, Theorem 1 and Theorem 2 are an investigation for the power of nondeterministic computation and co-nondeterministic computation.

The circuit model of this paper is U_2-circuits. The computational power by U_2-circuits is close to the computational power by the standard Turing machines, except that circuits are generally nonuniform computation models. More precisely, the class of decision problems solvable by a nonuniform family of polynomial-size deterministic U_2-circuits is equal to P/poly, and the class of decision problems solvable by a nonuniform family of polynomial-size nondeterministic U_2-circuits is equal to NP/poly. Lower bounds on the size of U_2-circuits have been well studied with a motivation as an approach to the P vs. NP problem [1,2,5].

Related Works. While both of nondeterministic computation and circuit complexity are central topics in computational complexity, the circuit complexity of nondeterministic circuits has not been relatively well studied. There are no published works for the separations on the power of deterministic circuits and nondeterministic circuits as far as we know.

The author proved a $3(n-1)$ lower bound for the size of nondeterministic U_2-circuits computing the parity function in his previous paper [3]. In the paper, the gate elimination method (See Sect. 2.2.) has been firstly applied to nondeterministic circuits. The idea (i.e., how we apply the method to nondeterministic circuits) is used also in this paper.

Proof Outlines. Once Theorem 1 is proved, Theorem 2 is obtained by the properties of nondeterministic circuits and co-nondeterministic circuits (and deterministic circuits). We describe the properties and the proof of Theorem 2 in Sect. 3. Thus, the main task of this paper is to prove Theorem 1.

To prove Theorem 1, we propose a simple proof strategy, and call the key idea *nondeterministic selecting*. In Sect. 4, we describe nondeterministic selecting and the proof idea using it. The proof of Theorem 1 is in Sect. 5.

2   Preliminaries

The parity function of n inputs $x_1, \ldots, x_n$, denoted by Parity_n, is 1 iff $\sum x_i \equiv 1 \pmod{2}$.

2.1   Circuits

Circuits are formally defined as directed acyclic graphs. The nodes of in-degree 0 are called *inputs*, and each one of them is labeled by a variable or by a constant 0 or 1. The other nodes are called *gates*, and each one of them is labeled by a Boolean function. The *fan-in* of a node is the in-degree of the node, and the *fan-out* of a node is the out-degree of the node. There is a single specific node called *output*. The *size* of a circuit is the number of gates in the circuit.

We denote by U_2 the set of all Boolean functions over two variables except for the XOR function and its complement. A Boolean function in U_2 can be represented as the following form:

$$f(x, y) = ((x \oplus a) \wedge (y \oplus b)) \oplus c,$$

where $a, b, c \in \{0, 1\}$. A U_2-*circuit* is a circuit in which each gate has fan-in 2 and is labeled by a Boolean function in U_2.

A *nondeterministic circuit* is a circuit with *actual inputs* $(x_1, \ldots, x_n) \in \{0, 1\}^n$ and some further inputs $(y_1, \ldots, y_m) \in \{0, 1\}^m$ called *guess inputs*. A nondeterministic circuit computes a Boolean function f as follows: For $x \in \{0, 1\}^n$, $f(x) = 1$ iff there exists a setting of the guess inputs $\{y_1, \ldots, y_m\}$ which makes the circuit output 1. A *co-nondeterministic circuit* is also a circuit with actual inputs $(x_1, \ldots, x_n) \in \{0, 1\}^n$ and guess inputs $(y_1, \ldots, y_m) \in \{0, 1\}^m$. A co-nondeterministic circuit computes a Boolean function f as follows: For $x \in \{0, 1\}^n$, $f(x) = 0$ iff there exists a setting of the guess inputs $\{y_1, \ldots, y_m\}$ which makes the circuit output 0. We call a circuit without guess inputs a *deterministic circuit* to distinguish it from a nondeterministic circuit or a co-nondeterministic circuit.

2.2   The Gate Elimination Method

In our proofs, we need the gate elimination method. In this subsection, we have a quick look at the method for the case of the parity function, which is relevant to our case.

Consider a gate g which is labeled by a Boolean function in U_2. Recall that any Boolean function in U_2 can be represented as the following form:

$$f(x, y) = ((x \oplus a) \wedge (y \oplus b)) \oplus c,$$

where $a, b, c \in \{0, 1\}$. If we fix one of two inputs of g so that $x = a$ or $y = b$, then the output of g becomes a constant c. In such case, we say that g is *blocked*.

Theorem 3 (Schnorr [4]).

$$size^{d}(\text{Parity}_n) = 3(n - 1).$$

Proof. Assume that $n \geq 2$. Let C be an optimal deterministic U_2-circuit computing Parity_n. Let g_1 be a top gate in C, i.e., whose two inputs are connected from two inputs x_i and x_j, $1 \leq i, j \leq n$. Then, x_i must be connected to another gate g_2, since, if x_i is connected to only g_1, then we can block g_1 by an assignment of a constant to x_j and the output of C becomes independent from x_i, which contradicts that C computes Parity$_n$. By a similar reason, g_1 is not the output of C. Let g_3 be a gate which is connected from g_1. See Fig. 1.

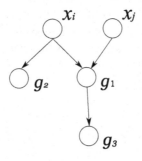

Fig. 1. Proof of Theorem 3

We prove that we can eliminate at least three gates from C by an assignment to x_i. We assign a constant 0 or 1 to x_i such that g_1 is blocked. Then, we can eliminate g_1, g_2 and g_3. If g_2 and g_3 are the same gate, then the output of g_2 ($= g_3$) becomes a constant, which means that g_2 ($= g_3$) is not the output of C and we can eliminate another gate which is connected from g_2 ($= g_3$). Thus, we can eliminate at least three gates and the circuit comes to compute Parity$_{n-1}$ or $\neg$Parity$_{n-1}$. For deterministic circuits, it is obvious that $size^{\mathrm{d}}($Parity$_{n-1}) = size^{\mathrm{d}}(\neg$Parity$_{n-1})$. Therefore,

$$size^{\mathrm{d}}(\text{Parity}_n) \geq size^{\mathrm{d}}(\text{Parity}_{n-1}) + 3$$

$$\vdots$$

$$\geq 3(n-1).$$

$x \oplus y$ can be computed with three gates by the following form:

$$(x \wedge \neg y) \vee (\neg x \wedge y).$$

Therefore, $size^{\mathrm{d}}($Parity$_n) \leq 3(n-1)$. □

3 Proof of Theorem 2

As mentioned in Sect. 1, Theorem 2 is immediately obtained by Theorem 1. In this section, we show the proof of Theorem 2, and we concentrate our main task (i.e., the proof of Theorem 1) in the rest of this paper.

Nondeterministic circuits and co-nondeterministic circuits have the following property. The property is a natural property which nondeterministic computation and co-nondeterministic computation have.

Proposition 1. *For every Boolean function* f, *if* $size^{nd}(f) \geq 1$ *and* $size^{cnd}(\neg f) \geq 1$, *then* $size^{nd}(f) = size^{cnd}(\neg f)$.

Proof. By the definitions of Nondeterministic circuits and co-nondeterministic circuits, a nondeterministic circuit C computing f and one NOT gate at the output of C is a co-nondeterministic circuit computing $\neg f$. Similarly, a co-nondeterministic circuit C' computing f' and one NOT gate at the output of C' is a nondeterministic circuit computing $\neg f'$. In U_2-circuits, the one NOT gate is not counted as the number of gates, since the NOT gate is merged to another gate. □

To prove Theorem 2, the following property of deterministic circuits is also needed.

Proposition 2. *For every Boolean function* f, *if* $size^d(f) \geq 1$ *and* $size^d(\neg f) \geq 1$, *then* $size^d(f) = size^d(\neg f)$.

Proof. A deterministic circuit C computing f and one NOT gate at the output of C is a deterministic circuit computing $\neg f$. In U_2-circuits, the one NOT gate is not counted as the number of gates, since the NOT gate is merged to another gate. □

By Proposition 1 and Proposition 2, once Theorem 1 is proved, Theorem 2 is immediately proved.

Proof (of Theorem 2). Let f_1 be the f in Theorem 1. We let f (in Theorem 2) be $\neg f_1$. □

4   Nondeterministic Selecting

In this section, we describe our idea for the proof of Theorem 1. We call the key idea nondeterministic selecting.

Let $f' : \{0,1\}^{\sqrt{n}} \to \{0,1\}$, and

$$f = \bigvee_{i=0}^{\sqrt{n}-1} f'(x_{\sqrt{n} \cdot i+1}, x_{\sqrt{n} \cdot i+2}, \ldots, x_{\sqrt{n} \cdot i+\sqrt{n}}).$$

Nondeterministic circuits can efficiently compute f. We construct a nondeterministic circuit C computing f as follows. Firstly, we select $\sqrt{n}$ inputs nondeterministically. More precisely, we construct a selector circuit C' which outputs $x_{\sqrt{n} \cdot i+1}, x_{\sqrt{n} \cdot i+2}, \ldots, x_{\sqrt{n} \cdot i+\sqrt{n}}$ for each i, $0 \leq i \leq \sqrt{n} - 1$, when guess inputs of C are assigned to an assignment. Then, one circuit C'' computing f' is enough in C. $\sqrt{n}$ variables of the output of C' are connected to the input of C''. It is not difficult to confirm that C computes f by the definition of nondeterministic circuits.

On the other hand, a trivial construction of deterministic circuits computing f uses $\sqrt{n}$ circuits computing f'. Note that it is a complicated problem (called a direct sum) whether $\sqrt{n}$ circuits are needed. In our proof of Theorem 1, we choose the parity function as f' so that we can prove the large lower bound on the circuit complexity of f for deterministic circuits and even co-nondeterministic circuits.

5   Proof of Theorem 1

To prove Theorem 1, we let

$$f = \bigvee_{i=0}^{\sqrt{n}-1} \text{Parity}_{\sqrt{n}}(x_{\sqrt{n}\cdot i+1}, x_{\sqrt{n}\cdot i+2}, \ldots, x_{\sqrt{n}\cdot i+\sqrt{n}}),$$

and prove some lemmas on the circuit complexity of f. Theorem 1 is immediately proved by the lemmas.

Proof (of Theorem 1). By the definitions of deterministic circuits and co-nondeterministic circuits, $size^{\text{cnd}}(f) \leq size^{\text{d}}(f)$. By Lemma 1, Lemma 2 and Lemma 4, the theorem holds. □

5.1   The Nondeterministic Circuit Complexity

Nondeterministic circuits can efficiently compute f as mentioned in Section 4.

Lemma 1. $size^{\text{nd}}(f) \leq 2n + o(n)$.

Proof. We construct a nondeterministic circuit computing f as mentioned in Sect. 4. For simplicity, we assume that $\sqrt{n}$ is a power of 2.

We use $\log\sqrt{n}$ guess inputs $y_1, \ldots, y_{\log\sqrt{n}}$, and compute z_i, $0 \leq i \leq \sqrt{n} - 1$, by $o(n)$ gates as follows.

$$z_i = 1 \text{ iff } i = (y_{\log\sqrt{n}} y_{\log\sqrt{n}-1} \cdots y_1)_2 \left(= \sum_{j=1}^{\log\sqrt{n}} y_j \cdot 2^{j-1}\right).$$

$\sqrt{n}$ inputs are nondeterministically selected by at most $2n$ gates as follows.

$$\bigvee_{i=0}^{\sqrt{n}-1} x_{\sqrt{n}\cdot i+1} \wedge z_i, \bigvee_{i=0}^{\sqrt{n}-1} x_{\sqrt{n}\cdot i+2} \wedge z_i, \ldots, \bigvee_{i=0}^{\sqrt{n}-1} x_{\sqrt{n}\cdot i+\sqrt{n}} \wedge z_i.$$

Finally, we construct a circuit computing $\text{Parity}_{\sqrt{n}}$, which can be constructed with $o(n)$ gates by Theorem 3. □

5.2   The Deterministic Circuit Complexity

We firstly show the upper bound.

Lemma 2. $size^d(f) \leq 3n - o(n)$.

Proof. By Theorem 3, the obvious construction of f satisfies the lemma.   □

The following lemma can be skipped for the proof of Theorem 1. If one hopes the separation on the power of deterministic circuits and nondeterministic circuits (i.e., a weaker result and an easier proof than Theorem 1), then Lemma 3 is useful. Lemma 1 and Lemma 3 imply the separation.

Lemma 3. $size^d(f) \geq 3n - o(n)$.

Proof. We refer the proof of Theorem 3. While we eliminate at least three gates from the circuit by an assignment to x_i as the proof of Theorem 3, we modify the proof as follows. If $x_{\sqrt{n} \cdot i+1}, x_{\sqrt{n} \cdot i+2}, \ldots, x_{\sqrt{n} \cdot i+\sqrt{n}}$ have been assigned except one variable for some i, $0 \leq i \leq \sqrt{n} - 1$, then we assign 0 or 1 to the variable so that $Parity_{\sqrt{n}}(x_{\sqrt{n} \cdot i+1}, x_{\sqrt{n} \cdot i+2}, \ldots, x_{\sqrt{n} \cdot i+\sqrt{n}}) = 0$ and we do not count the number of eliminated gates at the assignment. By the modification, we can eliminate at least 3 gates at $n - \sqrt{n}$ assignments.   □

5.3   The Co-Nondeterministic Circuit Complexity

We prove the lower bound by the gate elimination method. See Sect. 2.2 for the definition of "block".

Lemma 4. $size^{cnd}(f) \geq 3n - o(n)$.

Proof. Let C be an optimal co-nondeterministic U_2-circuit computing f. We prove that we can continuously eliminate at least 3 gates from C by an assignment of a constant 0 or 1 to an actual input.

In the continuous eliminations, if $x_{\sqrt{n} \cdot i+1}, x_{\sqrt{n} \cdot i+2}, \ldots, x_{\sqrt{n} \cdot i+\sqrt{n}}$ have been assigned except one variable for some i, $0 \leq i \leq \sqrt{n} - 1$, then we assign 0 or 1 to the variable so that $Parity_{\sqrt{n}}(x_{\sqrt{n} \cdot i+1}, x_{\sqrt{n} \cdot i+2}, \ldots, x_{\sqrt{n} \cdot i+\sqrt{n}}) = 0$ and we do not count the number of eliminated gates at the assignment. The number of such assignment is $\sqrt{n}$.

Case 1. There is an actual input x_i, $1 \leq i \leq n$, which is connected to at least two gates.

Let g_1 and g_2 be gates which are connected from x_i. Since we can block g_1 by an assignment of a constant to x_i, g_1 is not the output of C and there is a gate g_3 which is connected from g_1. See Fig. 2.

We prove that we can eliminate at least 3 gates from C by an assignment to x_i. We assign a constant 0 or 1 to x_i such that g_1 is blocked. Then, we can eliminate g_1, g_2 and g_3. If g_2 and g_3 are the same gate, then the output of g_2 $(= g_3)$ becomes a constant, which means that g_2 $(= g_3)$ is not the output of C

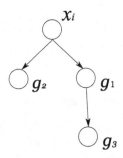

Fig. 2. Case 1

and we can eliminate another gate which is connected from g_2 $(= g_3)$. Thus, we can eliminate at least 3 gates.

Case 2. Every actual input is connected to at most one gates.

Let g_1 be a gate in C such that one of two inputs is connected from an actual input x_i and the other is connected from a node v whose output is dependent on only guess inputs and independent from actual inputs. (v may be a gate and may be a guess input.) Consider that an assignment to actual inputs and guess inputs is given. Then, if the value of the output of v blocks g_1 by the assignment, then the output of C must be 1, since, if the output of C is 0, then the value of the Boolean function which is computed by C becomes independent from x_i, which contradicts that C computes f. We use the fact above and reconstruct C as follows.

Let c be a constant 0 or 1 such that if the output of v is c, then g_1 is blocked. We fix the input of g_1 from v to $\neg c$ and eliminate g_1. We prepare a new output gate g_2 and connect the two inputs of g_2 from the old output gate and v. g_2 is labeled by a Boolean function in U_2 so that the output of g_2 is 0 iff the input from the old output gate is 0 and the input from v is $\neg c$. Let C' be the reconstructed circuit. See Fig. 3.

In the reconstruction, we eliminated one gate (g_1) and added one gate (g_2). Thus, the size of C' equals the size of C. In C', if the output of v is c, then the output of C' becomes 1 by g_2. If the output of v is $\neg c$, then the output of C' equals the output of the old output gate and g_1 has been correctly eliminated since we fixed the input of g_1 from v to $\neg c$ in the reconstruction. Thus, C and C' compute a same Boolean function. We continue such reconstruction until the reconstructed circuit satisfies the condition of Case 1. The reconstructions must end, since one reconstruction increases continuous gates whose one input is dependent on only guess inputs (i.e., g_2) at the output. Note that g_1 is not included in the continuous gates, since the output of C must depend on at least two actual inputs.

Thus, we can eliminate at least 3 gates at $n - \sqrt{n}$ assignments. ◻

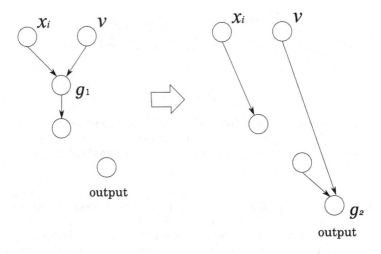

Fig. 3. Case 2

6 Concluding Remarks

In this paper, we considered the power of nondeterministic circuits and co-nondeterministic circuits. To prove our results, we proposed a simple proof strategy using nondeterministic selecting. It remains open that we use the strategy and prove a similar or improved result for U_2-circuits or other Boolean circuits.

Acknowledgement. The author would like to thank the anonymous reviewers for valuable and detailed comments.

References

1. Iwama, K., Morizumi, H.: An explicit lower bound of 5n - o(n) for boolean circuits. In: Proc. of MFCS. pp. 353–364 (2002)
2. Lachish, O., Raz, R.: Explicit lower bound of 4.5n - o(n) for boolean circuits. In: Proc. of STOC. pp. 399–408 (2001)
3. Morizumi, H.: Lower bounds for the size of nondeterministic circuits. In: Proc. of COCOON. pp. 289–296 (2015)
4. Schnorr, C.: Zwei lineare untere schranken für die komplexität boolescher funktionen. Computing **13**(2), 155–171 (1974)
5. Zwick, U.: A 4n lower bound on the combinational complexity of certain symmetric boolean functions over the basis of unate dyadic boolean functions. SIAM J. Comput. **20**(3), 499–505 (1991). https://doi.org/10.1137/0220032

On Hardest Languages for
One-Dimensional Cellular Automata

Mikhail Mrykhin[✉][iD] and Alexander Okhotin[iD]

Department of Mathematics and Computer Science, St. Petersburg State University,
7/9 Universitetskaya nab., Saint Petersburg 199034, Russia
mikhail.k.mrykhin@gmail.com, alexander.okhotin@spbu.ru

Abstract. Since the famous construction of "the hardest context-free language" by Greibach (1973), the existence of hardest languages under homomorphic reductions has been investigated for quite a few language families. This paper shows that for one-way real-time cellular automata, also known as trellis automata, there is no hardest language, whereas for linear-time cellular automata, the hardest language is constructed.

1 Introduction

The notion of a complete set for a family of formal languages is among the central concepts of theoretical computer science. Given a reduction mechanism, such as logarithmic-space Turing machines, a language, to which every language from a family is reducible, is called *hard* for that family. If, furthermore, this language belongs to the family, it is *complete* for that family.

For this definition to make sense, the reduction mechanism should be computationally weaker than the family itself. Furthermore, the weaker the reduction mechanism, the stronger are the results on the completeness of some language with respect to those reductions. The weakest reduction mechanism are *homomorphisms*, under which every symbol a from the alphabet of the given language is mapped to a substring $h(a)$ over the alphabet of the hard language. There is a notable result involving such reductions: Greibach's [8] "hardest context-free language", that is, a fixed language L_0 over an alphabet Σ_0 defined by a formal grammar, to which one can reduce every language L defined by a grammar over any alphabet Σ, by a suitable homomorphism $h \colon \Sigma^* \to \Sigma_0^*$, so that a nonempty string $w \in \Sigma^+$ belongs to L if and only if its image $h(w)$ is in L_0.

Greibach's result inspired a line of research on the existence of hardest languages under homomorphic reductions in various families of formal languages. Already Greibach [9] proved the first negative result, that for the family of languages described by *LR(1) grammars*, or, equivalently, recognized by *deterministic pushdown automata*, there cannot exist such a hardest language. For another important subfamily of grammars, the *linear grammars*, non-existence of hardest languages was established by Boasson and Nivat [2]. On the other hand, as

Research supported by Russian Science Foundation, project 18-11-00100.

A. Leporati et al. (Eds.): LATA 2021, LNCS 12638, pp. 118–130, 2021.
https://doi.org/10.1007/978-3-030-68195-1_10

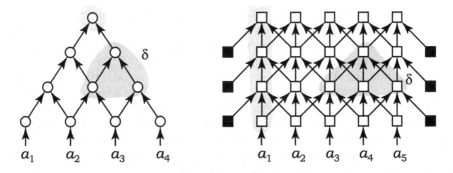

Fig. 1. Computation of (left) trellis automata, (right) cellular automata.

proved by Okhotin [20], two generalizations of ordinary ("context-free") grammars *do* have hardest languages: these are *conjunctive grammars* [15,19], with an explicit conjunction operation in the rules, and *Boolean grammars* [16], which are further equipped with a negation operation.

Hardest languages under homomorphic reductions have been investigated beyond formal grammars. For the family of *regular languages*, Čulík and Maurer [7] proved that a hardest language under homomorphic reductions cannot exist; a related stronger result was later established by Harju, Karhumäki and Kleijn [10]. Autebert [1] obtained the same negative result for one-counter automata. On the other hand, Greibach [8] proved a hardest language theorem for a certain restricted type of Turing machines, whereas Čulík and Maurer [7] found a hardest language in the complexity class NSPACE(n); their argument extends to NSPACE($f(n)$) and DSPACE($f(n)$) for every space-constructible function f.

The purpose of this paper is to investigate whether hardest languages exist for two basic kinds of one-dimensional cellular automata: *one-way real-time cellular automata*, also known under a proper name of *trellis automata*, and the more powerful *linear-time cellular automata*. The form of time-space diagrams for these automata is illustrated in Fig. 1: on an input of length n, a trellis automaton makes $n - 1$ parallel steps, with the value of a cell at each step depending on its value and the value of its right neighbour at the previous step; a linear-time cellular automaton makes at most $C \cdot n$ steps, for some constant C, and the next value of each cell depends on both its left and right neighbours, as well as on itself.

Both models were first studied by Smith [21], and have later received much attention. For **trellis automata**, already Smith [21] showed that they can accept the Dyck language [21]; Ibarra and Kim [12] proved that they can recognize the language $\{ a^n b^{2^n} \mid n \geqslant 1 \}$ and some P-complete language; Čulík [6] applied the famous Firing Squad Problem to construct a trellis automaton recognizing $\{ a^m b^{m+n} c^n \mid m, n \geqslant 1 \}$. Recently, Terrier [26] presented a sophisticated simulation of formal grammars describing bounded languages by trellis automata.

Okhotin [17] showed that trellis automata are equivalent to *linear conjunctive grammars*—a subclass of the earlier mentioned conjunctive grammars.

A variety of methods have been employed to show some limitations of this model: Yu [27] proved that the language $\{\, a^n b^{in} \mid i, n \geqslant 0 \,\}$ is not recognized by trellis automata; Terrier [23] presented a non-representability argument for $\{\, wxw \mid w, x \in \{a, b\}^+ \,\}$; Buchholz and Kutrib [4] characterized representable languages of the form $\{\, a^n b^{f(n)} \mid n \geqslant 0 \,\}$. A general lemma for proving non-representability of languages by trellis automata was established by Terrier [22], who used it to prove the non-closure of this family under concatenation.

The **linear-time cellular automaton** model is much more powerful: for instance, they can recognize the languages $\{\, a^{2^n} \mid n \geqslant 0 \,\}$ [5] and $\{\, a^n b^{in} \mid n, i \geqslant 1 \,\}$ [3]. There is an important theoretical result on *linear speed-up* for this model: every language recognized by a cellular automaton in time $C \cdot n$ can be recognized in time $(1 + \varepsilon)n$, see Ibarra et al. [13]. Whether these automata are equivalent to *real-time cellular automata* operating in time exactly n, is a long-standing open problem raised already by Smith [21] and addressed many times in the literature. For more details about cellular automata as language-recognition devices, the reader is directed to a survey by Terrier [25].

Two results on hardest languages for these models are established in this paper. The first result is that trellis automata cannot have a hardest language, which is proved in Sect. 3 by an argument based on state complexity: it is shown that, for every n, there is a language recognized by a trellis automaton, which is complicated enough not to be reducible to any n-state trellis automaton. The other result is that linear-time cellular automata *do* have a hardest language, which is constructed in Sect. 4. Whether a hardest language exists for real-time cellular automata, is left as yet another open problem concerning this elusive class, discussed in Sect. 5.

2   Definitions

Definition 1. *A trellis automaton is a 5-tuple $A = (\Sigma, Q, I, \delta, F)$, where*

- *Σ is a finite set of input symbols,*
- *Q is a finite set of states,*
- *I is a mapping from Σ to Q,*
- *δ is a mapping from $Q \times Q$ to Q called the transition function,*
- *$F \subseteq Q$ is a subset of accepting states.*

The initial configuration on an input string $w = a_1 \ldots a_n$, with $n \geqslant 1$, is $I(w) = I(a_1) \ldots I(a_n)$. Every next configuration is determined by a function $\Delta \colon Q^+ \to Q^+$, defined by $\Delta(q_1 \ldots q_n) = r_1 \ldots r_{n-1}$, where $r_i = \delta(q_i, q_{i+1})$.

After $|w| - 1$ such steps, the string of states shrinks into a single state, which determines the acceptance. In formal notation, the language recognized by the automaton is $L(A) = \{\, w \mid \Delta^{|w|-1}(I(w)) \in F \,\}$.

The following example of an automaton is given to illustrate the definition; it shall also be used in a proof later in Sect. 3.

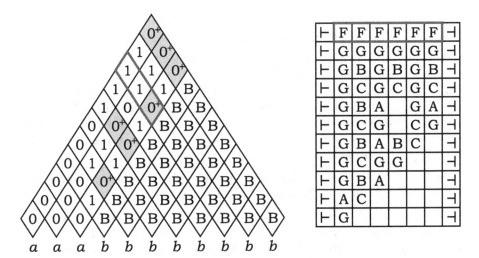

Fig. 2. (left) Trellis automaton for $\{a^n b^{i \cdot 2^n} \mid n, i \geqslant 1\}$, (right) cellular automaton for the FSP.

Example 1 [18]. For each $p \geqslant 2$, the language $L_p = \{a^n b^{i \cdot p^n} \mid n, i \geqslant 1\}$ is recognized by a trellis automaton $A = (\Sigma, Q, I, \delta, F)$ with $p + 3$ states, which computes, on each string $a^i b^j$, the i-th digit of the base-p representation of j, as illustrated for $p = 2$ in Fig. 2(left).

Its set of states is $Q = \{0, \ldots, p - 1, 0^+, B, D\}$, where $0, \ldots, p - 1$ are base-p digits and 0^+ is "zero with carry". The initial function sets $I(a) = 0$ and $I(b) = B$. The transition $\delta(B, B) = B$ sets the triangle on b^+. The leftmost B in each line increments the digits on the first diagonal to the left by $\delta(x, B) = x + 1$ for all $x \in \{0, \ldots, p - 2\}$, $\delta(p - 1, B) = 0^+$ and $\delta(0^+, B) = 1$. In every subsequent diagonal, the digits are incremented by zero with carry: $\delta(x, 0^+) = x + 1$, $\delta(p - 1, 0^+) = 0^+$. Otherwise, the digit is unchanged: $\delta(x, y) = x$ for all $x, y \in \{0, \ldots, p - 1\}$ and $\delta(0^+, y) = 0$ for all $y \in \{0, \ldots, p - 1\}$. For the remaining pairs of states, δ is defined as D. The state 0^+ is assumed on all strings $a^n b^{i \cdot p^n}$, and accordingly it is the only accepting state.

Definition 2. *A cellular automaton is a 5-tuple $A = (\Sigma, Q, I, \delta, F)$, where*

- *Σ is a finite set of terminal symbols,*
- *Q is a finite set of internal states, with $\vdash, \dashv \notin Q$,*
- *I is a mapping from Σ to Q,*
- *$\delta\colon (Q \cup \{\vdash\}) \times Q \times (Q \cup \{\dashv\}) \to Q$ is the transition function,*
- *$F \subseteq Q$ is a subset of accepting states.*

The initial configuration on $w = a_1 \ldots a_n$ is $I(a_1) \ldots I(a_n)$. Every next configuration is given by a function $\Delta\colon Q^+ \to Q^+$, defined by $\Delta(q_1 \ldots q_n) = r_1 \ldots r_n$, where $r_i = \delta(q_{i-1}, q_i, q_{i+1})$, assuming that $q_0 = \vdash$ and $q_{n+1} = \dashv$. A string is accepted if the leftmost cell ever enters a state from F.

A cellular automaton is linear-time, if, on an input of length n, it can enter a state from F in at most $C \cdot n$ steps, for some constant $C \geqslant 0$.

Among various interesting problems solvable by linear-time cellular automata, there is the famous *Firing Squad Problem* (FSP), stated as follows. Assume that each cell, except the leftmost one, begins in a so-called *quiescent state* $\square$, with $\delta(\square, \square, \square) = \delta(\square, \square, \dashv) = \square$. Then, in the beginning, only the leftmost cell knows that the computation has started. Its task is to communicate this news to all cells, and have them "fire" at once, that is, bring them to the same state at the same time.

Example 2 (Mazoyer [14]). There exists a 6-state cellular automaton solving the firing squad problem in minimal time $2n - 2$.

The automaton operates by sending signals at various speeds, which is achieved by using other signals to slow down the main signals. These signals form a complex geometric pattern that eventually leads to all cells firing at the same time.

A variant of the FSP investigated by Čulík [6] has both the leftmost and the rightmost states initiate the computation from both sides; then, all cells are synchronized in time $n - 1$. In this paper, Čulík's synchronization shall be used in the construction of a cellular automaton recognizing the hardest language.

3   No Hardest Language for Trellis Automata

Theorem 1. *There does not exist any "hardest" trellis automaton M_0, such that for every language L recognized by a trellis automaton there would be a homomorphism $h \colon \Sigma^* \to \Sigma_0^*$, such that $w \in L$ if and only if $h(w) \in L(A_0)$.*

The proof is by contradiction. If there exists a hardest language in this family, then it is recognized by a trellis automaton with t states. It turns out that for every fixed value of t there is a language that is complicated enough not to be reducible to the supposed hardest language.

Lemma 1. *For every $t \geqslant 1$, there is a language $L \subseteq \{a, b\}^*$ recognized by a trellis automaton, such that L is not an inverse homomorphic image of any language recognized by a t-state trellis automaton.*

Proof. Let p be any prime greater than t. Consider the language $L_p = \{ a^n b^{i \cdot p^n} \mid n, i \geqslant 1 \}$ from Example 1, recognized by a trellis automaton with $p + 3$ states. It is claimed that this language is not reducible to any trellis automaton with fewer than p states by any homomorphism.

Suppose the contrary, that L_p is reducible to some language recognized by some automaton $A = (\Sigma, Q, I, \delta, F)$, with $|Q| < p$, by a homomorphism $h \colon \{a, b\}^* \to \Sigma^*$.

For every $i \geqslant 0$, let u_i denote the last i symbols of $h(a)^n$, for n large enough. Similarly, for $j \geqslant 0$, denote the first j symbols of $h(b)^n$ by v_j. Let $v_{j,k}$, with $j \geqslant 0$ and $k \geqslant 1$, be the substring of $h(b)^n$ of length k beginning at the $(j+1)$-th symbol.

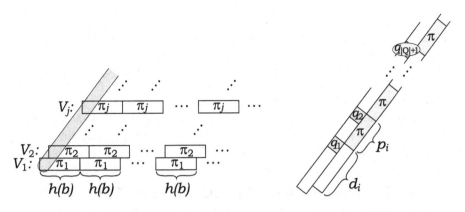

Fig. 3. Periodic sequences of states in the proof of Lemma 1.

Claim 1. *For each $k \geqslant 1$, the sequence of states $V_k = \{\Delta(I(v_{j,k}))\}_{j=0}^{\infty}$ is periodic with period $m = |h(b)|$.*

Proof. Indeed, $\Delta(v_{j+m,k}) = \Delta(v_{j,k})$, because $v_{j+m,k} = v_{j,k}$.

For each $j \geqslant 0$, let π_j be the periodic part of V_j, as illustrated in Fig. 3(left).

Claim 2. *For each $i \geqslant 0$, the sequence $\{\Delta(u_i v_j)\}_{j=1}^{\infty}$ is ultimately periodic, with a period of length $p_i \geqslant 1$ beginning from $d_i \geqslant 0$. Moreover, $p_{i+1} = c_i p_i$, for some $c_i \leqslant |Q|$.*

Proof. Induction on i.

Induction base: $i = 0$. Then the sequence $\{\Delta(v_j)\}_{j=1}^{\infty}$ is formed of the first elements of the sequences V_j. Each V_j has period $m = |h(b)|$, and the periodic part π_j of every next V_j depends only on π_{j-1}. Since there are at most $|Q|^m$ different periodic parts, the sequence $\{\pi_j\}_{j \geqslant 0}$ is ultimately periodic, and therefore so is the desired sequence of the first states in π_j.

Induction step: Assume that the sequence $\{\Delta(u_i v_j)\}_j$ is periodic beginning from d_i with period p_i. Denote its periodic part by $\pi = \Delta(u_i v_{d_i}), \Delta(u_i v_{d_i+1}), \ldots, \Delta(u_{i+1} v_{d_i+p_i-1})$.
Consider the following states in the next diagonal: $q_1 = \Delta(u_{i+1} v_{d_i})$, $q_2 = \Delta(u_{i+1} v_{d_i+p_i})$, $\ldots$, $q_{|Q|+1} = \Delta(u_{i+1} v_{d_i+|Q|p_i})$, as illustrated in Fig. 3(right). Since there are $|Q| + 1$ states in this sequence, two of them must coincide. Let these be $\Delta(u_{i+1} v_{d_i+sp_i}) = \Delta(u_{i+1} v_{d_i+tp_i})$, with $s < t$. This makes the $(i + 1)$-th diagonal periodic beginning with $d_i + sp_i$, with period $(t - s)p_i$.

Let $|h(a)| = \ell$ and $|h(b)| = m$. For each n, consider the n-th diagonal $D_n = \{\Delta(u_{\ell n} v_j)\}_j$. By Claim 2, each D_n is periodic, and the length $p_{\ell n}$ of its period divides $P_n = p_0 |Q|!^{\ell n}$. The number $|Q|!$ is coprime with p, because $|Q| < p$; therefore, for sufficiently large n, the number P_n is not divisible by p^n.

Now consider the sequence $D'_n = \{\Delta(h(a)^n h(b)^j)\}_j = \{\Delta(u_{\ell n} v_{mj})\}_j$ obtained from D_n by taking every m-th state. It is also periodic with period $p_{\ell n}$ beginning with $d_{\ell n}$.

For every $k \geqslant 0$, the string $h(a)^n h(b)^{kp^n}$ must be accepted by A as an image of a string in L_p. Let k be the smallest number with $mkp^n > d_{\ell n}$. Then the accepting state $\Delta(u_{\ell n} v_{mkp^n}) \in F$ is already in the periodic part of D'_n. Then, several periods later, the state $\Delta(u_{\ell n} v_{m(kp^n + P_n)}) = \Delta(u_{\ell n} v_{mkp^n})$ must be accepting as well. However, the string $u_{\ell n} v_{m(kp^n + P_n)} = h(a^n b^{kp^n + P_n})$ is an image of a string not in L_p, because P_n is not divisible by p^n. This is a contradiction. $\qquad\square$

4   The Hardest Language for Linear-Time Cellular Automata

In order to prove the hardest language theorem for linear-time cellular automata, one should construct a single automaton A_0 operating in time $C \cdot n$, for a fixed constant C, so that every linear-time cellular automaton could be homomorphically reduced to this one.

Theorem 2. *There exists a linear-time cellular automaton A_0, such that for every language L recognized by a linear-time cellular automaton there exists a homomorphism $h\colon \Sigma^* \to \Sigma_0^*$, such that $w \in L$ if and only if $h(w) \in L(A_0)$.*

The hardest language uses the following 18-symbol alphabet.

$$\Sigma_0 = \{ a_S \mid S \subseteq \{x_2''\} \} \cup \{ b_T \mid T \subseteq \{y_1, y_2, y_3, z\} \}$$

There are two symbols of the form a_S, namely, $a_{\{x_2''\}}$ and $a_\varnothing$, where the subscript is used to encode a single bit of the binary representation of some number. A symbol b_T encodes four such bits.

For each cellular automaton $A = (\Sigma, Q, I, \delta, F)$, its input strings shall be mapped to strings over the hardest language's alphabet by a homomorphism $h_A\colon \Sigma^* \to \Sigma_0^*$. Assume that $Q = \{1, \ldots, 2^k - 1\}$, for some $k \geqslant 1$, and that $F = \{2^{k-1}, \ldots, 2^k - 1\}$. Both end-markers are denoted by 0. Then, each state is encoded in k bits, and the acceptance status of each state is determined by the leading digit of its encoding.

The image $h(c)$ of each symbol $c \in \Sigma$ consists of two substrings. It begins with the binary encoding of the state $I(c)$, written down as k symbols of the form a_S, with $S \subseteq \{x_2''\}$, where the flag x_2'' denotes 1 in the corresponding position, whereas its absence denotes 0. The rest of the image $h(c)$ encodes the entire transition table δ, with each entry $\delta(p, q, r) = s$ encoded as k symbols of the form b_T, with $T \subseteq \{y_1, y_2, y_3, z\}$. This is a four-track encoding: every i-th symbol of this encoding contains four bits, and these are the i-th bits in the binary representations of the states p, q, r and s. Overall, the length of $h(c)$ is $m = k + k \cdot (2^k - 1) \cdot (2^k)^2 = \Theta(|Q|^3 \log |Q|)$.

For the desired hardest automaton A_0, it is only essential that it accepts a well-formed image $h(c_1 \ldots c_n)$, with $c_1, \ldots c_n \in \Sigma$, if and only if A accepts

$c_1 \ldots c_n$. For any ill-formed strings, it is irrelevant whether A_0 accepts or rejects them.

Given $h(c_1 \ldots c_n)$, the automaton A_0 simulates the computation of A on the string $c_1 \ldots c_n$, and accepts if and only if A accepts. Each step of A is simulated by A_0 in $2m$ steps, organized in two *phases*. During Phase I, A_0 communicates the encoded state in each cell to its two neighbours; at Phase II, A_0 has a triple of states (p, q, r) in each encoded cell, and it uses the encoded transition function of A to determine the new value $\delta(p, q, r)$ in this encoded cell.

The cells where the symbols a_S are originally placed are called *state-cells*: they come in blocks of k cells, and they store the state of a single cell in A. The cells originally containing b_T are the *rule-cells*: a block of k rule-cells stores a single value of the original automaton's transition function, and holds it throughout the computation.

At Phase I, the automaton A_0 copies the number of the state encoded in each block of state-cells to the two neighbouring blocks of state-cells, as illustrated in Fig. 4. Each bit is propagated in both directions by two signals. At the same time, a firing squad is used to synchronize the image of each symbol in exactly m steps. At the moment when the firing squad fires, the signals propagating each i-th bit of each state to the left and to the right will reach exactly the positions of the i-th bits in the two neighbouring blocks of state-cells. Thus, at the end of the phase, each block of state-cells knows the entire 3-symbol neighbourhood of the original automaton, represented as a 3-track encoding of binary numbers.

At Phase II, each block of state-cells contains the 3-track encoding of its neighbourhood. Now the task is to compare it with each transition in the rule-cells, encoded in 4 tracks. For exactly one transition, all three tracks shall match, and then, by the end of this phase, the state-cell should encode the resulting state of this transition. This is implemented as follows. In the beginning of Phase II, each rule-cell sends its contents towards the state-cells, as a left-bound signal. After k steps, the signals that comprise the first rule arrive at the corresponding positions in the state-cells; at the same time, the state-cells use a firing squad to count up to k. Then, at the time when the firing squad fires, each bit of the 3-state neighbourhood is compared to the corresponding bit of this transition.

If there is any mismatch, then the corresponding cell knows that. During the next k steps, the automaton communicates the data about the mismatches between the state-cells, and if no mismatch is found, then the state-cells change their value to the target state of this transition. At the same time, in the course of these k steps, the k signals comprising the next transition arrive at the state-cells, a firing squad again signals their arrival, and the matching goes on.

Thus, synchronization is used in two different places. At Phase I, as well as at Phase II, each block of m cells synchronizes in m steps. During Phase II, each block of k state-cells synchronizes in k steps once for each simulated transition. Both synchronizations are carried out using the minimal-time solution to the Firing Squad Problem with two generals, with the borders between state-cells and rule-cells reinitializing these two generals each time after the squad fires.

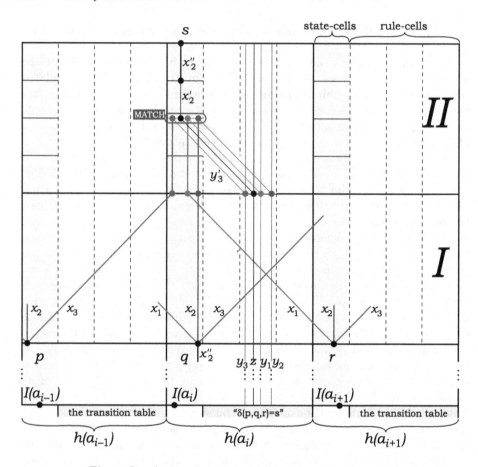

Fig. 4. Simulation of one step of A by the automaton A_0.

Let FSP_1 be the set of states of a cellular automaton that uses the border between rule-cells and state-cells (that is, the border between the images of the symbols) as both the left and the right general. This activates synchronization in m steps. Let FSP_2 be the states of an automaton that uses the border between rule-cells and state-cells as the left general, and the border between state-cells and rule-cells as the right general. This activaves the synchronization of a block of state-cells in k steps. These firing squads are embedded in the definition of the set of states given below.

In the full set of states of A_0, states of Phase I and of Phase II are specifically marked, and there is an attached firing squad that implements the alternation of phases.

$$Q_0 = FSP_1 \times \Big([\{I\} \times Q_I] \cup [\{II\} \times Q_{II}] \Big)$$

Both Phase-I states and Phase-II states remember whether the current cell is a state-cell or a rule-cell, and each contains individual bits of binary representations of the encoded states and transitions.

$$Q_{\mathrm{I}} = \left(\left[\{\text{state}\} \times 2^{\{x_2\}} \right] \cup \left[\{\text{rule}\} \times 2^{\{y_1, y_2, y_3, z\}} \right] \right) \times 2^{\{x_1, x_3\}}$$

$$Q_{\mathrm{II}} = \left(\left[\{\text{state}\} \times 2^{\{x_1, x_2, x_3, \#, x_2', x_2''\}} \mathrm{FSP}_2 \right] \cup \left[\{\text{rule}\} \times 2^{\{y_1, y_2, y_3, z\}} \right] \right) \times$$
$$\times 2^{\{y_1', y_2', y_3', z'\}}$$

In this definition, the variable x_2 represents a single bit of the number of the current state, encoded in a block of state-cells, and the notation $2^{\{x_2\}}$ means that the value of this bit is 0 or 1. Each rule-cell remembers 4 bits, denoted by y_1, y_2, y_3, z: these are digits of binary representations of three arguments to the transition function, along with a digit of its value. The variable x_1 denotes a bit in the number of the state being sent to the right neighbour; symmetrically, x_3 is a bit sent to the left neighbour.

The initialization function sets all cells in the final configuration of Phase II, so that the subsequent first transition starts Phase I. Symbols a_S initialize the state-cells with one bit each, symbols b_T similarly mark the rule-cells.

$$I_0(a_S) = (f, \mathrm{II}, \text{state}, S, f, \varnothing)$$
$$I_0(b_T) = (f, \mathrm{II}, \text{rule}, T, \varnothing)$$

Here f denotes the "firing" state in both firing squads.

The basic signals propagate as follows. At the start of Phase I, x_2'' sends x_3 to the left at unit speed, x_1 to the right and x_2 in place; once FSP_1 fires, all these signals stop in their positions (this is the same relative position within the images of three symbols, as exactly m steps have passed) until the next Phase I, at which point they are discarded and new ones are sent out. The bits y_1, y_2, y_2 and z always keep their positions, as the transition rules stay constant. At the start of Phase II, they send the corresponding signals y_1', y_2', y_3' and z' to the left at unit speed; these are also discarded at the start of a new Phase I.

Finally, the argument matching during Phase II proceeds as follows: at the moment when FSP_2 fires, all state-cells compare the bits of x_1, x_2 and x_3 with y_1, y_2 and y_3, respectively. If a mismatch is found, the cell sets a falsifying #-flag, which propagates in both directions at unit speed, and spreads over all neighbouring state-cells by the next firing moment. At the same time, state-cells containing z get the flag x_2' (unverified value), which stays in place until the next firing moment; unless it is overridden by the #-flag, it turns into x_2'' (verified value). A successful verification happens for exactly one transition: the true bits represented by x_2'' appear at that time and stay in place until Phase I.

The definition of the transition function is omitted due to space constraints.

Lemma 2. *For each $\ell \geqslant 0$, let $q_1 \ldots q_n$ be the configuration of A after ℓ steps starting with $I(w)$. Then, the automaton A_0, having started with $I_0(h_A(w))$, after $2\ell m$ steps, has the binary encoding of each state q_i, with $i \in \{1, \ldots, n\}$, in the cells numbered $(i-1)m + 1, \ldots, (i-1)m + k$.*

It remains to define the accepting states of A_0. Accepting states are effective at the first cell of a block of state-cells, which contains the leading digit of the state of A encoded in this block. Accordingly, a state of A_0 is deemed accepting if this is a Phase II state-cell, both firing squads fire, and the bit x_2'' is set; this is the following unique accepting state.

$$F_0 = \{(f, \text{II}, \text{state}, \{x_2''\}, f, \varnothing)\}$$

Proof (of Theorem 2). A linear-time automaton recognizing L is first transformed to work in time $2n$, by the method of Ibarra et al. [13]. Let A be the resulting automaton, operating on an n-symbol input. Then, A_0, as constructed above, reads a string of length mn, and completes the simulation in time $4mn$. Finally, A_0 is modified so that it rejects after this number of steps on all inputs.   □

5   On Real-Time vs Linear-Time Cellular Automata

In conclusion, one-way real-time cellular automata have no hardest languages, whereas linear-time cellular automata do have one. It is natural to ask whether a hardest language exists for the intermediate class of *real-time cellular automata*, that is, cellular automata that run for $n - 1$ steps.

Real-time cellular automata are known to be more powerful than trellis automata, but it is a long-standing open problem whether they are strictly weaker than linear-time cellular automata. Over half a century since the problem was first raised by Smith [21], several potential separating languages were conjectured, but eventually a way of recognizing each of them in real time was found.

A few theoretical characterizations of this problem were discovered. Ibarra and Jiang [11] showed that if the family accepted by real-time cellular automata is closed under reversal, then this model is equivalent in power to linear-time cellular automata. Terrier [24] proved that the closure under cyclic shift similarly implies equivalence to linear-time cellular automata.

Turning back to hardest languages, the kind of simulation used in this paper for linear-time cellular automata slows down the computation by a constant factor, and is accordingly unsuitable for establishing a hardest language for real-time cellular automata. If this family has no hardest language under homomorphic reductions, this would imply that real-time cellular automata and linear-time cellular automata define different families of languages. Any methods for proving or disproving this conjecture would be interesting to find and are suggested as a subject for future research.

References

1. Autebert, J.: Non-principalité du cylindre des langages à compteur. Math. Syst. Theory **11**, 157–167 (1977)
2. Boasson, L., Nivat, M.: Le cylindre des langages linéaires. Math. Syst. Theory **11**, 147–155 (1977)
3. Bucher, W., Culik, K.: On real time and linear time cellular automata. RAIRO Theor. Inform. Appl. **18**(4), 307–325 (1984)
4. Buchholz, T., Kutrib, M.: On time computability of functions in one-way cellular automata. Acta Informatica **35**(4), 329–352 (1998)
5. Choffrut, C., Culik II, K.: On real-time cellular automata and trellis automata. Acta Informatica **21**, 393–407 (1984)
6. Culik II, K.: Variations of the firing squad problem and applications. Inf. Process. Lett. **30**(3), 153–157 (1989)
7. Culik II, K., Maurer, H.A.: On simple representations of language families. RAIRO Theor. Informatics Appl. **13**(3), 241–250 (1979)
8. Greibach, S.A.: The hardest context-free language. SIAM J. Comput. **2**(4), 304–310 (1973)
9. Greibach, S.A.: Jump PDA's and hierarchies of deterministic context-free languages. SIAM J. Comput. **3**(2), 111–127 (1974)
10. Harju, T., Karhumäki, J., Kleijn, H.C.M.: On morphic generation of regular languages. Discrete Appl. Math. **15**(1), 55–60 (1986)
11. Ibarra, O.H., Jiang, T.: Relating the power of cellular arrays to their closure properties. Theoret. Comput. Sci. **57**, 225–238 (1988)
12. Ibarra, O.H., Kim, S.M.: Characterizations and computational complexity of systolic trellis automata. Theoret. Comput. Sci. **29**, 123–153 (1984)
13. Ibarra, O.H., Palis, M.A., Kim, S.M.: Some results concerning linear iterative (systolic) arrays. J. Parallel Distrib. Comput. **2**(2), 182–218 (1985)
14. Mazoyer, J.: A six-state minimal time solution to the firing squad synchronization problem. Theoret. Comput. Sci. **50**, 183–238 (1987)
15. Okhotin, A.: Conjunctive grammars. J. Autom. Lang. Comb. **6**(4), 519–535 (2001)
16. Okhotin, A.: Boolean grammars. Inf. Comput. **194**(1), 19–48 (2004)
17. Okhotin, A.: On the equivalence of linear conjunctive grammars and trellis automata. RAIRO Theor. Inform. Appl. **38**(1), 69–88 (2004)
18. Okhotin, A.: State complexity of linear conjunctive languages. J. Autom. Lang. Comb. **9**(2/3), 365–381 (2004)
19. Okhotin, A.: A tale of conjunctive grammars. In: Hoshi, M., Seki, S. (eds.) DLT 2018. LNCS, vol. 11088, pp. 36–59. Springer, Cham (2018). https://doi.org/10. 1007/978-3-319-98654-8_4
20. Okhotin, A.: Hardest languages for conjunctive and Boolean grammars. Inf. Comput. **266**, 1–18 (2019)
21. Smith III, A.R.: Real-time language recognition by one-dimensional cellular automata. J. Comput. Syst. Sci. **6**(3), 233–253 (1972)
22. Terrier, V.: On real time one-way cellular array. Theoret. Comput. Sci. **141**(1&2), 331–335 (1995)
23. Terrier, V.: Language not recognizable in real time by one-way cellular automata. Theoret. Comput. Sci. **156**(1&2), 281–287 (1996)
24. Terrier, V.: Closure properties of cellular automata. Theoret. Comput. Sci. **352**(1–3), 97–107 (2006)

25. Terrier, V.: Language recognition by cellular automata. In: Rozenberg, G., Bäck, T., Kok, J.N. (eds.) Handbook of Natural Computing, pp. 123–158. Springer, Heidelberg (2012). https://doi.org/10.1007/978-3-540-92910-9_4

26. Terrier, V.: Recognition of poly-slender context-free languages by trellis automata. Theoret. Comput. Sci. **692**, 1–24 (2017)

27. Yu, S.: A property of real-time trellis automata. Discrete Appl. Math. **15**(1), 117–119 (1986)

Usefulness of Information and Unary Languages

Giovanni Pighizzini[1], Branislav Rovan[2], and Šimon Sádovský[2]($\boxtimes$)

[1] Dipartimento di Informatica, Università degli studi di Milano, Via Celoria 18,
20133 Milan, Italy
pighizzini@di.unimi.it
[2] Department of Computer Science, Comenius University,
Mlynská Dolina, 842 48 Bratislava, Slovakia
{rovan,sadovsky}@dcs.fmph.uniba.sk

Abstract. In this paper we continue the research on usefulness of information examining the effect of supplementary information on the complexity of solving a problem (see Rovan and Sádovský [7] for an overview). We use deterministic finite automata for a formal setting. Given a problem (a regular language) L_{prob} we measure the complexity of its solution – a DFA A_{prob} such that $L_{prob} = L(A_{prob})$ – using the state complexity. A supplementary information (advice) L_{adv} given by A_{adv} is useful if a simpler problem L_{new} given by A_{new} exists such that $L_{prob} = L_{new} \cap L_{adv}$ and both L_{new} and L_{adv} are simpler than L_{prob}. This is formalized via the notion of decomposability of finite automata (see [1] for DFA case and [7] for NFA case). We address the problem of decomposability of unary regular languages and give a characterization of λ-cyclic languages upon deterministic decomposability.

Keywords: Descriptional complexity · Deterministic finite automaton · Regular language · Unary language · λ-cyclic language · Decomposability of regular languages · Supplementary information · Usefulness of information · State complexity

1 Introduction

In the early days of Shannon's theory of information the main concern was in transferring information reliably and fast over possibly noisy channels. The *amount* of information was the important attribute of information considered. Over half a century later we can identify additional aspects, like usefulness, timeliness, etc. Research focused on formalizing these aspects of information was initiated over fifteen years ago (see [7] for a brief overview and usefulness, also [1] and [4] for usefulness, and [8] for timeliness). This paper further elaborates the notion of *usefulness of information* in the regular languages setting.

This research has been supported in part by the grant 1/0601/20 of the Slovak Scientific Grant Agency VEGA.

A. Leporati et al. (Eds.): LATA 2021, LNCS 12638, pp. 131–142, 2021.
https://doi.org/10.1007/978-3-030-68195-1_11

The essence of our approach to the notion of usefulness of information can be described as follows: *Information is useful if it helps to solve a given problem easier.* A problem to be solved is formalized by a language L and its solution by an automaton accepting it. In this paper we limit our attention to unary regular languages and deterministic finite automata. Given a problem (a unary regular language) L_{prob}, we measure the complexity of its solution – a finite automaton A_{prob} such that $L_{prob} = L(A_{prob})$ – using state complexity. A supplementary information (advice) L_{adv} given by A_{adv} is useful if a problem L_{new} given by A_{new} exists such that $L_{prob} = L_{new} \cap L_{adv}$ and both L_{new} and L_{adv} are simpler than L_{prob}. One can interpret A_{new} as the simplification of A solving the original problem relying on the provided advice. Since we look for simple solutions it is natural to consider minimal automata. This is formalized via the notion of decomposability of a finite automaton into two smaller finite automata (see [1] for the DFA case and [7] for the NFA case). This notion is naturally extended to regular languages which are considered decomposable when minimal finite automata are decomposable. For a more detailed description of our approach and an overview of past results see [7].

Decompositions of DFAs to finitely many, but not necessarily two, smaller DFAs was recently elaborated in [2] and [3] using the notions of *composite* and *prime* DFA. A DFA A is said to be composite if there are DFAs $A_1, \ldots, A_t$ such that $L(A) = \bigcap_{i=1}^{t} L(A_i)$ and the size of every A_i is strictly smaller than the size of A. Otherwise, A is said to be prime. Authors refer to t as the *width* of decomposition. In [2] they address the problem for unary regular languages. They give the criterion of what they call decomposability for unary deterministic finite automata (UDFA, for short). They prove that in the case of UDFAs having tail of nonzero length it holds that if a UDFA is decomposable, then it is decomposable to two smaller UDFAs. However, they prove that this is not the case when one considers UDFAs consisting of one cycle only. In this case there exist decomposable UDFAs which are not decomposable to two smaller UDFAs.

In this paper we provide a characterization of the class of λ-cyclic languages upon deterministic decomposability. A λ-cyclic language is a unary regular language which can be accepted by a UDFA having its cycle of length λ without initial tail. We stress that we are interested in the decomposition of UDFA to exactly two smaller UDFAs.

In Sect. 2 we introduce our notation for standard notions, summarize past results needed and define the problem. Results about UDFAs, which are the main concern of this paper, can be found in Sect. 3. Section 4 contains main results. For the lack of space some of the proofs are omitted or just outlined.

2   Preliminaries, Notations and Definition of Problem

We use standard notation of formal languages and automata theory. The length of a word w is denoted by $|w|$. ε denotes the empty word. The cardinality of a set S is denoted by $|S|$.

We consider natural numbers including zero and use $\mathbb{N}^+$ to denote the set of all positive natural numbers. By $\mathbb{Z}_n$ we denote the set of all remainders modulo

n, $\oplus_n$ denotes addition modulo n, i.e., $a \oplus_n b = (a + b) \bmod n$ for any $a, b \in \mathbb{Z}$ and $n \in \mathbb{N}$. When we write $a \mid b$ we mean that a is a divisor of b and we use $a \nmid b$ to express that a is not a divisor of b for any $a, b \in \mathbb{Z}$.

We define the *deterministic finite automaton* to be a 5-tuple $(K, \Sigma, \delta, q_0, F)$ with the standard meaning of its components. We require the transition function $\delta : K \times \Sigma \to K$ to be complete. The language accepted by A is $L(A) = \{w \in \Sigma^* \mid (\exists q_F \in F)\, (q_0, w) \vdash_A^* (q_F, \varepsilon)\}$ where the relation 'step of computation' $\vdash_A$ on configurations is defined as usual. If it holds that $|\Sigma| = 1$, we say that A is *a unary deterministic finite automaton*, which we abbreviate by UDFA. In that case we use $\Sigma = \{a\}$. We use $\mathcal{R}$ to denote the class of regular languages.

Given a DFA A we denote by $sc(A)$ the number of its states. For any language $L \in \mathcal{R}$ we denote by $sc(L)$ the so called *state complexity of L*, namely the number of states of the minimal DFA accepting L.

A *bipartite graph* $G = (V_1, V_2, E)$ is a tuple where V_1 and V_2 are two sets of *vertices* and $E \subseteq V_1 \times V_2$ is a set of *edges*. For $v \in V_1$ we define a *degree of vertex v*, denoted $d(v)$, by $d(v) = |\{u \in V_2 \mid (v, u) \in E\}|$. Analogously we define a *degree of vertex $u \in V_2$*. If for $v \in V_1 \cup V_2$ it holds $d(v) = 0$, we say that v is an *isolated vertex*.

2.1 Number Theory

It is often the case that questions about UDFAs are related to number-theoretical problems. Also our proofs employ some number-theoretical results, which we summarize here.

Chinese Remainder Theorem is one of classical results of number theory. We provide its generalization to non-coprime moduli and the formulation for systems of two congruences which suits best for our purposes.

Theorem 1 (Generalized Chinese Remainder Theorem). *Let $m, n, a, b \in \mathbb{N}$. Consider following system of congruences.*

$$x \equiv a \pmod{m}$$
$$x \equiv b \pmod{n}$$

If $a \equiv b \pmod{\gcd(m, n)}$, then there exists the unique solution x of this system modulo $\operatorname{lcm}(m, n)$. Moreover if $y \equiv x \pmod{\operatorname{lcm}(m, n)}$, then y is also a solution. Otherwise no solution exists.

Two technical lemmas follow. We shall refer to them later in the paper. Due to the space constraints we provide them without proofs.

Lemma 1. *Let $a, c \in \mathbb{N}$ such that a is a divisor of c. Then for every $n \in \mathbb{N}$ it holds $n \bmod a = (n \bmod c) \bmod a$.*

Lemma 2. *Let $a, b \in \mathbb{N}^+$. Then for each $n_1, n_2 \in \mathbb{N}$ and $r \in \mathbb{Z}$:*

$$\{(ka + r) \bmod b \mid k \in \mathbb{N};\ k \geq n_1\} = \{(i \cdot \gcd(a, b) + r) \bmod b \mid i \in \mathbb{N};\ i \geq n_2\}.$$

2.2 Usefulness of Information

Our main interest is in the notion of *usefulness* of information. We consider information useful if it helps to solve some problem easier in some sense. This approach led our research group to define the notion of *decomposability of a regular language*. This notion was first defined for the deterministic finite automata setting in [1] and further elaborated for the nondeterministic automata setting in [7].

Definition 1. *Let A be a DFA. We say that two DFAs A_1 and A_2 form a decomposition of A if $L(A) = L(A_1) \cap L(A_2)$, $sc(A_1) < sc(A)$ and $sc(A_2) < sc(A)$. In case such decomposition of A exists we say that A is decomposable.*

We can interpret A_1 to be a simpler solution to the problem $L(A)$ which uses the additional information that the input word is accepted by A_2.

Definition 2. *Let L be in $\mathcal{R}$. We say that L is deterministically decomposable if the minimal DFA accepting L is decomposable.*

Notation 1. *We denote the family of all deterministically decomposable regular languages by $\mathcal{D}_{det}$.*

3 Unary Deterministic Finite Automata

When dealing with deterministic finite automata over a unary alphabet (UDFA), one can observe that all of them have a similar shape. Each UDFA is formed by a (possibly empty) initial path, which is also called a tail, which is followed by exactly one cycle (see Fig. 1). It is convenient to think about the size of a unary DFA in terms of the sizes of its tail and cycle.

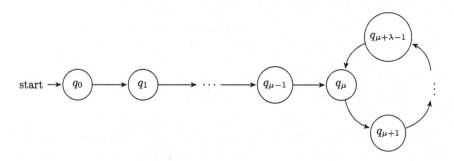

Fig. 1. Unary DFA of size (λ, μ)

Definition 3. *The size of a UDFA A is the pair (λ, μ) where λ is the number of states in the cycle of A and μ is the number of states in the tail of A.*

Notation 2. *When we say that we consider UDFA $A = (K, \{a\}, \delta, q[0], F)$ of size (λ, μ), then, if it is not stated otherwise, we implicitly mean that $K = \{q[i] \mid 0 \leq i < \lambda + \mu\}$ and for the transition function δ it holds that $(\forall i, 0 \leq i < \mu)$ $\delta(q[i], a) = q[i + 1]$ and $(\forall j \in \mathbb{Z}_\lambda)$ $\delta(q[\mu + j], a) = q[\mu + (j \oplus_\lambda 1)]$.*

Definition 4. *Let L be a unary infinite language and $\lambda \in \mathbb{N}$. We say that L is λ-cyclic if there exists a DFA A of the size $(\lambda, 0)$ such that $L(A) = L$. L is called properly λ-cyclic if it is λ-cyclic, but not λ'-cyclic for any $\lambda' < \lambda$.*

When asking questions about decomposability of a regular language, we need to deal with its minimal finite automaton. Here we provide a criterion of minimality for UDFAs.

Theorem 2 (Minimal UDFA characterization, [5,6]). *A UDFA $A = (K, \{a\}, \delta, p[0], F)$ of size (λ, μ) is minimal if and only if both the following conditions hold:*

(i) for any maximal proper divisor d of λ (i.e., $\lambda = \alpha \cdot d$ for some prime α) there exists an integer h, with $0 \leq h < \lambda$, such that $p[\mu + h] \in F$ if and only if $p[\mu + ((h + d) \bmod \lambda)] \notin F$, i.e., $a^{\mu+h} \in L$ if and only if $a^{\mu+h+d} \notin L$.
(ii) (If $\mu > 0$ then) $p[\mu - 1] \in F$ if and only if $p[\mu + \lambda - 1] \notin F$, i.e., $a^{\mu-1} \in L$ if and only if $a^{\mu+\lambda-1} \notin L$.

Informally the condition (ii) states that we cannot replace the cycle with a shorter one and condition (ii) states that we cannot 'roll the last state of the cycle into the last state of the tail.' Using Theorem 2 it is easy to prove the following lemma.

Lemma 3. *If L is a properly λ-cyclic language, then the minimal UDFA accepting L has size $(\lambda, 0)$.*

The decomposition of a DFA is defined in terms of intersection of two DFAs. We state a theorem providing a relationship between the sizes of two UDFAs and the UDFA accepting their intersection.

Theorem 3 ([6]). *Let L_1 and L_2 be two unary languages accepted by UDFAs A_1 and A_2 of size (λ_1, μ_1) and (λ_2, μ_2), respectively. The intersection of L_1 and L_2 is accepted by a UDFA of size $(\mathrm{lcm}(\lambda_1, \lambda_2), \max(\mu_1, \mu_2))$.*

3.1 Rolling Cycle of UDFA

We formulate our techniques of 'rolling-out' and 'rolling-in' cycles of UDFAs that we use in Sect. 4. These techniques come handy when examining the length of the tail in a decomposition of UDFA.

Lemma 4 (Roll-out Lemma). *Let $A = (K, \{a\}, \delta, q[0], F)$ be a UDFA of size (λ, μ). Then for any $\mu' \geq \mu$ and for the UDFA $A' = (K', \{a\}, \delta', q'[0], F')$ of size (λ, μ') where the set of accepting states is defined by*

$$F' = \{q'_F \in K' \mid (\exists w \in L(A)) \ (q'[0], w) \vdash_{A'}^* (q'_F, \varepsilon)\}$$

it holds $L(A') = L(A)$.

Lemma 5 (Roll-in Lemma). *Let* $A = (K, \{a\}, \delta, q[0], F)$ *be a UDFA of size* (λ, μ) *such that there exists* $\mu' \in \mathbb{N}$ *such that it holds:*

(i) $\mu' < \mu$

(i) For any $i, j \in \mathbb{N}$ *such that* $\mu' \le i, j < \lambda + \mu$ *and* $j - i \equiv 0 \pmod{\lambda}$ *it holds*
$q[i] \in F \Leftrightarrow q[j] \in F$.

Then for the UDFA $A' = (K', \{a\}, \delta', q'[0], F')$ *of size* (λ, μ') *where the set of accepting states is defined by*

$$F' = \{q'_F \in K' \mid (\exists w \in L(A)) \ (q'[0], w) \vdash^*_{A'} (q'_F, \varepsilon)\}$$

it holds $L(A') = L(A)$.

We state both Lemma 4 and Lemma 5 without presenting the proofs. However, both of them are quite intuitive. Lemma 4 states that if we have a UDFA A with a tail of length μ, we can 'roll-out' its cycle to get a UDFA with longer tail while accepting the same language. On the other hand, if the last state of the cycle and the last state of the tail of the given UDFA agree on their finality, then, by merging these states together, we can obtain a UDFA accepting the same language and having a shorter tail. We call this procedure 'rolling the last state of the cycle into the last state of the tail.' Let us consider a UDFA A of size (λ, μ). Lemma 5 states conditions which are sufficient in order to apply this 'roll-in' procedure several times to obtain a UDFA accepting $L(A)$ while having a tail of length $\mu' < \mu$.

4 Deterministic Decomposability of λ-cyclic Languages

Now we shall turn to proving our main result. When thinking about deterministic decomposability of a given properly λ-cyclic language, we found useful the point of view based on bipartite graphs and Extended Chinese Remainder Theorem (Theorem 1). We define notions based on this point of view and prove the criterion of the deterministic decomposability of properly λ-cyclic languages.

Definition 5. *Let* L *be a properly* λ-*cyclic language for some* $\lambda \in \mathbb{N}$ *and let* $\lambda_1, \lambda_2 \in \mathbb{N}$. *The bipartite graph induced by* L, λ_1 *and* λ_2 *is the bipartite graph* $G_{L,\lambda_1,\lambda_2} = (\mathbb{Z}_{\lambda_1}, \mathbb{Z}_{\lambda_2}, E)$ *where the set of edges* E *is defined as follows:*

$$E = \{(r_1, r_2) \mid r_1 \in \mathbb{Z}_{\lambda_1}; \ r_2 \in \mathbb{Z}_{\lambda_2};$$
$$(\exists m \in \mathbb{N}) \ m \equiv r_1 \pmod{\lambda_1} \ \wedge \ m \equiv r_2 \pmod{\lambda_2} \ \wedge \ a^m \in L\}.$$

Let $V_1' = \{r \in \mathbb{Z}_{\lambda_1} \mid d(r) > 0\}$ *and* $V_2' = \{r \in \mathbb{Z}_{\lambda_2} \mid d(r) > 0\}$ *be the sets obtained by removing all isolated vertices from* $G_{L,\lambda_1,\lambda_2}$. *We say that the graph* $G_{L,\lambda_1,\lambda_2}$ *decomposes* L *if for all* $(r_1, r_2) \in V_1' \times V_2'$ *it holds that*

$$(r_1, r_2) \in E \ \vee \ ((\nexists m \in \mathbb{N}) \ m \equiv r_1 \pmod{\lambda_1} \ \wedge \ m \equiv r_2 \pmod{\lambda_2}).$$

Intuitively Definition 5 states the following. Consider that we have a properly λ-cyclic language L and we want to decompose its minimal UDFA, which has size $(\lambda, 0)$, using UDFAs A_1 and A_2 of sizes $(\lambda_1, 0)$ and $(\lambda_2, 0)$ where moreover $\text{lcm}(\lambda_1, \lambda_2) = \lambda$. We shall show later that if A is decomposable, then also a decomposition of this type exists. The vertices of $G_{L,\lambda_1,\lambda_2}$ correspond to the remainders modulo λ_1 and λ_2, thus also to the states of A_1 and A_2. Observing the way E is defined one can see that if $(r_1, r_2) \in E$, we must mark the states corresponding to r_1 in A_1 and to r_2 in A_2 as accepting to ensure that $L(A) \subseteq L(A_1) \cap L(A_2)$. The condition in the last line of Definition 5 says that after this marking we accept nothing more than words from L in $L(A_1) \cap L(A_2)$ and thus $L(A_1) \cap L(A_2) = L$. For example consider $L_{12} = \{a^{12k+2}, a^{12k+11} \mid k \in \mathbb{N}\}$ with its minimal UDFA A_{12} of size $(12, 0)$. One can verify that the bipartite graph $G_{L_{12},4,6}$ (Fig. 2) induced by $L_{12}, 4$ and 6 decomposes L_{12}. This means that the UDFAs $A_4 = (K_4, \{a\}, \delta_4, q_4[0], \{q_4[2], q_4[3]\})$ of size $(4, 0)$ and $A_6 = (K_6, \{a\}, \delta_6, q_6[0], \{q_6[2], q_6[5]\})$ of size $(6, 0)$ form a decomposition of A_{12}.

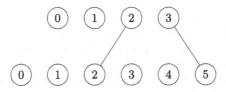

Fig. 2. Bipartite graph $G_{L_{12},4,6}$ induced by $L_{12}, 4$ and 6

Lemma 6. *Let L be a properly λ-cyclic language for some $\lambda \in \mathbb{N}$. If there exist $\lambda_1, \lambda_2 \in \mathbb{N}$ such that $\lambda_1, \lambda_2 < \lambda$, $\text{lcm}(\lambda_1, \lambda_2) = \lambda$ and the bipartite graph $G_{L,\lambda_1,\lambda_2}$ induced by L, λ_1 and λ_2 decomposes L, then $L \in \mathcal{D}_{det}$.*

Proof. Since L is a properly λ-cyclic language, the minimal UDFA accepting L has size $(\lambda, 0)$. Let us denote this UDFA $A = (K, \{a\}, \delta, q[0], F)$.

We shall construct a decomposition of A. To this end we define the UDFA $A_1 = (K_1, \{a\}, \delta_1, q_1[0], F_1)$ to be the λ_1-cycle with $F_1 = \{q_1[m \bmod \lambda_1] \mid m \in \mathbb{Z}_\lambda; q[m] \in F\}$. Similarly we define the UDFA $A_2 = (K_2, \{a\}, \delta_2, q_2[0], F_2)$ to be the λ_2-cycle with $F_2 = \{q_2[m \bmod \lambda_2] \mid m \in \mathbb{Z}_\lambda; q[m] \in F\}$. Since $\lambda_1, \lambda_2 < \lambda$, we have $sc(A_1) < sc(A)$ and $sc(A_2) < sc(A)$. So it remains to prove that $L(A) = L(A_1) \cap L(A_2)$.

- $L(A) \subseteq L(A_1) \cap L(A_2)$: Let $n \in \mathbb{N}$ be such that $a^n \in L(A)$. Let us denote $m = n \bmod \lambda$. It follows that $(q[0], a^n) \vdash_A^* (q[m], \varepsilon)$ and $q[m] \in F$. Clearly $(q_1[0], a^n) \vdash_{A_1}^* (q_1[n \bmod \lambda_1], \varepsilon)$. Since $\lambda = \text{lcm}(\lambda_1, \lambda_2)$, it holds that λ_1 is a divisor of λ. Then, following Lemma 1, we have $n \bmod \lambda_1 = (n \bmod \lambda) \bmod \lambda_1$. Therefore $q_1[n \bmod \lambda_1] = q_1[(n \bmod \lambda) \bmod \lambda_1] = q_1[m \bmod \lambda_1]$. Since $m \in \mathbb{Z}_\lambda$ and $q[m] \in F$, from definition of F_1 we can see that $q_1[m \bmod \lambda_1] \in F_1$. So we conclude $a^n \in L(A_1)$. One can prove that $a^n \in L(A_2)$ by a similar argumentation. Thus $a^n \in L(A_1) \cap L(A_2)$.

– $L(A_1) \cap L(A_2) \subseteq L(A)$: Let $n \in \mathbb{N}$ such that $a^n \in L(A_1) \cap L(A_2)$. Clearly

$$(q_1[0], a^n) \vdash^*_{A_1} (q_1[n \bmod \lambda_1], \varepsilon) \text{ where } q_1[n \bmod \lambda_1] \in F_1$$
$$(q_2[0], a^n) \vdash^*_{A_2} (q_2[n \bmod \lambda_2], \varepsilon) \text{ where } q_2[n \bmod \lambda_2] \in F_2$$

We denote $r_1 = n \bmod \lambda_1$ and $r_2 = n \bmod \lambda_2$. From the definition of F_1 it follows that there exists $m_1 \in \mathbb{Z}_\lambda$ such that $q[m_1] \in F$ and $m_1 \bmod \lambda_1 = r_1$. Similarly from the definition of F_2 it follows that there exists $m_2 \in \mathbb{Z}_\lambda$ such that $q[m_2] \in F$ and $m_2 \bmod \lambda_2 = r_2$. Since $q[m_1] \in F$, it follows that $a^{m_1} \in L(A) = L$. Let us denote $s_1 = m_1 \bmod \lambda_2$. By the definition of $G_{L,\lambda_1,\lambda_2}$ (Definition 5) we have that $(r_1, s_1) \in E$. Similarly from $q[m_2] \in F$ we have $a^{m_2} \in L(A) = L$. We denote $s_2 = m_2 \bmod \lambda_1$. From the definition of $G_{L,\lambda_1,\lambda_2}$ we obtain $(s_2, r_2) \in E$. This means that vertices $r_1 \in \mathbb{Z}_{\lambda_1}$ and $r_2 \in \mathbb{Z}_{\lambda_2}$ are not isolated in the graph $G_{L,\lambda_1,\lambda_2}$. Moreover, the system of congruences

$$r_1 \equiv x \pmod{\lambda_1}$$
$$r_2 \equiv x \pmod{\lambda_2}$$

has a solution, namely n. Therefore, since $G_{L,\lambda_1,\lambda_2}$ decomposes L, we obtain $(r_1, r_2) \in E$. Following the definition of the set of edges E of $G_{L,\lambda_1,\lambda_2}$ we conclude that there exists $m \in \mathbb{N}$ such that $a^m \in L = L(A)$ and m is a solution of the aforementioned system of congruences. Since $a^m \in L(A)$, we have $q[m \bmod \lambda] \in F$. Following Theorem 1 we have that all the solutions of the aforementioned system of modular equations are congruent modulo $\mathrm{lcm}(\lambda_1, \lambda_2) = \lambda$. Since both m and n are solutions of the aforementioned system of modular equations, it follows $m \bmod \lambda = n \bmod \lambda$. Therefore $q[n \bmod \lambda] \in F$. It is clear that $(q[0], a^n) \vdash^*_A (q[n \bmod \lambda], \varepsilon)$ and this computation is accepting, which completes the proof. □

The converse implication of the Lemma 6 also holds but to prove it we need some technical results first. We start with a result giving information about the length of a cycle in a decomposition of a given UDFA.

Lemma 7. *Let L be a unary properly λ-cyclic language such that $L \in \mathcal{D}_{det}$ and A be the minimal UDFA such that $L(A) = L$. Then there exists a decomposition of A to UDFAs A_1 and A_2 of sizes (λ_1, μ_1) and (λ_2, μ_2) such that $\lambda_1 \mid \lambda$ and $\lambda_2 \mid \lambda$.*

Proof. Consider L to be a unary properly λ-cyclic language such that $L \in \mathcal{D}_{det}$ and consider A to be its minimal UDFA of size $(\lambda, 0)$. Therefore there exists a decomposition of A, i.e., there exist UDFAs $A_1 = (K_1, \{a\}, \delta_1, q_1[0], F_1)$ and $A_2 = (K_2, \{a\}, \delta_2, q_2[0], F_2)$ of sizes (λ_1, μ_1) and (λ_2, μ_2), respectively, such that $sc(A_1) < sc(A), sc(A_2) < sc(A)$ and $L(A) = L(A_1) \cap L(A_2)$. W.l.o.g. we assume that A_1 and A_2 are minimal automata accepting their respective languages and

$F_i = \{q_F \in K_i \mid (\exists w \in L(A))\ (q_i[0], w) \vdash^*_{A_i} (q_F, \varepsilon)\}$ for $i \in \{1, 2\}$. We shall show that $\lambda_1 \mid \lambda$. Let us examine the set of accepting states F_1. We can write

$$F_1 = \{q_F \in K_1 \mid (\exists w \in L(A))\ (q_1[0], w) \vdash^*_{A_1} (q_F, \varepsilon)\} \tag{1}$$

$$= \{q_F \in K_1 \mid (\exists w \in L)\ (q_1[0], w) \vdash^*_{A_1} (q_F, \varepsilon)\} \tag{2}$$

$$= \{q_F \in K_1 \mid (\exists r \in \mathbb{Z}_\lambda)(\exists i \in \mathbb{N})\ a^r \in L;\ (q_1[0], a^{i\lambda + r}) \vdash^*_{A_1} (q_F, \varepsilon)\} \tag{3}$$

$$= \{q_1[k] \mid k \in \mathbb{N};\ k < \mu_1;\ a^k \in L\}\ \cup$$
$$\{q_1[\mu_1 + ((i\lambda + r - \mu_1) \bmod \lambda_1)] \mid r \in \mathbb{Z}_\lambda;\ a^r \in L;\ i \in \mathbb{N};\ i \geq \tfrac{\mu_1 - r}{\lambda}\} \tag{4}$$

$$= \{q_1[k] \mid k \in \mathbb{N};\ k < \mu_1;\ a^k \in L\}\ \cup$$
$$\{q_1[\mu_1 + ((i \cdot \gcd(\lambda_1, \lambda) + r - \mu_1) \bmod \lambda_1)] \mid r \in \mathbb{Z}_\lambda;\ a^r \in L;\ i \in \mathbb{N}\} \tag{5}$$

The equalities (1) and (2) are obvious. Since L is properly λ-cyclic, we have the equality (3). For the equality (4) we use the definition of the transition function δ_1 and we separate accepting states to those in the tail of A_1 and those in its cycle. Finally we use Lemma 2 to write the equality (5) which implies the following claim which we shall refer to as *Helpful Claim*:

$$(\forall r_1 \in \mathbb{Z}_{\lambda_1})\ q_1[\mu_1 + r_1] \in F_1 \Leftrightarrow (\forall i \in \mathbb{N})\ q_1[\mu_1 + ((i \cdot \gcd(\lambda_1, \lambda) + r_1) \bmod \lambda_1)] \in F_1$$

Now consider that $\lambda_1 \nmid \lambda$. Since $\lambda_1 < \lambda$, this means that $\gcd(\lambda_1, \lambda) < \lambda_1$. However, according to Theorem 2, *Helpful Claim* and $\gcd(\lambda_1, \lambda) \mid \lambda_1$ we obtain that the automton A_1 is not the minimal UDFA accepting $L(A_1)$ which is a contradiction. In fact, in that case we can replace the cycle of A_1 by a cycle of length $\gcd(\lambda_1, \lambda)$ and thus obtain a smaller UDFA accepting $L(A_1)$. Therefore it must hold $\lambda_1 \mid \lambda$. The proof of the fact that $\lambda_2 \mid \lambda$ is analogous. $\square$

We extend Lemma 7 using the observation that if a properly λ-cyclic language is deterministically decomposable, then there exists a decomposition in which both automata have no tails.

Lemma 8. *Let L be a unary properly λ-cyclic language such that $L \in \mathcal{D}_{det}$ and A be the minimal UDFA such that $L(A) = L$. Then there exists a decomposition of A to UDFAs A_1 and A_2 of sizes $(\lambda_1, 0)$ and $(\lambda_2, 0)$ such that $\mathrm{lcm}(\lambda_1, \lambda_2) = \lambda$.*

Proof. Let L be a unary properly λ-cyclic language such that $L \in \mathcal{D}_{det}$ and let A be the minimal UDFA such that $L(A) = L$. Obviously A has size $(\lambda, 0)$. Following Lemma 7 we obtain that there exists a decomposition of the automaton A to two UDFAs of sizes (λ_1, μ_1) and (λ_2, μ_2) such that $\lambda_1 \mid \lambda$ and $\lambda_2 \mid \lambda$. Let us denote these automata D_1 and D_2. If $\mu_1 = \mu_2 = 0$, then there is nothing to prove. So that for the rest of the proof we consider that $\mu_1 > 0$ or $\mu_2 > 0$. Let $k \in \mathbb{N}$ be an arbitrary number such that $k\lambda \geq \max(\mu_1, \mu_2)$. According to Rollout Lemma (Lemma 4) there exist UDFAs $C_1 = (K_1^C, \{a\}, \delta_1^C, q_1^C[0], F_1^C)$ and $C_2 = (K_2^C, \{a\}, \delta_2^C, q_2^C[0], F_2^C)$ of sizes $(\lambda_1, k\lambda)$ and $(\lambda_2, k\lambda)$, respectively, such that $L(C_1) = L(D_1)$ and $L(C_2) = L(D_2)$. Therefore it holds $L(C_1) \cap L(C_2) = L(A)$.

Now we, informally said, synchronize tails of C_1 and C_2 with their respective cycles and we prove that, after this, we still accept $L(A)$ in the intersection. We define UDFAs $B_1 = (K_1^B, \{a\}, \delta_1^B, q_1^B[0], F_1^B)$ and $B_2 = (K_2^B, \{a\}, \delta_2^B, q_2^B[0], F_2^B)$ of sizes $(\lambda_1, k\lambda)$ and $(\lambda_2, k\lambda)$ where sets of accepting states F_1^B and F_2^B are defined by

$$F_1^B = \{q_1^B[k\lambda + i] \mid i \in \mathbb{Z}_{\lambda_1}; \ q_1^C[k\lambda + i] \in F_1^C\} \cup$$
$$\{q_1^B[j] \mid j \in \mathbb{N}; \ j < k\lambda; \ q_1^C[k\lambda + (j \bmod \lambda_1)] \in F_1^C\}$$
$$F_2^B = \{q_2^B[k\lambda + i] \mid i \in \mathbb{Z}_{\lambda_1}; \ q_2^C[k\lambda + i] \in F_2^C\} \cup$$
$$\{q_2^B[j] \mid j \in \mathbb{N}; \ j < k\lambda; \ q_2^C[k\lambda + (j \bmod \lambda_2)] \in F_2^C\}.$$

We note that automata C_1 and B_1 are UDFAs of the same size, which means that they have the same transition function up to the names of states. The same holds for C_2 and B_2. Automata C_1 and B_1 resp. C_2 and B_2 may only differ in their respective sets of accepting states. We prove that $L(B_1) \cap L(B_2) = L(A)$.

- $L(B_1) \cap L(B_2) \subseteq L(A)$: Let $w \in L(B_1) \cap L(B_2)$. We divide the proof in two cases according to the length of w.
  - $|w| \geq k\lambda$: In this case w is accepted in the cycles of B_1 and B_2. Since C_1 and B_1 share the same cycle and the same holds for C_2 and B_2, and since $L(A) = L(C_1) \cap L(C_2)$, we have $w \in L(C_1) \cap L(C_2) = L(A)$.
  - $|w| < k\lambda$: Here we can write $w = a^t$ for some $t < k\lambda$. Examine accepting computations on a^t in both B_1 and B_2, which are $(q_1^B[0], a^t) \vdash_{B_1}^* (q_1^B[t], \varepsilon)$ and $(q_2^B[0], a^t) \vdash_{B_2}^* (q_2^B[t], \varepsilon)$ where $q_1^B[t] \in F_1^B$ and $q_2^B[t] \in F_2^B$. Following definitions of F_1^B and F_2^B we have $q_1^C[k\lambda + (t \bmod \lambda_1)] \in F_1^C$ and $q_2^C[k\lambda + (t \bmod \lambda_2)] \in F_2^C$. Therefore $a^{k\lambda + t} \in L(C_1) \cap L(C_2) = L(A)$. $L(A)$ is λ-cyclic, so that $a^t \in L(A)$.
- $L(A) \subseteq L(B_1) \cap L(B_2)$: Let $w \in L(A) = L(C_1) \cap L(C_2)$ and again divide the proof in two cases according to the length of w.
  - $|w| \geq k\lambda$: As above $w \in L(B_1) \cap L(B_2)$ follows from the fact that B_1 has the same cycle as C_1 and this is true also for B_2 and C_2.
  - $|w| < k\lambda$: Here we can write $w = a^t$ for some $t < k\lambda$. Since $a^t \in L(A)$ and the language $L(A)$ is λ-cyclic, we have that $a^{k\lambda + t} \in L(A) = L(C_1) \cap L(C_2)$. The word $a^{k\lambda + t}$ is accepted in the automata C_1 and C_2 by the states $q_1^C[k\lambda + (t \bmod \lambda_1)] \in F_1^C$ and $q_2^C[k\lambda + (t \bmod \lambda_2)] \in F_2^C$. Following the definitions of F_1^B and F_2^B we obtain $q_1^B[t] \in F_1^B$ and $q_2^B[t] \in F_2^B$. It means that the computations $(q_1^B[0], a^t) \vdash_{B_1}^* (q_1^B[t], \varepsilon)$ and $(q_2^B[0], a^t) \vdash_{B_2}^* (q_2^B[t], \varepsilon)$ are both accepting and we conclude $a^t \in L(B_1) \cap L(B_2)$ as desired.

Considering the definitions of F_1^B and F_2^B with $\lambda_1 \mid \lambda$ and $\lambda_2 \mid \lambda$ one can prove the following claims:

(C1) For all $i, j \in \mathbb{N}$ such that $0 \leq i, j < \lambda_1 + k\lambda$ and $i - j \equiv 0 \pmod{\lambda_1}$ it holds

$$q_1^B[i] \in F_1^B \iff q_1^B[j] \in F_1^B$$

(C2) For all $i, j \in \mathbb{N}$ such that $0 \leq i, j < \lambda_2 + k\lambda$ and $i - j \equiv 0 \pmod{\lambda_2}$ it holds

$$q_2^B[i] \in F_2^B \iff q_2^B[j] \in F_2^B$$

We define the UDFA $A_1 = (K_1, \{a\}, \delta_1, q_1[0], F_1)$ and the UDFA $A_2 = (K_2, \{a\}, \delta_2, q_2[0], F_2)$ of sizes $(\lambda_1, 0)$ and $(\lambda_2, 0)$, where $F_1 = \{q_F \in K_1 \mid (\exists w \in L(B_1))\ (q_1[0], w) \vdash^*_{A_1} (q_F, \varepsilon)\}$ and $F_2 = \{q_F \in K_2 \mid (\exists w \in L(B_2))\ (q_2[0], w) \vdash^*_{A_2} (q_F, \varepsilon)\}$.

The claims (C1) and (C2) show that the automata B_1 and B_2 fulfill the assumptions of Roll-in Lemma (Lemma 5) for $\mu' = 0$. So that from Roll-in Lemma it follows that $L(A_1) = L(B_1)$ and $L(A_2) = L(B_2)$. Therefore $L(A) = L(B_1) \cap L(B_2) = L(A_1) \cap L(A_2)$.

Since both λ_1 and λ_2 are divisors of λ, we have $\mathrm{lcm}(\lambda_1, \lambda_2) \leq \lambda$. According to Theorem 3 the language $L(A_1) \cap L(A_2) = L(A)$ can be accepted by a UDFA of size $(\mathrm{lcm}(\lambda_1, \lambda_2), 0)$. Therefore if $\mathrm{lcm}(\lambda_1, \lambda_2) < \lambda$, then the language $L(A)$ cannot be properly λ-cyclic, which is a contradiction. Thus it holds that $\mathrm{lcm}(\lambda_1, \lambda_2) = \lambda$. Moreover $\lambda_1 = sc(A_1) < \lambda = sc(A)$, $\lambda_2 = sc(A_2) < \lambda = sc(A)$. We have thus found the decomposition with desired properties. $\qquad\square$

Now we are ready to prove the opposite implication of Lemma 6.

Lemma 9. *Let L be a properly λ-cyclic language for some $\lambda \in \mathbb{N}$. If $L \in \mathcal{D}_{det}$ then there exist $\lambda_1, \lambda_2 \in \mathbb{N}$ such that $\lambda_1, \lambda_2 < \lambda$, $\mathrm{lcm}(\lambda_1, \lambda_2) = \lambda$ and the bipartite graph $G_{L,\lambda_1,\lambda_2}$ induced by L, λ_1 and λ_2 decomposes L.*

Proof. Let $L \in \mathcal{D}_{det}$ be a properly λ-cyclic language for some $\lambda \in \mathbb{N}$ and let A be the minimal UDFA such that $L(A) = L$. Following Lemma 8 there exists a decomposition of A to UDFAs $A_1 = (K_1, \{a\}, \delta_1, q_1[0], F_1)$ and $A_2 = (K_2, \{a\}, \delta_2, q_2[0], F_2)$ of sizes $(\lambda_1, 0)$ and $(\lambda_2, 0)$, such that $\mathrm{lcm}(\lambda_1, \lambda_2) = \lambda$. Let $G_{L,\lambda_1,\lambda_2} = (\mathbb{Z}_{\lambda_1}, \mathbb{Z}_{\lambda_2}, E)$ be the bipartite graph induced by L, λ_1 and λ_2 (Definition 5). We recall that the set of edges E is defined by:

$$E = \{(r_1, r_2) \mid r_1 \in \mathbb{Z}_{\lambda_1};\ r_2 \in \mathbb{Z}_{\lambda_2};$$
$$(\exists m \in \mathbb{N})\ m \equiv r_1 \ (\mathrm{mod}\ \lambda_1) \ \wedge \ m \equiv r_2 \ (\mathrm{mod}\ \lambda_2) \ \wedge \ a^m \in L\}.$$

We prove that $G_{L,\lambda_1,\lambda_2}$ decomposes L according to Definition 5. Let us assume for the contrary that $G_{L,\lambda_1,\lambda_2}$ does not decompose L. It means, following Definition 5, that there exist $s_1 \in \mathbb{Z}_{\lambda_1}$ and $s_2 \in \mathbb{Z}_{\lambda_2}$ such that both of them are not isolated vertices and moreover it holds that:

(1) $(s_1, s_2) \notin E$
(2) $(\exists m \in \mathbb{N})\ m \equiv s_1 \ (\mathrm{mod}\ \lambda_1) \wedge m \equiv s_2 \ (\mathrm{mod}\ \lambda_2)$

Since s_1 is not isolated, there exists $t_2 \in \mathbb{Z}_{\lambda_2}$ such that $(s_1, t_2) \in E$. Therefore there exists $m_1 \in \mathbb{N}$ such that $m_1 \equiv s_1 \ (\mathrm{mod}\ \lambda_1)$, $m_1 \equiv t_2 \ (\mathrm{mod}\ \lambda_2)$ and $a^{m_1} \in L$. The computation of A_1 on a^{m_1} is $(q_1[0], a^{m_1}) \vdash^*_{A_1} (q_1[s_1], \varepsilon)$. Following that $L = L(A) = L(A_1) \cap L(A_2)$ and $a^{m_1} \in L$ we get $q_1[s_1] \in F_1$. Analogously it also holds that $q_2[s_2] \in F_2$.

Now consider the computations of A_1 and A_2 on the word a^m. Following (2) they are $(q_1[0], a^m) \vdash^*_{A_1} (q_1[s_1], \varepsilon)$ and $(q_2[0], a^m) \vdash^*_{A_2} (q_2[s_2], \varepsilon)$. Since $q_1[s_1] \in F_1$ and $q_2[s_2] \in F_2$, we have $a^m \in L(A_1) \cap L(A_2) = L(A) = L$. From (2) and $a^m \in L$ we obtain that $(s_1, s_2) \in E$ which is a contradiction to (1). Therefore

$G_{L,\lambda_1,\lambda_2}$ decomposes L. Moreover we have that $\mathrm{lcm}(\lambda_1, \lambda_2) = \lambda$ and since A_1 and A_2 form the decomposition of A, we also have that $\lambda_1 = sc(A_1) < sc(A) = \lambda$ and $\lambda_2 = sc(A_2) < sc(A) = \lambda$ as desired. $\square$

Following Lemma 6 and Lemma 9 we obtain the main results of our paper.

Theorem 4 (Characterization of Properly λ-cyclic Languages upon Deterministic Decomposability). *Let L be a properly λ-cyclic language for some $\lambda \in \mathbb{N}$. $L \in \mathcal{D}_{det}$ if and only if there exist $\lambda_1, \lambda_2 \in \mathbb{N}$ such that $\lambda_1, \lambda_2 < \lambda$, $\mathrm{lcm}(\lambda_1, \lambda_2) = \lambda$ and the bipartite graph $G_{L,\lambda_1,\lambda_2}$ induced by L, λ_1 and λ_2 decomposes L.*

References

1. Gaži, P., Rovan, B.: Assisted problem solving and decompositions of finite automata. In: Geffert, V., Karhumäki, J., Bertoni, A., Preneel, B., Návrat, P., Bieliková, M. (eds.) SOFSEM 2008. LNCS, vol. 4910, pp. 292–303. Springer, Heidelberg (2008). https://doi.org/10.1007/978-3-540-77566-9_25
2. Jecker, I., Kupferman, O., Mazzocchi, N.: Unary prime languages. In: Esparza, J., Král', D. (eds.) 45th International Symposium on Mathematical Foundations of Computer Science, MFCS 2020, LIPIcs, Prague, Czech Republic, 24–28 August 2020, vol. 170, pp. 51:1–51:12. Schloss Dagstuhl - Leibniz-Zentrum für Informatik (2020). https://doi.org/10.4230/LIPIcs.MFCS.2020.51
3. Kupferman, O., Mosheiff, J.: Prime languages. Inf. Comput. **240**, 90–107 (2015). https://doi.org/10.1016/j.ic.2014.09.010
4. Labath, P., Rovan, B.: Simplifying DPDA using supplementary information. In: Dediu, A.-H., Inenaga, S., Martín-Vide, C. (eds.) LATA 2011. LNCS, vol. 6638, pp. 342–353. Springer, Heidelberg (2011). https://doi.org/10.1007/978-3-642-21254-3_27
5. Nicaud, C.: Average state complexity of operations on unary automata. In: Kutyłowski, M., Pacholski, L., Wierzbicki, T. (eds.) MFCS 1999. LNCS, vol. 1672, pp. 231–240. Springer, Heidelberg (1999). https://doi.org/10.1007/3-540-48340-3_21
6. Pighizzini, G., Shallit, J.: Unary language operations, state complexity and Jacobsthal's function. Int. J. Found. Comput. Sci. **13**(1), 145–159 (2002). https://doi.org/10.1142/S012905410200100X
7. Rovan, B., Sádovský, Š.: On usefulness of information: framework and NFA case. In: Adventures Between Lower Bounds and Higher Altitudes - Essays Dedicated to Juraj Hromkovič on the Occasion of His 60th Birthday pp. 85–99 (2018). https://doi.org/10.1007/978-3-319-98355-4_6
8. Rovan, B., Zeman, M.: Modeling time criticality of information. Inf. Process. Lett. **114**(3), 147–151 (2014). https://doi.org/10.1016/j.ipl.2013.10.008

Learning

Learnability and Positive Equivalence Relations

David Belanger[1], Ziyuan Gao[2(✉)], Sanjay Jain[3], Wei Li[2], and Frank Stephan[2,3]

[1] Department of Mathematics, Ghent University,
Krijgslaan 281, 9000 Ghent, Belgium
david.belanger@ugent.be
[2] Department of Mathematics, National University of Singapore, 10 Lower Kent Ridge Road, Singapore 119076, Republic of Singapore
{matgaoz,matliw}@nus.edu.sg
[3] School of Computing, National University of Singapore, Singapore 117417, Republic of Singapore
{sanjay,fstephan}@comp.nus.edu.sg

Abstract. Prior work of Gavryushkin, Khoussainov, Jain and Stephan investigated what algebraic structures can be realised in worlds given by a positive (= recursively enumerable) equivalence relation which partitions the natural numbers into infinitely many equivalence classes. The present work investigates the infinite one-one numbered recursively enumerable (r.e.) families realised by such relations and asks how the choice of the equivalence relation impacts the learnability properties of these classes when studying learnability in the limit from positive examples, also known as learning from text. For all choices of such positive equivalence relations, for each of the following entries, there are one-one numbered r.e. families which satisfy it: (a) they are behaviourally correctly learnable but not vacillatorily learnable; (b) they are explanatorily learnable but not confidently learnable; (c) they are not behaviourally correctly learnable. Furthermore, there is a positive equivalence relation which enforces that (d) every vacillatorily learnable one-one numbered family of languages closed under this equivalence relation is already explanatorily learnable and cannot be confidently learnable.

1 Introduction

Consider a learning scenario where all positive examples of a given target concept L belonging to a concept class $\mathcal{L}$ are shown sequentially to a learner M. After processing each example, M makes a conjecture as to the identity of the target concept, based on some fixed representation system of all concepts in $\mathcal{L}$. M is said to successfully identify L if its sequence of conjectures converges to a correct

D. Belanger (as RF), Z. Gao (as RF) and S. Jain (as Co-PI), F. Stephan (as PI) have been supported by the Singapore Ministry of Education Academic Research Fund grant MOE2016-T2-1-019/R146-000-234-112 and MOE2019-T2-2-121/R146-000-304-112. Furthermore, S. Jain is supported in part by NUS grant C252-000-087-001.

A. Leporati et al. (Eds.): LATA 2021, LNCS 12638, pp. 145–156, 2021.
https://doi.org/10.1007/978-3-030-68195-1_12

hypothesis describing L. This learning paradigm, due to Gold [18], is well-studied and has inspired the development of a large number of other learning models in inductive inference[1].

In this work, we study how the interrelations between the elements of a domain X influences the learnability of classes of languages defined over X. The domain of interest throughout this work is $\mathbb{N}$. We will be concerned with *recursively enumerable (r.e.)* equivalence relations defined on $\mathbb{N}$ that induce infinitely many equivalence classes. The main motivation for focussing on such relations comes from the study of r.e. structures. Here, r.e. structures are given by a domain, recursive functions representing basic operators in the structure, and some recursively enumerable predicates, among which there is a recursively enumerable equivalence relation η with infinitely many equivalence classes which plays the role of equality in the given structure. Such structures have been studied for a long time; for example, Novikov [25] constructed a finitely generated group with undecidable word-problem; in other words, there is a group which can be represented using an r.e. but nonrecursive equivalence relation (as equality of the group) but one cannot represent it using a recursive equivalence relation E. On the other hand, for Noetherian rings [24], Baur [5] showed that every r.e. Noetherian ring is a recursive ring, implying that the underlying equality η is always a recursive relation and that its equivalence classes are uniformly recursive. Another example of an r.e. equivalence relation is the relation of *provable equivalence* with respect to any formal system, say Peano Arithmetic (PA), where $x \sim_{\text{PA}} y$ holds iff $\alpha \leftrightarrow \beta$ is provable in PA, x and y being the Gödel numbers of α and β respectively according to some fixed Gödel numbering.

Fokina, Gavryushkin, Jain, Khoussainov, Semukhin, Stephan and Turetsky [13,16,17] focussed in a sequence of papers on the question of which type of structures could be realised by an r.e. equivalence relation on $\mathbb{N}$ with infinitely many equivalence classes and how different relations compare with respect to their ability to realise structures of certain type. Ershov [10,11], and following him Odifreddi [26], call r.e. equivalence relations *positive* equivalence relations. Fokina, Kötzing and San Mauro [14] studied Gold-style learnability of equivalence structures (with no computability restrictions on the learner), and gave a structural characterisation of families of equivalence structures that are learnable in the limit from informant.

Now a structure is realised by a positive equivalence relation η iff there is a bijection between the elements in the domain of the structure and the equivalence classes of η and all relations involved are recursively enumerable and all functions are realised by recursive functions which respect η. In the simplest case of functions from the domain to the domain, they respect η if they map η-equivalent numbers to η-equivalent numbers. This work focusses on the study of learnability within the framework of families realised by positive equivalence relations. In particular the topic of the investigations is to which extent separations between learning criteria known from inductive inference can be witnessed

---

[1] The reader is referred to [19,28] for an introduction to other basic learning notions in inductive inference and to [1–4,6–9,12,14,15,18,21–23] for further reading

by η-closed sets, that is, which of the positive equivalence relations witness a separation of two learning criteria or collapse them. Furthermore one asks, whether certain learning criteria can be void (non-existent) for certain equivalence relations η. The study of learning in a world given by some η is similar to that in complexity relative to an oracle; one wants to know how robust the results from the non-relativised world are and how much they generalise.

A full version of the paper is available at https://arxiv.org/pdf/2012.01466.pdf.

2   Preliminaries

Any unexplained recursion-theoretic notation may be found in [26,29,30]. We use $\mathbb{N} = \{0, 1, 2, \ldots\}$ to denote the set of all natural numbers. The set of all *partial recursive functions* and of all *recursive functions* of one, and two arguments over $\mathbb{N}$ is denoted by $\mathcal{P}$, $\mathcal{P}^2$, $\mathcal{R}$ and $\mathcal{R}^2$ respectively. Any function $\psi \in \mathcal{P}^2$ is called a *numbering of partial-recursive functions*—this numbering may or may not include all partial recursive functions. Moreover, let $\psi \in \mathcal{P}^2$, then we write ψ_e instead of $\lambda x. \psi(e, x)$ and set $\mathcal{P}_\psi = \{\psi_e \mid e \in \mathbb{N}\}$. A numbering $\varphi \in \mathcal{P}^2$ is said to be an *acceptable numbering* or *Gödel numbering* of all partial recursive functions if $\mathcal{P}_\varphi = \mathcal{P}$ and for every numbering $\psi \in \mathcal{P}^2$, there is a $c \in \mathcal{R}$ such that $\psi_e = \varphi_{c(e)}$ for all $e \in \mathbb{N}$ (see [29]). Throughout this paper, $\varphi_0, \varphi_1, \varphi_2, \ldots$ is a fixed acceptable numbering of all partial recursive functions and $W_0, W_1, W_2, \ldots$ is a fixed *numbering of all recursively enumerable sets* (abbr. r.e. sets) of natural numbers, where W_e is the domain of φ_e for all $e \in \mathbb{N}$.

Let $e, x \in \mathbb{N}$; if $\varphi_e(x)$ is defined then we say that $\varphi_e(x)$ *converges*. Otherwise, $\varphi_e(x)$ is said to *diverge*. Furthermore, if the computation of $\varphi_e(x)$ halts within s steps of computation then we write $\varphi_{e,s}(x) \downarrow = \varphi_e(x)$; otherwise $\varphi_{e,s}(x)$ diverges. For all $e, s \in \mathbb{N}$ the set $W_{e,s}$ is defined as the domain of $\varphi_{e,s}$.

Given any set S, S^* denotes the set of all finite sequences of elements from S. The symbol K denotes the *diagonal halting problem*, i.e., $K = \{e : e \in \mathbb{N}, \; \varphi_e(e) \text{ converges}\}$. For $\sigma \in (\mathbb{N} \cup \{\#\})^*$ and $n \in \mathbb{N}$ we write $\sigma(n)$ to denote the element in the nth position of σ. For any finite sequence σ we use $|\sigma|$ to denote the length of σ. Further, whenever $n \leqslant |\sigma|$, $\sigma[n]$ denotes the sequence $\sigma(0), \sigma(1), \ldots, \sigma(n-1)$. The concatenation of two sequences σ and τ is denoted by $\sigma \circ \tau$; for convenience, and whenever there is no possibility of confusion, this is occasionally denoted by $\sigma\tau$.

A class $\mathcal{L}$ is said to be *uniformly r.e.* (or just *r.e.*) if there is an r.e. set $S \subseteq \mathbb{N}$ such that $\mathcal{L} = \{W_i : i \in S\}$. A class is said to be *one-one r.e.*, if the r.e. set S as above additionally satisfies the condition that for $i, j \in S$, $W_i = W_j$ iff $i = j$. An r.e. class $\mathcal{L} = \{B_0, B_1, \ldots\}$ is said to be *uniformly recursive* or an *indexed family* if there exists a recursive function $f \in \mathcal{R}^2$ such that for all $i, x \in \mathbb{N}$, if $x \in B_i$ then $f(i, x) = 1$ else $f(i, x) = 0$.

3   Learnability

Background on inductive inference may be found in [19]. Let $\mathcal{L}$ be a class of r.e. languages. Throughout this paper, the mode of data presentation is that of a *text*. A text is any infinite sequence of natural numbers and the $\#$ symbol, where the symbol $\#$ indicates a pause in the data presentation. More formally, a *text* T_L for a language $L \in \mathcal{L}$ is any total mapping $T_L : \mathbb{N} \to \mathbb{N} \cup \{\#\}$ such that $L = \text{range}(T_L) - \{\#\}$. We use content$(T)$ to denote the set range$(T) - \{\#\}$, i.e., the content of a text T contains only the natural numbers appearing in T. Furthermore, for every $n \in \mathbb{N}$ we use $T[n]$ to denote the finite sequence $T(0), \ldots, T(n-1)$, i.e., the *initial segment* of length n of T. Analogously, for a finite sequence $\sigma \in (\mathbb{N} \cup \{\#\})^*$ we use content(σ) to denote the set of all numbers in the range of σ.

Description 1. Further basic ingredients of the notions considered in this paper are as follows.

(1) For each positive equivalence relation η, one can define an infinite sequence $a_0, a_1, \ldots$ of least representatives of the equivalence classes where each a_n is the ascending limit of approximations $a_{n,t}$ where $a_{n,t}$ is the least natural number which is not η_t-equivalent to any $a_{m,t}$ with $m < n$ (where η_t denotes the tth approximation to η). Alternatively, one can obtain η from a construction of such a sequence where the $a_{n,t}$ approximate the a_n from below and whenever an $a_{n,t+1} \neq a_{n,t}$ then $a_{n,t+1} = a_{m,t}$ for some $m > n$ and whenever $a_{n,t}$ is not in the list at $t+1$ then it is put into the equivalence class of some $a_{m,t}$ with $m < n$. Some algorithms to construct the equivalence relation η explain on how to update these approximations to $a_0, a_1, \ldots$ and one should note that (the construction implies) the limit satisfies $a_0 < a_1 < \ldots$ and that for each n there are only finitely many t with $a_{n,t} < a_{n,t+1}$.

(2) The classes whose learnability are considered are given by a uniformly r.e. one-one numbering of sets $B_0, B_1, \ldots$ where each set B_k is a union of η-equivalence classes; however, the indices k of B_k are usual natural numbers and not equivalence-classes of η. Such a family is called an η-*family* below and note that η-families are always infinite.

(3) Infinite indexed families as considered by Angluin [1] are too restrictive, as they might not exist for some η; however, every infinite indexed family has a one-one numbering and thus using the notion of infinite uniformly r.e. one-one numberings is the adequate choice for the present work.

(4) The learner sees an infinite sequence $x_0, x_1, \ldots$ of members of one set B_k (such sequences are called texts and can have pauses represented by a special pause symbol $\#$) and the learner has to find in the limit an r.e. index for B_k, which may not be equal to k.

(5) The hypotheses issued by the learners are always indices from a fixed acceptable numbering of all r.e. sets; without loss of generality one can assume that they incorporate the closure under the equivalence relation η and are thus indices for r.e. unions of equivalence classes of η.

(6) The present work focusses on the following learning criteria [3,7,8,12,18,19, 27,28]: *Explanatory learning*, where the hypotheses of the learner converge on every text for a set B_k to a single index of B_k; *Confident learning*, which is explanatory learning with the additional requirement that the learner also on texts not belonging to any language in the class has to converge

to some index; *Behaviourally correct learning*, which is more general than explanatory learning and where the learner is only required to output on any text for B_k almost always an index for B_k but these indices can all be different; *Vacillatory learning*, where a learner is vacillatory iff it is a behaviourally correct learner for the class, with the additional constraint that on every text for a language B_k in the class, the set of all hypotheses issued in response to this text is finite (thus, some of these hypotheses are output infinitely often).

We now provide formal definitions of these criteria as well as the criterion of *finite* learning (sometimes known as *one-shot learning* in the literature; see [18,31]), which is a more restrictive version of explanatory learning. In the following definitions, a learner M is a recursive function mapping $(\mathbb{N} \cup \{\#\})^*$ into $\mathbb{N} \cup \{?\}$; the ? symbol permits M to abstain from conjecturing at any stage. If M is presented with a text T for any η-closed set L, it is enough to assume that content(T) contains *at least one* element of each η-equivalence class contained in L; since η is r.e., M on T could simulate a complete text for L by enumerating at each stage s the sth approximation of the current input's η-closure.

Definition 2 (Angluin [1], Bārzdiņš [3], Case and Smith [9], Feldman [12], Gold [18], Osherson, Stob and Weinstein [28], Trakhtenbrot and Bārzdiņš [31]). Let $\mathcal{L}$ be any class of r.e. languages.

(1) M *explanatorily (Ex) learns* $\mathcal{L}$ if, for every L in $\mathcal{L}$ and each text T_L for L, there is a number n for which $L = W_{M(T_L[n])}$ and, for every $j \geqslant n$, $M(T_L[j]) = M(T_L[n])$.

(2) M *behaviourally correctly (BC) learns* $\mathcal{L}$ if, for every L in $\mathcal{L}$ and each text T_L for L, there is a number n for which $L = W_{M(T_L[j])}$ whenever $j \geqslant n$.

(3) M *finitely (Fin) learns* $\mathcal{L}$ if, for every L in $\mathcal{L}$ and each text T_L for L, there is a number n for which $L = W_{M(T_L[n])}$ and for every $m < n$, $M(T_L[m]) = ?$ and for every $j \geqslant n$, $M(T_L[j]) = M(T_L[n])$.

(4) M *confidently (Conf) learns* $\mathcal{L}$ if M Ex learns $\mathcal{L}$ and M converges on every text for any language, that is, for every $L \subseteq \mathbb{N}$ and text T_L for L, there is a number n such that for every $j \geqslant n$, $M(T_L[j]) = M(T_L[n])$.

(5) M *vacillatorily (Vac) learns* $\mathcal{L}$ if M BC learns $\mathcal{L}$ and for every L in $\mathcal{L}$ and each text T_L for L, $\{M(T_L[n]) : n \geqslant 1\}$ is finite.

Throughout this work, we only consider positive equivalence relations that induce *infinitely* many equivalence classes. For any positive equivalence relation η and $x \in \mathbb{N}$, let $[x]$ be $\{y : y \, \eta \, x\}$. Furthermore, for any $D \subseteq \mathbb{N}$, $[D]$ denotes $\bigcup_{x \in D}[x]$. For any finite $\{i_0, \ldots, i_n\} \subseteq \mathbb{N}$, the set $[\{a_{i_0}, a_{i_1}, \ldots, a_{i_n}\}]$ will simply be denoted by $[a_{i_0}, a_{i_1}, \ldots, a_{i_n}]$. An η-*family* $\mathcal{L}$ is a uniformly r.e. one-one infinite family, each of whose members is a union of η-equivalence classes. Note that uniformly recursive infinite families might not exist for some η and therefore an η-family is the nearest notion to a uniformly recursive family which exists for each positive equivalence relation η. A set is η-*infinite* (resp. η-finite) if it is equal to a union of infinitely (resp. finitely) many η-equivalence classes; note that an η-infinite set may not necessarily be recursively enumerable. A set is η-closed if it is either η-finite or η-infinite. In this paper, all families are assumed to consist of only η-closed sets (for some given η). For brevity's sake, we do not use any notation to indicate the dependence of a_n on η; the choice of η will always be clear from the context. A family $\mathcal{A}$ of sets is called a *superfamily* of another family $\mathcal{B}$ of sets iff $\mathcal{A} \supseteq \mathcal{B}$.

A useful notion that captures the idea of the learner converging on a given text is that of a *locking sequence*, or more generally that of a *stabilising sequence*. A sequence $\sigma \in (\mathbb{N} \cup \{\#\})^*$ is called a *stabilising sequence* [15] for a learner M on some language L if content$(\sigma) \subseteq L$ and for all $\tau \in (L \cup \{\#\})^*$, $M(\sigma) = M(\sigma \circ \tau)$. A sequence $\sigma \in (\mathbb{N} \cup \{\#\})^*$ is called a *locking sequence* [6] for a learner M on some language L if σ is a stabilising sequence for M on L and $W_{M(\sigma)} = L$. The following proposition due to Blum and Blum [6] will be occasionally useful.

Proposition 3 (Blum and Blum [6]**).** *If a learner M explanatorily learns some language L, then there exists a locking sequence for M on L. Furthermore, all stabilising sequences for M on L are also locking sequences for M on L.*

The following theorem due to Kummer [20] will be useful for showing that a given family of r.e. sets has a one-one numbering.

Theorem 4 (Kummer [20]**).** *Suppose $L_0, L_1, L_2, \ldots$ and $H_0, H_1, H_2, \ldots$ are two numberings such that (1) for all $i, j \in \mathbb{N}$, $L_i \neq H_j$; (2) $H_0, H_1, H_2, \ldots$ is a one-one numbering; (3) for all $i \in \mathbb{N}$ and all finite $D \subseteq L_i$, there are infinitely many j such that $D \subseteq H_j$. Then $\{L_i : i \in \mathbb{N}\} \cup \{H_j : j \in \mathbb{N}\}$ has a one-one numbering.*

4   Results for All Positive Equivalence Relations: Fin, Conf, Ex, Vac and BC Learning

In the present section, we investigate the relationship between the main learning criteria – namely, finite, confident, explanatory, vacillatory and behaviourally correct learning – with respect to families closed under any given positive equivalence relation. The first part of this section will study, for any general positive equivalence relation η, the learnability of a particular η-family known as the *ascending family for η*. As will be seen later, the ascending family provides a useful basis for constructing η-families that witness the separation of various learnability notions.

Definition 5. For all $n \in \mathbb{N}$, A_n denotes the set $[a_0, a_1, ..., a_{n-1}]$. The family $\{A_n : n \in \mathbb{N}\}$ will be denoted by $\mathcal{A}_\eta$, and is called the *ascending family for η*.

Note that each member of $\mathcal{A}_\eta$ is η-finite; furthermore, $\mathcal{A}_\eta$ is an η-family because η induces infinitely many equivalence classes and for all n, a_n can be approximated from below (c.f. Description 1, item (1)). For brevity's sake, we do not use any notation to indicate the dependence of A_n on η; the choice of η will always be clear from the context.

In the second part of this section, we study the question of whether the learning hierarchy

$$\text{Fin} \subset \text{Conf} \subset \text{Ex} \subset \text{Vac} \subset \text{BC}.$$

is strict for the class of η-families (for any given positive equivalence relation η). It turns out that while the two chains of inclusions $\text{Fin} \subset \text{Conf} \subset \text{Ex}$ and $\text{Vac} \subset \text{BC}$ hold for *all* positive equivalence relations, there is a positive equivalence relation ϑ for which every vacillatorily learnable ϑ-family is also explanatorily learnable. The construction of ϑ will be given in the next section. We begin with a few basic examples of η-families to illustrate some of the notions introduced so far.

Example 6 (Ershov, [10]). If A is a recursive and coinfinite set, then $x\,\eta_A\,y \Leftrightarrow (x = y \vee (x \in A \wedge\ y \in A))$ is a positive equivalence relation. $\mathcal{F} := \{A\} \cup \{\{x\} : x \notin A\}$ is an η_A-family since (1) every equivalence class of η_A is either A or a singleton $\{x\}$ with $x \notin A$, which implies that $\mathcal{F}$ is infinite and each member of $\mathcal{F}$ is η_A-closed, and (2) there is a uniformly recursive one-one numbering $\{F_i\}_{i\in\mathbb{N}}$ of $\mathcal{F}$; for example, one could set $F_0 = A$ and $F_{i+1} = \{x_i\}$ for all i, where $x_1, x_2, x_3, \ldots$ is a one-one recursive enumeration of $\mathbb{N} - A$. $\mathcal{F}$ is also finitely learnable via a learner that outputs ? until it sees the first number x in the input; if $x \in A$ then A is conjectured, and if $x \notin A$ then $\{x\}$ is conjectured.

Example 7 (Ershov, [10]). If R is an r.e. set and $D_0, D_1, D_2, \ldots$ is a one-one numbering of all finite sets, then $x\,\eta_R\,y \Leftrightarrow D_x \triangle D_y \subseteq R$ is a positive equivalence relation ($\triangle$ denotes the symmetric difference). If $S \cap R = \emptyset$, then $L_S := \{x : D_x \cap S \neq \emptyset\}$ is η_R-closed. Suppose $\mathbb{N} - R$ contains an infinite r.e. set C. Let $\mathcal{F}$ consist of all sets $L_{C'}$ such that $C' = C - F$ for some finite set F. Then $\mathcal{F}$ is an η_R-family that is not behaviourally correctly learnable.

The next theorem shows that for any positive equivalence relation η, the ascending family witnesses that explanatory learning is strictly more powerful than confident learning.

Theorem 8. *For every positive equivalence relation η, the ascending family $\mathcal{A}_\eta$ is explanatorily learnable but not confidently learnable. One can add the set $\mathbb{N}$ to $\mathcal{A}_\eta$ and obtain an η-family which is not behaviourally correctly learnable.*

The following proposition provides a method for establishing that a given uniformly r.e. family is an η-family.

Proposition 9. *Every uniformly r.e. superfamily of $\mathcal{A}_\eta$ that consists of η-closed sets is an η-family; in particular, the families of all η-finite sets and all η-closed r.e. sets are η-families.*

A minor modification of the proof of Proposition 9 reveals a slightly more general result: for any positive equivalence relation η and any strictly increasing recursive enumeration $e_0, e_1, e_2, \ldots$, every uniformly r.e. superfamily of $\{A_{e_i} : i \in \mathbb{N}\}$ is an η-family. This variant of Proposition 9 will be occasionally useful for showing that a given uniformly r.e. class is an η-family.

Proposition 10. *Let f be any strictly increasing recursive function. Then, for any given positive equivalence relation η, every uniformly r.e. superfamily of $\{A_{f(i)} : i \in \mathbb{N}\}$ consisting of η-closed sets is an η-family.*

The next result shows that for any given positive equivalence relation η, behaviourally correct learning is more powerful than explanatory learning with respect to the class of η-families.

Theorem 11. *For every positive equivalence relation η, there is an η-family which is behaviourally correctly learnable but not explanatorily learnable.*

Vacillatory learning, according to which a learner is allowed to switch between any finite number of correct indices in the limit, is known to be strictly weaker than behaviourally correct learning for general families of r.e. sets [7]. The next main result – Theorem 13 – asserts that for any given positive equivalence relation η, this relation between the two criteria holds even for certain η-families. We begin with the following proposition, from which the separation result may be deduced.

Proposition 12. *If the class of η-finite sets is vacillatorily learnable then one can relative to the halting problem K compute a sequence $e_0, e_1, \ldots$ of characteristic indices of η-finite and η-closed sets $E_0, E_1, \ldots$ which form a partition of $\mathbb{N}$.*

Theorem 13. *For every positive equivalence relation η, there is an η-family which is behaviourally correctly learnable but not vacillatorily learnable.*

Moving down the learning hierarchy given at the start of the present section, the following theorem shows that for any positive equivalence relation η, finite learning can be more restrictive than confident learning with respect to η-families.

Theorem 14. *Let η be any given positive equivalence relation such that there is at least one finitely learnable η-family. Then there is an η-family that is confidently learnable but not finitely learnable.*

As Gold [18] observed, the class consisting of $\mathbb{N}$ and all finite sets is not learnable in any sense considered in the present paper.[2] On the other hand, the class comprising only $\mathbb{N}$ and the class of all finite sets are both explanatorily learnable. Thus the union of two explanatorily learnable classes of r.e. languages may not even be behaviourally correctly learnable. Blum and Blum [6] noted that the family of explanatorily (resp. behaviourally correctly) learnable classes of recursive functions is also not closed under union. In the rest of this section, we investigate the question of whether the non-union property of explanatory (resp. vacillatory, behaviourally correct) learning holds for the class of η-families,

---

[2] However, there *are* many natural families of languages that are learnable in the limit, such as the class of non-erasing pattern languages (see [1, Example 1]).

where η is any given positive equivalence relation. For any learning criterion I and any positive equivalence relation η, say that I is *closed under union with respect to* η iff for any η-families $\mathcal{L}$ and $\mathcal{H}$ such that $\mathcal{L} \cup \mathcal{H}$ is an η-family, if $\mathcal{L}$ and $\mathcal{H}$ are I-learnable, then $\mathcal{L} \cup \mathcal{H}$ is I-learnable. Somewhat surprisingly, while explanatory and vacillatory learnability are not closed under union with respect to any η, the answer for behaviourally correct learning depends on whether or not there are at least two η-infinite r.e. sets.

Proposition 15. *Let η be any given positive equivalence relation. If $\mathbb{N}$ is the only η-infinite r.e. set, then $\mathbb{N}$ is not contained in any behaviourally correctly learnable η-family.*

Theorem 16. *Let η be any given positive equivalence relation. Then the following hold.*

(a) There are disjoint, explanatorily (resp. vacillatorily) learnable η-families $\mathcal{L}_1$ and $\mathcal{L}_2$ for which $\mathcal{L}_1 \cup \mathcal{L}_2$ is an η-family that is not explanatorily (resp. vacillatorily) learnable.

(b) Behaviourally correct learning is closed under union with respect to η iff $\mathbb{N}$ is the only η-infinite r.e. set.

We next establish the non-union theorem for finite learning of η-families, where η is any positive equivalence relation such that at least one finitely learnable η-family exists. It may be worth noting that, in contrast to explanatory learnability, there is a positive equivalence relation ϑ for which no ϑ-family is finitely (or even confidently) learnable, as will be seen in the subsequent section.

Theorem 17. *Let η be any given positive equivalence relation such that at least one η-family is finitely learnable. Then there are finitely learnable η-families $\mathcal{L}_1$ and $\mathcal{L}_2$ for which $\mathcal{L}_1 \cup \mathcal{L}_2$ is an η-family that is not finitely learnable.*

5   Learnability of Families Closed Under Special Positive Equivalence Relations

So the general results were that for every positive equivalence relation η, for each of the following conditions, there are η-families which satisfy it: (a) the family is explanatory learnable but not confidently learnable; (b) the family is behaviourally correctly learnable but not vacillatorily learnable; (c) the family is not behaviourally correctly learnable. The picture does not provide η-families which are confidently learnable and also not separate out the notion of vacillatory learning from explanatory learning. The first main result of this section is to construct a positive equivalence relation ϑ such that there is no confidently learnable ϑ-family and furthermore all vacillatorily learnable ϑ-families are explanatory learnable. Thus one cannot separate for all η the notions of vacillatory and explanatory learning and one also cannot show that every η has a confidently learnable η-family. The second main result shows that there is a positive equivalence relation ζ for which there are confidently learnable ζ-families but no finitely learnable ζ-families.

Theorem 18. *There is a positive equivalence relation ϑ satisfying:*

(1) There is only one ϑ-infinite r.e. set, namely $\mathbb{N}$.

(2) Every ϑ-family contains an infinite ascending chain $B_0 \subset B_1 \subset \cdots$ of ϑ-finite sets whose union is $\mathbb{N}$. In particular, no ϑ-family is confidently learnable; and every behaviorally-correctly learnable ϑ-family consists only of ϑ-finite languages.

(3) Every vacillatory learnable ϑ-family is explanatorily learnable.

According to Theorem 14, for every positive equivalence relation η such that there is at least one finitely learnable η-family, there is also an η-family that is confidently but not finitely learnable. The next main result complements this theorem by showing that there is a positive equivalence relation ζ for which no finitely learnable ζ-family exists even though there are confidently learnable ζ-families.

Description 19. One defines a positive equivalence relation ζ using a dense simple set Z with $0 \notin Z$ as below; recall for this that a set is dense simple iff it is recursively enumerable, coinfinite and the sequence $a_0, a_1, \ldots$ of its non-elements in ascending order grows faster than every recursive function. It is known that such sets Z exist [26].

Now one defines that $x \zeta y$ iff there is an n with $a_n \leqslant \min\{x, y\} \leqslant \max\{x, y\} < a_{n+1}$. This relation is positive, as $x \zeta y$ is equivalent to

$$\forall z \left[\min\{x, y\} < z \leqslant \max\{x, y\} \Rightarrow z \in Z \right]$$

which is an r.e. condition. Furthermore, in coincidence with the notation used in this paper, each a_n is the least element of its equivalence class and the a_n are the ascending limits of the approximations $a_{n,t}$ which are the non-elements (in ascending order) of the set Z_t of the first t elements enumerated into Z, so that $Z_0 = \emptyset$ and $a_{n,0} = n$. As Z is coinfinite, there are infinitely many a_n's and so ζ induces infinitely many equivalence classes.

Theorem 20. *There is a confidently learnable ζ-family but no finitely learnable ζ-family.*

6 Conclusion

The present work studied how the relations between the most basic inference criteria for learning from text are impacted when the only classes to be considered for learning are uniformly r.e. one-one families of sets which are closed under a given positive equivalence relation η. One considers the chain of implications finitely learnable $\Rightarrow$ confidently learnable $\Rightarrow$ explanatorily learnable $\Rightarrow$ vacillatorily learnable $\Rightarrow$ behaviourally correctly learnable which is immediate from the definitions. When choosing η as the explicitly constructed ϑ from Theorem 18, the implication from explanatorily learnable to vacillatorily learnable becomes an equivalence and the criterion of confidently learnable becomes void, that is, no

ϑ-family satisfies it. For the positive equivalence relation ζ from Description 19, there is a ζ-family which is confidently learnable, but none which is finitely learnable. Furthermore, in the case that a finitely learnable η-family exists for some η, then there is also a confidently learnable η-family which is not finitely learnable. For all choices of η, the implications from confident to explanatory learning and from vacillatory to behaviourally correct learning cannot be reversed and the class of all η-closed r.e. set is an η-family which cannot be learnt behaviourally correctly.

Besides investigating the situation for further learning criteria, future work can investigate to which extent the results generalise to arbitrary uniformly r.e. families of η-closed sets. Here one would get that the family of all η-singletons is r.e. and finitely, thus confidently learnable: the learner generates an index of the η-equivalence class to be learnt from the first data-item observed and keeps this hypothesis forever. So one has one more level in the learning hierarchy. However, the collapse of vacillatory learning to explanatory learning for the constructed equivalence relation ϑ generalises to uniformly r.e. families.

References

1. Angluin, D.: Inductive inference of formal languages from positive data. Inf. Control **45**, 117–135 (1980)
2. Baliga, G., Case, J., Jain, S.: The synthesis of language learners. Inf. Comput. **152**(1), 16–43 (1999)
3. Bārzdiņš, J.M.: Two theorems on the limiting synthesis of functions. In: Bārzdiņš, J.M. (ed.) Theory of Algorithms and Programs I, Proceedings of the Latvian State University, vol. 210, pp. 82–88. Latvian State University, Riga (1974). (in Russian)
4. Bārzdiņš, J.M.: Inductive inference of automata, functions and programs. In: American Mathematical Society Translations, pp. 107–122, 1977. Appeared Originally in the Proceedings of the 20-th International Congress of Mathematicians 1974, vol. 2, pp. 455–460 (1974). (in Russian)
5. Baur, W.: Rekursive Algebren mit Kettenbedingungen. Zeitschrift für mathematische Logik und Grundlagen der Mathematik **20**, 37–46 (1974). (in German)
6. Blum, L., Blum, M.: Toward a mathematical theory of inductive inference. Inf. Control **28**, 125–155 (1975)
7. Case, J.: The power of vacillation in language learning. SIAM J. Comput. **28**(6), 1941–1969 (1999)
8. Case, J., Lynes, C.: Machine inductive inference and language identification. In: Nielsen, M., Schmidt, E.M. (eds.) ICALP 1982. LNCS, vol. 140, pp. 107–115. Springer, Heidelberg (1982). https://doi.org/10.1007/BFb0012761
9. Case, J., Smith, C.: Comparison of identification criteria for machine inductive inference. Theoret. Comput. Sci. **25**, 193–220 (1983)
10. Ershov, Y.L.: Positive equivalences. Algebra Logic **10**(6), 378–394 (1974)
11. Ershov, Y.L.: Theory of Numberings. Nauka, Moscow (1977). (in Russian)
12. Feldman, J.A.: Some decidability results on grammatical inference and complexity. Inf. Control **20**(3), 244–262 (1972)
13. Fokina, E., Khoussainov, B., Semukhin, P., Turetsky, D.: Linear orders realized by c.e. equivalence relations. J. Symbol. Logic **81**(2), 463–482 (2016)

14. Fokina, E.B., Kötzing, T., Mauro, L.S.: Limit learning equivalence structures. In: Proceedings of the 30th International Conference on Algorithmic Learning Theory (ALT 2019), pp. 383–403 (2019)

15. Fulk, M.: A study of inductive inference machines. Ph.D. thesis, SUNY/Buffalo (1985)

16. Gavruskin, A., Jain, S., Khoussainov, B., Stephan, F.: Graphs realised by r.e. equivalence relations. Ann. Pure Appl. Logic **165**, 1263–1290 (2014)

17. Gavryushkin, A., Khoussainov, B., Stephan, F.: Reducibilities among equivalence relations induced by recursively enumerable structures. Theoret. Comput. Sci. **612**, 137–152 (2016)

18. Mark Gold, E.: Language identification in the limit. Inf. Control **10**, 447–474 (1967)

19. Jain, S., Osherson, D.N., Royer, J.S., Sharma, A.: Systems That Learn, 2nd edn. MIT Press, Cambridge (1999)

20. Kummer, M.: An easy priority-free proof of a theorem of Friedberg. Theoret. Comput. Sci. **74**, 249–251 (1990)

21. Lange, S., Zeugmann, T.: Types of monotonic language learning and their characterization. In: Haussler, D. (ed.) Proceedings of the Fifth Annual ACM Workshop on Computational Learning Theory, Pittsburgh, Pennsylvania, 27–29 July 1992, pp. 377–390. ACM Press, New York (1992)

22. Lange, S., Zeugmann, T.: Monotonic versus non-monotonic language learning. In: Brewka, G., Jantke, K.P., Schmitt, P.H. (eds.) NIL 1991. LNCS, vol. 659, pp. 254–269. Springer, Heidelberg (1993). https://doi.org/10.1007/BFb0030397

23. Mukouchi, Y.: Characterization of finite identification. In: Jantke, K.P. (ed.) AII 1992. LNCS, vol. 642, pp. 260–267. Springer, Heidelberg (1992). https://doi.org/10.1007/3-540-56004-1_18

24. Noether, E.: Idealtheorie in Ringbereichen. Mathematische Annalen **83**, 24–66 (1921)

25. Novikov, P.S.: On the algorithmic unsolvability of the word problem in group theory. Trudy Matematicheskogo Instituta imeni V.A. Steklova, Acad. Sci. USSR **44**, 3–143 (1955)

26. Odifreddi, P.: Classical Recursion Theory. North-Holland, Amsterdam (1989)

27. Odifreddi, P.: Classical Recursion Theory, vol. II. Elsevier, Amsterdam (1999)

28. Osherson, D., Stob, M., Weinstein, S.: Systems That Learn, An Introduction to Learning Theory for Cognitive and Computer Scientists. Bradford – The MIT Press, Cambridge (1986)

29. Rogers, H.: Theory of Recursive Functions and Effective Computability. McGraw-Hill, New York (1967)

30. Soare, R.: Recursively Enumerable Sets and Degrees. A Study of Computable Functions and Computably Generated Sets. Springer, Heidelberg (1987)

31. Trakhtenbrot, B.A., Bārzdiņš, J.M.: Konetschnyje awtomaty (powedenie i sinetez). Nauka, Moscow (1970). in Russian. English Translation: Finite Automata-Behavior and Synthesis, Fundamental Studies in Computer Science 1, North-Holland, Amsterdam (1975)

Learning Mealy Machines with One Timer

Frits Vaandrager[1(✉)], Roderick Bloem[2(✉)], and Masoud Ebrahimi[2(✉)]

[1] Radboud University, Nijmegen, Netherlands
f.vaandrager@cs.rul.nl
[2] Graz University of Technology, Graz, Austria
{roderick.bloem,masoud.ebrahimi}@iaik.tugraz.at

Abstract. We present Mealy machines with a single timer (MM1Ts), a class of models that is both sufficiently expressive to describe the real-time behavior of many realistic applications, and can be learned efficiently. We show how learning algorithms for MM1Ts can be obtained via a reduction to the problem of learning Mealy machines. We describe an implementation of an MM1T learner on top of LearnLib, and compare its performance with recent algorithms proposed by Aichernig et al. and An et al. on several realistic benchmarks.

1 Introduction

Model learning, also known as active automata learning, is a black-box technique for constructing state machine models of software and hardware components from information obtained through testing (i.e., providing inputs and observing the resulting outputs). Model learning has been successfully used in numerous applications, for instance for spotting bugs in implementations of major network protocols. e.g.. in [5–8,20]. We refer to [13,24] for surveys and further references.

Timing plays a crucial role in many applications. A TCP server, for instance, may retransmit packets if they are not acknowledged within a specified time. Also, a timeout will occur if a TCP server does not receive an acknowledgment after a number of retransmissions, or if it remains in certain states too long. Timing behavior cannot be captured using existing learning tools, which typically only support learning of deterministic finite automata (DFAs) and related models. In the case of TCP, previous work only succeeded to learn models of real implementations by having the network adaptor ignore all retransmissions, and by completing learning queries before the occurrence of certain timeouts [8]. All timing issues had to be artificially suppressed.

This work was supported by the Austrian Research Promotion Agency (FFG) through project TRUSTED (867558), Graz University of Technology's LEAD project "Dependable Internet of Things in Adverse Environments" and by Radboud University's NWO TOP project 612.001.852 "Grey-box learning of Interfaces for Refactoring Legacy Software (GIRLS)".

A. Leporati et al. (Eds.): LATA 2021, LNCS 12638, pp. 157–170, 2021.
https://doi.org/10.1007/978-3-030-68195-1_13

The challenge to extend model learning algorithms to a setting of timed systems has been addressed by several authors. Most proposals aim to develop learning algorithms for the popular framework of timed automata [2], which extends DFAs with clock variables. Transitions of timed automata may contain both guards that test the values of clocks, and resets that update the clocks. Since guards and resets are not directly observable in a black-box setting, this poses major challenges during learning. Grinchtein et al. [9,10] developed learning algorithms for deterministic event-recording automata (DERAs), which have a clock for each action in the alphabet, and where each transition resets the clock corresponding to its input action. This restriction makes resets observable, but the complexity of the resulting algorithms still appears to be prohibitively high, due to the difficulties of inferring guards. The restrictions of DERAs also make it hard to capture the timing behavior of common network protocols. For instance, a pattern that often occurs is that within t time units after an event a there should be an event b. (For instance, in TCP a SYN should be followed by a SYN-ACK within a specified time interval.) In a DERA, upon occurrence of two consecutive a's, the automaton no longer remembers when the first a has occurred, and can thus not ensure the occurrence of a timeout at the required moment in time. Recently, Henry et al. [11] proposed a learning algorithm for a slightly larger class of reset-free DERAs, where some transitions may reset no clocks. Even though this algorithm appears to be more efficient than those of [9,10], it still suffers from a combinatorial blow up because, for each transition, it has to guess whether this transition resets a clock. An et al. [3] developed a learning algorithm for deterministic one-clock timed automata (DOTAs), using a brute force approach to reset guessing, also leading to a combinatorial blow up. Entirely different, heuristic algorithms are proposed recently by Aichernig et al. [1,23], using genetic programming. They succeeded to learn timed automata models with one clock for several industrial benchmarks.

Given the difficulties to infer the guards and resets of timed automata, the question arises whether timed automata provide the right modeling framework to support learning algorithms. As an alternative, we propose to consider the use of *timers* instead of *clocks*. The difference is that the value of a timer decreases when time advances, whereas the value of a clock increases. In a setting with clocks, guards and invariants are required to constrain the timing of events, but a timer simply triggers a timeout whenever its value becomes 0. The absence of guards and invariants makes model learning much easier in a setting with timers. A learner still has to figure out which transitions set a timer, but this also becomes easier and does not create a combinatorial blow-up. If a transition sets a timer then slight changes in the timing of this transition will trigger corresponding changes in the timing of the resulting timeout, allowing a learner to figure out the exact cause of each timeout event. DFAs with timers are strictly less expressive than timed automata if we assume that timeout events can be observed. For many realistic applications, however, this reduced expressivity causes no problems. Kurose and Ross [15], for instance, use finite state machine models with timers to explain transport layer protocols. Caldwell et al. [4]

propose a learning algorithm for a simple class of automata with timers, which they call time delay Mealy machines. These machines have only a single timer, which is reset on every transition. As a result, time delay Mealy machines are not sufficiently expressive to capture the timing behavior of realistic network protocols.

In this paper, we present Mealy machines with a single timer (MM1Ts), a class of models that is both sufficiently expressive to describe the real-time behavior of many realistic applications, and can be learned efficiently. In an MM1T, the timer can be set to integer values on transitions, and may be stopped or time out in later transitions. Each timeout triggers an observable output, allowing a learner to observe the occurrence of timeouts. We show how learning algorithms for MM1Ts can be obtained via a reduction to the problem of learning Mealy machines. We describe an implementation of an MM1T learner on top of LearnLib, a state-of-the-art tool for learning Mealy machines [17], and compare its performance with the tools of Aichernig et al. [1] and An et al. [3] on several benchmarks: TCP connection setup, Android's Authentication and Key Management (AKM) service, and some industrial benchmarks taken from [1]. Our implementation outperforms the tool of [1] with several orders of magnitude in terms of the total number of input symbols required to learn a model. The tool of [3] is only able to learn the benchmarks with a "helpful" teacher that provides information about resets; without help, it is unable to learn the benchmarks.

2 Mealy Machines with a Single Timer

In this section, we introduce the notion of Mealy machines with a single timer (MM1T). We write $f : X \rightharpoonup Y$ to denote that f is a partial function from X to Y. We write $f(x) \downarrow$ to mean that the result is defined for x, that is, $\exists y : f(x) = y$, and $f(x) \uparrow$ if the result is undefined. We often identify a partial function f with the set of pairs $\{(x, y) \in X \times Y \mid f(x) = y\}$.

MM1Ts are just regular (deterministic) Mealy machines, augmented with a timer that can be switched on and off, a timeout input, and a function that specifies how transitions affect the timer. We view timeout's as input events, a choice that makes sense if we view the hardware clock (or whatever the device is that triggers timeout interrupts) as part of the environment of the machine.

Definition 1. A Mealy machine with a single timer (MM1T) is defined as a tuple $\mathcal{M} = (I, O, Q, q_0, \delta, \lambda, \tau)$, where

- I is a finite set of inputs, containing a special element timeout,
- O is a finite set of outputs,
- $Q = Q_{off} \cup Q_{on}$ is a finite set of states, partitioned into subsets where the timer is on and off, respectively; $q_0 \in Q_{off}$ is the initial state,
- $\delta : Q \times I \rightharpoonup Q$ is a transition function, satisfying

$$\delta(q, i) \uparrow \quad \Leftrightarrow \quad i = \text{timeout} \wedge q \in Q_{off} \tag{1}$$

(inputs are always defined, except for timeout in states where timer is off),

- $\lambda : Q \times I \rightharpoonup O$ is an output function, *satisfying*

$$\lambda(q, i) \downarrow \;\; \Leftrightarrow \;\; \delta(q, i) \downarrow \tag{2}$$

(each transition has both an input and an output),
- $\tau : Q \times I \rightharpoonup \mathbb{N}^{>0}$ *is a reset function, satisfying*

$$\tau(q, i) \downarrow \;\; \Rightarrow \;\; \delta(q, i) \in Q_{on} \tag{3}$$

$$q \in Q_{off} \wedge \delta(q, i) \in Q_{on} \Rightarrow \tau(q, i) \downarrow \tag{4}$$

$$\delta(q, \mathsf{timeout}) \in Q_{on} \;\; \Rightarrow \;\; \tau(q, \mathsf{timeout}) \downarrow \tag{5}$$

(when a transition (re)sets the timer, the timer is on in the target state; when it moves from a state where the timer is off to a state where the timer on, it sets the timer; if the timer stays on after a timeout, it is reset).

Let $\delta(q, i) = q'$ and $\lambda(q, i) = o$. We write $q \xrightarrow{i/o, n} q'$ if $\tau(q, i) = n \in \mathbb{N}^{>0}$, and $q \xrightarrow{i/o, \perp} q'$ or just $q \xrightarrow{i/o} q'$ if $\tau(q, i) \uparrow$.

Example 1. The MM1T shown in Fig. 1 is a simplified model of the sender from the alternating-bit protocol, adapted from [15, Figure 3.15]. We write *set-timer(n)* on the *i*-transition from state q to indicate that $\tau(q, i) = n$. The MM1T has four states, with $Q_{on} = \{q_1, q_3\}$ and $Q_{off} = \{q_0, q_2\}$. In the model, input *in* corresponds to a request from the upper layer to transmit data. Initially, upon receipt of such a request, the sender builds a packet from the data and a sequence number 0, sends this over the network (output *send0*), and starts the timer with timeout value 3. When the sender receives an acknowledgement with the correct sequence number 0 (input *ack0*) it stops the timer and jumps to state q_2 without generating visible output (*void*). Acknowledgement with the incorrect sequence number (input *ack1*) are ignored. Likewise, inputs *in* in state q_1 and acknowledgements in state q_0 are ignored (for readability, these transitions are not shown in the diagram). If no *ack0* input arrives within 3 timeunits, a timeout occurs and the same packet is retransmitted. The behavior in states q_2 and state q_3 is symmetric to that in states q_0 and q_1, respectively, except that the roles of sequence numbers 0 and 1 is swapped.

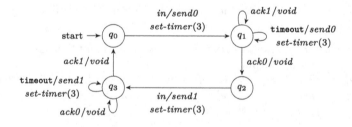

Fig. 1. MM1T model of alternating-bit protocol sender.

Semantics. We give two semantics for MM1Ts, an untimed and a timed one. In the untimed semantics, we just record the labels of sequences of transitions. Formally, an *untimed word* over inputs I and outputs O is a sequence

$$w = (i_0, o_0, n_0), (i_1, o_1, n_1) \cdots (i_k, o_k, n_k),$$

where each $i_j \in I$, each $o_j \in O$, and each $n_j \in \mathbb{N}^{>0} \cup \{\bot\}$ is a timer value. An *untimed run* of MM1T $\mathcal{M}$ over w is a sequence

$$\alpha = q_0 \xrightarrow{i_0/o_0, n_0} q_1 \xrightarrow{i_1/o_1, n_1} q_2 \cdots \xrightarrow{i_k/o_k, n_k} q_{k+1}$$

of transitions of $\mathcal{M}$ such that all q_j are states of $\mathcal{M}$ and q_0 is the initial state. Note that, since MM1Ts are deterministic, for each untimed word w there is at most one untimed run over w. We say that w is an untimed word of $\mathcal{M}$ iff $\mathcal{M}$ has an untimed run over w. MM1Ts $\mathcal{M}$ and $\mathcal{N}$ with the same set of inputs are *untimed equivalent*, $\mathcal{M} \approx_{untimed} \mathcal{N}$, iff they have the same untimed words.

The timed semantics, which is slightly more involved, describes the real-time behavior of a MM1T. It associates an infinite state transition system to a MM1T that describes all possible configurations and transitions between them. A *configuration* of a MM1T is a pair (q, t), where $q \in Q$ is a state and $t \in \mathbb{R}^{\geq 0} \cup \{\infty\}$ specifies the value of the timer. We require $t = \infty$ iff $q \in Q_{off}$. We refer to (q_0, ∞) as the *initial configuration*. Using four rules we define a transition relation that describes how one configuration may evolve into another. For all $q \in Q, r \in Q_{off}$, $s, s' \in Q_{on}, i \in I, o \in O, t \in \mathbb{R}^{\geq 0} \cup \{\infty\}, d \in \mathbb{R}^{\geq 0}$ and $n \in \mathbb{N}^{>0}$,

$$\frac{d \leq t}{(q, t) \xrightarrow{d} (q, t - d)} \quad (6) \qquad\qquad \frac{q \xrightarrow{i/o, n} s, \quad i = \mathsf{timeout} \Rightarrow t = 0}{(q, t) \xrightarrow{i/o} (s, n)} \quad (7)$$

$$\frac{q \xrightarrow{i/o} r, \quad i = \mathsf{timeout} \Rightarrow t = 0}{(q, t) \xrightarrow{i/o} (r, \infty)} \quad (8) \qquad \frac{s \xrightarrow{i/o} s', \quad i \neq \mathsf{timeout}}{(s, t) \xrightarrow{i/o} (s', t)} \quad (9)$$

Rule (6) states that the value of the timer decreases proportionally when time advances, until it becomes 0. Here we use the convention that $\infty - d = \infty$, for any $d \in \mathbb{R}^{>0}$. So when the timer is off, time may advance indefinitely. Rule (7) describes events where the timer is (re)set; a timeout may occur only when the timer has expired in the source state. Rule (8) describes events where the timer is off in the target state; again, a timeout may occur only when the timer has expired in the source state. Finally, rule (9) describes events where the timer remains on and is not reset.

A *timed word* over inputs I and outputs O is a sequence

$$w = (t_0, i_0, o_0), (t_1, i_1, o_1) \cdots (t_k, i_k, o_k),$$

where each $i_j \in I$, each $o_j \in O$, and each $t_j \in \mathbb{R}^{\geq 0}$. A timed word w describes a behavior that an experimenter may observe when interacting with an MM1T:

after an initial delay of t_0 time units, input i_0 is applies which triggers output o_0, after a subsequent delay of t_1 time units, input i_1 is applied, etc. For such a timed word w, a *timed run* of MM1T M over w is a sequence

$$\alpha = C_0 \xrightarrow{t_0} C_0' \xrightarrow{i_0/o_0} C_1 \xrightarrow{t_1} C_1' \xrightarrow{i_1/o_1} C_2 \cdots \xrightarrow{t_k} C_k' \xrightarrow{i_k/o_k} C_{k+1}$$

of transitions of M such that all C_j, C_j' are configurations of M and C_0 is the initial configuration. Since MM1Ts are deterministic, for each timed word w there exists at most one run over w. We say w is a timed word of M if there exists a run of M over w. MM1Ts M and N with the same set of inputs are *timed equivalent*, $M \approx_{timed} N$, iff they have the same sets of timed words.

Although the definitions are quite different, it turns out that timed and untimed equivalence coincide.

Theorem 1. $M \approx_{timed} N \Leftrightarrow M \approx_{untimed} N$

3   Learning MM1Ts

It will be useful to explore this connection between the timed and untimed semantics in some more detail, because this will allow us to reuse existing active learning algorithms for untimed systems [18,21] for learning MM1Ts.

3.1   From MM1Ts to Mealy Machines and Back

MM1Ts generalize the classical notion of a Mealy machine: essentially, a Mealy machine is just an MM1T in which the timer is off in all states. Conversely, each MM1T can be viewed as a Mealy machine of a special form.

Definition 2. *A Mealy machine is a tuple* $M = (I, O, Q, q_0, \delta, \lambda)$, *where* I *is a finite set of inputs,* O *a set of outputs,* Q *a finite set of states,* $q_0 \in Q$ *the initial state,* $\delta : Q \times I \to Q$ *a transition function, and* $\lambda : Q \times I \to O$ *an output function. We generalize the transition function to sequences of inputs as usual. Function* $mq_M : I^+ \to O$ *assigns to each sequence of inputs the final output:* $mq_M(\sigma i) = \lambda(\delta(q_0, \sigma), i)$. *Mealy machines* M *and* N *with the same set of inputs* I *are* equivalent, *denoted by* $M \approx N$, *if for all* $\sigma \in I^+$, $mq_M(\sigma) = mq_N(\sigma)$.

We associate a Mealy machine Mealy(M) to each MM1T M as follows. We keep the same states, inputs and transitions, but add timeout self-loops for each state in Q_{off} to make the Mealy machine input enabled. We introduce a fresh output nil and associate this special output to each new timeout self-loop. The outputs of the other transitions of Mealy(M) are pairs consisting of the output from M and the timer update.

Definition 3. *Let* $\mathcal{M} = (I, O, Q, q_0, \delta, \lambda, \tau)$ *be a MM1T. Then* Mealy$(\mathcal{M})$ *is the Mealy machine* $(I, (O \times (\mathbb{N}^{>0} \cup \{\bot\})) \cup \{\text{nil}\}, Q, q_0, \delta', \lambda')$, *where*

$$\delta'(q, i) = \begin{cases} \delta(q, i) & \text{if } \lambda(q, i) \downarrow \\ q & \text{otherwise} \end{cases}$$

$$\lambda'(q, i) = \begin{cases} (\lambda(q, i), \tau(q, i)) & \text{if } \tau(q, i) \downarrow \\ (\lambda(q, i), \bot) & \text{if } \lambda(q, i) \downarrow \text{ and } \tau(q, i) \uparrow \\ \text{nil} & \text{otherwise} \end{cases}$$

Conversely, suppose that $\mathcal{N} = (I, (O \times (\mathbb{N}^{>0} \cup \{\bot\})) \cup \{\text{nil}\}, Q, q_0, \delta', \lambda')$ *is a Mealy machine. Then we may reverse the above construction and define a tuple* MM1T$(\mathcal{N}) = (I, O, Q, q_0, \delta, \lambda, \tau)$ *in the obvious way.*

The following result, which follows from the definitions and Theorem 1, asserts that Mealy and MM1T act like adjoint operators.

Theorem 2. *Let* $\mathcal{M}$ *be a MM1T and let* $\mathcal{N}$ *be a Mealy machine such that* Mealy$(\mathcal{M}) \approx \mathcal{N}$. *Then* MM1T$(\mathcal{N})$ *is a MM1T and* $\mathcal{M} \approx_{timed}$ MM1T$(\mathcal{N})$.

Theorem 2 suggests that we can obtain a learner for MM1Ts from a learner for Mealy machines. To achieve this, we place an *adaptor* between a Mealy machine learner and a System Under Learning (SUL) that behaves like MM1T $\mathcal{M}$. From the perspective of the Mealy machine learner, the adaptor behaves like a teacher for Mealy$(\mathcal{M})$ that answers membership and equivalence queries. In order to answer these queries, the adaptor interacts with the SUL and observes timed words of $\mathcal{M}$. When the learner has succeeded to learn a Mealy machine $\mathcal{N}$ that is equivalent to Mealy$(\mathcal{M})$, we know by Theorem 2 that $\mathcal{M} \approx_{timed}$ MM1T$(\mathcal{N})$, and so we have learned a MM1T that is equivalent to $\mathcal{M}$. Effectively, the combination of the adaptor and the Mealy machine learner acts as an MM1T learner.

We implemented an adaptor that interacts with LearnLib [18] so we can benefit from all optimizations already integrated into this well maintained automata learning library. Our adaptor is available online[1]. Below we describe how to implement a membership oracle for learning MM1Ts. An equivalence oracle can be implemented in a similar manner, and is not discussed here for reasons of space.

3.2   Membership Queries

In order to answer membership queries, the adaptor maintains an observation tree defined as follows.

Definition 4. *Let* $\mathcal{M}$ *be a MM1T. An* observation tree *for* $\mathcal{M}$ *is a triple* $\mathcal{T} = (S, mq, timer)$, *where* $S \subset I^*$ *is a non empty, finite, prefix closed set of input sequences, referred to as* nodes, $mq : S \backslash \{\epsilon\} \to O \times (\mathbb{N}^{>0} \cup \{\bot\})$ *is a node labeling function, and* timer $: S \to \{on, off\}$ *is a function that specifies whether the timer is on or off in a node. We require that* timer$(\epsilon) = $ off *and* $\sigma \cdot$ timeout $\in S \Rightarrow$ timer$(\sigma) = on$.

---

[1] https://extgit.iaik.tugraz.at/scos/scos.sources/LearningMMTs.

Initially, the adaptor starts with a trivial observation tree with a single node ϵ and $timer(\epsilon) = off$. The observation tree is then extended one node at a time. For this, the adaptor maintains a maximum timer value Δ. Initially, Δ can be assigned some arbitrary value in $\mathbb{N}$. Suppose that $\sigma = i_1 \cdots i_{k-1}$ is a leaf node of observation tree $\mathcal{T}$, $i_k \in I$ and $i_k = \mathsf{timeout} \Rightarrow timer(\sigma) = on$. In order to add node $\sigma \cdot i_k$ to $\mathcal{T}$, the adaptor resets the SUL and then eagerly applies $\sigma \cdot i_k$. That is, for each $j \in [1, k]$, the adaptor processes input i_j as follows:

- if $i_j \neq \mathsf{timeout}$, the adaptor feeds the input to the SUL without any delay,
- otherwise, the oracle waits for the timeout event.

The immediate response o after feeding i_k accounts for the output value that will be recorded in $mq(\sigma \cdot i_k)$. Next the adaptor waits for Δ time units. If a timeout occurs after $n < \Delta$ time units then the value of $timer(\sigma \cdot i_k)$ is set to on, otherwise it is set to off. If $timer(\sigma \cdot i_k) = off$ then we set $mq(\sigma \cdot i_k) = (o, \bot)$. Otherwise, the adaptor performs another experiment to decide whether the clock was set on the last transition or before:

- it resets the SUL and eagerly applies σ,
- it waits for $1/2$ time unit and then applies input i_k,
- it then waits until a timeout event occurs at time $n' \leq n$.
- If $n' = n$ it sets $mq(\sigma \cdot i_k) = (o, n)$, otherwise it sets $mq(\sigma \cdot i_k) = (o, \bot)$.

Once the observation tree $\mathcal{T}$ is big enough, the adaptor can answer a membership query σ by computing the sequence σ' obtained by omitting spurious timeouts from σ, that is, timeouts from nodes of $\mathcal{T}$ where the timer is off. If σ ends with a spurious timeout then the response of the adaptor is nil, otherwise it is $mq(\sigma')$.

Query Complexity. Note that in order to add a new node to the observation tree, we need one or two experiments (membership queries) on the MM1T (SUL), depending whether the timer is on in the target node. Thus, starting from the trivial observation tree, we will need at most $2n$ membership queries on the MM1T to implement a single membership query with n input symbols by LearnLib, with a total number of inputs in $O(n^2)$. This way of learning MM1Ts has a higher query complexity than learning Mealy machines, but the growth of the total number of input symbols required is still polynomial. If the number of states where the timer is on is low, the query complexity is comparable.

Learning the Maximum Timer Value. If the maximum timer value Δ is greater than or equal to the SUL's maximum timer value, no timeout event will be missed during learning a hypothesis. Otherwise, the equivalence oracle will at some point return a counterexample containing a timeout event that is not present in the observation tree. Based on this counterexample, we then update Δ and start learning from scratch.

4   From MM1T to DOTA Learning

In order to compare our approach to those of [1,3], we translate MM1Ts to Deterministic One-Clock Timed Automata (DOTAS). In the interest of brevity, we will not formalize this transition, but rather illustrate it with an example. We will construct a DOTA of the alternating-bit protocol from Fig. 1; the result can be found in Fig. 2.

Edges of DOTAs are labeled with an action, a clock guard that is an interval on allowed clock values, and a Boolean that states whether to reset the clock to zero. The set of actions consists of all input labels of the MM1T (except timeout) prefixed with '?' and all output labels (except *void*) prefixed with '!'. In general, DOTAs have accepting and non-accepting states, we will construct DOTAs with only accepting states.

We split each transition of the MM1T into an input and an output transition. For instance, we encode the transition $q_0 \rightarrow q_1$ into transitions $q_0' \rightarrow l_0$ and $l_0 \rightarrow q_1'$. In this case, the input transition can be taken at any time and it resets the clock, causing the output transition to be taken immediately. For edges with a *void* output, we omit the output transition. Thus, the self loop on q_1 labeled *ack1/void* is represented by a self-loop on q_1' in the DOTA.

An MM1T transition that sets the timer is replaced by a DOTA transition that reset the clock and appropriate clock guards on subsequent states. For instance, the transition $q_0 \rightarrow q_1$ sets the timer to 3; thus, a timeout event will occur in q_1 at time 3, causing a *send*. In the DOTA, this is reflected by a clock reset on the transition $l_0 \rightarrow q_1'$ and a clock guard with value $[3, \infty)$ on the self loop on q_1' labeled *!send*.

The rest of the translation follows along the same lines.

5   Case Studies

5.1   Learning Setup

We instantiated L_M^* and TTT using the MM1T membership oracle. For counterexample processing we used *Rivest and Shapire's* method [19]. We close tables using *close shortest* strategy. Finally, we use a *random word* equivalence oracle with 1000 tests and word length of minimum 4 and of maximum 11. For further details on above terminologies, we refer readers to LearnLib's documentation.

5.2   Android Authentication and Key Management

To show that our algorithm can learn realistic Mealy machines with timers, we used our algorithm to learn the Authentication and Key Management of the WiFi implementation of a Huawei Mate10-lite running Android 8.0.0 (Kernel 4.4.23+) with a security patch dated July 5, 2019. The IEEE 802.11 standard gives an abstract automaton of an Authentication and Key Management (AKM) service in [14, p. 1643]. The automaton has a state that encapsulates a 4-way handshake mechanism granting access to the controlled port. Since learning the 4-way handshake mechanisms is already addressed in [22], we focus on learning the AKM service. We used the following management frames: Auth(Open), AssoReq, Deauth(leaving), Disas(leaving), ProbeReq, and timeout [14, p. 45–49].

Our learning experiments resulted in the MM1T shown in Fig. 3. The SUL deviates from the specified standards in the following ways.

Disassociation: The reference prescribes that a disassociation (Disas) terminates an established association but maintains authentication. In the learned model (state q_2), a disassociation instead drops both the established association and the authentication. To correct this, the access point should transit to q_1 when disassociating in q_2 (red transitions from q_2 must go to q_1).

Association Timeout: Along the red transition from q_1 to q_2, SUL does not include *BSS Max Idle Period* element in *AssoResp* frames. Yet, it implements an association timeout event, which violates the specification. To confirm this, we manually inspected the Android 8.0.0 (r39) source code, which excludes the element mentioned above except for access points of Wireless Mesh Networks.

5.3   Performance Comparison

We apply our learning method for MM1Ts to a set of real-world benchmarks. This demonstrates the expressiveness of MM1Ts, and shows the practicality of our implementation.

Benchmarks: Our benchmark set consists of the AKM (Sect. 5.2), the TCP Connection State Diagram ([16, p. 23]), a car alarm system (CAS) [1], and a particle counter (PC) [1]. For the TCP benchmark, we used the one timeout on the transmission control block indicated in the diagram in the RFC. See Table 1 for statistics on the size of the benchmarks.

Algorithms: Table 2 shows benchmark results for MM1T and DOTA learning algorithms. GTALearn represents the learning algorithm by Aichernig et al. [1]. OTALearn* represents the learning algorithm by An et al. [3] using a "smart" teacher that provides the clock reset information. Finally, OTALearn uses a normal teacher and timed out on all benchmarks.

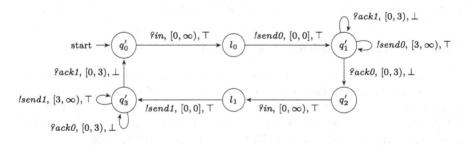

Fig. 2. DOTA model of alternating-bit protocol sender.

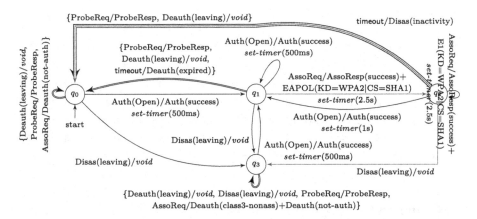

Fig. 3. MM1T of a Huawei Mate10-lite that captures granting uncontrolled port. Double and triple edges represent a set of transitions. We rounded timer values to the nearest 500 ms and marked specification violations with the color red.

Table 1. Benchmarks in terms of state-space ($|\mathbf{S}|$) and input size ($|\mathbf{\Sigma}|$).

Model	AKM		TCP		CAS		PC	
	$|\mathbf{S}|$	$|\mathbf{\Sigma}|$	$|\mathbf{S}|$	$|\mathbf{\Sigma}|$	$|\mathbf{S}|$	$|\mathbf{\Sigma}|$	$|\mathbf{S}|$	$|\mathbf{\Sigma}|$
MM1T	4	5	11	8	8	4	8	8
DOTA	15	12	20	13	14	10	26	14

Performance Metrics: Since OTALearn* implements its own equivalence checker and GTALearn is not an Angluin style algorithm, we report the total number of resets (**#R**), and inputs (**#I**) performed rather than the number of membership and equivalence queries. We believe this gives a fair comparison of the algorithms under the assumption that most time is spent executing the SUL.

AKM has a more sophisticated timed behavior than the other benchmarks, which explains the higher number of resets for MM1T learners. Meanwhile, if considering number of inputs, OTALearn* straggles by an order of magnitude. GTALearn shows a competitive performance in the number of inputs performed that indicates the potentials of this novel approach.

 TCP has only one timeout transition; thus, the learning algorithms do not need to reset the SUL as often. (With the exception of OTALearn.) L_M^* learns the MM1T for TCP in one round, while TTT requires 8 rounds, which justifies the better performance of L_M^*. MM1T learners outperform those for DOTAs by nearly an order of magnitude when considering the number of inputs performed.

 CAS and PC show a slightly more sophisticated timed behavior than TCP. For both, the MM1T algorithms also significantly outperform the algorithms for DOTA. Similarly, if considering the inputs performed, OTALearn* straggles by three orders of magnitude.

Table 2. Benchmark results in terms of total resets and total performed inputs.

Algorithm	AKM		TCP		CAS		PC	
	#R	#I	#R	#I	#R	#I	#R	#I
MM1T-L_M^*	5587	35002	413	2401	613	4822	408	2271
MM1T-TTT	5714	35948	640	4773	623	4978	369	2443
GTALEARN	2626	36411	1186	33779	1609	30870	3368	33824
OTALEARN*	2103	356762	2924	86880	1448	3791091	10003	3540458
OTALEARN	timeout		timeout		timeout		timeout	

6   Conclusion and Future Work

Timers are commonly used in software to enforce real-time behavior, and so it is natural to use them in formal models. We presented a framework of Mealy machines with a single timer and showed how a learning algorithm can be obtained via reduction to the problem of learning Mealy machines. Our approach assumes that timers are set when input events occur, and timeouts trigger instantaneous outputs. While these assumptions do not always hold, there are many real-time systems for which the delays between timer events and observable inputs and outputs are negligible, and the assumptions are justified. We evaluated our approach on a number of realistic applications, and showed that it outperforms the approaches of Aichernig [1] et al. and An et al. [3].

An obvious direction for future research is to extend our work to Mealy machines with multiple timers. We expect that a learning algorithm can be developed, but a simple reduction to Mealy machine learning is no longer possible. It would be interesting to apply the genetic programming approach of [1] in a setting of Mealy machines with timers. Since it no longer needs to learn transition guards, one may expect that a genetic algorithm will converge faster. Of course, as noted by [3], we may resort to grey-box techniques for model learning [12] to obtain efficient learning algorithms for real-time software. However, this forces us to deal with numerous programming language specific details. Black-box techniques can be applied without knowledge of the underlying hardware/software, which makes it important to push these techniques to their limits.

Acknowledgement. We would like to thank Andrea Pferscher and Miaomiao Zhang for help with running the benchmarks on their tools [1,3].

References

1. Aichernig, B.K., Pferscher, A., Tappler, M.: From passive to active: learning timed automata efficiently. In: Lee, R., Jha, S., Mavridou, A., Giannakopoulou, D. (eds.) NFM 2020. LNCS, vol. 12229, pp. 1–19. Springer, Cham (2020). https://doi.org/10.1007/978-3-030-55754-6_1

2. Alur, R., Dill, D.: A theory of timed automata. Theoret. Comput. Sci. **126**, 183–235 (1994)
3. An, J., Chen, M., Zhan, B., Zhan, N., Zhang, M.: Learning one-clock timed automata. TACAS 2020. LNCS, vol. 12078, pp. 444–462. Springer, Cham (2020). https://doi.org/10.1007/978-3-030-45190-5_25
4. Caldwell, B., Cardell-Oliver, R., French, T.: Learning time delay mealy machines from programmable logic controllers. IEEE Trans. Autom. Sci. Eng. **13**(2), 1155–1164 (2015)
5. Fiterău-Broştean, P., Howar, F.: Learning-based testing the sliding window behavior of TCP implementations. In: Petrucci, L., Seceleanu, C., Cavalcanti, A. (eds.) FMICS/AVoCS -2017. LNCS, vol. 10471, pp. 185–200. Springer, Cham (2017). https://doi.org/10.1007/978-3-319-67113-0_12
6. Fiterău-Broştean, P., Jonsson, B., Merget, R., de Ruiter, J., Sagonas, K., Somorovsky, J.: Analysis of DTLS implementations using protocol state fuzzing. In: USENIX Security Symposium. USENIX Association (2020)
7. Fiterău-Broştean, P., Lenaerts, T., et al.: Model learning and model checking of SSH implementations. In: SPIN Symposium. ACM (2017)
8. Fiterău-Broştean, P., Janssen, R., Vaandrager, F.: Combining model learning and model checking to analyze TCP implementations. In: Chaudhuri, S., Farzan, A. (eds.) CAV 2016. LNCS, vol. 9780, pp. 454–471. Springer, Cham (2016). https://doi.org/10.1007/978-3-319-41540-6_25
9. Grinchtein, O., Jonsson, B., Leucker, M.: Learning of event-recording automata. Theoret. Comput. Sci. **411**(47), 4029–4054 (2010)
10. Grinchtein, O., Jonsson, B., Pettersson, P.: Inference of event-recording automata using timed decision trees. In: Baier, C., Hermanns, H. (eds.) CONCUR 2006. LNCS, vol. 4137, pp. 435–449. Springer, Heidelberg (2006). https://doi.org/10.1007/11817949_29
11. Henry, L., Jéron, T., Markey, N.: Active learning of timed automata with unobservable resets. In: Bertrand, N., Jansen, N. (eds.) FORMATS 2020. LNCS, vol. 12288, pp. 144–160. Springer, Cham (2020). https://doi.org/10.1007/978-3-030-57628-8_9
12. Howar, F., Jonsson, B., Vaandrager, F.: Combining black-box and white-box techniques for learning register automata. In: Steffen, B., Woeginger, G. (eds.) Computing and Software Science. LNCS, vol. 10000, pp. 563–588. Springer, Cham (2019). https://doi.org/10.1007/978-3-319-91908-9_26
13. Howar, F., Steffen, B.: Active automata learning in practice. In: Bennaceur, A., Hähnle, R., Meinke, K. (eds.) Machine Learning for Dynamic Software Analysis: Potentials and Limits. LNCS, vol. 11026, pp. 123–148. Springer, Cham (2018). https://doi.org/10.1007/978-3-319-96562-8_5
14. IEEE: Std 802.11-2016 (Revision of IEEE Std 802.11-2012): Part 11: Wireless LAN Medium Access Control (MAC) and Physical Layer (PHY) Specifications (2016)
15. Kurose, J.F., Ross, K.W.: Computer Networking: A Top-Down Approach, 6th edn. Pearson, London (2013)
16. Postel, J.E.: Transmission Control Protocol. RFC 793, September 1981
17. Raffelt, H., Steffen, B., Berg, T.: LearnLib: a library for automata learning and experimentation. In: FMICS 2005. ACM Press (2005)
18. Raffelt, H., Steffen, B., Berg, T., Margaria, T.: LearnLib: a framework for extrapolating behavioral models. STTT **11**(5), 393–407 (2009)
19. Rivest, R., Schapire, R.: Inference of finite automata using homing sequences (extended abstract). In: ACM Symposium on Theory of Computing. ACM (1989)

20. Ruiter, J.d., Poll, E.: Protocol state fuzzing of TLS implementations. In: USENIX Security Symposium. USENIX (2015)
21. Steffen, B., Howar, F., Merten, M.: Introduction to active automata learning from a practical perspective. In: Bernardo, M., Issarny, V. (eds.) SFM 2011. LNCS, vol. 6659, pp. 256–296. Springer, Heidelberg (2011). https://doi.org/10.1007/978-3-642-21455-4_8
22. McMahon Stone, C., Chothia, T., de Ruiter, J.: Extending automated protocol state learning for the 802.11 4-way handshake. In: Lopez, J., Zhou, J., Soriano, M. (eds.) ESORICS 2018. LNCS, vol. 11098, pp. 325–345. Springer, Cham (2018). https://doi.org/10.1007/978-3-319-99073-6_16
23. Tappler, M., Aichernig, B.K., Larsen, K.G., Lorber, F.: Time to learn – learning timed automata from tests. In: André, É., Stoelinga, M. (eds.) FORMATS 2019. LNCS, vol. 11750, pp. 216–235. Springer, Cham (2019). https://doi.org/10.1007/978-3-030-29662-9_13
24. Vaandrager, F.: Model learning. Commun. ACM **60**(2), 86–95 (2017)

Logics and Languages

Finite-Word Hyperlanguages

Borzoo Bonakdarpour[1]($\boxtimes$) and Sarai Sheinvald[2]($\boxtimes$)

[1] Department of Computer Science and Engineering, Michigan State University,
East Lansing, USA
borzoo@msu.edu

[2] Department of Software Engineering, ORT Braude College, Karmiel, Israel
sarai@braude.ac.il

Abstract. *Formal languages* are in the core of models of computation and their behavior. A rich family of models for many classes of languages have been widely studied. *Hyperproperties* lift conventional trace-based languages from a set of execution traces to a set of sets of executions. Hyperproperties have been shown to be a powerful formalism for expressing and reasoning about information-flow security policies and important properties of cyber-physical systems. Although there is an extensive body of work on formal-language representation of trace properties, we currently lack such a general characterization for hyperproperties. We introduce *hyperlanguages* over finite words and models for expressing them. Essentially, these models express multiple words by using assignments to quantified *word variables*. Relying on the standard models for regular languages, we propose *hyperregular expressions* and *finite-word hyperautomata (NFH)*, for modeling the class of *regular hyperlanguages*. We demonstrate the ability of regular hyperlanguages to express hyperproperties for finite traces. We explore the closure properties and the complexity of the fundamental decision problems such as nonemptiness, universality, membership, and containment for various fragments of NFH.

1 Introduction

Formal languages along with the models that express them are in the core of modeling, specification, and verification of computing systems. Execution traces are formally described as words, and various families of automata are used for modeling systems of different types. *Regular languages* are a classic formalism for finite traces and when the traces are infinite, ω-*regular languages* are used.

There are well-known connections between specification logics and formal languages. For example, LTL [15] formulas can be translated to ω-regular expressions, and CTL* [8] formulas can be expressed using tree automata. Accordingly, many verification techniques that exploit these relations have been developed. For instance, in the automata-theoretic approach to verification [17,18], the model-checking problem is reduced to checking the nonemptiness of the product automaton of the model and the complement of the specification.

© Springer Nature Switzerland AG 2021
A. Leporati et al. (Eds.): LATA 2021, LNCS 12638, pp. 173–186, 2021.
https://doi.org/10.1007/978-3-030-68195-1_17

Hyperproperties [6] generalize the traditional trace properties [2] to *system properties*, i.e., a set of sets of traces. A hyperproperty prescribes how the system should behave in its entirety and not just based on its individual executions. Hyperproperties have been shown to be a powerful tool for expressing and reasoning about information-flow security policies [6] and important properties of cyber-physical systems [19] such as sensitivity and robustness, as well as consistency conditions in distributed computing such as linearizability [4]. While different types of logics have been suggested for expressing hyperproperties, their formal-language counterparts and the models that express them are currently missing.

In this paper, we establish a formal-language theoretical framework for *hyperlanguages*, that are sets of sets of words, which we term *hyperwords*. Our framework is based on an underlying standard automata model for formal languages, augmented with quantified *word variables* that are assigned words from a set of words in the hyperlanguage. This formalism is in line with logics for hyperproperties (e.g.., HyperLTL [5] and HyperPCTL [1]). These logics express the behavior of infinite trace systems. However, a basic formal model for expressing general hyperproperties for finite words has not been defined yet. Hyperlanguages based on finite words have many practical applications. For instance, path planning objectives for robotic systems often stipulate the *existence* of one or more *finite* paths that stand out from *all* other paths.

To begin with the basics, we focus this paper on a regular type of hyperlanguages of sets consisting of finite words, which we call *regular hyperlanguages*. The models we introduce and study are based on the standard models for regular languages, namely regular expressions and finite-word automata. We explain the idea with two examples.

Example 1. Consider the following *hyperregular expression* (HRE) over the alphabet $\{a\}$.

$$r_1 = \forall x. \exists y. \underbrace{\left(\{a_x, a_y\}^* \{\#_x, a_y\}^* \right)}_{\hat{r}_1}$$

The HRE r_1 uses two word variables x and y, which are assigned words from a hyperword. The HRE r_1 contains an underlying regular expression $\hat{r}_1$, whose alphabet is $(\{a\} \cup \{\#\})^{\{x,y\}}$, and whose (regular) language describes different word assignments to x and y, where $\#$ is used for padding at the end if the words assigned to x and y are of different lengths. In a word in the language of $\hat{r}_1$, the i'th letter describes both i'th letters in the words assigned to x and y. For example, the word $\{a_x, a_y\}\{a_x, a_y\}\{\#_x, a_y\}$ describes the assignment $x \mapsto aa, y \mapsto aaa$. The regular expression $\hat{r}_1$ requires that the word assigned to y be longer than the word assigned to x. The *quantification condition* $\forall x. \exists y$ of r_1 requires that for every word in a hyperword S in the hyperlanguage of r_1, there exists a longer word in S. This holds iff S contains infinitely many words. Therefore, the hyperlanguage of r_1 is the set of all infinite hyperwords over $\{a\}$.

□

Example 2. Robotics applications are often concerned with finding the shortest path that reaches a goal g, starting from an initial location i. The shortest path requirement can be expressed by the following HRE over an alphabet Σ:

$$r_2 = \exists x.\forall y.\{i_x, i_y\}\{\bar{g}_x, \bar{g}_y\}^* \Big(\{g_x, \bar{g}_y\} \mid \{g_x, g_y\}\Big)\{\#_x, \$_y\}^*$$

where $\bar{g} \in \Sigma - \{g\}$ and $\$ \in \Sigma$. That is, there exists a path x that is shorter than any other path y in reaching g.   □

Although there is an ongoing line of research on model-checking hyperproperties [3,7,11], the work on finite-trace hyperproperties is limited to [9], where the authors construct a finite-word representation for the class of regular k-safety hyperproperties. We make the following contributions:

Table 1. Summary of results on properties of hyperregular languages.

Property	Result	
Closure	Complementation, Union, Intersection (Theorem 1)	
Nonemptiness	$\forall\exists\exists$	Undecidable (Theorem 2)
	$\exists^* / \forall^*$	NL-complete (Theorem 2)
	$\exists^*\forall^*$	PSPACE-complete (Theorem 2)
Universality	$\exists\forall\forall$	Undecidable (Theorem 3)
	$\exists^* / \forall^*$	PSPACE-complete (Theorem 3)
	$\forall^*\exists^*$	EXPSPACE (Theorem 3)
Finite membership	NFH	PSPACE (Theorem 4)
	$O(\log(k))$ $\forall$	NP-complete (Theorem 4)
Regular membership	Decidable (Theorem 5)	
Containment	NFH	Undecidable (Theorem 6)
	$\exists^* \subseteq \forall^* / \forall^* \subseteq \exists^*$	PSPACE-complete (Theorem 7)
	$\exists^*\forall^* \subseteq \forall^*\exists^*$	EXPSPACE (Theorem 7)

- Introduce regular hyperlanguages and HREs, and demonstrate the ability of HREs to express important information-flow security policies such as different variations of *noninterference* [13] and *observational determinism* [20].
- Present *nondeterministic finite-word hyperautomata* (NFH), an automata-based model for expressing regular hyperlanguages.
- Conduct a comprehensive study of the properties of regular hyperlanguages (see Table 1). We show that regular hyperlanguages are *closed* under union, intersection, and complementation. We further prove that the *nonemptiness* problem is in general undecidable for NFH. However, for the alternation-free fragments (which only allow one type of quantifier), as well as for the $\exists\forall$ fragment (in which the quantification condition is limited to a sequence of $\exists$

quantifiers followed by a sequence of $\forall$ quantifiers), nonemptiness is decidable. We also study the *universality, membership* and *containment* problems. These results are aligned with the complexity of HyperLTL model checking for tree-shaped and general Kripke structures [3]. This shows that the complexity results in [3] mainly stem from the nature of quantification over finite words and depend on neither the full power of the temporal operators nor the infinite nature of HyperLTL semantics.

2   Preliminaries

An *alphabet* is a nonempty finite set Σ of *letters*. A *word* over Σ is a finite sequence of letters from Σ. The *empty word* is denoted by ϵ, and the set of all words is denoted by Σ^*. A *language* is a subset of Σ^*. We assume that the reader is familiar with the syntax and semantics of regular expressions (RE). We use the standard notations $\{\cdot, |, *\}$ for concatenation, union, and Kleene star, respectively, and denote the language of an RE r by $\mathcal{L}(r)$. A language L is *regular* if there exists an RE r such that $\mathcal{L}(r) = L$.

Definition 1. *A* nondeterministic finite-word automaton *(NFA) is a tuple* $A = \langle \Sigma, Q, Q_0, \delta, F \rangle$, *where* Σ *is an alphabet,* Q *is a nonempty finite set of states,* $Q_0 \subseteq Q$ *is a set of* initial *states,* $F \subseteq Q$ *is a set of* accepting *states, and* $\delta \subseteq Q \times \Sigma \times Q$ *is a transition relation.*

Given a word $w = \sigma_1 \sigma_2 \cdots \sigma_n$ over Σ, a *run of A on w* is a sequence of states $(q_0, q_1, \ldots q_n)$, such that $q_0 \in Q_0$, and for every $0 < i \leq n$, it holds that $(q_{i-1}, \sigma_i, q_i) \in \delta$. The run is *accepting* if $q_n \in F$. We say that A *accepts* w if there exists an accepting run of A on w. The *language* of A, denoted $\mathcal{L}(A)$, is the set of all words that A accepts. It holds that a language L is regular iff there exists an NFA A such that $\mathcal{L}(A) = L$.

3   Hyperregular Expressions

Definition 2. *A* hyperword *over Σ is a set of words over Σ and a* hyperlanguage *over Σ is a set of hyperwords over Σ.*

Before formally defining hyperregular expressions, we explain the idea behind them. A *hyperregular expression* (HRE) over Σ uses a set of *word variables* $X = \{x_1, x_2, \ldots, x_k\}$. When expressing a hyperword S, these variables are assigned words from S. An HRE r is composed of a *quantification condition* α over X, and an underlying RE $\hat{r}$, which represents word assignments to X. An HRE r defines a hyperlanguage $\mathfrak{L}(r)$. The condition α defines the assignments that should be in $\mathcal{L}(\hat{r})$. For example, $\alpha = \exists x_1.\forall x_2$ requires that there exists a word $w_1 \in S$ (assigned to x_1), such that for every word $w_2 \in S$ (assigned to x_2), the word that represents the assignment $x_1 \mapsto w_1, x_2 \mapsto w_2$, is in $\mathcal{L}(\hat{r})$. The hyperword S is in $\mathfrak{L}(r)$ iff S meets these conditions.

We represent an assignment $v : X \to S$ as a *word assignment* $\boldsymbol{w}_v$, which is a word over the alphabet $(\Sigma \cup \{\#\})^X$ (that is, assignments from X to $\Sigma \cup \{\#\}$), where the i'th letter of $\boldsymbol{w}_v$ represents the k i'th letters of the words $v(x_1), \ldots, v(x_k)$ (in case that the words are not of equal length, we "pad" the end of the shorter words with $\#$ symbols). We represent these k i'th letters as an assignment denoted $\{\sigma_{1x_1}, \sigma_{2x_2}, \ldots, \sigma_{kx_k}\}$, where x_j is assigned σ_j. For example, the assignment $v(x_1) = aa$ and $v(x_2) = abb$ is represented by the word assignment $\boldsymbol{w}_v = \{a_{x_1}, a_{x_2}\}\{a_{x_1}, b_{x_2}\}\{\#_{x_1}, b_{x_2}\}$.

Definition 3. *A* hyperregular expression *is a tuple* $r = \langle X, \Sigma, \alpha, \hat{r} \rangle$, *where* $\alpha = \mathbb{Q}_1 x_1 \cdots \mathbb{Q}_k x_k$, *where* $\mathbb{Q}_i \in \{\exists, \forall\}$ *for every* $i \in [1, k]$, *and where* $\hat{r}$ *is an RE over* $\hat{\Sigma} = (\Sigma \cup \{\#\})^X$.

Let S be a hyperword and let $v : X \to S$ be an assignment of the word variables of r to words in S. We denote by $v[x \mapsto w]$ the assignment obtained from v by assigning the word $w \in S$ to $x \in X$. We represent v by $\boldsymbol{w}_v$. We now define the membership condition of a hyperword S in the hyperlanguage of r. We first define a relation $\vdash$ for $S, \hat{r}$, a quantification condition α, and an assignment $v : X \to S$, as follows.

- For $\alpha = \epsilon$, define $S \vdash_v (\alpha, \hat{r})$ if $\boldsymbol{w}_v \in \mathcal{L}(\hat{r})$.
- For $\alpha = \exists x.\alpha'$, define $S \vdash_v (\alpha, \hat{r})$ if there exists $w \in S$ s.t. $S \vdash_{v[x \mapsto w]} (\alpha', \hat{r})$.
- For $\alpha = \forall x.\alpha'$, define $S \vdash_v (\alpha, \hat{r})$ if $S \vdash_{v[x \mapsto w]} (\alpha', \hat{r})$ for every $w \in S$.[1]

Since all variables are under the scope of α, membership is independent of v, and so if $S \vdash (\alpha, \hat{r})$, we denote $S \in \mathcal{L}(r)$. The hyperlanguage of r is $\mathcal{L}(r) = \{S \mid S \in \mathcal{L}(r)\}$.

Definition 4. *We call a hyperlanguage* $\mathcal{L}$ *a* regular hyperlanguage *if there exists an HRE* r *such that* $\mathcal{L}(r) = \mathcal{L}$.

Application of HRE in Information-flow Security

Noninterference [13] requires high-secret commands be removable without affecting observations of users holding low clearances:

$$\varphi_{\mathsf{ni}} = \forall x. \exists y \{l_x, l\lambda_y\}^*,$$

where l denotes a low state and $l\lambda$ denotes a low state such that all high commands are replaced by a dummy value λ.

Observational determinism [20] requires that if two executions of a system start with low-security-equivalent events, they should remain low equivalent:

$$\varphi_{\mathsf{od}} = \forall x. \forall y. \Big(\{l_x, l_y\}^+ \mid \{\bar{l}_x, \bar{l}_y\}\{\$_x, \$_y\}^* \mid \{l_x, \bar{l}_y\}\{\$_x, \$_y\}^* \mid \{\bar{l}_x, l_y\}\{\$_x, \$_y\}^* \Big)$$

---

[1] In case that α begins with $\forall$, membership holds vacuously with an empty hyperword. We restrict the discussion to nonempty hyperwords.

where l denotes a low event, $\bar{l} \in \Sigma \setminus \{l\}$, and $\$ \in \Sigma$. We note that similar policies such as *Boudol and Castellani's noninterference* [12] can be formulated in the same fashion.[2]

Generalized noninterference (GNI) [14] allows nondeterminism in the low-observable behavior, but requires that low-security outputs may not be altered by the injection of high-security inputs:

$$\varphi_{\mathsf{gni}} = \forall x.\forall y.\exists z. \left(\{h_x, l_y, hl_z\} \mid \{\bar{h}_x, l_y, \bar{h}l_z\} \mid \{h_x, \bar{l}_y, h\bar{l}_z\} \mid \{\bar{h}_x, \bar{l}_y, \bar{h}\bar{l}_z\} \right)^*$$

where h denotes the high-security input, l denotes the low-security output, $\bar{l} \in \Sigma \setminus \{l\}$, and $\bar{h} \in \Sigma \setminus \{h\}$.

Declassification [16] relaxes noninterference by allowing leaking information when necessary. Some programs must reveal secret information to fulfill functional requirements. For example, a password checker must reveal whether the entered password is correct or not:

$$\varphi_{\mathsf{dc}} = \forall x.\forall y.\{li_x, li_y\}\{pw_x, pw_y\}\{lo_x, lo_y\}^+$$

where li denotes low-input state, pw denotes that the password is correct, and lo denotes low-output states. We note that for brevity, φ_{dc} does not include behaviors where the first two events are not low or, in the second event, the password is not valid.

Termination-sensitive noninterference requires that for two executions that start from low-observable states, information leaks are not permitted by the termination behavior of the program (here, l denotes a low state and $\$ \in \Sigma$):

$$\varphi_{\mathsf{tsni}} = \forall x.\forall y. \left(\{l_x, l_y\}\{\$_x, \$_y\}^*\{l_x, l_y\} \mid \{\bar{l}_x, \bar{l}_y\}\{\$_x, \$_y\}^* \mid \right.$$
$$\left. \{l_x, \bar{l}_y\}\{\$_x, \$_y\}^* \mid \{\bar{l}_x, l_y\}\{\$_x, \$_y\}^* \right)$$

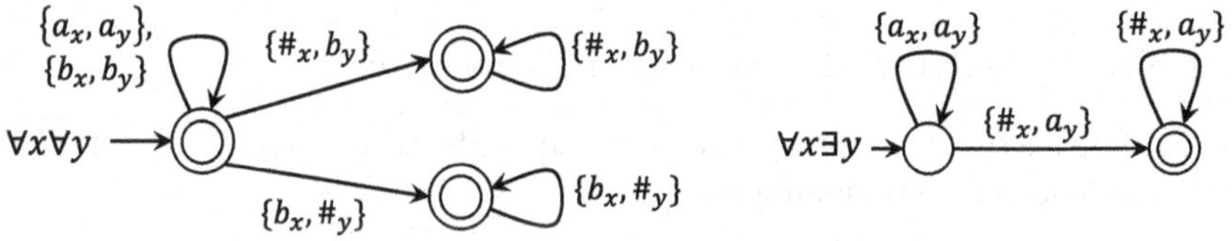

Fig. 1. The NFH $\mathcal{A}_1$ (left) and $\mathcal{A}_2$ (right).

4   Nondeterminsitic Finite-Word Hyperautomata

We now present a model for regular hyperlanguages, namely *finite-word hyperautomata*. A hyperautomaton is composed of a set X of word variables, a quantification condition, and an underlying finite-word automaton that accepts representations of assignments to X.

---

[2] This policy states that every two executions that start from bisimilar states (in terms of memory low-observability), should remain bisimilarly low-observable.

Definition 5. *A* nondeterministic finite-word hyperautomaton *(NFH) is a tuple* $\mathcal{A} = \langle \Sigma, X, Q, Q_0, F, \delta, \alpha \rangle$, *where* Σ, X *and* α *are as in Definition 3, and where* $\langle \hat{\Sigma}, Q, Q_0, F, \delta \rangle$ *forms an underlying NFA over* $\hat{\Sigma} = (\Sigma \cup \{\#\})^X$.

The acceptance condition for NFH, as for HRE, is defined with respect to a hyperword S, the NFH $\mathcal{A}$, the quantification condition α, and an assignment $v : X \to S$. For the base case of $\alpha = \epsilon$, we define $S \vdash_v (\alpha, \mathcal{A})$ if $\hat{\mathcal{A}}$ accepts $\boldsymbol{w}_v$. The cases where α is of the type $\exists x.\alpha'$ and $\forall x.\alpha'$ are defined similarly as for HRE, and if $S \vdash (\alpha, \mathcal{A})$, we say that $\mathcal{A}$ *accepts* S.

Definition 6. *Let* $\mathcal{A}$ *be an NFH. The* hyperlanguage of $\mathcal{A}$, *denoted* $\mathfrak{L}(\mathcal{A})$, *is the set of all hyperwords that* $\mathcal{A}$ *accepts.*

Example 3. Consider the NFH $\mathcal{A}_1$ in Fig. 1 (left), whose alphabet is $\Sigma = \{a, b\}$, over two word variables x and y. The NFH $\mathcal{A}_1$ contains an underlying standard NFA $\hat{\mathcal{A}}_1$. For two words w_1, w_2 that are assigned to x and y, respectively, $\hat{\mathcal{A}}_1$ requires that (1) w_1, w_2 agree on their a (and, consequently, on their b) positions, and (2) once one of the words has ended (denoted by $\#$), the other must only contain b letters. Since the quantification condition of $\mathcal{A}_1$ is $\forall x_1.\forall x_2$, in a hyperword S that is accepted by $\mathcal{A}_1$, every two words agree on their a positions. As a result, all the words in S must agree on their a positions. The hyperlanguage of $\mathcal{A}_1$ is then all hyperwords in which all words agree on their a positions.

Example 4. The NFH $\mathcal{A}_2$ of Fig. 1 (right) depicts the translation of the HRE of Example 1 to an NFH.

Since regular expressions are equivalent to NFA, we can translate the underlying regular expression $\hat{r}$ of an HRE r to an equivalent NFA, and vice versa – translate the underlying NFA $\hat{\mathcal{A}}$ of an NFH $\mathcal{A}$ to a regular expression. It is then easy to see that every HRE has an equivalent NFH over the same set of variables with the same quantification condition.

We consider several fragments of NFH, which limit the structure of the quantification condition α. HRE$_\forall$ is the fragment in which α contains only $\forall$ quantifiers, and similarly, in HRE$_\exists$, α contains only $\exists$ quantifiers. In the fragment HRE$_{\exists\forall}$, α is of the form $\exists x_1 \cdots \exists x_i \forall x_{i+1} \cdots \forall x_k$.

5   Properties of Regular Hyperlanguages

5.1   Closure Properties

We now consider closure properties of regular hyperlanguages. We show, via constructions on NFH, that regular hyperlanguages are closed under all the Boolean operations.

Theorem 1. *Regular hyperlanguages are closed under union, intersection, and complementation.*

Proof Sketch. Complementing an NFH $\mathcal{A}$ amounts to dualizing its quantification condition by replacing $\exists$ with $\forall$ and vice versa, and complementing $\hat{\mathcal{A}}$ via the standard construction for NFA. Since complementing $\hat{\mathcal{A}}$ is exponential in its state space, $\overline{\mathcal{A}}$ is exponential in the size of $\mathcal{A}$.

Now, let $\mathcal{A}_1$ and $\mathcal{A}_2$ be two NFH over Σ, with the variables X and Y, respectively. The NFH $\mathcal{A}_\cap$ for $\mathcal{L}(\mathcal{A}_1) \cap \mathcal{L}(\mathcal{A}_2)$ is based on the product construction of $\hat{\mathcal{A}}_1$ and $\hat{\mathcal{A}}_2$. The quantification condition of $\mathcal{A}_\cap$ is $\alpha_1 \cdot \alpha_2$. The underlying NFA $\hat{\mathcal{A}}_\cap$ advances simultaneously on both $\mathcal{A}_1$ and $\mathcal{A}_2$: when $\hat{\mathcal{A}}_1$ and $\hat{\mathcal{A}}_2$ run on word assignments $\boldsymbol{w_1}$ and $\boldsymbol{w_2}$, respectively, $\hat{\mathcal{A}}_\cap$ runs on a word assignment $\boldsymbol{w_1} \cup \boldsymbol{w_2}$, which represents both assignments $\boldsymbol{w_1}$ and $\boldsymbol{w_2}$, and accepts only if both $\hat{\mathcal{A}}_1$ and $\hat{\mathcal{A}}_2$ accept. To run on both assignments simultaneously, every letter in $\mathcal{A}_\cap$ is of the type $f_1 \cup f_2$, where $f_1 : X \rightarrow (\Sigma \cup \{\#\})$ is a letter in $\hat{\Sigma}_1$, and $f_2 : Y \rightarrow (\Sigma \cup \{\#\})$ is a letter in $\hat{\Sigma}_2$. This construction is polynomial in the sizes of $\mathcal{A}_1$ and $\mathcal{A}_2$.

Similarly, the NFH $\mathcal{A}_\cup$ for $\mathcal{L}(\mathcal{A}_1) \cup \mathcal{L}(\mathcal{A}_2)$ is based on the union construction of $\hat{\mathcal{A}}_1$ and $\hat{\mathcal{A}}_2$. The quantification condition of $\mathcal{A}_\cup$ is again $\alpha_1 \cdot \alpha_2$. The underlying NFA $\hat{\mathcal{A}}_\cup$ advances either on $\mathcal{A}_1$ or $\mathcal{A}_2$. For every word assignemt $\boldsymbol{w}$ read by $\hat{\mathcal{A}}_1$, the NFH $\hat{\mathcal{A}}_\cup$ reads $\boldsymbol{w} \cup \boldsymbol{w}'$, for every $\boldsymbol{w}' \in \hat{\Sigma}_2^*$, and dually, for every word $\boldsymbol{w}$ read by $\hat{\mathcal{A}}_2$, the NFH $\hat{\mathcal{A}}_\cup$ reads $\boldsymbol{w}' \cup \boldsymbol{w}$, for every $\boldsymbol{w}' \in \hat{\Sigma}_1^*$. The state space of $\mathcal{A}_\cup$ is linear in the state spaces of $\mathcal{A}_1, \mathcal{A}_2$. However, the size of the alphabet of $\mathcal{A}_\cup$ may be exponentially larger than that of $\mathcal{A}_1$ and $\mathcal{A}_2$. $\qquad\square$

5.2   Decision Procedures

We now turn to study several decision problems for the various fragments of NFH. Throughout this section, $\mathcal{A}$ is an NFH $\langle \Sigma, X, Q, Q_0, \delta, F, \alpha \rangle$, where $X = \{x_1, \dots x_k\}$.

Nonemptiness. The *nonemptiness problem* is to decide, given an NFH $\mathcal{A}$, whether $\mathcal{L}(\mathcal{A}) = \emptyset$. In [10], a reduction from the *Post correspondence problem* is used for proving the undecidability of HyperLTL satisfiability. A roughly similar reduction shows that the nonemptiness problem for NFH is, in general, undecidable. However, nonemptiness is decidable for the fragments we consider, with varying complexities.

For the alternation-free fragments, we show that a simple reachability test on their underlying automata suffices to verify nonemptiness.

For NFH$_{\exists\forall}$, we show that the problem is decidable, by checking the nonemptiness of an exponentially larger equi-empty NFA. To summarize, we have the following.

Theorem 2. *The nonemptiness problem for*

1. *NFH$_\exists$ and NFH$_\forall$ is NL-complete,*
2. *NFH$_{\exists\forall}$ is PSPACE-complete, and*
3. *NFH is undecidable.*

Proof Sketch. **NFH$_\forall$ and NFH$_\exists$.** The lower bound follows from the NL-hardness of NFA nonemptiness. For the upper bounds, an NFH$_\exists$ $\mathcal{A}_\exists$ is nonempty iff $\hat{\mathcal{A}}_\exists$ accepts some word assignment $\boldsymbol{w}_v$. Indeed, any hyperword that contains the words in v is accepted by $\mathcal{A}_\exists$. We can therefore run a restricted reachability test on $\hat{\mathcal{A}}_\exists$, that considers only consecutive transitions in which for every $x \in X$, a letter σ_x never follows $\#_x$, which guarantees a run on a legal word assignment.

We can show that an NFH$_\forall$ $\mathcal{A}_\forall$ is nonempty iff $\mathcal{A}_\forall$ accepts a hyperword of size 1. Accordingly, $\mathcal{A}_\forall$ is nonempty iff $\hat{\mathcal{A}}$ accepts a word that represents an assignment that assigns all variables the same word. We thus restrict the transitions of $\hat{\mathcal{A}}_\forall$ to fixed functions, and check the nonemptiness of the restricted NFA.

NFH$_{\exists\forall}$. We begin with a PSPACE upper bound. Let $\mathcal{A}$ be an NFH$_{\exists\forall}$ with m existential quantifiers, and let $S \in \mathfrak{L}(\mathcal{A})$. Then, there exist $w_1, \ldots, w_m \in S$, such that for every assignment $v : X \to S$ in which $v(x_i) = w_i$ for every $1 \leq i \leq m$, we have that $\hat{\mathcal{A}}$ accepts $\boldsymbol{w}_v$. In particular, $\hat{\mathcal{A}}$ accepts every assignment that agrees with v on $x_1, \ldots x_m$, and assigns only words from $\{w_1, \ldots, w_m\}$. Therefore, $\hat{\mathcal{A}}$ accepts the hyperword $\{w_1, \ldots, w_m\}$. That is, $\mathcal{A}$ is nonempty iff it accepts a hyperword of size at most m. We can construct an NFA A based on $\hat{\mathcal{A}}$ that is nonempty iff $\hat{\mathcal{A}}$ accepts all appropriate assignments of a hyperword of size m. The size of A is exponential in the size of $\hat{\mathcal{A}}$, and the result follows from the NL upper bound for NFA nonemptiness.

Next, we prove the lower-bound for NFH$_{\exists\forall}$ by a reduction from a polynomial version of the *corridor tiling problem*, defined as follows. We are given a finite set T of tiles, two relations $V \subseteq T \times T$ and $H \subseteq T \times T$, an initial tile t_0, a final tile t_f, and a bound $n > 0$. We have to decide whether there is some $m > 0$ and a tiling of a $n \times m$-grid such that (1) The tile t_0 is in the bottom left corner and the tile t_f is in the top right corner, (2) Every pair of horizontal neighbors is in H, and (3) Every pair of vertical neighbors is in V. When n is given in unary notation, the problem is known to be PSPACE-complete.

Given an instance C of the tiling problem, we construct an NFH$_{\exists\forall}$ $\mathcal{A}$ that is nonempty iff C has a solution. We encode a solution to C as a word $w_{sol} = w_1 \cdot w_2 \cdot w_m\$$ over $\Sigma = T \cup \{1, 2, \ldots n, \$\}$, where the word w_i, of the form $1 \cdot t_{1,i} \cdot 2 \cdot t_{2,i}, \ldots n \cdot t_{n,i}$, describes the contents of row i. To check that w_{sol} indeed encodes a solution, we need to make sure that: (1) w_1 begins with t_0 and w_m ends with $t_f\$$, (2) Every w_i is of the correct form, (3) Within every w_i, it holds that $(t_{j,i}, t_{j+1,i}) \in H$, and (4) For w_i, w_{i+1}, it holds that $(t_{j,i}, t_{j,i+1}) \in V$ for every $j \in [1, n]$.

Verifying conditions (1) $-$ (3) above is easy via an NFA of size $O(n|H|)$. The main obstacle is condition (4). We describe an NFH$_{\exists\forall}$ $\mathcal{A} = \langle T \cup \{0, 1, \ldots n, \$\}, \{y_1, y_2, y_3, x_1, \ldots x_{\log(n)}\}, Q, \{q_0\}, \delta, F, \alpha = \exists y_1 \exists y_2 \exists y_3 \forall x_1 \ldots \forall x_{\log(n)} \rangle$ that is nonempty iff there exists a word that satisfies conditions (1) $-$ (4). The NFH $\mathcal{A}$ only proceeds on letters whose assignments to y_1, y_1, y_3 is $r, 0, 1$, respectively, where $r \in T \cup \{1, \ldots n, \$\}$. Then $\mathcal{A}$ requires the existence of the words $0^{|w_{sol}|}$ and $1^{|w_{sol}|}$ (the 0 word and 1 word, henceforth). $\mathcal{A}$ makes sure that the word assigned to y_1 matches a correct solution w.r.t. conditions (1) $-$ (3) above. Now, we need to make sure that for every position j in a row, the tile in position j in the next row matches the current one w.r.t. V. We can use a state q_j to remember the tile in position j, and compare it to the tile in the next occurrence of j. To avoid checking all positions simultaneously (which would require exponentially many states), we use $\log(n)$ copies of the 0 and 1 words to encode j. The $\log(n)$ $\forall$ conditions make sure that every position within $1 - n$ is checked.

We limit the checks to words in which $x_1, \ldots x_{\log(n)}$ are the 0 or 1 words, by having $\hat{A}$ accept every word in which some x variable is not assigned 0 or 1. This accepts all cases in which the word assigned to y_1 is also assigned to one of the x variables.

To check that $x_1, \ldots x_{\log(n)}$ are the 0 or 1 words, $\hat{A}$ checks that the letter assignments to these variables remain constant throughout the run. In these cases, upon reading the first letter, $\hat{A}$ remembers the value j that is encoded by the assignments to $x_1, \ldots x_{\log(n)}$ in a state, and makes sure that throughout the run, the tile that occurs in the assignment to y_1 in position j in the current row matches the tile in position j in the next row.

We construct a similar reduction for the case that the number of $\forall$ quantifiers is fixed: instead of encoding the position by $\log(n)$ bits, we can directly specify the position by a word of the form j^*, for every $j \in [1, n]$, and we construct a matching NFH$_{\exists\forall}$ over $O(n)$ variables under $\exists$, and a single variable under $\forall$. □

Universality. The *universality problem* is to decide whether a given NFH A accepts every hyperword over Σ. Notice that A is universal iff $\overline{A}$ is empty. Since complementing an NFH involves an exponential blow-up, we conclude the following from the results in Sect. 5.2, combined with the PSPACE lower bound for the universality of NFA.

Theorem 3. *The universality problem for*

1. *NFH is undecidable,*
2. *NFH$_\exists$ and NFH$_\forall$ is PSPACE-complete, and*
3. *NFH$_{\forall\exists}$ is in EXPSPACE.*

Membership. We turn to study the membership problem for NFH: given an NFH A and a hyperword S, is $S \in \mathfrak{L}(A)$? When S is finite, so is the set of assignments from X to S, and so the problem is decidable. We call this case the *finite membership problem*.

Theorem 4. – *The finite membership problem for NFH is in PSPACE.*
 – *The finite membership problem for NFH with $O(\log(k))$ $\forall$ quantifiers is NP-complete.*

Proof. We can decide the membership of a finite hyperword S in $\mathfrak{L}(A)$ by iterating over all relevant assignments from X to S, and for every such assignment v, checking on-the-fly whether $\boldsymbol{w}_v \in \mathcal{L}()A$. The space size of this algorithm is polynomial in k and logarithmic in $|A|$ and in $|S|$.

When the number of $\forall$ quantifiers in $\mathcal{A}$ is $|O(\log(k))|$, we can iterate over all assignments to the $\forall$ variables in polynomial time, while guessing assignments to the variables under $\exists$. Thus, membership in this case is in NP.

We show NP-hardness for this case by a reduction from the Hamiltonian cycle problem. Given a graph $G = \langle V, E \rangle$ where $V = \{v_1, \ldots, v_n\}$ and $|E| = m$, we construct an $\text{NFH}_\exists$ $\mathcal{A}$ over $\{0, 1\}$ with n states, n variables, δ of size m, and a hyperword S of size n, as follows. $S = \{w_1, \ldots, w_n\}$, where $w_i = 0^{i-1} \cdot 1 \cdot 0^{n-i}$. The structure of $\hat{\mathcal{A}}$ is identical to that of G, and we set $Q_0 = F = \{v_1\}$. For every $(v_i, v_j) \in E$, we have $(v_i, f_i, v_j) \in \delta$, where $f_i(x_i) = 1$ and $f_i(x_j) = 0$ for every $x_j \neq x_i$. Intuitively, the i'th letter in an accepting run of $\hat{\mathcal{A}}$ marks traversing v_i. Assigning w_j to x_i means that the j'th step of the run traverses v_i. Since the words in w make sure that every $v \in V$ is traversed exactly once, and are all of length n, we have that $\mathcal{A}$ accepts S iff there exists some ordering of the words in S that matches a Hamiltonian cycle in G.

Note: For a hyperword of size ≥ 2, the size of δ must be exponential in the number k' of $\forall$ quantifiers, to account for all the assignments to these variables. Thus, if $k = O(k')$, an algorithm that uses a space of size k is in fact logarithmic in the size of $\mathcal{A}$. □

When S is infinite, it may still be finitely represented, allowing for algorithmic membership testing. We now address the problem of deciding whether a regular language $\mathcal{L}$ (given as an NFA) is accepted by an NFH. We call this *the regular membership problem for NFH*. We show that this problem is decidable for the entire class of NFH.

Theorem 5. *The regular membership problem for NFH is decidable.*

Proof. Let $A = \langle \Sigma, P, P_0, \rho, F \rangle$ be an NFA with n states.

We first extend the alphabet of A to $\Sigma \cup \{\#\}$, and extend its language to $\mathcal{L}(A) \cdot \{\#\}^*$. We next describe a procedure for deciding whether $\mathcal{L}(A) \in \mathfrak{L}(\mathcal{A})$.

For the case that $k = 1$, if $\alpha = \exists x_1$, then $\mathcal{L}(A) \in \mathfrak{L}(\mathcal{A})$ iff $\mathcal{L}(A) \cap \mathcal{L}(\hat{\mathcal{A}}) \neq \emptyset$. Otherwise, if $\alpha = \forall x_1$, then $\mathcal{L}(A) \in \mathfrak{L}(\mathcal{A})$ iff $\mathcal{L}(A) \notin \mathfrak{L}(\overline{\mathcal{A}})$, where $\overline{\mathcal{A}}$ is the NFH for $\overline{\mathfrak{L}(\mathcal{A})}$. The quantification condition for $\overline{\mathcal{A}}$ is $\exists x_1$, conforming to the base case.

For $k > 1$, we construct a sequence of NFA $A_k, A_{k-1} \ldots, A_1$ as follows. Initially, $A_k = \hat{\mathcal{A}}$. Let $A_i = \langle \Sigma_i, Q_i, Q_i^0, \delta_i, \mathcal{F}_i \rangle$. If $\mathbb{Q}_i = \exists$, then we construct A_{i-1} as follows. The set of states of A_{i-1} is $Q_i \times P$, and the set of initial states is $Q_i^0 \times P_0$. The set of accepting states is $\mathcal{F}_i \times F$. For every $(q \xrightarrow{f} q') \in \delta_i$ and every $(p \xrightarrow{f(x_i)} p') \in \rho$, we have $((q, p) \xrightarrow{f \setminus \{\sigma_{i_{x_i}}\}} (q', p')) \in \delta_{i-1}$. We denote this construction by $A \cap_{x_i} A_i$. Then, A_{i-1} accepts a word assignment $\boldsymbol{w}_v$ iff there exists a word $u \in \mathcal{L}(A)$, such that A_i accepts $\boldsymbol{w}_{v \cup \{x_i \mapsto u\}}$.

If $\mathbb{Q}_i = \forall$, then we set $A_{i-1} = \overline{A \cap_{x_i} \overline{A_i}}$. Notice that A_{i-1} accepts a word assignment $\boldsymbol{w}_v$ iff for every $u \in \mathcal{L}(A)$, it holds that A_i accepts $\boldsymbol{w}_{v \cup \{x_i \mapsto u\}}$.

For $i \in [1, k]$, let $\mathcal{A}_i$ be the NFH whose quantification condition is $\alpha_i = \mathbb{Q}_1 x_1 \cdots \mathbb{Q}_i x_i$, and whose underlying NFA is A_i. Then, according to the construction of A_{i-1}, we have that $\mathcal{L}(A) \in \mathfrak{L}(\mathcal{A}_i)$ iff $\mathcal{L}(A) \in \mathfrak{L}(\mathcal{A}_{i-1})$.

The NFH $\mathcal{A}_1$ has a single variable, and we can now apply the base case.

Every $\forall$ quantifier requires complementation, which is exponential in n. Therefore, in the worst case, the complexity of this algorithm is $O(2^{2^{\cdots^{|Q||A|}}})$, where the tower is of height k. If the number of $\forall$ quantifiers is fixed, then the complexity is $O(|Q||A|^k)$.

Containment. The *containment problem* is to decide, given NFH $\mathcal{A}_1$ and $\mathcal{A}_2$, whether $\mathfrak{L}(\mathcal{A}_1) \subseteq \mathfrak{L}(\mathcal{A}_2)$. Since we can reduce the nonemptiness problem to the containment problem, we have the following as a result of Theorem 2.

Theorem 6. *The containment problem for NFH is undecidable.*

However, the containment problem is decidable for various fragments of NFH.

Theorem 7. *The containment problem of $NFH_\exists \subseteq NFH_\forall$ and $NFH_\forall \subseteq NFH_\exists$ is PSPACE-complete. The containment problem of $NFH_{\exists\forall} \subseteq NFH_{\forall\exists}$ is in EXPSPACE.*

Proof. A lower bound for all cases follows from the PSPACE-hardness of the containment problem for NFA. For the upper bound, for two NFH $\mathcal{A}_1$ and $\mathcal{A}_2$, we have that $\mathfrak{L}(\mathcal{A}_1) \subseteq \mathfrak{L}(\mathcal{A}_2)$ iff $\mathfrak{L}(\mathcal{A}_1) \cap \overline{\mathfrak{L}(\mathcal{A}_2)} = \emptyset$. We can compute an NFH $\mathcal{A} = \mathcal{A}_1 \cap \overline{\mathcal{A}_2}$ (Theorem 1), and check its nonemptiness. Complementing $\mathcal{A}_2$ is exponential in its number of states, and the intersection construction is polynomial.

If $\mathcal{A}_1 \in NFH_\exists$ and $\mathcal{A}_2 \in NFH_\forall$ or vice versa, then $\mathcal{A}$ is an $NFH_\exists$ or $NFH_\forall$, respectively, whose nonemptiness can be decided in space that is logarithmic in $|\mathcal{A}|$.

The quantification condition of an NFH for the intersection may be any interleaving of the quantification conditions of the two intersected NFH (Theorem 1). Therefore, for the rest of the fragments, we can construct the intersection such that $\mathcal{A}$ is an $NFH_{\exists\forall}$. The exponential blow-up in complementing $\mathcal{A}_2$, along with The PSPACE upper bound of Theorem 2 gives an EXPSPACE upper bound for the rest of the cases. $\qquad\square$

6   Discussion and Future Work

We have introduced and studied *hyperlanguages* and a framework for their modeling, focusing on the basic class of regular hyperlanguages, modeled by HRE and NFH. We have shown that regular hyperlanguages are closed under set operations and are capable of expressing important hyperproperties for information-flow security policies over finite traces. We have also investigated fundamental decision procedures for various fragments of NFH. Some gaps, such as the precise lower bound for the universality and containment problems for $\text{NFH}_{\exists\forall}$, are left open.

Since our framework does not limit the type of underlying model, it can be lifted to handle hyperwords consisting of infinite words, with an underlying model designed for such languages, such as *nondeterministic Büchi automata*, which model ω-regular languages. Just as Büchi automata can express LTL, such a model can express the entire logic of HyperLTL [5].

As future work, we plan on studying non-regular hyperlanguages (e.g., context-free), and object hyperlanguages (e.g., trees). Another direction is designing learning algorithms for hyperlanguages, by exploiting known canonical forms for the underlying models, and basing on existing learning algorithms for them. The main challenge would be handling learning sets and a mechanism for learning word variables and quantifiers.

References

1. Ábrahám, E., Bonakdarpour, B.: HyperPCTL: a temporal logic for probabilistic hyperproperties. In: QEST, pp. 20–35 (2018)
2. Alpern, B., Schneider, F.: Defining liveness. Inf. Process. Lett. **24**, 181–185 (1985)
3. B. Bonakdarpour, B., Finkbeiner, B.: The complexity of monitoring hyperproperties. In: CSF, pp. 162–174 (2018)
4. Bonakdarpour, B., Sánchez, C., Schneider, G.: Monitoring hyperproperties by combining static analysis and runtime verification. In: ISoLA, pp. 8–27 (2018)
5. Clarkson, M., Finkbeiner, B., Koleini, M., Micinski, K., Rabe, M., Sánchez, C.: Temporal logics for hyperproperties. In: POST, pp. 265–284 (2014)
6. Clarkson, M., Schneider, F.: Hyperproperties. J. Comput. Securi. **18**, 1157–1210 (2010)
7. Coenen, N., Finkbeiner, B., Sánchez, C., Tentrup, L.: Verifying hyperliveness. In: Dillig, I., Tasiran, S. (eds.) CAV 2019, Part I. LNCS, vol. 11561, pp. 121–139. Springer, Cham (2019). https://doi.org/10.1007/978-3-030-25540-4_7
8. Emerson, E.A., Halpern, J.: "Sometimes" and "not never" revisited: on branching versus linear time temporal logic. J. ACM **33**, 151–178 (1986)
9. Finkbeiner, B., Haas, L., Torfah, H.: Canonical representations of k-safety hyperproperties. In: CSF 2019, pp. 17–31 (2019)
10. Finkbeiner, B., Hahn, C.: Deciding hyperproperties. In: CONCUR, pp. 13:1–13:14 (2016)
11. Finkbeiner, B., Rabe, M.N., Sánchez, C.: Algorithms for model checking hyperLTL and hyperCTL*. In: Kroening, D., Păsăreanu, C.S. (eds.) CAV 2015, Part I. LNCS, vol. 9206, pp. 30–48. Springer, Cham (2015). https://doi.org/10.1007/978-3-319-21690-4_3
12. Boudol, G., Castellani, I.: Noninterference for concurrent programs and thread. In: TCS 2002, pp. 109–130 (2002)
13. Goguen, J., Meseguer, J.: Security policies and security models. In: IEEE Symposium on Security and Privacy, pp. 11–20 (1982)
14. McCullough, D.: Noninterference and the composability of security properties. In: Proceedings of the 1988 IEEE Symposium on Security and Privacy, pp. 177–186 (1988)

15. Pnueli, A.: The temporal logic of programs. In: FOCS, pp. 46–57 (1977)
16. Sabelfeld, A., Sands, D.: Probabilistic noninterference for multi-threaded programs. In: CSFW, pp. 200–214 (2000)
17. Vardi, M., Wolper, P.: Automata theoretic techniques for modal logic of programs. J. Comput. Syst. Sci. **32**, 183–221 (1986)
18. Vardi, M., Wolper, P.: Reasoning about infinite computations. Inf. Comput. **115**, 1–37 (1994)
19. Wang, Y., Zarei, M., Bonakdarpour, B., Pajic, M.: Statistical verification of hyperproperties for cyber-physical systems. ACM Trans. Embed. Comput. Syst. (TECS) **18**, 92:1–92:23 (2019)
20. Zdancewic, S., Myers, A.: Observational determinism for concurrent program security. In: CSFW, p. 29 (2003)

Temporal Logics with Language Parameters

Jens Oliver Gutsfeld$^{(\boxtimes)}$, Markus Müller-Olm, and Christian Dielitz

Institut für Informatik, Westfälische Wilhelms-Universität Münster,
Einsteinstraße 62, 48149 Münster, Germany
{jens.gutsfeld,markus.mueller-olm,christian.dielitz}@wwu.de

Abstract. We develop a generic framework to extend the logics LTL, CTL$^+$ and CTL* by automata-based connectives from formal language classes and analyse this framework with regard to regular languages, visibly pushdown languages and (deterministic) context-free languages. More precisely, we consider how the use of different automata classes changes the expressive power of the logics and provide algorithms for the satisfiability and model checking problems induced by the use of different automata. For the model checking problem, we treat not only finite Kripke transition systems, but also visibly pushdown systems and pushdown systems. We provide completeness or undecidability results in almost all cases and show that the extensions we consider can formulate properties not expressible in classical temporal logics or regular extensions thereof.

1 Introduction

Temporal logics like LTL, CTL and CTL* are widely used for verification purposes through satisfiability checking and model checking. Their usefulness is based on their simple logical structure, their ability to capture useful safety and liveness properties and their low (polynomial) model checking complexity for fixed formulae. However, as noticed already in the seminal paper by Wolper [15], classical temporal logics are unable to express even basic regular properties like *Every other index fulfills* φ for a specification φ. Several temporal logics were developed to express (subsets of) ω-regular properties [8,10,15]. Nevertheless, these formalisms are not sufficient to specify properties like *A sequence of n requests should be followed by n acknowledgements* that are not ω-regular. While new temporal logics have also been developed for these types of properties [1], they are usually ad hoc constructions for a single class of languages or properties in the sense that there is no straightforward way to extend them to even larger language classes or to alternatively restrict their expressive power to a fragment related to a smaller language class in order to improve the model checking complexity (or obtain decidability). In this paper, we study a generic

This work was partially funded by the DFG under project MoNaLog (MU 1508/3).

A. Leporati et al. (Eds.): LATA 2021, LNCS 12638, pp. 187–199, 2021.
https://doi.org/10.1007/978-3-030-68195-1_14

approach for equipping classical temporal logics with automata-based connectives. More precisely, we take the logics LTL, CTL$^+$ and CTL* and annotate
the Until- and Release-operator with automata in order to restrict the analysis of the paths under consideration in accordance with the language specified
by the automata. We instantiate this framework with regular languages, visibly
pushdown languages and (deterministic) context-free languages and consider the
satisfiability problem and the differences in expressive power between the different instantiations. Furthermore, we study the model checking problem for these
instantiations not only against finite Kripke transition systems, but also against
(Visibly) Pushdown Systems. For both the satisfiability and the model checking
problem, we obtain a fairly complete picture, providing completeness results in
almost all cases. Our approach is inspired by the work of Axelsson et al. [3]
which introduces a similar automata-based framework for the logic CTL named
Extended CTL. While our results and algorithms are largely complementary to
theirs, some decision problems turn out not to be more expensive when switching
from Extended CTL to Extended CTL$^+$ or Extended CTL*. Thus, we obtain
increased expressivity for free.

The paper is structured as follows: in Sect. 2, we introduce some automata
models. Section 3 introduces our extended variants of LTL, CTL* and CTL$^+$
and establishes basic properties of them. Afterwards, we characterise LTL[U]
formulae by alternating automata (Sect. 4). In Sect. 5, we establish inclusion
and separation theorems for several logics. Finally, in Sect. 6 and Sect. 7, we
address the satisfiability and model checking problems for our logics.

2   Preliminaries

Let AP be a finite set of atomic propositions and Γ be a set of actions. A Kripke
Transition System (KTS) is a tuple $\mathcal{K} = (S, \rightarrow, \lambda)$ where S is a set of states,
$\rightarrow \subseteq S \times \Gamma \times S$ is a total transition relation and $\lambda : S \rightarrow 2^{AP}$ is a labelling
function. We also write $s \xrightarrow{a} s'$ to denote $(s, a, s') \in \rightarrow$. Every KTS has an initial
state $s_0 \in S$. A path is an infinite sequence of alternations of states and actions
$\pi = s_0 a_0 s_1 a_1 \ldots$ such that $s_i \xrightarrow{a_i} s_{i+1}$. We denote the suffix $s_i a_i s_{i+1} a_{i+1} \ldots$ by
π^i and the state s_i by $\pi(i)$. For every state, $Paths(s)$ denotes the paths starting
in s and $Paths(\mathcal{K})$ denotes $Paths(s_0)$. For a path $\pi = s a_0 s_1 a_1 \ldots$ starting in a
state s, we define the corresponding trace to be $\hat{\pi} = (\lambda(s)a_0)(\lambda(s_1)a_1)\ldots$. We
lift $Paths(s)$ to $Traces(s)$ and $Paths(\mathcal{K})$ to $Traces(\mathcal{K})$ in the obvious manner.
We denote the action sequence $a_0 \ldots a_n$ by $Actions(\pi, n)$.

A *Pushdown System* is a tuple $\mathcal{P} = (P, \Gamma, \Delta, \lambda, I)$ consisting of a finite,
non-empty set of control locations P, a finite, non-empty stack alphabet Γ,
a transition relation $\Delta \subseteq (P \times \Gamma) \times \Sigma \times (P \times \Gamma^*)$, a labelling function $\lambda :
P \times \Gamma \rightarrow 2^{AP}$, and a non-empty set of initial configurations $I \subseteq P \times \Gamma$. We write
$(p, \gamma) \xrightarrow{\sigma} (p', \omega)$ for $((p, \gamma), \sigma, (p', \omega)) \in \Delta$. We also write $(p, \gamma) \xrightarrow{\sigma} (p', \omega) \in \Delta$. A
configuration of a pushdown system is an element $c \in P \times \Gamma^+$. A rule $(p, \gamma) \xrightarrow{\sigma}$
$(p', \omega) \in \Delta$ is a *call* rule if $|\omega| = 2$, an *internal* rule if $|\omega| = 1$ and a *return* rule if
$|\omega| = 0$. We assume the standard semantics for PDS [3] and identify them with
their configuration graph (as a KTS).

A PDS is called a *Visibly Pushdown System* (VPS) [2] if its input alphabet Σ can be partitioned into disjoint sets Σ_{call}, Σ_{int} and Σ_{ret} such that every rule for an input symbol from the set Σ_{call} is a call rule (and analogously for the other two sets). An alphabet thus partitioned is called a *pushdown alphabet*. A Büchi Pushdown System (BPDS) $\mathcal{BP}$ is a PDS with an additional set $F \subseteq P$ of accepting control locations. A path of a BPDS $\mathcal{BP}$ is accepting if it visits some state $p \in F$ infinitely often. A word is accepted if there is an accepting path for it. We denote by $\mathcal{L}(\mathcal{BP})$ the set of words accepted by $\mathcal{BP}$. Pushdown Automata and Visibly Pushdown Automata, which we abbreviate by PDA and VPA respectively, are defined analogously, with the difference that we only consider finite runs over finite input words and a word is accepted if an accepting control location is reached after the last input symbol. We call an automaton *deterministic* (signified by prepending a D to the model class) if there is a unique transition for every input symbol, top of stack symbol and control location. We also consider deterministic and non-deterministic finite automata (DFA and NFA) which we define as PDA with only internal transitions. For all these models, the size of an automaton $\mathcal{A}$, denoted $|\mathcal{A}|$, is the sum of its number of states and number of transitions. For BPDS we make use of the following theorem:

Theorem 1 [13]. *It can be checked in time polynomial in $|\mathcal{BP}|$ whether a BPDS $\mathcal{BP}$ has an accepting path.*

We call the languages accepted by DFA *regular languages* (REG), those accepted by VPA *visibly pushdown languages* (VPL), those accepted by DPDA *deterministic context-free languages* (DCFL) and those accepted by PDA *context-free languages* (CFL). Following [3], we call a class U of finite word automata *reasonable* iff it contains automata recognising the languages Σ and Σ^* and for any $\mathcal{A} \in U$ and $\mathcal{B}$ of the same automaton type with $\mathcal{L}(\mathcal{A}) = \mathcal{L}(\mathcal{B})$, $\mathcal{B} \in U$ (closure under *equivalences*). All types of finite word automata considered are reasonable classes of automata.

Let $DIR = \{\downarrow, \downarrow_a\}$. An Alternating Jump Automaton (AJA) [4] is a 5-Tuple $(Q, \Sigma, \delta, q_0, \Omega)$ where Q is a finite set of states, Σ is a finite pushdown alphabet, $\delta : Q \times \Sigma \to \mathbb{B}^+(DIR \times Q \times Q)$ is a transition function, $q_0 \in Q$ is an initial state, and $\Omega : Q \to \mathbb{N}$ is a colouring function. Here $\mathbb{B}^+(X)$ denotes the set of positive boolean formulae over X. Let $\mathcal{A} = (Q, \Sigma, \delta, q_0, \Omega)$ be an AJA and $w = \alpha_1 \alpha_2 \cdots \in \Sigma^\omega$ an infinite word. We define two types of successor relations of indices on w. The direct successor is simply $succ(\downarrow, i) = i + 1$. The *abstract* successor $succ(\downarrow_a, i)$ is the index of the next input symbol to be read on the same recursion level for a call or internal action, if it exists. Otherwise, it is $\top$. Formally, $succ(\downarrow_a, i) = min\{j > i \mid |v|_{\Sigma_{call}} = |v|_{\Sigma_{ret}}$ for $v = \alpha_i \ldots \alpha_{j-1}\}$ if it exists and $\alpha_i \in \Sigma_{call} \cup \Sigma_{int}$, where $|v|_X$ is the number of occurences of symbols from X in v. Otherwise, $succ(\downarrow_a, i) = \top$. An AJA $\mathcal{A}$ processes the input word w by reading it from left to right starting in its initial state q_0. Whenever $\mathcal{A}$ is in a state q and reads a symbol α_i, it guesses a set of targets $T \subseteq (DIR \times Q \times Q)$ satisfying $\delta(q, \alpha_i)$. Then, it creates a copy for each $(d, q', q'') \in T$, which moves to the state q' if $j = succ(d, \alpha_i) \neq \top$ and reads the symbol α_j next or otherwise moves to state q'' and reads the next symbol α_{i+1}. Formally, an execution of $\mathcal{A}$ on w is an infinite tree $T(\mathcal{A}, w)$ with root ε and the following properties: 1) each node of $T(\mathcal{A}, w)$ is associated with a pair $(i, q) \in \mathbb{N} \times Q$ which indicates that a copy of $\mathcal{A}$ is currently in state q and reads the input symbol α_i next. For a node v, this pair is denoted by $p(v)$ and $p(\varepsilon) = (1, q_0)$. 2) For each node v in the tree with $p(v) = (i, q)$, there is a set of targets $T = \{(d_1, q_1, q_1') \ldots (d_k, q_k, q_k')\} \subseteq DIR \times Q \times Q$ satisfying $\delta(q, \alpha_i)$ such that for each child node v_h of v with $1 \le h \le k$ we have $p(v_h) = (i+1, q_h')$ if $succ(d_h, w_i) = \top$ and $p(v_h) = (succ(d_h, w_i), q_h)$ otherwise. A branch $\beta = v_0 v_1 \ldots$ of $T(\mathcal{A}, w)$ is an infinite sequence of nodes such that $v_0 = \varepsilon$ and v_i is the parent of v_{i+1} for each $i \in \mathbb{N}_0$. The set of colours appearing infinitely often in a branch is defined as $\mathcal{C}(\beta) = \{\Omega(q) \mid$ for infinitely many $j \in \mathbb{N}_0$ there is $i \in \mathbb{N}_0$ with $p(v_j) = (i, q)\}$. The execution tree $T(\mathcal{A}, w)$ is accepting iff for each branch β the smallest colour in $\mathcal{C}(\beta)$ is even (parity condition).

Theorem 2 [4]. *For every AJA $\mathcal{A}$, one can construct in exponential time a BVPS $\mathcal{A}'$ with size exponential in the size of $\mathcal{A}$ such that $\mathcal{L}(A) = \mathcal{L}(\mathcal{A}')$.*

As a special case of AJA, we define Alternating Büchi Automaton (ABA) and non-deterministic Büchi Automaton (BA) in the obvious way [5]. For ABA, we can compute BA of exponential size and perform a PSPACE emptiness test [5].

3   Logics

3.1   Extended LTL

Let U be a reasonable class of finite word automata and AP be a set of atomic propositions. For $ap \in AP$ and $\mathcal{A} \in U$, an LTL[U] formula is given by the grammar $\varphi ::= ap \mid \neg\varphi \mid \varphi \wedge \varphi \mid \varphi \mathcal{U}^{\mathcal{A}} \varphi$. The semantics of an LTL[U] formula for a path π is standard for atomic propositions, negations and boolean connectives. For the Until-operator, we set $\pi \models \varphi_1 \mathcal{U}^{\mathcal{A}} \varphi_2$ iff $\exists k : \pi^k \models \varphi_2$ and $\forall j < k : \pi^j \models \varphi_1$ and $Actions(\pi, k) \in \mathcal{L}(\mathcal{A})$. We denote by $\mathcal{L}(\varphi)$ the set of traces of paths fulfilling φ. For a KTS $\mathcal{K}$, we define $\mathcal{K} \models \varphi$ iff $Traces(\mathcal{K}) \subseteq \mathcal{L}(\varphi)$. For LTL[U] (and later logics), we denote by $|\varphi|$ the sum of the number of operators in the syntax tree of φ and the size of the automata $\mathcal{A}$ occurring in φ. For simplicity, we often write $\mathcal{U}^{\mathcal{L}}$ for some language $\mathcal{L}$ and use this as an abbreviation for $\mathcal{U}^{\mathcal{A}}$ where $\mathcal{A}$ is an automaton with $\mathcal{L}(\mathcal{A}) = \mathcal{L}$ of minimal size. We also write LTL[REG] and LTL[VPL] to emphasise that it is not relevant whether a deterministic or non-deterministic automaton is used for the respective language class, while we use the specific class of automata (e.g.. LTL[DVPA] or LTL[NFA]) otherwise. If no parameter is given for a modality, the language Σ^* is assumed. We can define $\vee$, $\Rightarrow$, *true* and *false* in the obvious manner. We define $\varphi_1 \mathcal{R}^{\mathcal{A}} \varphi_2 \equiv \neg(\neg\varphi_1 \mathcal{U}^{\mathcal{A}} \neg\varphi_2)$, $\mathcal{X}\varphi \equiv true \, \mathcal{U}^{\Sigma}\varphi$, $\mathcal{G}^{\mathcal{A}}\varphi \equiv false \mathcal{R}^{\mathcal{A}}\varphi$ and $\mathcal{F}^{\mathcal{A}}\varphi \equiv true \, \mathcal{U}^{\mathcal{A}}\varphi$. The explicit semantics of $\mathcal{R}^{\mathcal{A}}$ is as follows: $\pi \models \varphi_1 \mathcal{R}^{\mathcal{A}}\varphi_2$ iff $\forall i : Actions(\pi, i) \notin \mathcal{L}(\mathcal{A})$ or $\pi^i \models \varphi_2$ or $\exists j < i : \pi^j \models \varphi_1$.

An LTL[U] formula is in *Negation Normal Form* (NNF) if negations occur only in front of atomic propositions. In order to convert LTL[U] formulae to

NNF, we assume $\mathcal{R}^{\mathcal{A}}$ to be a first class modality and $\vee$ to be a first class operator of LTL[U]. The equivalence $\neg(\varphi_1\,\mathcal{U}^{\mathcal{A}}\,\varphi_2) \equiv (\neg\varphi_1)\,\mathcal{R}^{\mathcal{A}}(\neg\varphi_2)$ and the classical equivalences for LTL can be used to convert every LTL[U] formula into an equivalent LTL[U] formula in NNF with linear blowup.

In LTL[U], we can express the example properties mentioned in the introduction in a straightforward manner after choosing an appropriate language class and automaton: given a DFA $\mathcal{A}$ accepting all finite words of even length, the property *Every other index fulfills* φ can be expressed by the LTL[DFA] formula $\mathcal{G}^{\mathcal{A}}\varphi$. Likewise, for a VPA $\mathcal{A}$ that recognises the language $req^n grant^n$ (if req is marked a call and $grant$ a return symbol), the LTL[VPA] formula *work* $\mathcal{U}^{\mathcal{A}}$ *complete* states that an agent takes n requests and then grants the same number before completing its work. LTL[U] can also express the classical fairness property $\mathcal{G}\mathcal{F}q$ which, using additional automata, can also be refined by requirements that certain prefixes of a fair path must be contained in a language.

3.2   Extended CTL* and CTL⁺

Let U be a reasonable class of automata. For a set AP of atomic propositions and automata $\mathcal{A} \in U$, a CTL*[U] formula φ is given by the grammar $\varphi ::= ap \mid \neg\varphi \mid \varphi \wedge \varphi \mid \varphi\,\mathcal{U}^{\mathcal{A}}\,\varphi \mid E\varphi$. A CTL⁺[U] formula ψ is given by the grammar $\psi ::= ap \mid \neg\psi \mid \psi \wedge \psi \mid E\varphi$ where φ is constructed according to the grammar $\varphi ::= \psi\,\mathcal{U}^{\mathcal{A}}\,\psi \mid \neg\varphi \mid \varphi \wedge \varphi$.

The semantics of all operators except existential quantification is inherited from LTL[U] in the obvious manner. For the quantifier, we define: $\pi \models E\varphi$ iff $\exists\pi' \in Paths(\pi(0)) : \pi' \models \varphi$. We define universal quantifiers by $A\varphi \equiv \neg E\neg\varphi$. In order to compare our logics to the logics in [3], we define CTL[V, W] to be the sub-logic of CTL*[V \cup W] in which all modalities are quantified, quantifiers appear only in front of modalities and we have $\mathcal{A} \in V$ for $\mathcal{U}^{\mathcal{A}}$ modalities and $\mathcal{B} \in W$ for $\mathcal{R}^{\mathcal{B}}$ modalities. For $c \in \{+, *\}$, we use the notation CTL^c to discuss both logics simultaneously.

4   Automata for *LTL[U]*

In this section, we show how LTL[REG] formulae can be translated to ABA. Our translation is inspired by the classical translation of LTL into ABA [5].

Theorem 3. *For every LTL[REG] formula φ, there is an ABA $\mathcal{A}$ with $\mathcal{L}(\mathcal{A}) = \mathcal{L}(\varphi)$ with size quadratic in $|\varphi|$.*

Proof. Without loss of generality, φ is given in NNF. The proof is by structural induction over φ with the induction hypothesis that for every subformula ψ of φ, there is an ABA $\mathcal{A}$ with $\mathcal{L}(\mathcal{A}) = \mathcal{L}(\psi)$ and $|\mathcal{A}| \in \mathcal{O}(|\psi|)$. (Negated) Atomic propositions and logical connectives are handled in the obvious way. We consider the case $\varphi = \psi_1\mathcal{U}^{\mathcal{X}}\psi_2$ for an automaton $\mathcal{X} = (Q', \Sigma, \delta', q_0', F')$. By assumption, we have ABA $\mathcal{A}_i = (Q^i, 2^{AP} \times \Sigma, \delta^i, q_0^i, F^i)$ with $\mathcal{L}(\mathcal{A}_i) = \mathcal{L}(\psi_i)$ for $i \in \{1,2\}$. Let Q^1, Q^2 and Q' be pairwise disjoint. Intuitively, we translate the formula just like an LTL formula, but also run $\mathcal{X}$ to guess an accepting path, requiring φ_1 to hold until an accepting state of $\mathcal{X}$ is hit and then forcing φ_2 to hold. Let $\mathcal{A}_\varphi = (Q, \hat{\Sigma}, \delta, F)$ where $Q = Q_1 \cup Q_2 \cup Q'$, $\delta = \delta_1 \cup \delta_2 \cup \hat{\delta}'$, $q_0 = q_0'$ and $F = F_1 \cup F_2$. We set $\hat{\delta}'(q', (X, \sigma)) = (\delta^1(q_0^1, (X, \sigma)) \wedge \bigvee_{q'' \in \delta'(q', \sigma)} q'')$ with an additional disjunct $\delta^2(q_0^2, (X, \sigma))$ if $q' \in F'$ for all $q' \in Q'$.

Consider finally the case $\varphi = \psi_1 R^{\mathcal{X}} \psi_2$. Again, the ABA $\mathcal{A}_i = (Q^i, 2^{AP} \times \Sigma, \delta^i, q_0^i, F^i)$ with $\mathcal{L}(A_i) = \mathcal{L}(\psi_i)$ for $i \in \{1, 2\}$ are given by the induction hypothesis. Let $\mathcal{X} = (Q', \Sigma, \delta', q_0', F')$ be the NFA used in the formula and let Q^1, Q^2 as well as Q' be pairwise disjoint. Let $\mathcal{A}_\varphi = (Q, 2^{AP} \times \Sigma, \delta, F)$ where $Q = Q_1 \cup Q_2 \cup Q'$, $\delta = \delta_1 \cup \delta_2 \cup \hat{\delta}'$, $q_0 = q_0'$ and $F = F_1 \cup F_2 \cup Q'$. Finally, we set $\hat{\delta}'(q', (X, \sigma)) = (\delta_1(q_0^1, (X, \sigma)) \vee \bigwedge_{q'' \in \delta'(q', \sigma)} q'') \wedge \delta_2(q_0^2, (X, \sigma))$ for final states $q' \in F'$. For non-final states $q' \notin F'$, we set $\hat{\delta}'(q', (X, \sigma)) = \delta^1(q_0^1, (X, \sigma)) \vee (\bigwedge_{q'' \in \delta'(q', \sigma)} q'')$. For this modality, we have to distinguish several cases: if $\varphi_1 \wedge \varphi_2$ is fulfilled at a final state, φ is fulfilled there. Likewise, it suffices to require only φ_1 to hold if the current state in the NFA is non-final. If φ_1 holds, we require φ_2 to hold unless the current state is non-accepting, in which case we do not force any formulae to hold. However, unless φ_2 is released from holding by one of these cases, we always have to pursue all possible successors in the NFA lest we miss paths which reach an accepting state and thus require φ_2 to hold. All accepting states of the automata $\mathcal{A}_i$ stay accepting in order to reflect the semantics of ψ_i. Moreover, all states of the NFA $\mathcal{X}$ are declared accepting as paths on which φ_1 never holds, but φ_2 holds on any accepting state of $\mathcal{X}$ are allowed by the semantics of the $\mathcal{R}^{\mathcal{X}}$ operator. □

Using a similar approach and handling the Until-modality similar to [14], we obtain the following theorem for VPL:

Theorem 4. *Every LTL[VPL] formula φ can be translated into an AJA $\mathcal{A}$ with $\mathcal{L}(\varphi) = \mathcal{L}(\mathcal{A})$ and $|\mathcal{A}| \in \mathcal{O}(|\varphi|^2)$.*

5   Expressivity

In this section, we compare the expressive power of different logics. For two logics L, L', we write $L \leq L'$ if every formula φ in L has a matching formula φ' in L' such that for all KTS $\mathcal{K}$, $\mathcal{K} \models \varphi$ iff $\mathcal{K} \models \varphi'$. The relation $<$ is derived in the obvious way. Furthermore, we write $L \leq_{lin} L'$ (resp. $L \leq_{exp} L'$ or $L \leq_{poly} L'$) to denote that the translation of formulae in L to formulae in L' is possible in linear (resp. exponential or polynomial) time. By $L \leq_{comp} L'$, we denote that there is a computable translation from L to L'. The following is immediate:

Theorem 5

1. *LTL[U] $\leq_{lin}$ CTL*[U]*
2. *CTL[U, V] $\leq_{lin}$ CTL$^+$[W] if U $\subseteq$ W and V $\subseteq$ W*
3. *CTL$^+$[U] $\leq_{lin}$ CTL*[U]*

It was shown in [3] that CTL[U, V] cannot express the fairness property $EG\mathcal{F}q$ regardless of the automata classes chosen, which implies:

Theorem 6. $CTL^*[W] \not\leq CTL[U, V]$ *for arbitrary classes U, V, W.*

In order to use model theoretic properties needed for further separation results, we introduce the logic PDL-Δ[U] [3,12]. For a set Π of atomic programs, PDL-Δ[U] is defined as follows: every atomic proposition ap is a formula and every atomic program $\pi \in \Pi$ is a program. If φ_1 and φ_2 are formulae, then so are $\varphi_1 \wedge \varphi_2$ and $\neg\varphi_1$. If φ is a formula, then φ? is a test where the set of tests is denoted $Test$. A regular expression over $\Pi \cup Test$ is a program. If α is a program and φ is a formula, then $\langle\alpha\rangle\varphi$ is a formula. Finally, a Büchi automaton $\mathcal{A}$ of type U over the alphabet $\Sigma \cup Test$ is an ω-program and for every ω-program $\mathcal{A}$, $\Delta\mathcal{A}$ is a formula.

A formula of PDL-Δ[U] is evaluated over a structure $\mathcal{M} = (S, R, v)$ where S is a set of states, $R : \Pi \to 2^{S \times S}$ is a transition relation that assigns to atomic programs the enabled state transitions, and $v : S \to 2^{AP}$ assigns atomic propositions to states. We can interpret $\mathcal{M}$ as a KTS by interpreting atomic programs as actions. The relation R is extended to tests as follows: $R(\varphi?) = \{(s, s) \mid M, s \models \varphi\}$. Moreover, for programs α, $R(\alpha) = \{(s, s') \mid \exists w = w_1 \ldots w_m \in \mathcal{L}(\alpha) : \exists s_0 \ldots s_m \in S : s = s_0 \wedge s' = s_m \wedge (s_{i-1}, s_i) \in R(w_i)$ for all $1 \leq i \leq m\}$ where $w_i \in \Pi \cup Test$. For each Büchi automaton $\mathcal{A}$ of type U over the alphabet $\Pi \cup Test$, there is a unary relation $R_\omega(\mathcal{A})$ such that $s \in R_\omega(\mathcal{A})$ iff there is an infinite word $w = w_0 w_1 \cdots \in \mathcal{L}(\mathcal{A})$ and a sequence of states $s_0 s_1 \ldots$ such that $s_0 = s$ and $(s_i, s_{i+1}) \in R(w_i)$ for all $i \geq 0$. If $\mathcal{A}$ is a VPS, we assume $Test \subseteq \Sigma_{int}$.

The semantics of PDL-Δ[U] for a structure $\mathcal{M}$ and a state $s \in S$ is as usual for the standard connectives. Additionally, we set $\mathcal{M}, s \models \langle\alpha\rangle\varphi$ iff there is $s' \in S$ such that $(s, s') \in R(\alpha)$ and $s' \models \varphi$. We define $\mathcal{M}, s \models \Delta\mathcal{A}$ iff $s \in R_\omega(\mathcal{A})$.

Theorem 7

1. $CTL^c[REG] \leq_{exp} PDL\text{-}\Delta[REG]$.
2. $CTL^c[VPL] \leq_{exp} PDL\text{-}\Delta[VPL]$.

Proof. For the innermost existential formula $E\psi$ of a CTL^c[REG] formula φ, we build the ABA $\mathcal{A}_\psi$ and dealternate it to obtain an equivalent BA $\mathcal{A}'$. We can thus replace $E\psi$ by the formula $\psi' \equiv \Delta\mathcal{A}'$. Inductively, we always temporarily replace such a formula by a fresh atomic proposition and then, after the dealternation (with exponential blowup), replace this proposition by the test $(\psi'?)$ in the new automaton. We then integrate the BA into each other. For $CTL^c[VPL]$, the proof is analogous, replacing ABA by AJA. $\square$

For PDL-Δ[U], we make use of the following model-theoretic results:

Theorem 8 [3]

1. *Every satisfiable PDL$-\Delta$[REG] has a finite model (finite model property).*

	REG	VPL	DCFL
LTL	PSPACE-complete	EXPTIME-complete	undecidable
CTL^c	2EXPTIME-complete	3EXPTIME-complete	undecidable

Fig. 1. The complexity of the satisfiability problem for different logics.

2. *Every satisfiable* $PDL-\Delta[VPL]$ *formula has a VPS model (VPS model property).*

Due to the embedding of Theorem 7 and the fact that there is already a satisfiable CTL[VPA, VPA] formula with no finite model [3], we obtain:

Corollary 1. *1.* $CTL^c[REG]$ *has the finite model property.*
2. There is a satisfiable $CTL^c[VPL]$ *formula with no finite model.* $CTL^c[VPL]$ *has the VPS model property.*
3. $CTL^c[REG] < CTL^c[VPL]$

Theorem 9. $LTL < LTL[REG] < LTL[VPL] < LTL[DCFL]$.

Proof. In every case, the relation $\leq$ is clear. The inequality LTL < LTL[REG] follows from Wolper's argument [15] that LTL cannot express that a proposition occurs on every other index. For every formula φ of LTL[REG], there is an ABA recognising $\mathcal{L}(\varphi)$. Thus, $\mathcal{L}(\varphi)$ is ω-regular, while the language of the LTL[DCFL] formula $\mathcal{F}^{a^n b^n} true$ is not. The last strict inclusion follows from the fact that languages definable in LTL[VPL] are ω-VPLs, but the language of the formula $\mathcal{F}^{\{a^n ba^n\}} true$ is not [2].   □

By considering the class of linear KTS (which only have one path), we get:

Corollary 2. $CTL^c[VPL] < CTL^c[DCFL]$.

Finally, we can separate LTL[U] and all variants of extended CTL since LTL[U] formulae cannot distinguish between non-bisimilar, but trace equivalent KTS and CTL[V,W— cannot express the property $\psi \equiv E\mathcal{G}\mathcal{F}q$ [3]:

Theorem 10. *LTL[U] and CTL[V, W] have incomparable expressivity for arbitrary classes U, V and W.*

6   Satisfiability

In this section, we tackle the satisfiability problem for our logics. For this purpose, we call a formula φ of a temporal logic *satisfiable* if there is a KTS $\mathcal{K}$ such that $\mathcal{K} \models \varphi$. We obtain exhaustive decidability and complexity classifications in all cases. An overview of our results can be found in Fig. 1. Unfortunately, unlike for Extended CTL, satisfiability is undecidable for all our logics for language classes going beyond VPL[1] since LTL[DCFL] can express the emptiness of the intersection of two DCFL:

---

[1] Indeed, as the undecidability results for DCFL carry over to CFL, we do not consider the latter explicitly in the following.

Theorem 11

1. *LTL[DCFL] satisfiability checking is undecidable.*
2. *CTL^c[DCFL] satisfiability checking is undecidable.*

For regular languages, the complexity of the satisfiability problem for LTL[REG] does not increase beyond the complexity of the satisfiability problem for LTL since we can test an automaton of polynomial size for emptiness in PSPACE:

Theorem 12. *LTL[REG] satisfiability checking is PSPACE-complete.*

For VPL, the problem remains decidable but the complexity increases because we can express hard problems about VPA.

Theorem 13. *LTL[VPL] satisfiability checking is EXPTIME-complete.*

Proof. The emptiness test on the AJA for a formula φ can be done in EXP-TIME and thus establishes membership. For showing hardness, we reduce from the LTL[VPL] model checking problem for a formula ϕ against a finite KTS $\mathcal{K} = (S, \rightarrow, \lambda)$ which is shown to be EXPTIME-complete later in this paper. For any $s \in S$, we introduce a fresh atomic proposition a_s and write $\psi_s \equiv \bigvee_{(s,t,s') \in \delta} (\mathcal{X}^{\{t\}}(a_{s'} \wedge \bigwedge_{a \in \lambda(s')} a \wedge \bigwedge_{a' \notin \lambda(s')} \neg a'))$. Then, the formula $\varphi' = a_{s_0} \wedge \bigwedge_{a \in \lambda(s_0)} a \wedge \bigwedge_{a' \notin \lambda(s_0)} \neg a' \wedge \mathcal{G}(\bigwedge_{s \in S}(a_s \Rightarrow \psi_s)) \wedge \neg \varphi$ is satisfiable iff there is a path in $\mathcal{K}$ violating φ. □

For the satisfiability problem of branching time logics, we follow the approach of [3] and use established results for the satisfiability problem of PDL-Δ[U].

Theorem 14 [7,12]

1. *PDL$-\Delta$[REG] satisfiability is EXPTIME-complete.*
2. *PDL$-\Delta$[VPL] satisfiability is 2EXPTIME-complete.*

Using our embeddings into PDL-Δ[U], we then obtain the following result:

Theorem 15

1. *CTL^c[REG] satisfiability checking is 2EXPTIME-complete.*
2. *CTL^c[VPL] satisfiability checking is 3EXPTIME-complete.*

Proof. By Theorem 7, we have an exponential translation of CTL^c[U] into PDL-Δ[U] for $U \in \{REG, VPL\}$. Furthermore, PDL-Δ[REG] satisfiability checking is EXPTIME-complete, implying a 2EXPTIME upper bound for CTL^c[REG]. The lower bound already holds for ordinary CTL$^+$. Since the satisfiability test for PDL$-\Delta$[VPL] is possible in 2EXPTIME, we obtain a 3EXPTIME upper bound and the lower bound follows from the corresponding lower bound for CTL[DVPA,DVPA∪ NFA] [3]. □

The last theorem shows that parametrisation by regular languages does not increase the complexity of the satisfiability problem for either CTL* or CTL+. While the use of VPL increases the complexity exponentially, this already holds for Extended CTL. Finally, the lower bound for CTL^c satisfiability and the complexity of PDL-Δ[U] satisfiability imply that there is no polynomial-time translation of CTL^c[U] into that logic. Since there is a linear translation of CTL[U,V] into PDL-Δ[$U \cup V$], a possible translation of Extended CTL+ into Extended CTL (if any) must involve an exponential overhead:

Corollary 3. *Let* $U \in \{REG, VPL\}$.

1 CTL^c[U] $\not\leq_{poly}$ PDL$-\Delta$[U].
2 CTL+[U] $\not\leq_{poly}$ CTL[U, U].

	PSPACE	EXPTIME	2EXPTIME	undecidable
KTS	LTL[REG]	LTL[VPL]		LTL[DCFL]
	CTL^c[REG]	CTL^c[VPL]		CTL^c[DCFL]
VPS		LTL[REG/VPL]	CTL+[REG/VPL](i)	LTL[DCFL]
		CTL+[REG/VPL](h)	CTL*[REG/VPL]	CTL^c[DCFL]
PDS		LTL[REG]	CTL+[REG](i)	LTL[VPL/DCFL]
		CTL+[REG](h)	CTL*[REG]	CTL^c[VPL/DCFL]

Fig. 2. The model checking complexity for different logics and models. (h) indicates hardness, (i) indicates inclusion and no annotation indicates completeness.

7 Model Checking

We discuss the model checking problem for extended logics and different model classes. An overview of our results can be found in Fig. 2. We begin with finite KTS and DCFL. The model checking problem also is undecidable in this case since LTL[DCFL] formulae can encode the intersection emptiness problem for DCFL:

Theorem 16. *Both LTL[DCFL] and CTL^c[DCFL] model checking for KTS are undecidable.*

As a consequence, since CTL[DCFL, DCFL] has a decidable model checking problem [3], there can be no computable embedding of CTL+ equipped with DCFL into corresponding CTL variants, which is surprising because the basic logics CTL+ and CTL have the same expressive power [6]:

Corollary 4. *CTL+[DCFL] $\not\leq_{comp}$ CTL[DCFL,DCFL].*

The last result shows that there is no generic translation from Extended CTL+ into Extended CTL that is uniform for all types of automata.

As for satisfiability, LTL[REG] model checking is not more costly than LTL model checking:

Theorem 17. *LTL[REG] model checking for finite KTS is PSPACE-complete.*

Again, switching to VPL increases the complexity to EXPTIME:

Theorem 18. *LTL[VPL] model checking for finite KTS is EXPTIME-complete.*

Proof. We reduce the problem of model checking an LTL formula φ against a VPS $\mathcal{A} = (Q, \Gamma, \Delta, \lambda)$ which is EXPTIME-complete [13]. For this, we define a KTS $\mathcal{K}_{\mathcal{A}} = (Q \times \Gamma, \rightarrow, \lambda)$ as a particular regular overapproximation of $\mathcal{A}$. The KTS $\mathcal{K}_{\mathcal{A}}$ follows the evolution of the configuration heads (q, γ) in evolutions of $\mathcal{A}$. It can do so precisely for call and internal steps but guesses the topmost stack symbol in the configuration reached after a return step. In order to recover the actual executions of $\mathcal{A}$, an adequately defined DVPA $\mathcal{A}'$ is used in the formula. For this purpose, the KTS $\mathcal{K}_{\mathcal{A}}$ makes visible in the actions the symbol pushed onto the stack for a call rule in a corresponding call-symbol and the stack symbol guessed as target of a return step in a corresponding return-symbol. Then, for $\psi_{\mathcal{A}'} \equiv \mathcal{G}^{\overline{\mathcal{A}'}} false$ and $\varphi' \equiv \psi_{\mathcal{A}'} \Rightarrow \varphi$, we have $\mathcal{K}_{\mathcal{A}} \models \varphi'$ iff $\mathcal{A} \models \varphi$ since $\psi_{\mathcal{A}'}$ holds iff a path of $\mathcal{K}_{\mathcal{A}}$ corresponds to a proper path of $\mathcal{A}$. The upper bound follows from Theorem 2 since we can build an AJA for $\neg\varphi'$ and test it for emptiness. □

Our results for LTL[REG] can be used to derive a model checking algorithm for CTL^c[REG]:

Theorem 19. *CTL^c[REG] model checking for finite KTS is PSPACE-complete.*

Proof. For hardness, a classical result of complexity theory states that the problem of deciding for n different DFAs whether their intersection is non-empty is PSPACE-complete [9]. This problem can be reduced to CTL^c[DFA] model checking in polynomial time: for DFA $\mathcal{A}_1 \ldots \mathcal{A}_n$, the formula $E(\mathcal{F}^{\mathcal{A}_1} end \wedge \cdots \wedge \mathcal{F}^{\mathcal{A}_n} end)$ can be checked against a structure generating all finite words followed by end markers to test the intersection of the $\mathcal{A}_i$ for emptiness. For model checking, we use the classical CTL* [5] algorithm by iteratively applying the LTL[REG] algorithm and replacing existentially quantified LTL[REG] formulae by atomic propositions accordingly. □

Note that CTL* model checking against finite KTS is already PSPACE-complete, in contrast to CTL$^+$ model checking which is Δ_2^p-complete [11].

For CTL^c[VPL], we again make use of our results for the corresponding LTL variant to derive:

Corollary 5. *CTL^c[VPL] model checking for finite KTS is EXPTIME-complete.*

Proof. We reduce from the model checking problem for LTL against VPS $\mathcal{A}$ which is EXPTIME complete for the fixed formula $\varphi \equiv \mathcal{G}(\neg fin)$ if we use *checkpoints*, i.e. DFA accepting configurations of $\mathcal{A}_{(q,\gamma)}$ for every head (q, γ) [13]. Each transition for (q, γ) is conditional on the DFA $A_{(q,\gamma)}$ accepting the current configuration. We model this restriction by a DVPA representing $\mathcal{A}_{(q,\gamma)}$ and check it by a $\mathcal{G}$-formula as before, encoding the transitions in the transition labels of $\mathcal{K}_\mathcal{A}$. The resulting universally quantified conjunction of $\mathcal{G}^\mathcal{B} false$ formulae and φ is a CTL$^+$[DVPA] formula and completes the reduction.   □

It is important to notice that in all cases heretofore discussed, the complexity becomes polynomial if we fix the formula.

For model checking against VPS, we again make use of our approach based on alternating automata, but lift it by employing classical results of pushdown model checking. More concretely, we dealternate the ABA/AJA for a LTL[REG/VPL] formula φ, build a product with the VPS and apply an emptiness test to obtain EXPTIME-completeness which also holds for LTL [13]:

Theorem 20. *LTL[U] model checking against a VPS is EXPTIME-complete and in P for a fixed formula for $U \in \{REG, VPL\}$.*

Our AJA for LTL[VPL] can be used to extend the classical CTL* model checking algorithm for PDS [13] to CTL*[VPL]:

Theorem 21. *CTL*[U] model checking for a VPS is 2EXPTIME-complete and EXPTIME-complete for a fixed formula for $U \in \{REG, VPL\}$.*

Naturally, our algorithm can also be used for our variants of CTL$^+$ to obtain the same upper bound. However, we do not obtain completeness since the complexity of CTL$^+$ model checking on VPS (and PDS) - even without language parameters - is an open question. For PDS, we can use the same argument as for VPS when it comes to regular languages to obtain the same completeness results. However, for VPL, we can express the formula $E\mathcal{F}^\mathcal{A} true$ for some VPA $\mathcal{A}$ and model checking this formula against a VPS is undecidable [3]:

Theorem 22. *Both LTL[VPL] model checking and CTL^c[VPL] model checking for a PDS are undecidable.*

References

1. Alur, R., Etessami, K., Madhusudan, P.: A temporal logic of nested calls and returns. In: Jensen, K., Podelski, A. (eds.) TACAS 2004. LNCS, vol. 2988, pp. 467–481. Springer, Heidelberg (2004). https://doi.org/10.1007/978-3-540-24730-2_35
2. Alur, R., Madhusudan, P.: Visibly pushdown languages. In: STOC 2004, pp. 202–211. ACM (2004)
3. Axelsson, R., Hague, M., Kreutzer, S., Lange, M., Latte, M.: Extended computation tree logic. In: Fermüller, C.G., Voronkov, A. (eds.) LPAR 2010. LNCS, vol. 6397, pp. 67–81. Springer, Heidelberg (2010). https://doi.org/10.1007/978-3-642-16242-8_6
4. Bozzelli, L.: Alternating automata and a temporal fixpoint calculus for visibly pushdown languages. In: Caires, L., Vasconcelos, V.T. (eds.) CONCUR 2007. LNCS, vol. 4703, pp. 476–491. Springer, Heidelberg (2007). https://doi.org/10.1007/978-3-540-74407-8_32

5. Demri, S., Goranko, V., Lange, M.: Temporal Logics in Computer Science: Finite-State Systems. Cambridge Tracts in Theoretical Computer Science. Cambridge University Press, Cambridge (2016)
6. Emerson, E.A., Halpern, J.Y.: Decision procedures and expressiveness in the temporal logic of branching time. In: STOC 1982, pp. 169–180 (1982)
7. Emerson, E.A., Jutla, C.S.: The complexity of tree automata and logics of programs. SIAM J. Comput. **29**(1), 132–158 (1999)
8. Fischer, M.J., Ladner, R.E.: Propositional dynamic logic of regular programs. J. Comput. Syst. Sci. **18**(2), 194–211 (1979)
9. Kozen, D.: Lower bounds for natural proof systems. In: FOCS 1977, pp. 254–266 (1977)
10. Lange, M.: Model checking propositional dynamic logic with all extras. J. Appl. Logic **4**(1), 39–49 (2006)
11. Laroussinie, F., Markey, N., Schnoebelen, P.: Model checking CTL$^+$ and FCTL is hard. In: FOSSACS 2001, pp. 318–331 (2001)
12. Löding, C., Lutz, C., Serre, O.: Propositional dynamic logic with recursive programs. J. Log. Algebr. Program. **73**(1–2), 51–69 (2007)
13. Schwoon, S. Model checking pushdown systems. Ph.D. thesis, Technical University Munich, Germany (2002)
14. Weinert, A., Zimmermann, M.: Visibly linear dynamic logic. Theoret. Comput. Sci. **747**, 100–117 (2018)
15. Wolper, P. Temporal logic can be more expressive. Inf. Control **56**(1/2), 72–99 (1983)

Commutative Rational Term Rewriting

Mamoru Ishizuka[1], Takahito Aoto[1(✉)] [iD], and Munehiro Iwami[2] [iD]

[1] Niigata University, Niigata, Japan
ishizuka@nue.ie.niigata-u.ac.jp, aoto@ie.niigata-u.ac.jp
[2] Simane University, Shimane, Japan
munehiro@cis.shimane-u.ac.jp

Abstract. Term rewriting for rational terms, i.e. infinite terms with a finite number of different subterms, has been considered e.g. in Corradini & Gadducci (1998) and Aoto & Ketema (2012). In this paper, we consider rational term rewriting by a set of commutativity rules i.e. rules of the form $f(x, y) \rightarrow f(y, x)$, based on the framework of Aoto & Ketema (2012). A rewrite step with a commutativity rule is specified via a regular set of redex positions, thus via a finite automaton. We present some finite automata constructions that correspond to (in particular) taking inverse rewrite steps, merging two branching rewrite steps, and merging two consecutive rewrite steps. As a corollary, we show that rational rewrite steps by the commutativity rules are closed under taking equivalence of the rewrite steps.

Keywords: Rational term rewriting · Commutativity · Finite automata

1 Introduction

Term rewriting systems (TRSs) is a computational model based on equational logic [3]. Besides the standard rewriting formalism, many variations and extensions have been considered. One direction of such extensions is towards incorporating infinitary phenomena for various aspects of computation. In particular, there is a long history of investigations on infinitary rewriting where (infinitary long) rewriting of infinite terms is considered, and that on graph rewriting where rewriting of (cyclic or acyclic) term graphs is considered (see e.g. [2,10,11]). In this paper, we consider yet another such a formalism of rewriting dealing with infinitary phenomena, *rewriting of rational terms*.

Rational terms are infinite terms with a finite number of different subterms [1,5,6,8]. Unraveling a cyclic term graph into an infinite term yields a term that is rational, and rational terms are represented finitely [6–8]. In [1], a framework of rational term rewriting has been considered, and some basic decidability results concerning computations of the rewriting are given. In this framework, a rewrite step is specified by a rewrite rule and regular set of redex positions; the reduct is obtained by simultaneously rewriting at the redex positions. In this paper,

ⓒ Springer Nature Switzerland AG 2021
A. Leporati et al. (Eds.): LATA 2021, LNCS 12638, pp. 200–212, 2021.
https://doi.org/10.1007/978-3-030-68195-1_15

we consider (a variant of) rational rewriting by *commutativity rules*—rewrite rules of the form $f(x, y) \rightarrow f(y, x)$. Commutative rewriting is a basis of the *C*-unification and *AC*-unification which have been well-studied in the case of the standard rewriting [4]. To the best of our knowledge, however, commutative rewriting has been yet beyond the scope of the study in rational term rewriting.

We present some finite automata constructions that correspond to (in particular) taking inverse rewrite steps, merging two branching rewrite steps, and merging two consecutive rewrite steps of rational term rewriting by the commutativity rules. It seems such constructions have not been studied in literature. As a corollary, we show that rational rewrite steps by the commutativity rules are closed under taking equivalence of the rewrite steps.

2 Preliminaries

In this section, we explain notions and notations that will be used in this paper. Our definitions and notation follow [1].

2.1 Finite Automata

Let Σ be a finite set of symbols. An *empty* sequence is denoted by ε and the *concatenation* of finite sequence $p, q \in \Sigma^*$ is denoted by $p.q$. A *deterministic finite automaton* (*DFA* for short) is a tuple $M = \langle Q, \Sigma, \delta, q_0, F \rangle$ where Q is a set of states, Σ is a set of input symbols, $\delta : Q \times \Sigma \rightarrow Q$ is a transition function, $q_0 \in Q$ is an initial state and $F \subseteq Q$ is a set of final states. The homomorphic extension of δ is denoted by $\hat{\delta} : Q \times \Sigma^* \rightarrow Q$. Let $\mathcal{L}(M, q_i) \subseteq \Sigma^*$ ($q_i \in Q$) be the smallest set such that (i) $\varepsilon \in \mathcal{L}(\mathcal{M}, q_0)$, and (ii) if $p \in \mathcal{L}(\mathcal{M}, q)$ and $\delta(q, a) = q_i$ then $p.a \in \mathcal{L}(\mathcal{M}, q_i)$. The *language* of a DFA M is given by $\mathcal{L}(M) = \bigcup_{q_i \in F} \mathcal{L}(M, q_i)$. Let $M_1 = \langle Q_1, \Sigma, \delta_1, q_1, F_1 \rangle, M_2 = \langle Q_2, \Sigma, \delta_2, q_2, F_2 \rangle$ be DFAs and suppose $\approx \subseteq Q_1 \times Q_2$. The relation $\approx$ is a *bisimulation relation* if (i) $q_1 \approx q_2$ (ii) $p \approx q$ implies $\delta_1(p, a) \approx \delta_2(q, a)$ for any $a \in \Sigma$ and (iii) if $p \approx q$, then $p \in F_1$ iff $q \in F_2$. Two DFAs M_1, M_2 are *bisimilar* ($M_1 \approx M_2$) if there exists a bisimulation relation. The following property is known (e.g. [9]).

Proposition 1. *Let* M_1, M_2 *be DFAs. Then,* $M_1 \approx M_2$ *iff* $\mathcal{L}(M_1) = \mathcal{L}(M_2)$.

2.2 Rational Terms

We denote a set of arity-fixed *function symbols* by $\mathcal{F}$ and a countably infinite set of *variables* by $\mathcal{V}$, where $\mathcal{F} \cap \mathcal{V} = \emptyset$. The *arity* of function symbol $f \in \mathcal{F}$ is denoted by $\text{arity}(f)$. Let $\mathcal{F}_n = \{f \in \mathcal{F} \mid \text{arity}(f) = n\}$. Function symbols in $\mathcal{F}_0$ are called *constants*. We assume there exists some $n \geq 0$ such that $\text{arity}(f) \leq n$ for all $f \in \mathcal{F}$. We denote the set of positive integers by $\mathbb{N}_+$, and the set of finite sequence of positive integers by $\mathbb{N}_+^*$. An *infinite term* t over $\mathcal{F}$ and $\mathcal{V}$ is a partial function from $\mathbb{N}_+^*$ to $\mathcal{F} \cup \mathcal{V}$ such that (i) $t(\varepsilon)$ is defined, and (ii) $t(p.i)$ ($i \in \mathbb{N}$) is defined iff $t(p) \in \mathcal{F}_n$ and $1 \leq i \leq n$ for some n. The set of infinite terms is denoted by $\mathcal{T}_{inf}(\mathcal{F}, \mathcal{V})$. Infinite terms are often abbreviated as terms below. The set $\mathrm{Pos}(t)$ of *positions* of a term t is the domain of the partial function t. In particular, ε is called the *root* position. A term t is a *finite* term if $\mathrm{Pos}(t)$ is a finite set. The symbol $t(p) \in \mathcal{F} \cup \mathcal{V}$ is called *the symbol at the position* p. $\mathcal{V}(t)$ is the set of variables appearing in t, that is $\mathcal{V}(t) = \{t(p) \in \mathcal{V} \mid p \in \mathrm{Pos}(t)\}$. A *subterm* $t|_p$ of t at the position $p \in \mathrm{Pos}(t)$ is a mapping given by $t|_p(q) = t(p.q)$. A term $t \in \mathcal{T}_{inf}(\mathcal{F}, \mathcal{V})$ is *rational* if the set of subterms $\{t|_p \mid p \in \mathrm{Pos}(t)\}$ of t is finite. Clearly, finite terms are always rational.

Example 1. Let $\mathsf{g}, \mathsf{h} \in \mathcal{F}_1$. Let s be a partial mapping $\{\varepsilon \mapsto \mathsf{g}, 1 \mapsto \mathsf{h}, 1.1 \mapsto x\}$. It is easy to see that s is a term; furthermore, the domain of s is a finite set $\{\varepsilon, 1, 1.1\}$, and thus s is a finite term. In usual notation, $s = \mathsf{g}(\mathsf{h}(x))$. Let t be a partial mapping given by $t(1^n) = \mathsf{g}$ for any $n \geq 0$, and undefined otherwise. Here, 1^n is the sequence of 1's of length n. Intuitively, t is an infinite term $t = \mathsf{g}(\mathsf{g}(\mathsf{g}(\cdots)))$. In fact, the set of subterm of t is given by $\{t\}$ (i.e. all subterms are equal to t), thus t is a rational term. Similarly, if we take $u = \mathsf{g}(\mathsf{h}(\mathsf{g}(\mathsf{h}(\cdots))))$, then the set of subterms of u equals to $\{u, \mathsf{h}(u)\}$, and hence u is a rational term. Clearly, $u|_{1^{2n}} = u$ and $u|_{1^{2n+1}} = \mathsf{h}(u)$ hold for each $n \geq 0$. Now, let $\mathsf{f} \in \mathcal{F}_2$ in addition, and let v be a mapping $v = \{1^i \mapsto \mathsf{f} \mid i \geq 0\} \cup \{1^i.2.1^j \mapsto \mathsf{g} \mid i \geq 0, j < i\} \cup \{1^i.2.1^i \mapsto x \mid i \geq 0\}$. Then v is an infinite term that is not rational. □

A *substitution* is a mapping $\sigma : \mathcal{V} \to \mathcal{T}_{inf}(\mathcal{F}, \mathcal{V})$ such that its domain $\mathrm{dom}(\sigma) = \{x \mid \sigma(x) \neq x\}$ is finite. A substitution is identified with its homomorphic extension; as usual, $\sigma(t)$ is rewritten as $t\sigma$.

A *regular system* is a finite set $E = \{x_1 = t_1, \ldots, x_n = t_n\}$ of equations such that the left hand sides $x_1, \ldots, x_n$ are mutually distinct variables and t_i is a finite term for all $1 \leq i \leq n$. We set its domain as $\mathcal{D}om(E) = \{x_1, \ldots, x_n\}$ and its range as $\mathcal{R}an(E) = \{t_1, \ldots, t_n\}$. We write $E(y) = t$ if $y = t \in E$. A variable $x_i \in \mathcal{D}om(E)$ is *looping* if the exists $1 \leq i_1, \ldots, i_k \leq n$ such that $x_i = t_{i_1}$, and for each $1 \leq j \leq k$, $t_{i_j} = x_{i_{(j \bmod k)+1}}$ holds. Otherwise, $x_i \in \mathcal{D}om(E)$ is *non-looping*. Let $\bot$ be a new constant and $\mathcal{F}_\bot = \mathcal{F} \cup \{\bot\}$. We define a term $E^*(x_i) \in \mathcal{T}_{inf}(\mathcal{F}_\bot, \mathcal{V})$ for each $x_i \in \mathcal{D}om(E)$ as follows:

$$E^*(x_i)(p) = \begin{cases} t_i(p) & \text{if } p \in \mathrm{Pos}(t_i) \text{ and } t_i(p) \notin \mathcal{D}om(E) \\ \bot & \text{if } t_i(p) = x_j \in \mathcal{D}om(E) \text{ and } x_j \text{ is looping} \\ E^*(x_j)(q) & \text{if there exists } p' \text{ such that } p = p'.q \\ & \text{and } t_i(p') = x_j \in \mathcal{D}om(E) \text{ and } x_j \text{ is non-looping} \\ \text{undefined} & \text{otherwise} \end{cases}$$

If $E^*(x) = t$ then the pair $\langle E, x \rangle$ (E_x in short) is called a *representation* of t.

Example 2. Let s, u be terms in Example 1. $\{y = \mathsf{g}(z), z = \mathsf{h}(x)\}_y$ and $\{y = \mathsf{g}(\mathsf{h}(x))\}_y$ are representations of s. Let $E = \{x = \mathsf{g}(y), y = \mathsf{h}(x)\}$. Then $u = E^*(x)$ and E_x is a representation of u. If we identify a mapping E with its homomorphic extension, then we have $E^0(x) = x$, $E^1(x) = \mathsf{g}(y)$, $E^2(x) = E(E(x)) = E(\mathsf{g}(y)) = \mathsf{g}(E(y)) = \mathsf{g}(\mathsf{h}(x))$, $E^3(x) = \mathsf{g}(\mathsf{h}(\mathsf{g}(y)))$, $\cdots$ whose limit will be u. On the other hand, if we set $F = \{x = y, y = x\}$, then

$F^0(x), F^1(x), F^2(x), F^3(x), \ldots$ are $x, y, x, y, \ldots$, which does not converge. Note we obtain $F^\star(x) = \bot$. Note that a non-looping regular system can be obtained by replacing every equation $x = t$ with x looping by $x = \bot$. $\qquad\Box$

Henceforth, we assume $\mathcal{F}$ contains the constant $\bot$.

The following proposition on regular systems will be used later.

Proposition 2 (Lemma 3.3 of [1]). *Let E and F be regular systems and suppose there exists a surjection $\delta : \mathcal{D}om(E) \to \mathcal{D}om(F)$ such that $\delta(y) = \delta(s) \in F$ for every $y = s \in E$, where δ is homomorphically extended to a substitution on terms in the usual way. Then, $E^\star(y) = F^\star(\delta(y))$ for every $y \in \mathcal{D}om(E)$.*

Let E be a regular system and $x \in \mathcal{D}om(E)$. Then, define $\mathcal{U}_E(x)$ as the smallest set satisfying: (1) $x \in \mathcal{U}_E(x)$, and (2) if $y \in \mathcal{U}_E(x)$ and $y = t \in E$ then $\mathcal{V}(t) \cap \mathcal{D}om(E) \subseteq \mathcal{U}_E(x)$. We write $y \sqsubseteq_E x$ if $y \in \mathcal{U}_E(x)$. If E is obvious from the context, the subscript $_E$ may be omitted. Next, for each $y \sqsubseteq x$ we define $SP_{E_x}(y)$ as the smallest set satisfying: (1) $\varepsilon \in SP_{E_x}(x)$ and (2) if $p \in SP_{E_x}(z)$ and there exists $z = t \in E$ such that $t|_q = y$, then $p.q \in SP_{E_x}(y)$. We also define $SP_{E_x}(y) = \emptyset$ for $y \not\sqsubseteq x$. Intuitively, $SP_{E_x}(y)$ denotes the set of positions in $E^\star(x)$ corresponding to $y \in \mathcal{D}om(E)$. Finally, we put for any set $W \subseteq \mathcal{U}_E(x)$, $SP_{E_x}(W) = \bigcup_{y \in W} SP_{E_x}(y)$; note $SP_{E_x}(U \cup W) = SP_{E_x}(U) \cup SP_{E_x}(W)$ and $SP_{E_x}(U \setminus W) = SP_{E_x}(U) \setminus SP_{E_x}(W)$ follow from the definition.

Example 3. Let $E = \{x = \mathsf{f}(y, x), y = \mathsf{g}(z), z = \mathsf{h}(y)\}$ be a regular system. Then $\mathcal{U}_E(x) = \{x, y, z\}, \mathcal{U}_E(y) = \mathcal{U}_E(z) = \{y, z\}$. If we put $E^\star(y) = \mathsf{g}(\mathsf{h}(\mathsf{g}(\mathsf{h}(\cdots)))) = s$, then $E^\star(x) = \mathsf{f}(s, \mathsf{f}(s, \mathsf{f}(\cdots)))$. Now, $SP_{E_x}(x) = \{2^n \mid n \geq 0\}$, $SP_{E_x}(y) = \{2^n.1^{2m+1} \mid n, m \geq 0\}$ and $SP_{E_x}(z) = \{2^n.1^{2m+2} \mid n, m \geq 0\}$. $\qquad\Box$

A regular system $E = \{x_1 = t_1, \ldots, x_n = t_n\}$ is *canonical* if E satisfies the condition: for each $1 \leq i \leq n$, either (i) $t_i \in \mathcal{V} \setminus \mathcal{D}om(E)$, or (ii) $t_i = f(y_1, \ldots, y_m)$ for some $f \in \mathcal{F}_m$ and $y_1, \ldots, y_m \in \mathcal{D}om(E)$. We say a representation $\langle E, x \rangle$ (or E_x) is canonical if so is E. It is known that from any regular system E one can construct a canonical regular system F such that (i) $\mathcal{D}om(E) \subseteq \mathcal{D}om(F)$ (ii) $E^\star(x) = F^\star(x)$ for all $x \in \mathcal{D}om(E)$, and (iii) $SP_{E_x}(y) = SP_{F_x}(y)$ for all $x, y \in \mathcal{D}om(E)$ such that $y \sqsubseteq x$.

2.3   Rational Term Rewriting

A pair $\langle l, r \rangle$, written also as $l \to r$, of finite terms l and r is a *rewrite rule* if $l \notin \mathcal{V}$ and $\mathcal{V}(l) \supseteq \mathcal{V}(r)$. A *term rewriting system* (*TRS* for short) is a finite set of rewrite rules. A TRS $\mathcal{R}$ is said to be *orthogonal* if l is linear term (any variable occurs at most once) for any $l \to r \in \mathcal{R}$, and there is no overlaps between rules, i.e. $l|_p$ and l' does not unify (w.l.o.g. assuming variables are disjoint) for rewrite rules $l \to r, l' \to r' \in \mathcal{R}$ and for each non-variable position p in l (when $l \to r = l' \to r'$, we moreover assume $p \neq \varepsilon$).

Definition 1. *Let $\mathcal{R}$ be an orthogonal TRS and s, t be rational terms. We have a development rewrite step $s \twoheadrightarrow_{\mathcal{R}} t$ if there exist representations E_x and F_x of s and t, resp., such that $\mathcal{D}om(E) = \mathcal{D}om(F)$, and a set $W \subseteq \mathcal{D}om(E)$ such that (1) $E(y) = F(y)$ for any $y \in \mathcal{D}om(E) \setminus W$ and (2) for any $y \in W$, there exist a rewrite rule $l \to r \in \mathcal{R}$ and a substitution ρ such that $E(y) = l\rho$ and $F(y) = r\rho$.*

We say that the rewrite step $s \twoheadrightarrow t$ is specified by $\langle E_x, F_x, W \rangle$, or $s \twoheadrightarrow t$ is a rewrite step obtained by applying the rewrite rules on W of E_x. If $\mathcal{R}$ is clear from the context, $s \twoheadrightarrow_{\mathcal{R}} t$ is abbreviated as $s \twoheadrightarrow t$. The set of *redex positions* of the rewrite step is given by $\Delta = SP_{E_x}(W)$, and we write $s \twoheadrightarrow^{\Delta} t$ to make the redex positions explicit. Note that a rewrite step may be specified by multiple representations.

Example 4. Let $\mathcal{F} = \{f, g, h, \perp\}$ and $\mathcal{R} = \{f(x, y) \to f(y, x), g(x, y) \to g(y, x)\}$. Let $E = \{x = f(x, y), y = g(y, y)\}, F = \{x = f(y, x), y = g(y, y)\}$ be regular systems. Let $W = \{x\}$. Then, we have a rewrite step $s \twoheadrightarrow^{\Delta} t$, where $s = E^{\star}(x), t = F^{\star}(x)$ and $\Delta = SP_{E_x}(W) = \{1^n \mid n \geq 0\}$. This rewrite step is specified by $\langle E_x, F_x, W \rangle$. Let $E' = \{x = f(z, y), z = f(x, y), y = g(y, y)\}, F' = \{x = f(z, y), z = f(y, x), y = g(y, y)\}$ be regular systems. Then we have $E'^{\star}(x) = s$. Thus, by applying the rewrite rule on $W' = \{z\}$ of E'_x, we obtain a rewrite step $s \twoheadrightarrow^{\Gamma} u$, where $u = F'^{\star}(x)$ and $\Gamma = SP_{E'_x}(W') = \{1^{2n+1} \mid n \geq 0\}$. Lastly, suppose $G = \{x = f(z, y), y = g(z, x), z = h(z)\}$ and $H = \{x = f(y, z), y = g(x, z), z = h(z)\}$. Then, we have $G^{\star}(x) \twoheadrightarrow H^{\star}(x)$. The step $G^{\star}(x) \twoheadrightarrow H^{\star}(x)$ is specified by $\langle E_x, F_x, \{x, y\} \rangle$. As in the last example, different rewrite rules can be employed in a single development rewrite step. □

Remark 1. In [1], a (standard) rewrite step $s \to t$ is defined in such a way that a single rewrite rule is allowed to use in a rewrite step; the restriction is needed to deal with rewriting of possibly non-orthogonal TRS in general (see Remarks 4.3 and 4.4 in [1]). Contrast to this, in the development rewrite step $s \twoheadrightarrow t$, different rewrite rules $l \to r \in \mathcal{R}$ can be employed depending on $y \in W$. Note, however, because of the orthogonality, there can not be multiple candidates for such a rewrite rule for each $y \in W$.

In this paper, we focus on development rewrite steps by a set of commutativity rewrite rules, i.e. rules of the form $f(x, y) \to f(y, x)$. It should be also clear that any development rewrite step can be specified on canonical representations because of the form of the commutativity rules. Thus, *we will w.l.o.g. specify a rewrite step via canonical representations.*

2.4   Products of Canonical Regular Systems

In this subsection, we present some basic properties of the product construction of canonical regular systems, which will be used in the subsequent proofs.

Definition 2 (product of canonical regular systems). *Let E, F be canonical regular systems. We define the product $E \times F$ of E and F as follows.*

$E \times F = \{\langle x, y \rangle = f(\langle x_1, y_1 \rangle, \ldots, \langle x_n, y_n \rangle) \mid x = f(x_1, \ldots, x_n) \in E$ and $y = f(y_1, \ldots, y_n) \in F\} \cup \{\langle x, y \rangle = z \mid x = z \in E, y = z \in F$ and $z \notin \mathcal{D}om(E) \cup \mathcal{D}om(F)\}$. Now, by regarding the pairs of variables as variables, we treat $E \times F$ as a canonical regular system.

The following lemmas characterizes the term represented by $(E \times F)_{\langle x,y \rangle}$ and the positions in it, in terms of those in E_x.

Lemma 1. Let E, F be canonical regular systems and $x \in \mathcal{D}om(E)$, $y \in \mathcal{D}om(F)$ such that $E^\star(x) = F^\star(y)$. Then, $E^\star(x) = (E \times F)^\star(\langle x, y \rangle)$.

Lemma 2. Let E, F be canonical regular systems and $x \in \mathcal{D}om(E)$, $y \in \mathcal{D}om(F)$ such that $E^\star(x) = F^\star(y)$. Let $W \subseteq \mathcal{D}om(E)$. Then, $SP_{E_x}(W) = SP_{(E \times F)_{\langle x,y \rangle}}(W \times \mathcal{D}om(F))$.

Using these lemmas, we can characterize rewrite steps of the products.

Lemma 3. Let $\mathcal{R}$ be a TRS and E_x a canonical representation of s. Suppose a rewrite step $s \xrightarrow{\Gamma}_{\mathcal{R}} t$ is obtained by applying the rewrite rules on $W \subseteq \mathcal{D}om(E)$ of E_x. Let F be a canonical regular system such that $s = F^\star(y)$. Then, $(E \times F)_{\langle x,y \rangle}$ is a representation of s, and the rewrite step $s \xrightarrow{\Gamma}_{\mathcal{R}} t$ is obtained by applying the rewrite rules on $W \times \mathcal{D}om(F)$ of $(E \times F)_{\langle x,y \rangle}$.

Proof. By the assumption, $s = E^\star(x)$ and $\Gamma = SP_{E_x}(W)$. Then by Lemma 1, we have $s = (E \times F)^\star(\langle x, y \rangle)$. Moreover, by Lemma 2, $\Gamma = SP_{E_x}(W) = SP_{(E \times F)_{\langle x,y \rangle}}(W \times \mathcal{D}om(F))$. Thus the claim follows. $\square$

3 Automata for Inverse Rewrite Steps

In what follows, we consider rewrite steps by commutativity rules and characterize the set of redex positions of rewrite steps via automata. For this, several conventions, which are going to be introduced now, are useful.

First, we assume $n = \max_{f \in \mathcal{F}} \text{arity}(f) \geq 2$; as, otherwise, one does not have any rewrite step by commutativity rules. And, for the automata characterizing the redex positions, we use DFAs over the signature $\Sigma = \{1, \ldots, n\}$; we put them as *position* automaton.

Definition 3 (position automata). A DFA $M = \langle Q, \Sigma, \delta, q_0, F \rangle$ is said to be a position DFA if $\Sigma = \{1, \ldots, n\}$.

Now, to work with position DFAs, it is useful to identify each rational term as a complete n-tree, i.e., an infinite tree where all nodes have n-children. Let us assume $\text{arity}(f) = n$ for any $f \in \mathcal{F}$ (including the case $f = \bot$). The rationale for this convention is that we encode $t = f(t_1, \ldots, t_l)$ $(l \leq n)$ over the original signature by $t^\circ = f(t_1^\circ, \ldots, t_l^\circ, t_\bot, \ldots, t_\bot)$, where $t_\bot = \{x_\bot = \bot(x_\bot, \ldots, x_\bot)\}^\star(x_\bot)$ and $x_\bot$ is a special variable reserved for this equation. Thus, *we assume an equation* $x_\perp = \perp(x_\perp, \ldots, x_\perp)$ *is (implicitly[1]) included to any regular system E. Moreover, we also identify each equation* $x = z \in E$ *where* $z \in \mathcal{V} \setminus Dom(E)$ *with the equation* $x = z(x_\perp, \ldots, x_\perp)$. Using these conventions, each rational term is identified with a complete *n*-tree labelled by $f \in \mathcal{F}$ or $z \in \mathcal{V}$.

Example 5. Let $\mathcal{F} = \{f, g, \perp\}$, $E = \{x = f(x, y, z), y = g(y), z = w\}$. We identify E with $E' = \{x = f(x, y, z), y = g(y, x_\perp, x_\perp), z = w(x_\perp, x_\perp, x_\perp), x_\perp = \perp(x_\perp, x_\perp, x_\perp)\}$. □

Let $\mathcal{F}_C \subseteq \mathcal{F}$ *and* $C = \{f(x_1, x_2, x_3, \ldots, x_n) \to f(x_2, x_1, x_3, \ldots, x_n) \mid f \in \mathcal{F}_C\}$. *This* C *is the TRS that we will consider henceforth.*

We now show that a DFA that recognized the set of redex positions of a rewrite step can be constructed via canonical regular system that specify that rewrite step.

Definition 4 (canonical DFA). *Let* E *be a canonical regular system and* $W \subseteq Dom(E)$, $x \in Dom(E)$. *Then the* canonical DFA *for* $\langle E_x, W \rangle$ *is a position DFA* $\mathcal{M}(E_x, W)$ *given by* $\langle Dom(E), \Sigma, \delta, x, W \rangle$, *where* $\delta : Dom(E) \times \Sigma \to Dom(E)$ *is defined as* $\delta(y, i) = E(y)|_i$.

Example 6. Let $\mathcal{F} = \{f, g\}$, $E = \{x = f(y, x), y = g(y, y)\}$ and $F = \{x = f(x, y), y = g(y, y)\}$. By applying commutativity rule $f(x, y) \to f(y, x)$ to $W = \{x\}$ on E_x we have $s \to^\Delta t$ where $\Delta = \{2^n \mid n \geq 0\}$, $s = E^\star(x)$ and $t = F^\star(x)$. Now, the DFA recognizing Δ is obtained as $\mathcal{M}(E_x, W) = \langle Dom(E)(= \{x, y\}), \Sigma(= \{1, 2\}), \delta, x, W \rangle$, where $\delta(z, i) = E(z)|_i$. □

Lemma 4 (redex positions and the language of canonical DFAs). *Let* $s \multimap^\Delta t$ *be a rewrite step specified by* $\langle E_x, F_x, W \rangle$. *Then* $\Delta = \mathcal{L}(\mathcal{M}(E_x, W))$.

Since commutativity rules are symmetric, the rewrite steps by commutativity rules are symmetric. From our definition and the previous lemma, the set of redex positions of the inverse rewrite step also becomes clear.

Lemma 5 (positions of inverse rewrite step). *Let* $s \multimap t$ *be a rewrite step specified by* $\langle E_x, F_x, W \rangle$. *Then we have a rewrite step* $t \multimap^\Lambda s$ *specified by* $\langle F_x, E_x, W \rangle$, *where* $\Lambda = \mathcal{L}(\mathcal{M}(F_x, W))$.

Now, what is the relation between the set Δ in $s \multimap^\Delta t$ and the set Λ in $t \multimap^\Lambda s$? Since these set Δ and Λ are regular sets, the relation should be also characterized via automata. This motives us to define an "inverse" automaton.

The following convention is very useful hereafter: for $i \in \Sigma$, we let $\bar{1} = 2, \bar{2} = 1, \bar{i} = i$ $(3 \leq i \leq n)$.

---

[1] To ease the readability, however, we omit below the equation $x_\perp = \perp(x_\perp, \ldots, x_\perp)$ if the equation is not necessary, i.e. if there is no equation in E such that its right hand side is a variable or all $f \in \mathcal{F}$ originally have the same arity.

Definition 5 (inverse automata). *Let* $M = \langle Q, \Sigma, \delta, q_0, F \rangle$ *be a position DFA. Then we define the* inverse automaton *of* M *by* $M^{-1} = \langle Q, \Sigma, \delta', q_0, F \rangle$ *where*

$$\delta'(q, i) = \begin{cases} \delta(q, \bar{i}) & \text{if } q \in F \\ \delta(q, i) & \text{otherwise.} \end{cases}$$

We remark that M^{-1} is a position DFA and $(M^{-1})^{-1} = M$.

First, we consider automata that recognize Δ and Λ of a rewrite step $s \multimap^{\Delta} t$ and its inverse $t \multimap^{\Lambda} s$ obtained by the triple $\langle E_x, F_x, W \rangle$ that specifies these rewrite step. We show that the automaton for the latter is the inverse of the one for the former.

Lemma 6 (inverse of canonical DFA). *Let* $s \multimap t$ *be a rewrite step specified by* $\langle E_x, F_x, W \rangle$. *Then we have* $\mathcal{M}(E_x, W)^{-1} = \mathcal{M}(F_x, W)$.

We now show that the inverse operation preserves the equivalence of the languages.

Lemma 7 (language preservation of inverse). *Let* M_1, M_2 *be position DFAs. If* $\mathcal{L}(M_1) = \mathcal{L}(M_2)$ *then* $\mathcal{L}(M_1^{-1}) = \mathcal{L}(M_2^{-1})$.

Based on our preparations so far, we are now going to show that regardless of the specification of rewrite steps, inverse rewrite steps are given by reducing the redex positions of the inverse automaton.

Theorem 1 (inverse rewrite steps and inverse automaton). *Let* M *be a position DFA and suppose* $s \multimap^{\Delta}_C t$ *where* $\Delta = \mathcal{L}(M)$. *For* $\Lambda = \mathcal{L}(M^{-1})$, *we have* $t \multimap^{\Lambda}_C s$.

Proof. Suppose $s \multimap^{\Delta} t$ is specified by $\langle E_x, F_x, W \rangle$. Then, by Lemma 4, we have $\Delta = \mathcal{L}(\mathcal{M}(E_x, W))$. Hence, $\mathcal{L}(M) = \Delta = \mathcal{L}(\mathcal{M}(E_x, W))$ is obtained. Then, by Lemma 7, $\mathcal{L}(M^{-1}) = \mathcal{L}(\mathcal{M}(E_x, W)^{-1})$. On the other hand, by Lemma 5, we have $t \multimap^{\Gamma} s$ where $\Gamma = \mathcal{L}(\mathcal{M}(F_x, W))$. Furthermore, by Lemma 6, $\mathcal{M}(E_x, W)^{-1} = \mathcal{M}(F_x, W)$. Thus, $\Lambda = \mathcal{L}(M^{-1}) = \mathcal{L}(\mathcal{M}(E_x, W)^{-1}) = \mathcal{L}(\mathcal{M}(F_x, W)) = \Gamma$. Therefore, from $t \multimap^{\Gamma} s$, we obtain $t \multimap^{\Lambda} s$. $\square$

Before ending this section, we remark that the results in this section hold not only for the development rewrite step $\multimap$ but also for the rewrite step $\rightarrow$, i.e. $s \rightarrow^{\Delta}_C t$ implies $t \rightarrow^{\Gamma}_C s$. The situation, however, becomes different in the next section.

4   Automata for Join of Branching Steps

From this section, we consider automata constructions that arise from branching development rewrite steps, i.e. rewrite steps of the form $t \leftomultimap s \multimap u$.

The first operation we consider is called *join* of branching steps. Let us explain the intuition of the join of rewrite steps informally. Suppose we have branching rewrite steps from s as $s \multimap^{\Gamma} t$ and $s \multimap^{\Delta} u$. The join of two rewrite steps expresses the effect of doing these two reductions *simultaneously*. However, this does not mean rewriting all the positions in $\Gamma \cup \Delta$, that is, for $p \in \Gamma \cap \Delta$, we consider applying the commutativity rule twice has an effect same as $s|_p = f(s_1, s_2) \to f(s_2, s_1) \to f(s_1, s_2) = s|_p$. That is, we regard that the one rewrite step at $s|_p$ is cancelled by the other. Thus, the join of the redex positions is defined as follows.

Definition 6 (join of position sets). *Let $\Gamma, \Delta \subseteq \mathrm{Pos}(s)$. The* join *of Γ and Δ is defined as $\Gamma \oplus \Delta = \{p \in \Gamma \mid p \notin \Delta\} \cup \{p \in \Delta \mid p \notin \Gamma\}$.*

Example 7. Let $\mathcal{F}_C = \{f, g\}$ and $s = \{x = f(y, z), y = g(y, w), z = g(w, z), w = h(w, w)\}^\star(x)$. Let $\Gamma = \{1^n \mid n \geq 0\}$ and $\Delta = \{2^n \mid n \geq 0\}$. We have $s \multimap^{\Gamma \oplus \Delta} t$, where $t = \{x = f(y, z), y = g(w, y), z = g(z, w), w = h(w, w)\}^\star(x)$.   □

We now want to achieve the effect of doing reduction at $\Gamma \oplus \Delta$ on regular systems. Note that two rewrite steps $s \multimap^\Gamma t$ and $s \multimap^\Delta u$ may be achieved using different regular systems. To synchronize two regular systems, we use the product construction.

We now introduce a notation that is used in the lemma below. Let E, E' be regular systems and $W \subseteq \mathcal{D}om(E), W' \subseteq \mathcal{D}om(E')$. We put $W \oplus W' = (W \times W'^c) \cup (W^c \times W')$. Here, $W^c = \mathcal{D}om(E) \setminus W$ and $W'^c = \mathcal{D}om(E') \setminus W'$.

Lemma 8 (join of branching steps). *Let $E_x, E'_{x'}$ be regular representations of s. Let $s \multimap^\Gamma t$ ($s \multimap^\Delta u$) be the rewrite step obtained by applying the rewrite rules on $W \subseteq \mathcal{D}om(E)$ of E_x ($W' \subseteq \mathcal{D}om(E')$ of $E'_{x'}$, respectively). Then, $(E \times E')_{\langle x, x' \rangle}$ is a regular representation of s, and by applying the rewrite rules on $W \oplus W'$ of $(E \times E')_{\langle x, x' \rangle}$, one obtains a rewrite step $s \multimap^{\Gamma \oplus \Delta} v$ for some v. (Hence, $\Gamma \oplus \Delta = \mathcal{L}(\mathcal{M}((E \times E')_{\langle x, x' \rangle}, W \oplus W')).$)*

The previous lemma motivates us to introduce the following automata construction.

Definition 7 (join automata). *We define the* join automaton $M_1 \oplus M_2$ *of two position DFAs $M_1 = \langle Q_1, \Sigma, \delta_1, q_1, F_1 \rangle$ and $M_2 = \langle Q_2, \Sigma, \delta_2, q_2, F_2 \rangle$ as follows: $M_1 \oplus M_2 = \langle Q_1 \times Q_2, \Sigma, \delta, \langle q_1, q_2 \rangle, F_1 \oplus F_2 \rangle$, where*

- *δ is given like this: $\delta(\langle x, y \rangle, i) = \langle \delta_1(x, i), \delta_2(y, i) \rangle$ and*
- *$F_1 \oplus F_2 = \{\langle x, y \rangle \mid x \in F_1, y \in Q_2 \setminus F_2\} \cup \{\langle x, y \rangle \mid x \in Q_1 \setminus F_1, y \in F_2\}$.*

Next lemmas are easily obtained.

Lemma 9 (join of canonical DFAs). *Let $E_x, E'_{x'}$ be regular representations of s, $W \subseteq \mathcal{D}om(E)$, and $W' \subseteq \mathcal{D}om(E')$. Then, $\mathcal{M}(E_x, W) \oplus \mathcal{M}(E'_x, W') = \mathcal{M}((E \times E')_{\langle x, x' \rangle}, W \oplus W')$.*

Lemma 10 (language preservation of join). *Suppose that M_1, M_2, M'_1, M'_2 are position DFAs. If $\mathcal{L}(M_1) = \mathcal{L}(M'_1)$ and $\mathcal{L}(M_2) = \mathcal{L}(M'_2)$ then $\mathcal{L}(M_1 \oplus M_2) = \mathcal{L}(M'_1 \oplus M'_2)$.*

We now arrive the main theorem of this section.

Theorem 2 (join rewrite steps and join automata). *Let M_1, M_2 be position DFAs. Suppose $s \multimap\!\!\twoheadrightarrow_C^\Gamma t$ and $s \multimap\!\!\twoheadrightarrow_C^\Delta u$, where $\Gamma = \mathcal{L}(M_1)$ and $\Delta = \mathcal{L}(M_2)$. Then, $s \multimap\!\!\twoheadrightarrow_C^{\Gamma \oplus \Delta} v$ and $\Gamma \oplus \Delta = \mathcal{L}(M_1 \oplus M_2)$ for some v.*

Proof. Suppose that the rewrite step $s \multimap\!\!\twoheadrightarrow^\Gamma t$ ($s \multimap\!\!\twoheadrightarrow^\Delta u$) is obtained by applying the rewrite rules on W of E_x (W' of $E'_{x'}$, respectively). Then $\Gamma = \mathcal{L}(M_1) = \mathcal{L}(\mathcal{M}(E_x, W))$ and $\Delta = \mathcal{L}(M_2) = \mathcal{L}(\mathcal{M}(E'_{x'}, W'))$. Then, by Lemma 10, we have $\mathcal{L}(M_1 \oplus M_2) = \mathcal{L}(\mathcal{M}(E_x, W) \oplus \mathcal{M}(E'_{x'}, W'))$. By Lemmas 8 and 9, $s \multimap\!\!\twoheadrightarrow^{\Gamma \oplus \Delta} v$ and $\Gamma \oplus \Delta = \mathcal{L}(\mathcal{M}((E \times E')_{\langle x,x'\rangle}, W \oplus W')) = \mathcal{L}(\mathcal{M}(E_x, W) \oplus \mathcal{M}(E'_{x'}, W')) = \mathcal{L}(M_1 \oplus M_2)$. □

Remark 2. For branching (standard) steps $s \to_C^\Gamma t_1$ and $s \to_C^\Delta t_2$, we obtain $s \multimap\!\!\twoheadrightarrow_C^{\Gamma \oplus \Delta} v$, as $s \to_C t_i$ implies $s \multimap\!\!\twoheadrightarrow_C t_i$. However, because the employed rules in $s \to_C^\Gamma t_1$ and $s \to_C^\Delta t_2$ may be different, it is not always the case $s \to_C^{\Gamma \oplus \Delta} v$. This is why we had to introduce the development rewrite step $\multimap\!\!\twoheadrightarrow$.

5 Automata for Difference of Branching Steps

Suppose that we have $s \multimap\!\!\twoheadrightarrow_C^\Gamma t$, $s \multimap\!\!\twoheadrightarrow_C^\Delta u$ and $s \multimap\!\!\twoheadrightarrow_C^{\Gamma \oplus \Delta} v$. Then, naturally there would be a rewrite step that will close the gap between t and v (u and v)—we will call rewrite steps such as $t \multimap\!\!\twoheadrightarrow v$ and $u \multimap\!\!\twoheadrightarrow v$ 'difference' of that branching rewrite steps. Below we present an automata construction that capture taking the difference of that branching rewrite steps.

Below, we put $(f(t_1, t_2, t_3, \ldots t_n))^C = f(t_2, t_1, t_3, \ldots, t_n)$.

Lemma 11 (difference of branching steps). *Let $E_x, E'_{x'}$ be regular representations of s. Let $s \multimap\!\!\twoheadrightarrow^\Gamma t$ ($s \multimap\!\!\twoheadrightarrow^\Delta u$) be the rewrite step obtained by applying the rewrite rules on $W \subseteq \mathcal{D}om(E)$ of E_x ($W' \subseteq \mathcal{D}om(E')$ of $E'_{x'}$, respectively). Suppose $s \multimap\!\!\twoheadrightarrow^{\Gamma \oplus \Delta} v$.*

1. *Let $F = \{\langle y, y'\rangle = w^C \mid \langle y, y'\rangle = w \in E \times E', y \in W\} \cup \{\langle y, y'\rangle = w \in E \times E' \mid y \notin W\}$. Then, $F_{\langle x,x'\rangle}$ is a regular representation of t and one obtains a rewrite step $t \multimap\!\!\twoheadrightarrow v$ by applying the rewrite rules on $\mathcal{D}om(E) \times W'$ of $F_{\langle x,x'\rangle}$.*
2. *Let $F' = \{\langle y, y'\rangle = w^C \mid \langle y, y'\rangle = w \in E \times E', y' \in W'\} \cup \{\langle y, y'\rangle = w \in E \times E' \mid y' \notin W'\}$. Then, $F'_{\langle x,x'\rangle}$ is a regular representation of u and one obtains a rewrite step $u \multimap\!\!\twoheadrightarrow v$ by applying the rewrite rules on $W \times \mathcal{D}om(E')$ of $F'_{\langle x,x'\rangle}$.*

The characterization of the previous lemma motivates us to define the difference automata as follows.

Definition 8 (difference automata). *Let $M_1 = \langle Q_1, \Sigma, \delta_1, q_1, F_1\rangle$, $M_2 = \langle Q_2, \Sigma, \delta_2, q_2, F_2\rangle$ be position DFAs. We define the difference automaton by $M_2 \setminus M_1 = \langle Q_1 \times Q_2, \Sigma, \eta, \langle q_1, q_2\rangle, Q_1 \times F_2\rangle$, where η is given like this:*

$$\eta(\langle x, y\rangle, i) = \begin{cases} \langle \delta_1(x, \bar{i}), \delta_2(y, \bar{i})\rangle & \text{if } x \in F_1 \\ \langle \delta_1(x, i), \delta_2(y, i)\rangle & \text{otherwise.} \end{cases}$$

The next lemma is shown using Lemma 11.

Lemma 12 (difference of canonical DFAs). *Let* $s \rightarrowtail^\Gamma t$ *(*$s \rightarrowtail^\Delta u$*) be obtained by applying the rewrite rules on* W *of* E_x *(on* W' *of* $E'_{x'}$*, respectively). Suppose* $s \rightarrowtail^{\Gamma \oplus \Delta} v$*. Then (1)* $t \rightarrowtail^\Lambda v$ *where* $\Lambda = \mathcal{L}(\mathcal{M}(E'_{x'}, W') \setminus \mathcal{M}(E_x, W))$*, and (2)* $u \rightarrowtail^\Pi v$ *where* $\Pi = \mathcal{L}(\mathcal{M}(E_x, W) \setminus \mathcal{M}(E'_{x'}, W'))$*.*

Lemma 13 (language preservation of difference). *Let* M_1, M'_1, M_2, M'_2 *be position DFAs such that* $\mathcal{L}(M_1) = \mathcal{L}(M'_1)$ *and* $\mathcal{L}(M_2) = \mathcal{L}(M'_2)$*. Then,* $\mathcal{L}(M_2 \setminus M_1) = \mathcal{L}(M'_2 \setminus M'_1)$*.*

Thus, we are ready to show that the difference of branching rewrite steps is characterized by the difference automata.

Theorem 3 (difference rewrite steps and difference automata). *Let* M_1, M_2 *be position DFAs. Let* $s \rightarrowtail^\Gamma_C t$ *and* $s \rightarrowtail^\Delta_C u$*, where* $\Gamma = \mathcal{L}(M_1)$ *and* $\Delta = \mathcal{L}(M_2)$*. Suppose* $s \rightarrowtail^{\Gamma \oplus \Delta}_C v$*. Then, (1)* $t \rightarrowtail^\Lambda_C v$*, where* $\Lambda = \mathcal{L}(M_2 \setminus M_1)$*, and (2)* $u \rightarrowtail^{\Lambda'}_C v$*, where* $\Lambda' = \mathcal{L}(M_1 \setminus M_2)$*.*

Proof. We here only show (1), as (2) can be shown in the symmetric way. Suppose that the rewrite step $s \rightarrowtail^\Gamma t$ ($s \rightarrowtail^\Delta u$) is obtained by applying the rewrite rules on W of E_x (W' of $E'_{x'}$, respectively). Then $\Gamma = \mathcal{L}(M_1) = \mathcal{L}(\mathcal{M}(E_x, W))$ and $\Delta = \mathcal{L}(M_2) = \mathcal{L}(\mathcal{M}(E'_{x'}, W'))$. Then, it follows from Lemma 13 that $\mathcal{L}(M_2 \setminus M_1) = \mathcal{L}(\mathcal{M}(E'_{x'}, W') \setminus \mathcal{M}(E_x, W))$. By Lemma 12, $t \rightarrowtail^\Lambda v$ by taking $\Lambda = \mathcal{L}(\mathcal{M}(E'_{x'}, W') \setminus \mathcal{M}(E_x, W)) = \mathcal{L}(M_2 \setminus M_1)$. □

6   Closure Under Equivalence

In this section, we give an application of the results in previous three sections. Namely, we show that development rewrite step $\rightarrowtail_C$ is closed under taking equivalence. It is clear from the definition that $\rightarrowtail_C$ is reflexive, and in Theorem 1 we have already shown that $\rightarrowtail_C$ is symmetric. Thus, only transitivity is yet to be shown.

We need one lemma for this.

Lemma 14. *For any position DFAs* M_1, M_2*, we have* $\mathcal{L}((M_2 \setminus M_1^{-1}) \setminus M_1) = \mathcal{L}(M_2)$*.*

Theorem 4 (merging of consecutive steps). *Let* M_1, M_2 *be position DFAs. Let* $s \rightarrowtail^\Delta_C t$ *and* $t \rightarrowtail^\Gamma_C u$*, where* $\Delta = \mathcal{L}(M_1)$ *and* $\Gamma = \mathcal{L}(M_2)$*. Then,* $s \rightarrowtail^\Lambda_C u$*, where* $\Lambda = \mathcal{L}((M_2 \setminus M_1^{-1}) \oplus M_1)$*.*

Proof. From $s \rightarrowtail^\Delta t$ and Theorem 1, we have $t \rightarrowtail^{\Delta'} s$, where $\Delta' = \mathcal{L}(M_1^{-1})$. Thus, from $t \rightarrowtail^{\Delta'} s$ and $t \rightarrowtail^\Gamma u$, we obtain by Theorem 2 that $t \rightarrowtail^{\Delta' \oplus \Gamma} v$ for some v. Furthermore, $s \rightarrowtail^\Pi v$ by the Theorem 3, where $\Pi = \mathcal{L}(M_2 \setminus M_1^{-1})$. Now we have $s \rightarrowtail^\Pi v$ and $s \rightarrowtail^\Delta t$. Thus, from Theorem 2, we have $s \rightarrowtail^\Lambda u'$ for some u', where $\Lambda = \Pi \oplus \Delta = \mathcal{L}((M_2 \setminus M_1^{-1}) \oplus M_1)$. Furthermore, we have

$t \twoheadrightarrow^{\Gamma'} u'$ by Theorem 3, where $\Gamma' = \mathcal{L}((M_2 \setminus M_1^{-1}) \setminus M_1)$. From Lemma 14, $\Gamma' = \mathcal{L}((M_2 \setminus M_1^{-1}) \setminus M_1) = \mathcal{L}(M_2) = \Gamma$. Thus, since we have $t \twoheadrightarrow^{\Gamma} u$ by our assumption, we obtain $u = u'$ from $t \twoheadrightarrow^{\Gamma'} u'$. As we have $s \twoheadrightarrow^{\Lambda} u'$, we conclude $s \twoheadrightarrow^{\Lambda} u$. $\qquad\square$

The following is an immediate corollary of Theorems 1 and 4.

Corollary 1 (closure under equivalence). *Equivalence closure of development rewrite steps is identical to a single development rewrite step in rational term rewriting of commutativity rules, i.e.* $\twoheadleftrightarrow^*_C = \twoheadrightarrow_C$ *in rational term rewriting for any set C of commutativity rules.*

7   Conclusion

We have studied development rewrite steps $\twoheadrightarrow_C$ of rational term rewriting by commutativity rules C, where each rewrite step $s \twoheadrightarrow^{\Gamma}_C t$ is specified by a regular set Γ of positions (hence by a finite automaton) in the rational term s. We have shown the inverse automata construction $()^{-1}$ such that $s \twoheadrightarrow^{\mathcal{L}(M)}_C t$ give rise to $t \twoheadrightarrow^{\mathcal{L}(M^{-1})}_C s$. We have also given the constructions of join $M_1 \oplus M_2$ and difference $M_1 \setminus M_2$ of automata M_1 and M_2 specifying branching steps $s \twoheadrightarrow^{\mathcal{L}(M_1)}_C t_1$ and $s \twoheadrightarrow^{\mathcal{L}(M_2)}_C t_2$. Then, consecutive steps $s \twoheadrightarrow^{\mathcal{L}(M_1)}_C t \twoheadrightarrow^{\mathcal{L}(M_2)}_C u$ give rise to $s \twoheadrightarrow^{\mathcal{L}(M')}_C u$ with $M' = (M_2 \setminus M_1^{-1}) \oplus M_1$. As a corollary, it has been shown that the equivalence closure $\twoheadleftrightarrow^*_C$ of development rewrite steps is identical to a single development rewrite step $\twoheadrightarrow_C$ for any set C of commutativity rules.

A possible future work would be the commutative unification in the setting of rational term rewriting. It would be also an interesting question how one can obtain the automata constructions for showing reversibility (i.e. $s \twoheadrightarrow^* t$ implies $t \twoheadrightarrow^* s$) of associative-commutative rational term rewriting. Another possible future work would be to generalize our constructions to deal with any *flat* rules.

Acknowledgement. Thanks are due to anonymous referees and Akihisa Yamada for helpful comments. This work was partially supported by a grant from JSPS No. 18K11158.

References

1. Aoto, T., Ketema, J.: Rational term rewriting revisited: decidability and confluence. In: Ehrig, H., Engels, G., Kreowski, H.-J., Rozenberg, G. (eds.) ICGT 2012. LNCS, vol. 7562, pp. 172–186. Springer, Heidelberg (2012). https://doi.org/10.1007/978-3-642-33654-6_12
2. Ariola, Z.M., Klop, J.W.: Equational term graph rewriting. Fundam. Informaticae **26**, 207–240 (1996)
3. Baader, F., Nipkow, T.: Term Rewriting and All That. Cambridge University Press, Cambridge (1998)
4. Baader, F., Snyder, W.: Unification theory. In: Handbook of Automated Reasoning vol. 1, pp. 445–533. Elsevier (2001)

5. Corradini, A.: Term rewriting in CT_Σ. In: Gaudel, M.-C., Jouannaud, J.-P. (eds.) CAAP 1993. LNCS, vol. 668, pp. 468–484. Springer, Heidelberg (1993). https://doi.org/10.1007/3-540-56610-4_83
6. Corradini, A., Gadducci, F.: Rational term rewriting. In: Nivat, M. (ed.) FoSSaCS 1998. LNCS, vol. 1378, pp. 156–171. Springer, Heidelberg (1998). https://doi.org/10.1007/BFb0053548
7. Courcelle, B.: Fundamental properties of infinite trees. Theor. Comput. Sci. **25**, 95–169 (1983)
8. Inverardi, P., Zilli, M.V.: Rational rewriting. In: Prívara, I., Rovan, B., Ruzička, P. (eds.) MFCS 1994. LNCS, vol. 841, pp. 433–442. Springer, Heidelberg (1994). https://doi.org/10.1007/3-540-58338-6_90
9. Kozen, D.C.: Automata and Computability. Springer, New York (1997)
10. Plump, D.: Term graph rewriting. In: Handbook of Graph Grammars and Computing by Graph Transformation Volume 2: Applications, Languages and Tools, pp. 3–61. World Scientific (1999)
11. Terese (ed.): Term Rewriting Systems, Cambridge Tracts in Theoretical Computer Science, vol. 55. Cambridge University Press (2003)

Context-Free Grammars with Lookahead

Takayuki Miyazaki$^{(\boxtimes)}$ and Yasuhiko Minamide

Tokyo Institute of Technology, Tokyo, Japan
miyazaki.t.af@m.titech.ac.jp, minamide@is.titech.ac.jp

Abstract. We introduce context-free grammars with lookahead. The grammars are an extension of both context-free grammars and parsing expression grammars, hence we can handle the two grammars in a unified way. To accommodate lookahead, we use a language with lookahead, which is a set of string pairs. We considered the grammar as a system of equations and give the language with lookahead by the limit of iterations from the empty set. The language class is closed under union, intersection, complement, and a weak version of concatenation and Kleene star.

Keywords: Context-free grammars · Parsing expression grammars · Lookahead · Syntactic predicates · Regular expressions · Formal language theory · Parsing · Denotational semantics

1 Introduction

We introduce context-free grammars with lookahead (CFGLa). Lookahead is used in the traditional theory of parsing and also in other areas of formal language theory [8,9,15,17,18]. It is a constraint on the following string. A positive lookahead $\&e$ and a negative lookahead $!e$ indicate that the following string starts with e and does not start with e, respectively. For example, if the languages of X and Y are $\{a^n b^n c^m \mid n, m \geq 1\}$ and $\{a^n b^m c^m \mid n, m \geq 1\}$, respectively, then the language of $(\&X)Y$ is $\{a^n b^n c^n \mid n \geq 1\}$. We can represent intersection and complement by lookahead. Besides, lookahead can be used to remove ambiguity and suppress branching. Parsing expression grammars (PEG) [9] are grammars with lookahead and the ordered choice instead of the alternation. They are widely used for parsing and have linear time parser, unambiguity, and expressiveness beyond deterministic context-free languages. CFGLa are an extension of both context-free grammars (CFG) and PEG, hence we can handle the two grammars in a unified way.

We develop the theory of CFGLa. We consider CFGLa as a system of equations over languages with lookahead [14]. Languages with lookahead are introduced to develop the theory of regular expressions with lookahead (RELa) [14,15], which are common in regular expression libraries. They are the set of pairs of strings, for example, a RELa $a(\&b)$ represents a language with lookahead $\{(a, bx) \mid x \in \Sigma^*\}$. We can naturally define union, concatenation, Kleene star, and lookahead on the set of languages with lookahead. Next, CFG can be

© Springer Nature Switzerland AG 2021
A. Leporati et al. (Eds.): LATA 2021, LNCS 12638, pp. 213–225, 2021.
https://doi.org/10.1007/978-3-030-68195-1_16

regarded as a system of language equations and the language of CFG is the least solution of the system [2,7]. Similarly, a CFGLa can be regarded as a system of equations over languages with lookahead. However, since CFGLa have negative lookahead, some grammars have no solution such as $X = !X$. Hence, we partially give the semantics by the naturally reachable solution [16], which is the limit of iterations from the empty set. This technique is derived from Boolean grammars [16], which are CFG with intersection and complement. A CFGLa is called valid if the CFGLa has the naturally reachable solution.

This study is closely related to grammars with one-sided contexts [3]. The grammars essentially support positive lookahead by intersection and right context. Their semantics is given by two approaches: a deductive system and a system of language equations. This study differs significantly in its support for negative lookahead and differs in several points such as the way to give semantics and simulation of PEG.

In Sect. 2, we define strings with lookahead as pairs of strings and languages with lookahead as sets of strings with lookahead. In Sect. 3.1, we introduce CFGLa by the naturally reachable solution on languages with lookahead. We observe some examples, for $k \geq 1$, the language $\{a^{k^n} \mid n \geq 0\}$ is represented by CFGLa. In Sect. 3.2, we show some closure properties. The recognition algorithm is omitted due to space limitations. In Sect. 3.3, we introduce operational semantics and show that CFGLa is an extension of PEG. In Sect. 4, we refer to related work and discuss the differences of expressiveness between the language class of CFGLa and other language classes.

2   Languages with Lookahead

In this section, we introduce strings with lookahead and languages with lookahead [14]. Strings with lookahead are pairs of strings with partial concatenation and positive lookahead. This is the same concept as strings with contexts [3,4]. Languages with lookahead are sets of pairs of strings with union, concatenation, Kleene star, and negative lookahead. We consider a limit of a sequence of languages with lookahead and show some basic properties.

Definition 1 (Strings with lookahead). *A string with lookahead is a pair of strings. The concatenation $\cdot$ is the partial function defined as $(x, yz) \cdot (y, z) = (xy, z)$. The positive lookahead $\&$ is defined as $\&(x, y) = (\varepsilon, xy)$.*

A string with lookahead (x, y) is a string x with a constraint that imposes the following string is y. For concatenation, the first elements of pairs are concatenated as usual: $x \cdot y = xy$. Considering the second elements of pairs, the operation is defined only when the second operand (y, z) satisfies the condition yz of the first operand. Positive lookahead creates a constraint. For any string with lookahead r, we have $\&r \cdot r = r$. The set of strings with lookahead forms a small category. For $(x, yz) : xyz \to yz$ and $(y, z) : yz \to z$, the concatenation of the two is $(xy, z) : xyz \to z$. A string with lookahead $(\varepsilon, x) : x \to x$ is the identity morphism.

Definition 2 (Languages with Lookahead). *A language with lookahead R over an alphabet Σ is a set of strings with lookahead. For two languages with lookahead R and S, the concatenation $R \cdot S$ is defined as $\{r \cdot s \mid r \in R,\ s \in S,\ r \cdot s\ \text{is defined}\}$. The power of R is defined as $R^0 = \{\varepsilon\} \times \Sigma^*$ and $R^{n+1} = R \cdot R^n$. The Kleene star R^* is defined as $\bigcup_{n \geq 0} R^n$ and R^+ is defined as $\bigcup_{n \geq 1} R^n$. The positive lookahead $\&R$ is defined as $\{\&r \mid r \in R\}$ and the negative lookahead $!R$ is defined as $(\{\varepsilon\} \times \Sigma^*) \setminus \&R$.*

Positive lookahead $\&R$ can be expressed as $!(!R)$ using negative lookahead. The operations $\cup$, $\cdot$, *, and $\&$ are monotonically increasing functions. For concatenation, that is if $R_1 \subseteq R_2$ and $S_1 \subseteq S_2$ then $R_1 \cdot S_1 \subseteq R_2 \cdot S_2$. The negative lookahead $!$ is a monotonically decreasing function, i.e. if $R_1 \subseteq R_2$ then $!R_1 \supseteq !R_2$. The set of languages with lookahead forms a Kleene algebra with tests [11,14]. The set of constraints $\mathcal{P}(\{\varepsilon\} \times \Sigma^*)$ with $\cup$, $\cdot$, and $!$ forms Boolean algebra.

Let R_n be a sequence over sets. The *limit superior* $\varlimsup_{n \to \infty} R_n$ and the *limit inferior* $\varliminf_{n \to \infty} R_n$ are defined as follows.

$$\varlimsup_{n \to \infty} R_n = \bigcap_{i \geq 0} \bigcup_{j \geq i} R_j \qquad\qquad \varliminf_{n \to \infty} R_n = \bigcup_{i \geq 0} \bigcap_{j \geq i} R_j$$

If the limit superior and the limit inferior coincide, then R_n *converges* and the *limit* $\lim_{n \to \infty} R_n$ is equal to the common value.

It is a standard result that union is continuous, i.e. $\lim_{n \to \infty} (R_n \cup S_n) = \lim_{n \to \infty} R_n \cup \lim_{n \to \infty} S_n$ for convergent sequences R_n and S_n. We show that concatenation and negative lookahead are also continuous.

Lemma 1. *For convergent sequences R_n and S_n, $\lim_{n \to \infty} (R_n \cdot S_n) = \lim_{n \to \infty} R_n \cdot \lim_{n \to \infty} S_n$.*

Proof. The key is that for a given string with lookahead r, there is a finite number of combinations of r_1 and r_2 such that $r = r_1 \cdot r_2$. First, if $r \in \varlimsup_{n \to \infty} (R_n \cdot S_n)$, then there are infinitely many n such that $r \in R_n \cdot S_n$. By the finiteness of combinations, there are r_1, r_2 and infinitely many n such that $r = r_1 \cdot r_2, r_1 \in R_n$, and $r_2 \in S_n$. Therefore, $\varlimsup_{n \to \infty} (R_n \cdot S_n) \subseteq (\varlimsup_{n \to \infty} R_n) \cdot (\varlimsup_{n \to \infty} S_n)$. Easily, $\varliminf_{n \to \infty} (R_n \cdot S_n) \supseteq (\varliminf_{n \to \infty} R_n) \cdot (\varliminf_{n \to \infty} S_n)$. Thus, $\lim_{n \to \infty} (R_n \cdot S_n) = (\lim_{n \to \infty} R_n) \cdot (\lim_{n \to \infty} S_n)$. $\square$

Lemma 2. *For convergent sequences R_n, $\lim_{n \to \infty} !R_n = ! \lim_{n \to \infty} R_n$.*

Proof. We can show this using the following equations.

$$!\left(\bigcup_{n \in I} R_n\right) = \bigcap_{n \in I} !R_n \qquad\qquad !\left(\bigcap_{n \in I} R_n\right) = \bigcup_{n \in I} !R_n$$

$\square$

The following lemma is useful to prove convergence of a sequence over sets.

Lemma 3. *Let R_n and M_n be sequences of sets, if $R_n \cap M_n$ converges and M_n converges to the universal set, then R_n converges and $\lim_{n \to \infty} R_n = \lim_{n \to \infty} (R_n \cap M_n)$.*

Proof. In general, $\varlimsup_{n \to \infty} (R_n \cap M_n) = \varlimsup_{n \to \infty} R_n \cap \varlimsup_{n \to \infty} M_n$ holds. If M_n converges to the universal set, then $\varlimsup_{n \to \infty} (R_n \cap M_n) = \varlimsup_{n \to \infty} R_n$. Therefore, R_n converges and $\lim_{n \to \infty} R_n = \lim_{n \to \infty} (R_n \cap M_n)$. □

Let B_n be a sequence over functions to sets. If for all X, the limit $\lim_{n \to \infty} B_n(X)$ converges, then B_n *converges* and the *limit* $(\lim_{n \to \infty} B_n)(X) = \lim_{n \to \infty} B_n(X)$.

3   Context-Free Grammars with Lookahead

In this section, we introduce CFGLa and illustrate some examples. We discuss closure properties and show CFGLa is an extension of PEG [9].

3.1   Definition of CFGLa

First, we introduce RELa including variables and their interpretation. Next, we introduce CFGLa and define the language with lookahead of CFGLa by the naturally reachable solution, which is the limit of iterations from the empty set. This technique is also used in Boolean grammars [16] and partial fixed-point logic [20]. A CFGLa is called valid if it has the naturally reachable solution. Lastly, we see that CFGLa is an extension of CFG and illustrate some examples.

Definition 3 (Regular expressions with lookahead). *Let Σ be an alphabet and V be a set of variables disjoint from Σ. The regular expressions with lookahead including variables $RELa(\Sigma, V)$ are given by the following syntax.*

$$e ::= \emptyset \mid \varepsilon \mid a \mid X \mid e|e \mid ee \mid !e \quad (a \in \Sigma,\ X \in V)$$

The lookahead has the highest priority, then concatenation and then alternation. For example, a RELa $!XY \mid \varepsilon$ is equal to $((!X)Y) \mid \varepsilon$. The positive lookahead $\&e$ is an abbreviation for $!(!e)$. Any character "." is an abbreviation for $a_1 \mid \ldots \mid a_n$ when $\Sigma = \{a_1, \ldots, a_n\}$. The end of a string $\$$ is an abbreviation for "$!.$".

A function $B : RELa(\Sigma, V) \to \mathcal{P}(\Sigma^* \times \Sigma^*)$ is called an *interpretation* if the following hold.

$$B(\emptyset) = \emptyset,\ B(\varepsilon) = \{\varepsilon\} \times \Sigma^*,\ B(a) = \{a\} \times \Sigma^*,$$
$$B(e_1|e_2) = B(e_1) \cup B(e_2),\ B(e_1 e_2) = B(e_1) \cdot B(e_2),\ B(!e) = !B(e)$$

The *empty interpretation* $B_\emptyset$ is an interpretation generated by $B_\emptyset(X) = \emptyset$ for all $X \in V$. For example, $B_\emptyset(a\,\&b) = \{(a, bx) \mid x \in \Sigma^*\}$ and $B_\emptyset(!X) = !\emptyset = \{\varepsilon\} \times \Sigma^*$. For two interpretations B_1 and B_2, we define $B_1 \subseteq B_2$ by $B_1(X) \subseteq B_2(X)$ for all

X. For an interpretation B and $e \in \mathrm{RELa}(\Sigma, V)$, the *language* $L_B(e)$ is defined as $\{x \mid (x, \varepsilon) \in B(e)\}$. The language is the set of strings that the end of a string can follow. We have $B(e\$) = L_B(e) \times \{\varepsilon\}$ and $B(\&(e\$)) = \{\varepsilon\} \times L_B(e)$. If e does not contain lookahead, $B_\emptyset(e) = L_{B_\emptyset}(e) \times \Sigma^*$.

Definition 4 (Context-free grammars with lookahead). *Let Σ be an alphabet and V be a set of variables disjoint from Σ. A CFGLa G is a tuple (V, Σ, P, S) where*

- *$P : V \to RELa(\Sigma, V)$ is a production function,*
- *$S \in V$ is a start variable.*

A production function P is extended to a function from $\mathrm{RELa}(\Sigma, V)$.

$$P(\emptyset) = \emptyset, \ P(\varepsilon) = \varepsilon, \ P(a) = a,$$
$$P(e_1 | e_2) = P(e_1) | P(e_2), \ P(e_1 e_2) = P(e_1) P(e_2), \ P(!e) = !P(e)$$

The power of a production function is defined as $P^0(e) = e$ and $P^{n+1}(e) = P(P^n(e))$. For example, if $P(X) = aXb|\varepsilon$, then $P^2(X) = a(aXb|\varepsilon)b|\varepsilon$.

A production function can be regarded as a system of equations. An interpretation B is called a *solution* of P if B satisfies $B = B \circ P$. Note that $B = B \circ P$ represents $B(X) = B(P(X))$ for any X. We write $\llbracket P \rrbracket(B)$ for $B \circ P$.

Lemma 4. *For an interpretation B, if a sequence $B \circ P^n$ converges, then the limit $\lim_{n \to \infty} (B \circ P^n)$ is a solution of P.*

Proof. First, we show $\lim_{n \to \infty} (B \circ P^n)$ is an interpretation. We show the case of concatenation.

$$\lim_{n \to \infty} B(P^n(e_1 e_2)) = \lim_{n \to \infty} B(P^n(e_1) P^n(e_2)) \qquad \text{(induction on } n)$$
$$= \lim_{n \to \infty} (B(P^n(e_1)) \cdot B(P^n(e_2))) \quad (B \text{ is an interpretation})$$
$$= \lim_{n \to \infty} B(P^n(e_1)) \cdot \lim_{n \to \infty} B(P^n(e_2)) \qquad \text{(Lemma 1)}$$

Next, we have $\lim_{n \to \infty} (B \circ P^n) \circ P = \lim_{n \to \infty} (B \circ P^{n+1}) = \lim_{n \to \infty} (B \circ P^n)$. Thus, $\lim_{n \to \infty} (B \circ P^n)$ is a solution. $\qquad \square$

The *naturally reachable solution* $B_{P,nat}$ is defined as $\lim_{n \to \infty} (B_\emptyset \circ P^n)$. A CFGLa is called *valid* if $B_{P,nat}$ exists. A CFGLa is called *uniquely convergent* if for all interpretation B, the limit $\lim_{n \to \infty} (B \circ P^n)$ converges to the same value. A uniquely convergent CFGLa is valid and has a unique solution.

Definition 5 (Language with lookahead). *The language with lookahead of a valid CFGLa $G = (V, \Sigma, P, S)$ is the set $\mathcal{B}(G) = B_{P,nat}(S)$ and the language of G is the set $\mathcal{L}(G) = L_{B_{P,nat}}(S)$.*

For a CFGLa without lookahead $G = (V, \Sigma, P, S)$, the function $[\![P]\!]$ is a monotone function, hence $\mathcal{B}(G) = \bigcup_{n \geq 0}(B_\emptyset \circ P^n)$. In addition, $\mathcal{B}(G) = \mathcal{L}(G) \times \Sigma^*$ holds. Therefore, the language of CFGLa without lookahead and the language of CFG coincide. Thus, CFGLa is an extension of CFG.

Example 1 (Valid Grammars).

1. A CFGLa $X = aXb \mid \varepsilon$ is uniquely convergent. The naturally reachable solution is calculated as follows.

$$B_\emptyset(P^0(X)) = B_\emptyset(X) = \emptyset,$$
$$B_\emptyset(P^1(X)) = B_\emptyset(aXb|\varepsilon) = \{\varepsilon\} \times \Sigma^*,$$
$$B_\emptyset(P^2(X)) = B_\emptyset(a(aXb|\varepsilon)b|\varepsilon) = \{\varepsilon, ab\} \times \Sigma^*,$$
$$B_\emptyset(P^3(X)) = B_\emptyset(a(a(aXb|\varepsilon)b|\varepsilon)b|\varepsilon) = \{\varepsilon, ab, aabb\} \times \Sigma^*.$$

Therefore, $B_{P,nat}(X) = \lim_{n \to \infty} B_\emptyset(P^n(X)) = \{a^n b^n \mid n \geq 0\} \times \Sigma^*$.

2. A CFGLa $X = X$ is valid and $B_{P,nat}(X) = \emptyset$. It is the same as CFG.
3. A CFGLa "$X = !Y, Y = \&Y$" is valid. For any language L, an interpretation $B(X) = \{\varepsilon\} \times L^c$, $B(Y) = \{\varepsilon\} \times L$ is a solution. Unlike in the case of CFG, there is no least solution. The naturally reachable solution is $B_{P,nat}(X) = \{\varepsilon\} \times \Sigma^*$, $B_{P,nat}(Y) = \emptyset$ and it is the desired result.
4. A CFGLa $X = !(aX)$ is valid. When $\Sigma = \{a\}$, we have the following.

$$B_\emptyset(P^n(X)) = \begin{cases} \{\varepsilon\} \times \{a^i \mid i < n, \ i \text{ is even}\} & (n \text{ is even}) \\ \{\varepsilon\} \times (\{a\}^* \setminus \{a^i \mid i < n, \ i \text{ is odd}\}) & (n \text{ is odd}) \end{cases}$$

Therefore, $B_{P,nat}(X) = \{(\varepsilon, a^{2n}) \mid n \geq 0\}$. A proper right recursion in negative lookahead is allowed.

5. A CFGLa G defined by

$$S = a\&(Y(a|\$))XSb \mid ab, \quad X = bX \mid b, \quad Y = bYb \mid a$$

is valid and $\mathcal{L}(G) = \{(ab^n)^n \mid n \geq 1\}$. If you ignore the lookahead part, the language is $\{ax_1 \ldots ax_{n-1}ab^n \mid n \geq 1, \ x_1, \ldots, x_{n-1} \in \{b\}^*\}$. Lookahead requires that all lengths of x_i be equal.

6. For $k \geq 1$, let G be $(\{X\}, \{a\}, P, X)$, where $P(X) = a\$|\&(X\$)|a^{k-1}Xa$. The CFGLa G is valid, $\mathcal{B}(G) = \{(a^n, a^m) \mid \exists l. \ n + km = k^l\}$, and $\mathcal{L}(G) = \{a^{k^n} \mid n \geq 0\}$. This grammar takes advantage of the property of lookahead and contains only positive lookahead without the end of a string. This example is essentially the same as Example 4 in [4].

Example 2 (Invalid Grammars).

1. A CFGLa $X = !X$ has no solution, hence it is invalid. Substitutions are correct for valid grammars. Contrary, substitutions do not preserve invalidity because a substitution yields a valid grammar $X = !(!X)$.

2. A CFGLa $X = !(Xx)$ has no solution because $(\varepsilon, x) \in B(X)$ if and only if $(\varepsilon, x) \notin B(!(Xx))$. A left recursion in negative lookahead is not allowed.
3. A CFGLa "$S = !X!S$, $X = \&X$" is invalid but has a unique solution $B(S) = \emptyset$, $B(X) = \{\varepsilon\} \times \Sigma^*$. Thus, validity and the existence of solutions are different. Also, uniquely convergence and the existence of a unique solution are different.
4. A CFGLa "$X = !Y, Y = !X$" is invalid. For any language L, an interpretation $B(X) = \{\varepsilon\} \times L^c$, $B(Y) = \{\varepsilon\} \times L$ is a solution. However, iterations from the empty set oscillate between $\emptyset$ and $\{\varepsilon\} \times \Sigma^*$ and do not converge. This invalid CFGLa is obtained by introducing a variable Y from a valid grammar $X = !(!X)$. The introduction of new variables does not change the set of solutions but change the convergence. This is one of the reasons we do not discuss normal forms of CFGLa in this paper.

What kind of grammars are valid? We conjecture that a CFGLa is valid if there is no left recursion through negative lookahead. Intuitively, there are no such patterns $X = !(Xx)$ or "$X = !Y, Y = !X$". It is easy to see that a CFGLa with only positive lookahead is valid because the function $[\![P]\!]$ is a monotone function.

3.2   Closure Properties

We write $\mathcal{B}(\text{CFGLa})$ for the class of languages with lookahead of CFGLa and $\mathcal{L}(\text{CFGLa})$ for the class of languages of CFGLa. First, we show that $\mathcal{B}(\text{CFGLa})$ is closed under union, concatenation, lookahead, and Kleene star. Next, we show that $\mathcal{L}(\text{CFGLa})$ is closed under union, intersection, complement, and the weak version of concatenation and Kleene star. Lastly, we show undecidability of emptiness checking.

Lemma 5. *For two valid CFGLa $G_1 = (V_1, \Sigma, P_1, S_1)$ and $G_2 = (V_2, \Sigma, P_2, S_2)$, if $V_1 \subseteq V_2$ and $P_1(X) = P_2(X)$ for all $X \in V_1$, then $B_{P_1,nat}(X) = B_{P_2,nat}(X)$ for any $X \in V_1$.*

Proof. For any $e \in \text{RELa}(\Sigma, V_1)$, $P_1(e) = P_2(e)$ by induction on e. For any $X \in V_1$, $P_1^n(X) = P_2^n(X)$ by induction on n, thus $B_{P_1,nat}(X) = B_{P_2,nat}(X)$. □

Corollary 6. $\mathcal{B}(\textit{CFGLa})$ *is closed under union, concatenation, and lookahead.*

For Kleene star, we first show a restricted version.

Proposition 7. *If $R \subseteq \Sigma^+ \times \Sigma^*$ is a language with lookahead of CFGLa, then R^* is a language with lookahead of CFGLa.*

Proof. Let G_1 be (V, Σ, P_1, X) where $\mathcal{B}(G_1) = R$ and G_2 be $(V \cup \{Y\}, \Sigma, P_1 \cup P_2, Y)$ where $Y \notin V$ and $P_2(Y) = XY|\varepsilon$. Let R_n be $B_\emptyset(P_1^n(X))$ and S_n be $\bigcup_{0 \le k \le n} R_n \cdots R_{n-k}$. Since $\mathcal{B}(G_2) = \lim_{n \to \infty} S_n \cup (\{\varepsilon\} \times \Sigma^*)$, it is sufficient that we show $\lim_{n \to \infty} S_n = R^+$. First, for any k, $R^{k+1} = \lim_{n \to \infty} (R_n \cdots R_{n-k}) = \lim_{n \to \infty} (R_n \cdots R_{n-k})$ by Lemma 1. Therefore $R^{k+1} \subseteq \lim_{n \to \infty} S_n$, thus $R^+ \subseteq \lim_{n \to \infty} S_n$.

Second, if $r \in \varprojlim_{n \to \infty} S_n$, then there exists infinitely many n_i and k_i such that $r \in R_{n_i} \cdots R_{n_i - k_i}$. There exists $r_{i0}, \ldots, r_{ik_i}$ such that $r = r_{i0} \cdots r_{ik_i}$, $r_{i0} \in R_{n_i}$, $\ldots$, $r_{ik_i} \in R_{n_i - k_i}$. By assumption, for $x \in \Sigma^*$, there are only finitely many n such that $(\varepsilon, x) \in R_n$. Therefore, the set $\{k_i \mid i \geq 1\}$ is finite. Hence, there exists $k \geq 0$ and infinitely many n_i' such that $r \in R_{n_i'} \cdots R_{n_i' - k}$. That is $r \in \varprojlim_{n \to \infty} (R_n \cdots R_{n-k}) = R^{k+1}$. Thus, $\varprojlim_{n \to \infty} S_n \subseteq R^+$. $\qquad\square$

Proposition 8. *If R is a language with lookahead of CFGLa, then $R \backslash (\{\varepsilon\} \times \Sigma^*)$ is a language with lookahead of CFGLa.*

Proof. Let V' be a copy of V. A function $f : \mathrm{RELa}(\Sigma, V) \to \mathrm{RELa}(\Sigma, V')$ is defined as follows.

$$f(\emptyset) = \emptyset, \ f(\varepsilon) = \emptyset, \ f(a) = a, \ f(X) = X'$$
$$f(e_1 | e_2) = f(e_1) | f(e_2), \ f(e_1 e_2) = f(e_1) e_2 | e_1 f(e_2), \ f(!e) = \emptyset$$

We have $B_\emptyset(f(e)) = B_\emptyset(e) \backslash (\{\varepsilon\} \times \Sigma^*)$ by induction on $e \in \mathrm{RELa}(\Sigma, V)$. For a given valid CFGLa $G_1 = (V, \Sigma, P, S)$, we consider $G_2 = (V \cup V', \Sigma, P_2, S')$ where $P_2(X) = P(X)$ for any $X \in V$ and $P_2(X') = f(P(X))$ for any $X' \in V'$. For any $e \in \mathrm{RELa}(\Sigma, V)$, we have $P_2(f(e)) = f(P(e))$, hence $P_2^n(f(e)) = f(P^n(e))$. Thus, $B_\emptyset(P_2^n(S')) = B_\emptyset(P_2^n(f(S))) = B_\emptyset(f(P^n(S))) = B_\emptyset(P^n(S)) \backslash (\{\varepsilon\} \times \Sigma^*)$. Therefore, G_2 is valid and $\mathcal{B}(G_2) = \mathcal{B}(G_1) \backslash (\{\varepsilon\} \times \Sigma^*)$. $\qquad\square$

By two propositions, we obtain the following result.

Corollary 9. $\mathcal{B}(CFGLa)$ *is closed under Kleene star.*

It is open whether $\mathcal{B}(\mathrm{CFGLa})$ is closed under intersection, complement.

Next, we show closure properties of $\mathcal{L}(\mathrm{CFGLa})$.

Proposition 10. $\mathcal{L}(CFGLa)$ *is closed under union, intersection, and complement.*

Proof. For two variables X and Y, $B(\&(X\$)Y\$) = (L_B(X) \cap L_B(Y)) \times \{\varepsilon\}$. If $B(Y) = \Sigma^* \times \Sigma^*$, then $B(!(X\$)Y\$) = (\Sigma^* \backslash L_B(X)) \times \{\varepsilon\}$. Thus, the statement holds by Lemma 5. $\qquad\square$

Hence, CFGLa can represent $\{a^n b^n c^n \mid n \geq 0\}$ and $\{xx \mid x \in \Sigma^*\}$. It is open whether $\mathcal{L}(\mathrm{CFGLa})$ is closed under concatenation and Kleene star. However, a weak version holds.

Lemma 11. *Let $L \subseteq \Sigma^*$ be a language of CFGLa and $\# \notin \Sigma$ be a letter. $L\# \times (\Sigma \cup \{\#\})^*$ is a language with lookahead of CFGLa.*

Proof. Let G_1 be (V, Σ, P_1, X) where $\mathcal{L}(G_1) = L$. Let $G_2 = (V \cup \{Y\}, \Sigma \cup \{\#\}, P_2, Y)$ where $P_1 \subseteq P_2$ and $P_2(Y) = X\#$. We define a predicate $p(R)$ as "$\forall x, y \in \Sigma^*, w \in (\Sigma \cup \{\#\})^*. \ (x, y) \in R \iff (x, y\#w) \in R$". If $p(B_{P_2, nat}(X))$ holds, then $\mathcal{B}(G_2) = L\# \times (\Sigma \cup \{\#\})^*$. For $e \in \mathrm{RELa}(\Sigma, V)$, $p(B_\emptyset(e))$ holds by induction on e. Since $P_2^n(X) \in \mathrm{RELa}(\Sigma, V)$, $p(B_{P_2, nat}(X))$ holds. $\qquad\square$

Corollary 12. *Let $L_1, L_2 \subseteq \Sigma^*$ be languages of CFGLa and $\# \notin \Sigma$ be a letter. $L_1 \# L_2$ and $(L_1 \#)^*$ are languages of CFGLa.*

Lastly, it is undecidable whether the intersection of two context-free languages is empty. For two context-free languages L_1 and L_2, $(L_1 \cap L_2) \times \{\varepsilon\}$ is a language with lookahead of CFGLa. Therefore, the following holds.

Proposition 13. *For a valid CFGLa G, emptiness $\mathcal{B}(G) = \emptyset$ and $\mathcal{L}(G) = \emptyset$ are also undecidable.*

For a given valid CFGLa $G_1 = (V, \Sigma, P_1, X)$, we consider a CFGLa $G_2 = (V \cup \{Y\}, \Sigma, P_1 \cup P_2, Y)$ with $P_2(Y) = \&(X\$)!Y$. The CFGLa G_2 is valid if and only if $\mathcal{L}(G_1) = \emptyset$. Therefore, validity is undecidable.

For a valid CFGLa G, membership $(x, y) \in \mathcal{B}(G)$ is decidable. A recognition algorithm is obtained by Brzozowski derivatives [5,13,14]. However, it is omitted in this paper due to space limitations.

3.3   Operational Semantics and Parsing Expression Grammars

The semantics by the naturally reachable solution can be considered denotational semantics. We introduce operational semantics of CFGLa and show that two semantics are consistent for a limited range of valid CFGLa. The results show that PEG can be regarded as a subclass of CFGLa.

For an expression $e \in \text{RELa}(\Sigma, V)$, we define the derivation $(e, x) \Rightarrow O$ where x is an input string and O is a set of the rest of the input string.

$$\overline{(\emptyset, x) \Rightarrow \emptyset} \qquad \overline{(\varepsilon, x) \Rightarrow \{x\}} \qquad \overline{(a, ax) \Rightarrow \{x\}} \qquad \overline{(a, bx) \Rightarrow \emptyset} \qquad \overline{(a, \varepsilon) \Rightarrow \emptyset}$$

$$\frac{(e_1, x) \Rightarrow O_1 \quad (e_2, x) \Rightarrow O_2}{(e_1 | e_2, xy) \Rightarrow O_1 \cup O_2}$$

$$\frac{(e_1, x) \Rightarrow \{y_1, \ldots, y_n\} \quad (e_2, y_1) \Rightarrow O_1 \quad \ldots \quad (e_2, y_n) \Rightarrow O_n}{(e_1 e_2, x) \Rightarrow O_1 \cup \cdots \cup O_n}$$

$$\frac{(e, x) \Rightarrow \emptyset}{(!e, x) \Rightarrow \{x\}} \qquad \frac{(e, x) \Rightarrow \{y\} \cup O}{(!e, x) \Rightarrow \emptyset} \qquad \frac{(P(X), x) \Rightarrow O}{(X, x) \Rightarrow O}$$

For example, $(a|ab, abc) \Rightarrow \{bc, c\}$.

For $e \in \text{RELa}(\Sigma, V)$, the *operational semantics* $B_{P,op}(e)$ is defined as $\{(x, y) \mid (e, xy) \Rightarrow O, y \in O\}$. A CFGLa G is *operationally complete* if, for any string x and any variable X, there exists O such that $(X, x) \Rightarrow O$. We consider that the semantics is given only if G is operationally complete. In general, $(e, x) \Rightarrow O$ if and only if $(P(e), x) \Rightarrow O$, hence $B_{P,op} = B_{P,op} \circ P$. For an operationally complete CFGLa, $B_{P,op}$ is an interpretation and a solution of P.

We show if a CFGLa is operationally complete then the CFGLa is uniquely convergent. The converse does not hold. We write $(e, x) \Rightarrow_0 O$ for a derivation that the rule of variables is not used. We can define $\Rightarrow_0$ inductively as $\Rightarrow$.

Lemma 14. *If* $(e, x) \Rightarrow_0 O$, *then* $B_{P,op}(e) \cap M_x = B(e) \cap M_x$ *for all interpretation* B *where* $M_x = \{(y, z) \mid x = yz\}$.

Proof. By induction on $\Rightarrow_0$. We show the case of concatenation. First, we have $(e_1, x) \Rightarrow_0 \{y_1, \ldots, y_n\}$, $(e_2, y_1) \Rightarrow_0 O_1$, $\ldots$, $(e_2, y_n) \Rightarrow_0 O_n$. By the induction hypothesis, $B_{P,op}(e_1) \cap M_x = B(e_1) \cap M_x$, $B_{P,op}(e_2) \cap M_{y_1} = B(e_2) \cap M_{y_1}$, $\ldots$, $B_{P,op}(e_2) \cap M_{y_n} = B(e_2) \cap M_{y_n}$. Thus, $B_{P,op}(e_1 e_2) \cap M_x = B(e_1 e_2) \cap M_x$. □

Lemma 15. *If* $(e, x) \Rightarrow O$, *then there exists* n *such that* $(P^m(e), x) \Rightarrow_0 O$ *for any* $m \geq n$.

Proof. By induction on $\Rightarrow$. We show the case of union. We have $(e_1, x) \Rightarrow_0 O_1$, $(e_2, x) \Rightarrow_0 O_2$, and $O = O_1 \cup O_2$. By the induction hypothesis, there exist n_1 and n_2 such that for all $m_1 \geq n_1, m_2 \geq n_2$, $(P^{m_1}(e_1), x) \Rightarrow_0 O_1$, $(P^{m_2}(e_2), x) \Rightarrow_0 O_2$. Let $n = \max(n_1, n_2)$, for all $m \geq n$, $(P^m(e_1), x) \Rightarrow_0 O_1$, $(P^m(e_2), x) \Rightarrow_0 O_2$. We have $P^m(e_1 e_2) = P^m(e_1) P^m(e_2)$. Therefore, $(P^m(e_1 e_2), x) \Rightarrow_0 O$. □

Proposition 16. *An operationally complete CFGLa is uniquely convergent.*

Proof. Let $M(e) = \{(x, y) \mid (e, xy) \Rightarrow O\}$. By Lemma 15 and operational completeness, we have $\lim_{n \to \infty} M(P^n(X)) = \Sigma^* \times \Sigma^*$.

$$
\begin{aligned}
B_{P,op}(X) &= \lim_{n \to \infty} (B_{P,op}(X) \cap M(P^n(X))) \\
&= \lim_{n \to \infty} (B_{P,op}(P^n(X)) \cap M(P^n(X))) && (B_{P,op} \text{ is a solution}) \\
&= \lim_{n \to \infty} (B(P^n(X)) \cap M(P^n(X))) && (\text{Lemma 14}) \\
&= \lim_{n \to \infty} B(P^n(X)) && (\text{Lemma 3})
\end{aligned}
$$

□

Two semantics coincide for an operationally complete CFGLa, i.e. $B_{P,nat} = B_{P,op}$. The result shows that PEG can be regarded as a subclass of CFGLa.

Definition 6 (Parsing expression grammars). *We define an ordered choice* e_1/e_2 *is an abbreviation for* $e_1 \| ! e_1 e_2$. *An expression* $e \in RELa(\Sigma, V)$ *is called a parsing expression if* e *can be written using only* $/$ *without using* $\|$. *A PEG* $G = (V, \Sigma, P, S)$ *is a CFGLa that* $P(X)$ *is a parsing expression for any* X.

The original definition of the language of PEG [9] is the set of successful input strings and for the derivation $(e, x) \Rightarrow o$, o is a consumed string or fail. We can easily show the language of PEG coincides with our definition.

PEG can represent all deterministic context-free languages and $\mathcal{L}(\text{PEG})$ is closed under union, intersection, and complement [9]. For PEG's derivation $(e, x) \Rightarrow O$, there exists y such that $O = \{y\}$ or $O = \emptyset$. That is if $(x_1, y_1), (x_2, y_2) \in B_{P,op}(e)$ and $x_1 y_1 = x_2 y_2$, then $y_1 = y_2$. Hence, the class $\mathcal{B}(\text{CFGLa})$ is a proper superset of $\mathcal{B}(\text{PEG})$.

Example 3.

1. A CFGLa $X = Xa \mid \varepsilon$ is valid and uniquely convergent but is not operationally complete.
2. A PEG $X = Xa/\varepsilon$ is equivalent to $X = Xa \mid !(Xa)$ and it is invalid. The reason is the same as Example 2.2.
3. A PEG $X = aXa \mid !(aXa)$ is uniquely convergent. $B(X) = \{(a^{2n}, a^m) \mid n + m = 2^k - 1\}$ is the solution and $\mathcal{L}(G) = \{a^{2^n - 2} \mid n \geq 1\}$.

Let $\Sigma = \{a, b\}$. We conjecture that the languages $\{xy \mid y \text{ is reverse of } x\}$ and $\{xay \mid y \text{ has the same length as } x\}$ cannot be represented by PEG because $X = aXa/bXb/\varepsilon$ and $Y = .Y./a$ do not represent them. It is an open problem whether there are context-free languages that cannot be represented by PEG.

4   Related Work and Discussion

Finite lookahead is used in the traditional theory of parsing. Regular lookahead is used in various domains of formal language theory, e.g.. RELa [14,15], LL(*) [17], transducers with lookahead [18], and tree transducers with lookahead [8]. A language with lookahead of RELa is a finite union of the form $A \times B$ where A and B are regular languages [14], hence it is expected that a language with lookahead of CFG with regular lookahead is a finite union of the form $A \times B$ where A is a context-free language and B is a regular language.

PEG [9] have non-regular lookahead. For the relationship between PEG and CFG, it is shown how to convert subclasses of CFG, such as LL, into PEG [12]. PEG with unordered choices [6] are an extension of PEG and can handle grammars without left recursion. Grammars with one-sided contexts [3] are an extension of CFG and have non-regular positive lookahead.

Regular expressions with complement and CFG with complement are called extended regular expressions (ERE) [5,19] and Boolean grammars [16], respectively. Lookahead differs from complement in the scope of influence and non-consumption of letters. RELa are more suitable for parsing than ERE [15].

The relationships of the language class are summarized below where an arrow means proper inclusion, a dashed arrow means inclusion.

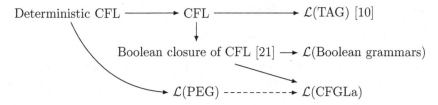

The class $\mathcal{L}(\text{CFGLa})$ is a proper superset of Boolean closure of context-free languages [21]. A context-free language over one letter alphabet is regular, hence the Boolean closure of unary context-free languages is also regular. On the other hand, a non-regular unary language $\{a^{2^n} \mid n \geq 1\}$ is represented by CFGLa.

The class $\mathcal{L}$(CFGLa) is not a subset of $\mathcal{L}$(TAG) because of the decidability of emptiness. It is expected that $\mathcal{L}$(CFGLa) is not a superset of $\mathcal{L}$(TAG) because the complexity of recognition of TAG is relatively large. A similar argument applies to some supersets of $\mathcal{L}$(TAG) such as indexed languages [1].

We consider it is difficult to prove of inclusion between $\mathcal{L}$(CFGLa) and $\mathcal{L}$(Boolean grammars). It is because of the difference in the properties between lookahead and complement. For example, the class $\mathcal{L}$(Boolean grammars) is closed under concatenation and Kleene star.

Acknowledgment. We are grateful to anonymous reviewers for introducing us to the important references of the grammars with one-sided contexts. This work was supported by JSPS KAKENHI Grant Number 20J23184 and 19K11899.

References

1. Aho, A.V.: Indexed grammars—an extension of context-free grammars. J. ACM **15**(4), 647–671 (1968)
2. Autebert, J.-M., Berstel, J., Boasson, L.: Context-free languages and pushdown automata. In: Rozenberg, G., Salomaa, A. (eds.) Handbook of Formal Languages, pp. 111–174. Springer, Heidelberg (1997). https://doi.org/10.1007/978-3-642-59136-5_3
3. Barash, M., Okhotin, A.: An extension of context-free grammars with one-sided context specifications. Inf. Comput. **237**, 268–293 (2014)
4. Barash, M., Okhotin, A.: Linear grammars with one-sided contexts and their automaton representation. RAIRO Theor. Inf. Appl. **49**(2), 153–178 (2015)
5. Brzozowski, J.A.: Derivatives of regular expressions. J. ACM **11**(4), 481–494 (1964)
6. Chida, N., Kuramitsu, K.: Parsing expression grammars with unordered choices. J. Inf. Process. **25**, 975–982 (2017)
7. Chomsky, N., Schützenberger, M.: The algebraic theory of context-free languages. In: Computer Programming and Formal Systems, vol. 35, pp. 118–161. Elsevier (1963)
8. Engelfriet, J.: Top-down tree transducers with regular look-ahead. Math. Syst. Theory **10**(1), 289–303 (1977)
9. Ford, B.: Parsing expression grammars: a recognition-based syntactic foundation. In: Proceedings of the 31st ACM SIGPLAN-SIGACT Symposium on Principles of Programming Languages, pp. 111–122. ACM (2004)
10. Joshi, A.K., Levy, L.S., Takahashi, M.: Tree adjunct grammars. J. Comput. Syst. Sci. **10**(1), 136–163 (1975)
11. Kozen, D.: Kleene algebra with tests. ACM Trans. Program. Lang. Syst. **19**(3), 427–443 (1997)
12. Mascarenhas, F., Medeiros, S., Ierusalimschy, R.: On the relation between context-free grammars and parsing expression grammars. Sci. Comput. Program. **89**, 235–250 (2014)
13. Might, M., Darais, D., Spiewak, D.: Parsing with derivatives: a functional pearl. In: Proceedings of the 16th ACM SIGPLAN International Conference on Functional Programming, pp. 189–195. ACM (2011)
14. Miyazaki, T., Minamide, Y.: Derivatives of regular expressions with lookahead. J. Inf. Process. **27**, 422–430 (2019)

15. Morihata, A.: Translation of regular expression with lookahead into finite state automaton. Comput. Softw. **29**(1), 147–158 (2012)
16. Okhotin, A.: Boolean grammars. Inf. Comput. **194**(1), 19–48 (2004)
17. Parr, T., Fisher, K.: LL(*) the foundation of the ANTLR parser generator. In: Proceedings of the 32nd ACM SIGPLAN Conference on Programming Language Design and Implementation, pp. 425–436 (2011)
18. Sakuma, Y., Minamide, Y., Voronkov, A.: Translating regular expression matching into transducers. J. Appl. Logic **10**(1), 32–51 (2012)
19. Stockmeyer, L.J., Meyer, A.R.: Word problems requiring exponential time (preliminary report). In: Proceedings of the 5th Annual ACM Symposium on Theory of Computing, pp. 1–9. ACM (1973)
20. Vardi, M.Y.: The complexity of relational query languages. In: Proceedings of the 14th Annual ACM Symposium on Theory of Computing, pp. 137–146. ACM (1982)
21. Wotschke, D.: The Boolean closures of the deterministic and nondeterministic context-free languages. In: Brauer, W. (ed.) GI Gesellschaft für Informatik e. V. LNCS, pp. 113–121. Springer, Heidelberg (1973). https://doi.org/10.1007/978-3-662-41148-3_11

Tree-Like Unit Refutations in Horn Constraint Systems

K. Subramani$^{(\boxtimes)}$ and Piotr Wojciechowski

LDCSEE, West Virginia University, Morgantown, WV, USA
k.subramani@mail.wvu.edu, pwojciec@mail.wvu.edu

Abstract. In this paper, we examine the problem of finding unit refutations of Horn constraint systems (HCSs). Recall that a Horn constraint is a linear constraint in which every coefficient belongs to the set $\{0, 1, -1\}$ and in which at most one coefficient is positive. In the current work, we extend the notion of unit refutations from CNF formulas to systems of linear constraints. Recall that for CNF formulas a unit resolution refutation is one in which every resolution step uses a one-literal (unit) clause. The equivalent notion in linear systems requires every inference step to use a one-variable (absolute) constraint. We analyze two problems associated with unit refutations of Horn constraint systems. In the length-bounded tree-like unit refutation (TLUR$_D$) problem, we ask if a given Horn constraint system has a tree-like unit refutation using at most L inference steps. In the optimal tree-like unit refutation (TLUR$_{Opt}$) problem, we ask for a tree-like unit refutation with the fewest inference steps. We show that the former problem is **NP-complete** and the latter is **NPO-complete**. We also show that the TLUR$_D$ problem does not admit a polynomial size kernel with respect to a natural output parameter under some well-accepted complexity theoretic assumptions.

1 Introduction

This paper analyzes the problem of finding unit refutations of Horn constraint systems (HCSs). Recall that a Horn constraint is a linear constraint such that every coefficient belongs to the set $\{0, 1, -1\}$ with at most one positive coefficient. That is, a Horn constraint is a constraint of the form $x_i - \sum x_j \geq b$ or of the form $-\sum x_j \geq b$.

Unit refutations have been extensively studied for CNF formulas. In CNF formulas, unit resolution is a restricted form of resolution in which each resolution step must use a one-literal (unit) clause. Note that in general CNF formulas unit resolution is an incomplete proof system. However, unit resolution is complete for Horn formulas. In this paper, we extend the concept of unit resolution to systems of linear constraints, in particular to HCSs. Thus, a unit refutation

K. Subramani—This research was supported in part by the Air-Force Office of Scientific Research through Grant FA9550-19-1-0177 and in part by the Air-Force Research Laboratory, Rome through Contract FA8750-17-S-7007.

A. Leporati et al. (Eds.): LATA 2021, LNCS 12638, pp. 226–237, 2021.
https://doi.org/10.1007/978-3-030-68195-1_18

of a system of linear constraints is one in which each inference step must use a one-variable (absolute) constraint.

Note that unit refutation is incomplete for HCSs. That is, not every HCS has a unit refutation (see Example 1). This is in stark contrast to unit resolution, which is complete for Horn formulas. Recently, we showed that the problem of checking whether a Horn constraint system has a unit tree-like refutation is in **P** [19].

In this paper, we analyze the length-bounded tree-like unit refutation (TLUR$_D$) problem and the optimal tree-like unit-refutation (TLUR$_{Opt}$) problem. Let **H** be an HCS and let L be a positive integer. The TLUR$_D$ problem asks if **H** has a tree-like unit refutation with at most L inference steps. The TLUR$_{Opt}$ problem asks for a tree-like unit refutation of **H** with the fewest inference steps.

The principal contributions of this paper are as follows:

1. Establishing that the TLUR$_D$ problem for HCSs is **NP-complete** (see Sect. 4).
2. Establishing that the TLUR$_{Opt}$ problem for HCSs is **NPO-complete** (see Sect. 5).
3. Establishing that the TLUR$_D$ problem for HCSs does not have a kernel whose size is polynomial in the length of the shortest tree-like unit refutation (see Sect. 6).

The rest of this paper is organized as follows: In Sect. 2, we introduce the problems being studied. Section 3 provides motivation for studying these problems as well as related work in the literature. In Sect. 4, we study the TLUR$_D$ problem for HCSs. Section 5 examines the TLUR$_{Opt}$ problem for HCSs. In Sect. 6, we prove our kernelization result for the TLUR$_D$ problem. We conclude in Sect. 7 by summarizing our contributions.

2   Statement of Problems

In this section, we define the problems under consideration in this paper.

Definition 1. *A system of constraints* $\mathbf{A} \cdot \mathbf{x} \geq \mathbf{b}$ *is a Horn Constraint System (HCS) if: 1. The entries in* $\mathbf{A}$ *belong to the set* $\{0, 1, -1\}$. *2. Each row of* $\mathbf{A}$ *contains at most one positive entry. 3.* $\mathbf{x}$ *is a real valued vector. 4.* $\mathbf{b}$ *is an integral vector.*

In a Horn constraint $l_k : \mathbf{a} \cdot \mathbf{x} \geq b_k$, b_k is called the defining constant. For a Horn constraint l_k, let $P(l_k)$ denote the number of positive coefficients in l_k.

If a Horn constraint has only one non-zero coefficient, then it is called an absolute constraint. If that coefficient is 1, then it is called a positive absolute constraint.

We are interested in certificates of infeasibility. Such certificates are called refutations. A refutation consists of a sequence of inference steps that results in a contradiction. Each inference step in a refutation consists of an application of an inference rule. One such inference rule, used for refuting the linear feasibility of systems of linear constraints, is known as the **ADD rule**. This inference rule derives a new constraint by summing a pair of constraints (either from the original system or derived by previous inference steps) and is defined as follows:

$$\text{ADD}: \frac{\sum_{i=1}^{n} a_i \cdot x_i \geq b_1 \qquad \sum_{i=1}^{n} a'_i \cdot x_i \geq b_2}{\sum_{i=1}^{n}(a_i + a'_i) \cdot x_i \geq b_1 + b_2} \tag{1}$$

This inference rule plays a role similar to the role played by resolution in clausal formulas.

Using Rule (1), we can now define a linear refutation.

Definition 2. *A linear refutation is a sequence of applications of the ADD rule that results in a contradiction of the form $0 \geq b$, $b > 0$.*

The form of refutation defined in Definition 2 is both **sound** and **complete** when used as a proof system for linear feasibility. It is **sound** since any assignment that satisfies the constraints used by an application of the ADD rule also satisfies the constraint derived by that application. It is **complete** since repeated application of the ADD rule will eventually result in a contradiction of the form: $0 \geq b$, $b > 0$ for any linearly infeasible system. The completeness of ADD rule based linear refutations was established by Farkas [7], in a lemma that is famously known as Farkas' Lemma for systems of linear inequalities [15].

We now formally define the type of refutation discussed in this paper.

Definition 3. *A **tree-like** refutation is a refutation in which each derived constraint can be used at most once.*

Note that if a derived constraint needs to be reused, then it can be re-derived.

In proof theory, there are many ways to measure the length of a refutation. In this paper, we define the length of a refutation as the number of inference steps in that refutation.

Definition 4. *The **length** of a refutation is the number of inference steps (applications of the ADD rule) in the refutation. The length of a refutation R is denoted as $|R|$.*

Let $\mathbf{H} : \mathbf{A} \cdot \mathbf{x} \geq \mathbf{b}$ be an HCS with m constraints over n variables. Using Farkas' Lemma [7], we can represent a tree-like refutation R of $\mathbf{H}$ using a vector $\mathbf{y} \in \mathbb{Z}^m$, $\mathbf{y} \geq 0$ such that $\mathbf{y} \cdot \mathbf{A} = \mathbf{0}$ and $\mathbf{y} \cdot \mathbf{b} > 0$. We now show that $\mathbf{y}$ corresponds to a tree-like linear refutation of $\mathbf{H}$ of length $(\sum_{k=1}^{m} y_k - 1)$.

Theorem 1. *Let $\mathbf{H} : \mathbf{A} \cdot \mathbf{x} \geq \mathbf{b}$ be an HCS with m constraints over n variables. If there exists a vector $\mathbf{y} \in \mathbb{Z}^m$, $\mathbf{y} \geq 0$ such that $\mathbf{y} \cdot \mathbf{A} = \mathbf{0}$ and $\mathbf{y} \cdot \mathbf{b} > 0$, then $\mathbf{H}$ has a tree-like linear refutation of length $(\sum_{k=1}^{m} y_k - 1)$.*

Proof. Let $\mathbf{H} : \mathbf{A} \cdot \mathbf{x} \geq \mathbf{b}$ be an HCS with m constraints over n variables. Let $\mathbf{y} \in \mathbb{Z}^m$, $\mathbf{y} \geq \mathbf{0}$ be a vector such that $\mathbf{y} \cdot \mathbf{A} = \mathbf{0}$ and $\mathbf{y} \cdot \mathbf{b} > 0$.

Note that $\mathbf{y}$ represents a weighted sum of the constraints in $\mathbf{H}$ where each constraint l_k is used y_k times. This weighted sum results in the constraint $0 \geq \mathbf{y} \cdot \mathbf{b}$. Since $\mathbf{y} \cdot \mathbf{b} > 0$, this is a contradiction.

Constraint summation is associative and commutative. Thus, the weighted sum corresponding to $\mathbf{y}$ can be expressed as a sequence of applications of the ADD rule that does not reuse derived constraints. This is a tree-like refutation R of $\mathbf{H}$.

Note that each application of the ADD rule derives one constraint from a pair of constraints. Initially, the weighted sum has $\sum_{k=1}^{m} y_k$ constraints which are used to derive a single contradiction. Thus, R has $(\sum_{k=1}^{m} y_k - 1)$ applications of the ADD rule. Consequently $|R| = (\sum_{k=1}^{m} y_k - 1)$.   □

In this paper, we study a restricted version of the ADD rule, known as the **unit-ADD rule**. In the unit-ADD rule, at least one of the constraints must be an absolute constraint. In HCSs, this rule has the following form:

$$\text{unit-ADD} : \frac{a_i \cdot x_i \geq b_1 \qquad a_j \cdot x_j - \sum_{k \in S} x_k \geq b_2}{a_j \cdot x_j + a_i \cdot x_i - \sum_{k \in S} x_k \geq b_1 + b_2} \tag{2}$$

A linear refutation using only the unit-ADD rule is called a unit refutation. Note that unit refutation is *not* a complete proof system. This can be seen in the following example.

Example 1. Consider the HCS H in System (3).

$$x_1 \geq -1 \qquad x_1 - x_2 \geq 1 \qquad x_2 - x_1 \geq 1 \tag{3}$$

System (3) has the following refutation:

1. ADD $x_1 - x_2 \geq 1$ and $x_2 - x_1 \geq 1$ to get $0 \geq 2$.

However, HCS $\mathbf{H}$ does not have a unit refutation. Observe that $x_1 \geq -1$ is the only absolute constraint in $\mathbf{H}$, thus it must be used in any unit refutation of $\mathbf{H}$. It is easy to see that a unit refutation of $\mathbf{H}$, if one existed, would have the following form:

1. ADD $x_1 \geq -1$ and $x_2 - x_1 \geq 1$ to get $x_2 \geq 0$.
2. ADD $x_2 \geq 0$ and $x_1 - x_2 \geq -1$ to get $x_1 \geq 1$.

3. ⋮

$\mathbf{H}$ does not have any unit constraints which cancel either x_1 from the constraint $x_1 \geq 1$ or which cancel x_2 from the constraint $x_2 \geq 0$. Thus, there is no way to complete this unit refutation.

Since unit refutation is a restriction of linear refutation, it is clearly a **sound** proof system. However, as just demonstrated, it is not a **complete** proof system.

As shown in Theorem 1, a tree-like linear refutation can be represented by a vector $\mathbf{y}$. We now prove a similar result for tree-like unit refutations.

Theorem 2. *Let* $\mathbf{H} : \mathbf{A} \cdot \mathbf{x} \geq \mathbf{b}$ *be an HCS with* m *constraints over* n *variables. If there exists a vector* $\mathbf{y} \in \mathbb{Z}^m$, $\mathbf{y} \geq \mathbf{0}$ *such that* $\mathbf{y} \cdot \mathbf{A} = \mathbf{0}$, $\mathbf{y} \cdot \mathbf{b} > 0$, *and* $\sum_{k=1}^m y_k \cdot P(l_k) = \sum_{k=1}^m y_k - 1$, *then* $\mathbf{H}$ *has a tree-like unit refutation of length* $(\sum_{k=1}^m y_k - 1)$.

Proof. Let $\mathbf{H} : \mathbf{A} \cdot \mathbf{x} \geq \mathbf{b}$ be an HCS with m constraints over n variables. Let $\mathbf{y} \in \mathbb{Z}^m$, $\mathbf{y} \geq \mathbf{0}$ be a vector such that $\mathbf{y} \cdot \mathbf{A} = \mathbf{0}$, $\mathbf{y} \cdot \mathbf{b} > 0$ and $\sum_{k=1}^m y_k \cdot P(l_k) = \sum_{k=1}^m y_k - 1$. Recall that $P(l_k)$ is the number of positive coefficients in the constraint l_k.

From Theorem 1, we know that $\mathbf{y}$ corresponds to a tree-like linear refutation of $\mathbf{H}$. We now show that the additional restriction, $\sum_{k=1}^m y_k \cdot P(l_k) = \sum_{k=1}^m y_k - 1$ ensures that $\mathbf{y}$ corresponds to a tree-like unit refutation of $\mathbf{H}$.

Since constraint summation is both associative and commutative, we can assume without loss of generality that each application of the ADD results in the cancellation of at least one variable. Initially, there are $\sum_{k=1}^m y_k \cdot P(l_k)$ total variables with a coefficient of 1 in the constraints in the weighted sum corresponding to $\mathbf{y_k}$. From Theorem 1, $\mathbf{y}$ corresponds to a tree-like linear refutation with $\sum_{k=1}^m y_k - 1 = \sum_{k=1}^l y_k \cdot P(l_k)$ applications of the ADD rule. Thus, each application of the ADD rule cancels exactly one variable.

Consider the ADD rule applied to constraints l_{k_1} and l_{k_2}. Assume without loss of generality that this summation cancels a variable x_i that appears with positive coefficient in l_{k_1} and negative coefficient in l_{k_2}. No other variable can be canceled. Thus, no variable with positive coefficient in l_{k_2} appears with negative coefficient in l_{k_1}. It follows that all variables with negative coefficient in l_{k_1} must be canceled by other applications of the ADD rule.

Recall that constraint summation is both associative and commutative, thus we can assume without loss of generality that all the applications of the ADD rule which canceled variables with negative coefficient from l_{k_1} occurred earlier in the refutation. This means that, l_{k_1} is a constraint of the form $x_i \geq b_{k_1}$. Consequently, this is a unit refutation of $\mathbf{H}$. □

This paper examines tree-like refutations using only the unit-ADD rule. We refer to these refutations as tree-like unit refutations. For these refutations, we are also interested in the problem of finding the length of the shortest refutation.

Thus, we focus on the following problems:

1. TLUR$_D$: Given an HCS $\mathbf{H}$ and an integer L, does $\mathbf{H}$ have a tree-like unit refutation with at most than L inference steps?
2. TLUR$_{Opt}$: Given an HCS $\mathbf{H}$, what is the number of inference steps in a shortest tree-like unit refutation of $\mathbf{H}$?

3   Motivation and Related Work

Horn constraint systems are a more general form of Difference Constraint Systems (DCSs), since each constraint in an HCS is allowed to have multiple variables with coefficient -1. As in difference constraint systems, the linear feasibility and integer feasibility problems coincide for Horn constraint systems [3]. Veinott devised a non-polynomial algorithm for the linear feasibility problem for Horn type programs where the positive and negative elements can take any value [18].

Horn constraint systems have been used as domains in abstract interpretation [4]. They are also used in Satisfiability Modulo Theory (SMT) solvers, which in turn are being increasingly used in program verification procedures [5,9]. These solvers are also part of procedures for bounded model checking, infinite-state systems, and test-case generation [6]. Additionally, Horn systems find applications in declarative programming [10]. The applications of Horn constraints to program verification have been discussed extensively [12].

In this paper, we focus on a restriction to general linear refutation. Specifically, we examine refutations in which each application of the ADD rule must use an absolute constraint. Note that such a restriction can cause the original refutation system to become **incomplete**. However, it is still necessary to study this type of restricted proof system. Observe that:

1. Restricted proofs tend to be compact (polynomial in the size of the input). For instance, read-once unit refutations are at most linear in the size of the input.
2. For specific constraint systems, the existence of these restricted refutations can be checked efficiently. For example, [16] showed that every infeasible DCS has a read-once refutation and that such a refutation can be found in polynomial time. While systems of Unit Two Variable per Inequality (UTVPI) constraints do not always have read-once refutations, these refutations can still be found in polynomial time [17].

The focus on unit refutations in this paper stems from a fundamental difference between absolute constraints and non-absolute constraints in HCSs. Since absolute constraints bound only a single variable, they can be used to define the domain over which feasibility is considered. Meanwhile, non-absolute constraints define the relationship between variables and can be considered domain agnostic. This difference in the two types of constraints carries over to create a difference between unit and non-unit refutations.

A unit refutation relies on the absolute constraints in the underlying HCS. Thus, a unit refutation serves as a domain specific refutation. This is in contrast to an unrestricted refutation which may be domain agnostic. It follows that a study of unit refutations is important since it reveals the structure of such domain specific refutations.

Several papers have studied unit refutations of Horn clauses [2,14].

4   Length-Bounded Tree-Like Unit Refutations

In this section, we study the computational complexity of the TLUR_D problem for HCSs.

We show that the problem of checking if an HCS has a tree-like unit refutation of length at most L is **NP-complete**. This is done by a reduction from the Integer Knapsack problem.

Definition 5. *The Integer Knapsack problem: Given n items $i = 1 \ldots n$, each with weight w_i and profit p_i, maximum capacity W, and target profit P, can you select items with total weight at most W such that the total profit is at least P.*

Note that items can be selected more than once.

This variant of knapsack is known to be **NP-complete** [13].

Consider an instance K of the Integer Knapsack problem as described above. We construct the HCS $\mathbf{H}_K$ as follows:

1. For each item $i = 1 \ldots n$, create the variables x_i and z_i. Additionally create the constraint $x_{i+1} - x_i \geq 0$ for each $i = 1 \ldots n - 1$.
2. Create the constraints $x_1 \geq -P + 1$ and $-x_n \geq 0$.
3. Let w_{max} be the largest weight of any item and let $l = \lfloor \log w_{max} \rfloor$. Create the variables y_0 through y_l and the constraint $y_j - \sum_{i=0}^{j-1} y_i \geq 0$ for each $j = 0 \ldots l$.
4. For each item i, let S_i be a set of integers such that $\sum_{j \in S_i} 2^j = w_i - 1$. Create the constraints $x_i - z_i - \sum_{j \in S_i} y_j \geq p_i$ and $z_i - x_i - \sum_{j \in S_i} y_j \geq 0$.

Using this reduction we have the following theorem.

Theorem 3. *Let K be an instance of integer knapsack and let $\mathbf{H}_K$ be the corresponding HCS. A total profit of at least P can be obtained from the items in K if and only if $\mathbf{H}$ has a tree-like unit refutation of length at most $L = (2 \cdot W + n)$*

Proof. Let R be a tree-like unit refutation of $\mathbf{H}_K$. Observe the following properties of R:

1. R uses the constraint $x_i - z_i - \sum_{j \in S_i} y_j \geq p_i$ if and only if it uses the constraint $z_i - x_i - \sum_{j \in S_i} y_j \geq 0$ – By construction, these are the only constraints with the variable z_i. Thus, any refutation which uses one must use the other. In fact, these two constraints must be used an equal number of times.
2. For each j, deriving the constraint $y_j \geq 0$ takes $(2^j - 1)$ applications of the ADD rule – This is obviously true for $y_0 \geq 0$. To derive $y_j \geq 0$, the constraint $y_j - \sum_{i=0}^{j-1} y_i \geq 0$ must be used. The cancellation of each $-y_i$ from this constraint takes 2^i applications of the ADD rule, $(2^i - 1)$ to derive $y_i \geq 0$ and an additional application to perform the cancellation. Thus, the derivation of $y_j \geq 0$ takes a total of $\sum_{i=1}^{j-1} 2^i = 2^j - 1$ inference steps as desired.
3. The cancellation of the y_js from the constraint $x_i - z_i - \sum_{j \in S_i} y_j \geq p_i$ takes a total of $(w_i - 1)$ inference steps – As before, the cancellation of each $-y_j$ from this constraint takes a total of 2^j inference steps. Thus the cancellation of all $-y_j$s from this constraint takes a total of $\sum_{j \in S_i} 2^j = w_i - 1$ inference steps as desired. The same is true for the constraint $z_i - x_i - \sum_{j \in S_i} y_j \geq 0$.
4. Using the constraint $x_i - z_i - \sum_{j \in S_i} y_j \geq p_i$ adds $2 \cdot w_i$ inference steps to R – Using this constraint adds the following inference steps to R: The inference step using $x_i - z_i - \sum_{j \in S_i} y_j \geq p_i$, the inference step using $z_i - x_i - \sum_{j \in S_i} y_j \geq 0$, and the $2 \cdot (w_i - 1)$ inference steps canceling the $-y_j$s from both constraints. This is a total of $2 \cdot w_i$ inference steps as desired.

Consider the constraints $x_1 \geq -P + 1$, $x_2 - x_1 \geq 0$, ..., $x_n - x_{n-1} \geq 0$, and $-x_n \geq 0$. Recall that R is a unit refutation. Thus, for each $i = 1 \ldots n$, the variable z_i cannot be canceled by summing the constraints $x_i - z_i - \sum_{j \in S_i} y_j \geq p_i$ and $z_i - x_i - \sum_{j \in S_i} y_j \geq 0$ unless the variable x_i is canceled from one of these constraints. It follows that R must use the constraint $x_{i+1} - x_i \geq 0$ (or $-x_n \geq 0$ if $i = n$) or the constraint $x_i - x_{i-1} \geq 0$ (or $x_1 \geq -P + 1$ if $i = 1$). Since R uses the constraints $x_i - z_i - \sum_{j \in S_i} y_j \geq p_i$ and $z_i - x_i - \sum_{j \in S_i} y_j \geq 0$ an equal number of times, R must use both of the constraints $x_{i+1} - x_i \geq 0$ and $x_i - x_{i-1} \geq 0$. Thus, R must use all of the constraints $x_1 \geq -P + 1$, $x_2 - x_1 \geq 0$, ..., $x_n - x_{n-1} \geq 0$.

Note that the sum of the defining constants in these constraints is $(-P + 1)$. For the defining constant of the final constraint derived by R to be positive, the sum of the defining constants of the remaining constraints in R must be least P.

Observe that this is possible using at most $(2 \cdot W + n)$ inference steps if and only if a total profit of at least P is obtainable from K with total weight at most W. □

From Theorem 3, we have the following result.

Theorem 4. *The TLUR$_D$ problem for HCSs is* **NP-complete**.

The complete proof will be in the journal version of this paper.

5 Optimal Length Tree-Like Unit Refutations

In this section, we study the approximation complexity of the TLUR$_{Opt}$ problem for HCSs.

We show that the TLUR$_{Opt}$ problem for HCSs is **NPO-complete**. Any problem that is **NPO-complete** cannot be approximated to within a polynomial factor of the optimum unless **P = NP** [11].

We establish **NPO-completeness** by a reduction from the Weighted Minimum Ones problem.

Definition 6. *Weighted Minimum Ones: Given a 3CNF formula Φ and non-negative variable weight function w, what is the minimum weight of variables assigned to* **true** *in any satisfying assignment to Φ?*

The Weighted Minimum Ones problem is known to be **NPO-complete** [1].

Let Φ be a 3CNF formula with n variables and m clauses and let w be a weight function on the variables of Φ. From Φ, we construct the HCS **H** as follows:

1. Let w_{max} be the largest weight of any variable and let $l = \lfloor \log w_{max} \rfloor$. Create the variables w_0 through w_l and the constraint $w_j - \sum_{i=0}^{j-1} w_i \geq 0$ for each $j = 0 \ldots l$.
2. For each variable x_i in Φ:
   (a) let W_i be a set of integers such that $\sum_{j \in W_i} 2^j = w(i) - 1$.

(b) Create the variables p_0 through $p_{\lfloor \log m \rfloor}$.

(c) Let d_i be the number of clauses in which x_i appears.

(d) Create the variables x_i, $y_{i,0}$ through $y_{i,\lfloor \log d_i \rfloor}$, and $z_{i,0}$ through $z_{i,\lfloor \log d_i \rfloor}$.

(e) For each subset S of $\{0, \ldots, \lfloor \log d_i \rfloor\}$, create the constraints
$$x_i - \sum_{j \in S} y_{i,j} - \sum_{j \in W_i} w_j \geq 0 \text{ and } x_i - \sum_{j \in S} z_{i,j} \geq 0.$$

(f) For each $j = 1 \ldots \lfloor \log d_i \rfloor$, create the constraints
$$y_{i,j} - \sum_{k=j+1}^{\lfloor \log d_i \rfloor} y_{i,k} \geq 0 \text{ and } z_{i,j} - \sum_{k=j+1}^{\lfloor \log d_i \rfloor} z_{i,k} \geq 0.$$

3. For each clause ϕ_l in Φ:

(a) Let $S_l \subseteq \{0, \ldots, \lfloor \log m \rfloor\}$ be such that $\sum_{j \in S_l} 2^j = l$.

(b) If ϕ_l contains the literal x_i, create the constraint $y_{i,\lfloor \log d_i \rfloor} - \sum_{j \in S_l} p_j \geq 1$.

(c) If ϕ_l contains the literal $\neg x_i$, create the constraint $z_{i,\lfloor \log d_i \rfloor} - \sum_{j \in S_l} p_j \geq 1$.

4. For each $j \in \{0, \ldots, \lfloor \log m \rfloor\}$, create the constraint $p_j \geq 1$.

5. Create the constraint $-\sum_{i=1}^{n} x_i \geq 1 - m - \sum_{l=1}^{m} |S_l|$.

We now show that **H** has a tree-like unit refutation of length $m \cdot (W + 1)$ if and only if Φ has a satisfying assignment for which the weights of the variables set to **true** sum to W.

Lemma 1. *H has a tree-like unit refutation of length $m \cdot (W + 1)$ if and only if Φ has a satisfying assignment for which the weights of the variables set to **true** sum to W.*

Proof. Let $\mathbf{x}^*$ be a satisfying assignment for Φ. We can construct a tree-like unit refutation R of **H** as follows:

1. Add the constraint $-\sum_{i=1}^{n} x_i \geq 1 - m - \sum_{l=1}^{m} |S_l|$ to R.

2. For each clause ϕ_l, at least one literal in ϕ_1 is assigned **true**. Let $Lit(\phi_l)$ be this literal. If $Lit(\phi_l) = x_i$, then add the constraint $y_{i,\lfloor \log d_i \rfloor} \geq 1$ to R. If $Lit(\phi_l) = \neg x_i$, then add the constraint $z_{i,\lfloor \log d_i \rfloor} \geq 1$ to R.

3. For each variable x_i:

(a) Let d_i^+ be the number of clauses ϕ_l such that $Lit(\phi_l) = x_i$ and let d_i^- be the number of clauses such that $Lit(\phi_l) = \neg x_i$.

(b) Note that if $\mathbf{x}^*$ assigns **true** to x_i, then $d_i^- = 0$ and if $\mathbf{x}^*$ assigns **false** to x_i, then $d_i^+ = 0$. Thus, at most one of d_i^+ and d_i^- is non-zero.

(c) If d_i^+ is non-zero, let $S_i \subseteq \{0, \ldots, \lfloor \log d_i \rfloor\}$ be such that $\sum_{j \in S^i} 2^j = d_i^+$. Then add the constraint $x_i - \sum_{j \in S_i} y_{i,j} - \sum_{j \in W_i} w_j \geq 0$. Additionally, add enough constraints of the form $y_{i,j} - \sum_{k=j+1}^{\lfloor \log d_i \rfloor} y_{i,k} \geq 0$ to derive $x_i - d_i^+ \cdot y_{i,\lfloor \log d_i \rfloor} \geq 0$.

(d) As in the proof of Theorem 3, cancellation of the w_js from this constraint takes a total of $(w(i) - 1)$ inference steps.

(e) If d_i^- is non-zero, let $S_i \subseteq \{0, \ldots, \lfloor \log d_i \rfloor\}$ be such that $\sum_{j \in S^i} 2^j = d_i^-$. Then add the constraint $x_i - \sum_{j \in S_i} z_{i,j} \geq 0$. Additionally, add enough constraints of the form $y_{i,j} - \sum_{k=j+1}^{\lfloor \log d_i \rfloor} y_{i,k} \geq 0$ to derive $x_i - d_i^- \cdot z_{i,\lfloor \log d_i \rfloor} \geq 0$.

(f) If $d_i^+ = d_i^- = 0$, then add the constraint $x_i \geq 0$ to R.

Note that for each x_i, R can derive either $x_i - d_i^+ \cdot y_{i, \lfloor \log d_i \rfloor} \geq 0$ or $x_i - d_i^- \cdot z_{i, \lfloor \log d_i \rfloor} \geq 0$.

For each clause ϕ_l, we can derive a constraint of the form $y_{i, \lfloor \log d_i \rfloor} \geq 1 + |S_l|$. Namely we can derive this constraint for $x_i = Lit(\phi_l)$. Similarly, we can derive the constraint $z_{i, \lfloor \log d_i \rfloor} \geq 1 + |S_l|$ for $\neg x_i = Lit(\phi_l)$.

Thus, R can derive the constraint $\sum_{i=1}^{n} x_i \geq m + \sum_{l=1}^{m} |S_l|$. Together with the constraint $-\sum_{i=1}^{m} x_i \geq 1 - m - \sum_{l=1}^{m} |S_l|$, R derives the contradiction $0 \geq 1$. It follows that R is a tree-like unit refutation.

Let R be a tree-like unit refutation of $\mathbf{H}$ such that $|R| \leq m \cdot (W + 1)$. We construct a satisfying assignment $\mathbf{x}^*$ to Φ as follows: For each x_i, if R uses a constraint with the variable $y_{i, \lfloor \log d_i \rfloor}$ then set x_i to **true**. Otherwise set x_i to **false**.

Note that without the constraint $-\sum_{i=1}^{n} x_i \geq 1 - m - \sum_{l=1}^{m} |S_l|$, $\mathbf{H}$ is feasible. Thus, this constraint must be in R. Since R derives a contradiction, the right hand side of the constraint derived by R must be positive. It follows that R must use at least $\sum_{l=1}^{m} |S_l|$ constraints of the form $p_j \geq 1$ and at least m constraints of the form $y_{i, \lfloor \log d_i \rfloor} - \sum_{j \in S_l} p_j \geq 1$ or $z_{i, \lfloor \log d_i \rfloor} - \sum_{j \in S_l} p_j \geq 1$. Each of these constraints corresponds to a clause satisfied by the assignment $\mathbf{x}^*$. Note that the p_js ensure that at most one constraint per clause is used. Thus, $\mathbf{x}^*$ must satisfy at least m clauses of Φ. Consequently $\mathbf{x}^*$ satisfies Φ.

If R uses a constraint with the variable $y_{i, \lfloor \log d_i \rfloor}$, then R must use the constraint $x_i - \sum_{j \in S} y_{i,j} - \sum_{j \in W_i} w_j \geq 0$. As before, to cancel the w_j variables from this constraint, R must use $(w(i) - 1)$ inference steps. Since the length of R is at most $m \cdot (W + 1)$ the total weight of variables set to **true** cannot be more than W.   □

From Lemma 1, we have the following result.

Theorem 5. *The TLUR$_{Opt}$ problem for HCSs is* **NPO-complete**.

The complete proof will be in the journal version of this paper.

6   Lower Bounds on Kernel Size for TLUR$_D$

In this section, we provide a lower bound on the size of the kernel for the TLUR$_D$ problem for HCSs.

We now show that that the TLUR$_D$ problem for HCSs does not have a kernel whose size is polynomial in L, the length of the refutation. This is done through the use of a t-bounded OR-distillation [8].

Definition 7. *Let P and Q be a pair of problems and let $t : \mathbb{N} \to \mathbb{N} \setminus \{0\}$ be a polynomially bounded function. Then a t-**bounded OR-distillation** from P into Q is an algorithm that for every s, given as input $t(s)$ strings $x_1, \ldots, x_{t(s)}$ with $|x_j| = s$ for all j:*

1. Runs in polynomial time, and

2. *Outputs a string y of length at most $t(s) \cdot \log s$ such that y is a* **yes** *instance of Q if and only if x_j is a* **yes** *instance of P for some $j \in \{1, \ldots, t(s)\}$.*

If any **NP-hard** problem has a t-bounded OR-distillation, then **coNP** $\subseteq$ **NP/poly** [8]. If **coNP** $\subseteq$ **NP/poly**, then $\Sigma_3^P = \Pi_3^P$ [20]. Thus, the polynomial hierarchy would collapse to the third level.

Theorem 6. *The $TLUR_D$ problem for HCSs does not have a kernel, whose size is polynomial in the length of the refutation, unless* **coNP** $\subseteq$ **NP/poly**.

Proof. We will prove this by showing that if the $TLUR_D$ problem for HCSs has a polynomial sized kernel, then there exists a t-bounded OR-distillation from the $TLUR_D$ problem for HCSs into itself.

For each j, let $\mathbf{H}_j$ be an HCS with m constraints over n variables such that, for each constraint $l_k \in \mathbf{H}_j$, the defining constant b_k satisfies $|b_k| \leq b_{max}$ for a fixed integer b_{max}. We can encode $\mathbf{H}_j$ as an $m \times (n+1)$ matrix where each row of the matrix corresponds to a constraint. In each row, the first n values store the coefficients of the constraint and the last value stores the defining constant. Using this representation, we have that $s = |\mathbf{H}_j| = m \cdot (n + \lceil \log b_{max} \rceil)$.

From Lemma 1, we have that the problem of determining if an HCS has a tree-like unit refutation of length at most L is still **NP-hard** when $L \leq m \cdot (n+1)$. Thus, we can assume without loss of generality that $L \leq s$.

Assume that for some constant c, the $TLUR_D$ problem has a kernel of size L^c. Let $t(s) = s^c$. Note that $t(s)$ is a polynomial.

For each $j = 1 \ldots t(s)$, let $\mathbf{H}_j$ be an HCS with m constraints over n variables such that $|\mathbf{H}_j| = s$. From, these instances we can create a new HCS $\mathbf{H}$ with $t(s) \cdot m$ constraints over $t(s) \cdot n$ variables such that: For each $j = 1 \ldots t(s)$ constraints $l_{1+m \cdot (j-1)}$ through $l_{m \cdot j}$ use variables $x_{1+n \cdot (j+1)}$ through $x_{n \cdot j}$ and correspond to the constraints in HCS $\mathbf{H}_j$.

Note that no constraint in $\mathbf{H}$ corresponding to a constraint in $\mathbf{H}_j$ shares variables with a constraint in $\mathbf{H}$ corresponding to a constraint in $\mathbf{H}_{j'}$, $j' \neq j$. Thus, any refutation of $\mathbf{H}$ corresponds to a refutation of the HCS $\mathbf{H}_j$ for some $j \in \{1, \ldots, t(s)\}$. Consequently, $\mathbf{H}$ has a tree-like unit refutation of length L if and only if $\mathbf{H}_j$ has a tree-like unit refutation of length L for some $j \in \{1, \ldots, t(s)\}$.

Let $\mathbf{H}'$ be a kernel of $\mathbf{H}$ such that $|\mathbf{H}'| \leq L^c$. Since we can assume that $L \leq s$, we have that $|\mathbf{H}'| \leq t(s)$. Additionally, $\mathbf{H}'$ has a tree-like unit refutation of length L if and only if $\mathbf{H}_j$ has a tree-like unit refutation of length L for some $j \in \{1, \ldots, t(s)\}$. Thus, we have a t-bounded OR-distillation from the $TLUR_D$ problem for HCSs to itself. This cannot happen unless **coNP** $\subseteq$ **NP/poly**. $\square$

7   Conclusion

In this paper, we examined the problem of finding unit refutations of HCSs. This extends the study of unit refutations from CNF formulas to linear systems. In particular, we analyzed the tree-like unit refutation problem for HCSs. This paper establishes that the $TLUR_D$ problem for HCSs is **NP-hard**. Furthermore, the TLUR$_{Opt}$ problem for HCSs is **NPO-complete**. We also showed that the TLUR$_D$ problem for HCSs cannot (under current complexity theoretic assumptions) admit a kernel whose size is polynomial in the length of the refutation.

References

1. Ausiello, G., D'Atri, A., Protasi, M.: Lattice theoretic ordering properties for NP-complete optimization problems. Fundam. Informaticae **4**(1), 83–94 (1981)
2. Baumgartner, P.: Linear and unit-resulting refutations for horn theories. J. Autom. Reason. **16**(3), 241–319 (1996)
3. Chandrasekaran, R., Subramani, K.: A combinatorial algorithm for Horn programs. Discrete Optim. **10**, 85–101 (2013)
4. Cousot, P., Cousot, R.: Abstract interpretation: a unified lattice model for static analysis of programs by construction or approximation of fixpoints. In: POPL, pp. 238–252 (1977)
5. de Moura,L., Owre, S., Ruess, H., Rushby, J.M., Shankar, N.: The ICS decision procedures for embedded deduction. In: IJCAR, pp. 218–222 (2004)
6. Duterre, B., de Moura, L.: The yices SMT solver. Technical report, SRI International (2006)
7. Farkas, G.: Über die Theorie der Einfachen Ungleichungen. Journal für die Reine und Angewandte Mathematik **124**(124), 1–27 (1902)
8. Fomin, F.V., Lokshtanov, D., Saurabh, S., Zehavi, M.: Theory of Parameterized Preprocessing. Cambridge University Press, Kernelization (2019)
9. Ford, J., Shankar, N.: Formal verification of a combination decision procedure. In: CADE, pp. 347–362 (2002)
10. Jaffar, J., Maher, M.: Constraint logic programming: a survey. J. Log. Program. **19–20581**, 503–581 (1994)
11. Kann, V.: On the Approximability of NP-complete Optimization Problems. Ph.D. thesis, Royal Institute of Technology Stockholm (1992)
12. Komuravelli, A., Bjørner, N., Gurfinkel, A., McMillan, K.L.: Compositional verification of procedural programs using Horn clauses over integers and arrays. In: Formal Methods in Computer-Aided Design, FMCAD 2015, Austin, Texas, USA, 27–30 September 2015, pp. 89–96 (2015)
13. Lueker, G.S.: Two NP-complete Problems in Nonnegative Integer Programming. Princeton University, Department of Electrical Engineering (1975)
14. Neiman, V.S.: Refutation search for horn sets by a subgoal-extraction method. J. Log. Program. **9**(2&3), 267–284 (1990)
15. Schrijver, A.: Theory of Linear and Integer Programming. John Wiley and Sons, New York (1987)
16. Subramani, K.: Optimal length resolution refutations of difference constraint systems. J. Autom. Reason. (JAR) **43**(2), 121–137 (2009)
17. Subramani, K., Wojciechowki, P.: A polynomial time algorithm for read-once certification of linear infeasibility in UTVPI constraints. Algorithmica **81**(7), 2765–2794 (2019)
18. Veinott, A.F., LiCalzi, M.: Subextremal functions and lattice programming, July 1992. Unpublished Manuscript
19. Wojciechowski, P., Subramani, K.: A certifying algorithm for checking for the presence of unit refutations in horn constraint systems. European Symposium on Programming (Submitted)
20. Yap, C.K.: Some consequences of non-uniform conditions on uniform classes. Theor. Comput. Sci. **26**(3), 287–300 (1983)

Trees and Graphs

Homomorphic Characterization of Tree Languages Based on Comma-Free Encoding

Stefano Crespi Reghizzi and Pierluigi San Pietro$^{(\boxtimes)}$

Dipartimento di Elettronica, Informazione e Bioingegneria (DEIB),
Politecnico di Milano, Piazza Leonardo da Vinci 32, Milano 20133, Italy
{stefano.crespireghizzi,pierluigi.sanpietro}@polimi.it

Abstract. A classic result in language theory is Medvedev's theorem for trees, stating that any regular tree language can be defined by the projection of a *local* tree language, i.e., of a language defined by its tiles of height 2, a.k.a. di-grams. The simple proof of the statement is based on a local alphabet whose size (linearly) depends on the number of states of the tree automaton recognizing the original language. We prove a new extended version of Medvedev's theorem for trees, by using a k-locally testable tree language defined by tiles of height $k \geq 2$ (k-grams). The size of the local alphabet is just the double of the original one, hence it is independent from the complexity of the tree automaton. This result relies on an encoding of the states traversed by a tree automaton, by means of binary comma-free codes carefully laid on tree paths. We thus generalize from words to trees our recent extended Medvedev's theorem for word languages that was based on strictly locally testable word languages. By applying the result to the syntax trees of context-free grammars, we characterize them by k-locally testable, binary-labeled trees.

1 Introduction

This study is about the theory of (regular) tree languages, a well-known domain (see, e.g., [3,6,7]) that keeps attracting much attention for fundamental research and application in many sectors. Early studies have shown that many formal properties generalize from finite-state, i.e., regular, word languages, to tree languages. Thus, tree languages admit equivalent characterizations based on finite tree automata (TA), on tree grammars (closely related to the context-free ones for words), on logic formulas, and, what is more relevant for us, on homomorphism of *local* tree languages.

The latter characterization is also called Medvedev's theorem [10]. In fact, in the case of words, Medvedev's original theorem asserts that a language R over a finite terminal alphabet Σ is regular if, and only if, there exists another finite alphabet Λ, a *local* word language L over Λ and a letter-to-letter homomorphism h from Λ^* to Σ^* such that R is the image $h(L)$. The reformulation of Medvedev's theorem for tree languages (available in the references cited above), uses a *local*

© Springer Nature Switzerland AG 2021
A. Leporati et al. (Eds.): LATA 2021, LNCS 12638, pp. 241–254, 2021.
https://doi.org/10.1007/978-3-030-68195-1_19

tree language. Intuitively, a local word language is defined by the set of substrings of length two (2-factors) occurring in language sentences. Similarly, a local tree language is defined by the set of subtrees of height two (also called 2-*grams*) that occur in the valid trees. Local languages have a local testability property, in the sense that membership of a word/tree can be checked by repeated local tests: respectively, on a word by means of a sliding window of width two, and on a tree by a sliding frame of height two and width sufficient to enclose the subtree.

The size of Medvedev's local word language alphabet Λ depends on the size of the original alphabet Σ and on the complexity of the language, measured by the number of states, $|Q|$, of the recognizer. More precisely, the size of the local alphabet in the original statement of the theorem is $|\Sigma| \cdot |Q|^2$, but is easily lowered to $|\Sigma| \cdot |Q|$. We may say that the *alphabetic ratio* of the original theorem is $|Q|$. A corresponding alphabetic ratio holds in the analogous theorem for tree languages.

Local word languages are at the lowest level of an infinite language hierarchy called *k-strictly locally testable* (slt) [9], each level $k \geq 2$ being characterized by the use of a sliding window of width k. Such a hierarchy has been extended to tree languages, see (see, e.g., [12]), using k-grams instead of k-factors.

The question we address is: if in Medvedev's theorem for trees we allow slt languages of order greater then 2, can we reduce the alphabetic ratio, and how much? In the case of word languages, our answer, referred to as the Extended Medvedev's Th. [5], says that the minimal alphabetic ratio is 2 and can be always achieved by using an slt language with slt constant of the order of the logarithm of the recognizer size. We hint to its proof that is the starting point of the present development for trees. Given a finite automaton (FA), the technique samples in each computation the sub-sequences made by k state-transitions, where k is in $O(\lg |Q|)$, so that a binary code of length k suffices to encode all the states of the FA. To prevent mistakes, we resort to codes that can be decoded without synchronisation, using a $2k$-slt DFA as decoder. The family of *comma-free* codes [2] has such a property and is the one we use also here. A similar approach was used in the proof of Theorem 5.2 of [13], to give a logical characterization of regular languages, which could be used as the starting point for an alternative proof of the Extended Medveded Theorem.

Moving from word to tree languages often complicates matters, and we had to examine and discard several possible ways of encoding the computation of a TA by means of comma-free codes laid on a tree, before we found a successful one. The result is an exact extension of the Extended Medvedev's Th. from word languages to regular tree languages, and says that any tree language is the projection, by means of a letter-to-letter homomorphism, of an slt tree language over an alphabet with size double of the original alphabet size. The alphabetic ratio 2 is the minimal possible.

Since the syntax trees of context free grammars are regular (indeed, local) trees, the Extended Medvedev's Th. is easily transposed as follows. A word language is generated by a context-free grammar G if, and only if, the language is the yield (a.k.a. frontier) of an slt tree language having the internal nodes labels in $\{0, 1\}$ and such that its trees are structurally identical to the syntax trees of G.

Section 2 introduces the notation and preliminary facts. Section 3 defines the tree transformation that encodes on tree paths the states traversed by sampled TA computations. Then, it shows that the encoding trees have the slt property and proves the Extended Medvedev's Th. for trees. It ends with a corollary for the syntax trees of context-free grammars.

2 Basic Definitions

Tree languages. We refer to [6] for the notions and notations not defined below. A *ranked alphabet* is a finite alphabet Δ with the property that there is a finite number of integers $k \geq 0$ such that a nonempty subset $\Delta_k \subseteq \Delta$ is given. If $a \in \Delta_k$ then k is one of the ranks of $a \in \Delta$. The Δ_k sets need not be disjoint.

A *tree* over a ranked alphabet Δ is an element of the set T_Δ defined by induction as:

- if $a \in \Delta_0$ then $a \in T_\Delta$;
- for all $k \geq 1$, if $a \in \Delta_k$ and $t_1, t_2, \ldots, t_k \in T_\Delta$ then $a[t_1, t_2, \ldots, t_k] \in T_\Delta$.

A *tree language* T over Δ is a subset of T_Δ.

Such a tree can be represented by a (root to leaf) directed (left to right) oriented acyclic graph with node labels: each node has a distinct identifier x, and has a *label* in Δ. We can assume for convenience that the identifier x depends only on the structure of the tree and not on its labels, so that the same identifier permits to individuate two nodes located in the same position, but belonging to different superposable trees.

The *label of a path* $x_1 \ldots x_m$ is the concatenation of the labels of nodes $x_1, \ldots, x_m$. Let t be a tree and x, y be two of its nodes. We denote by $t_{|x}$ the *subtree* of t at node x. The *distance* between nodes x and y is the length of the path $x \to y$. The *height* of t is the maximal distance from the root to a leaf (hence, a tree with one node has height 0). The *yield* (a.k.a. frontier) of a tree is a mapping from T_Δ into $\Delta_0{}^*$ returning the leaf labels in left to right order.

Let Λ, Δ be two ranked alphabets and let $\eta : \Lambda \to \Delta$ be a mapping. A *projection* $\eta : T_\Lambda \to T_\Delta$, is a map associating every tree $t \in T_\Lambda$ with the tree $\eta(t) \in T_\Delta$ such that $\eta(t) = \eta(a)$ if t is a single node labeled $a \in \Lambda_0$, and $\eta(t) = \eta(a)[\eta(t_1), \ldots, \eta(t_n)]$ if t is $a[t_1, \ldots, t_k]$, $k \geq 1$.

Let N be a symbol. Two trees $t \in T_\Lambda, t' \in T_\Delta$ are *structurally equivalent* if there exist two projections $\eta : T_\Lambda \to \{N\}, \eta' : T_\Delta \to \{N\}$ such that $\eta(t) = \eta'(t')$. A *tree transformation* from T_Δ into T_Λ is any subset of $T_\Delta \times T_\Lambda$.

The family of regular tree languages, henceforth tree languages, is characterized by their recognizer, the *bottom-up tree automaton* which may be assumed to be deterministic; the latter is equivalent to the nondeterministic top-down tree automaton. Moreover, the bottom-up version admits an equivalent nondeterministic top-down version that is *associated* (e.g., see Theorem 2.2.10 of [7]);

the associated automata have identical states, with starting and final states interchanged. Since we only need *nondeterministic top-down tree automata*, for brevity we just call them tree automata (TA).

Definition 1 (tree automaton). *A tree automaton is a 5-tuple* $M = (Q, \Delta, \delta, S, F)$ *where* Q *is the set of states,* $S \subseteq Q$ *is the set of starting states,* F *is a family* $\{F_a\}_{a \in \Delta_0}$ *of sets* $F_a \subseteq Q$ *of final states and* δ *is a family* $\{\delta_a^k\}_{k \geq 1, a \in \Delta_k}$ *of mappings* $\delta_a^k : Q \to \mathcal{P}(Q)^k$.
The mapping $\tilde{\delta} : \mathcal{T}_\Delta \to \mathcal{P}(Q)$ *is defined recursively as follows:*

$$\begin{cases} for\ a \in \Delta_0,\ \tilde{\delta}(a) = F_a; \\ for\ k \geq 1, a \in \Delta_k,\ and\ t_1, \ldots, t_k \in \mathcal{T}_\Delta, \\ \quad \tilde{\delta}(a[t_1, \ldots, t_k]) = \left\{ q \mid \exists (q_1, \ldots, q_k) \in \delta_a^k(q) : q_i \in \tilde{\delta}(t_i)\ for\ all\ 1 \leq i \leq k \right\}. \end{cases}$$

Define the tree language recognized by M *as* $T(M) = \{t \in \mathcal{T}_\Delta \mid \tilde{\delta}(t) \cap S \neq \emptyset\}$.

Given a computation by a TA on a tree $t \in T(M)$, to represent it, it is customary to additionally label each tree node with the TA state, to the effect that the tree thus obtained is in $\mathcal{T}_{\Delta \times Q}$, and it is called a *state-labeled* tree (S-tree). When useful to prevent confusion, we denote an S-tree by $\hat{t}$; also, $\widehat{T(M)} = \{\hat{t} \mid \hat{t}\ \text{is an S-tree computed by M for some}\ t \in T(M)\}$ denotes the set of all S-trees for $T(M)$. See Example 1 below.

Example 1 (a counting language). Consider the tree language (from [6]) $T \subseteq \mathcal{T}_\Delta$ with $\Delta_0 = \{0, 1, \ldots, 9\}$ and $\Delta_2 = \{+, *\}$, that includes the arithmetic expressions such that their result modulo 3 is 1. T is defined by the top-down TA $M = (\{q_0, q_1, q_2\}, \Delta, \delta, \{q_1\}, F)$ where for $0 \leq h \leq 2$,
$$\begin{cases} \delta_+(q_h) = \{(q_i, q_j) \mid i + j \mod 3 = h\} \\ \delta_*(q_h) = \{(q_i, q_j) \mid i * j \mod 3 = h\} \end{cases}$$
and $F_0 = F_3 = F_6 = F_9 = \{q_0\}$, $F_1 = F_4 = F_7 = \{q_1\}$, $F_2 = F_5 = F_8 = \{q_2\}$. A tree $t \in T(M)$ is shown in Fig. 1 as well as its S-tree $\hat{t}$.

Tree Languages Defined by Local Tests. We introduce our notation and terminology for strictly locally testable tree languages. The following definition of h-grams corresponds to the definition of $(h-1)$-type of [12].

Definition 2. (h-gram). *Let* $h \geq 2$, *let* $t \in \mathcal{T}_\Delta$ *and* x *a node of* t. *The restriction of the subtree* $t_{|x}$ *to the set of nodes at distance less than* h *from* x *is called a* h-gram *of* t *with root* x; *if* x *is the root of* t, *it is called a* root h-gram. *The set of* h-grams *of* t *is denoted by* $\langle\!\langle t \rangle\!\rangle_h$; *such a definition is naturally extended to the* h-grams $\langle\!\langle T \rangle\!\rangle_h$ *of a language* T.

Notice that the yield of a h-gram, unlike the one of a tree, may include symbols not belonging to Δ_0.

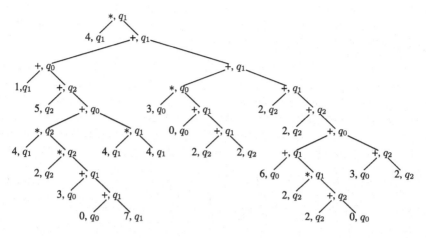

Fig. 1. The tree (labels in $\Delta_0 \cup \Delta_2$) and the corresponding S-tree (labels in $(\Delta_0 \cup \Delta_2) \times \{q_0, q_1, q_2\}$) for Example 1.

Definition 3 (strictly locally testable (tree) language). *Let $h \geq 2$, let Γ_h be a set of h-grams over the alphabet Δ and let $\Theta_h \subseteq \Gamma_h$ be the set of root h-grams. The two sets jointly define the h-strictly locally testable (h- slt) language, denoted by $T = T(\Gamma_h, \Theta_h) \subseteq \mathcal{T}_\Delta$, such that the inclusions hold:*

$$\langle\!\langle T \rangle\!\rangle_h \subseteq \Gamma_h \quad \text{and} \quad \{y \mid y \text{ is a root } h\text{-gram of } T\} \subseteq \Theta_h.$$

A language is called strictly locally testable *(slt) if it is h-slt for some h; in particular, if $h = 2$ it is also called* local.

A well-known example of local language are the S-trees $\widehat{T(M)}$ of a language $T(M) \subseteq \mathcal{T}_\Delta$. See for instance the S-trees of language $T(M)$ in Example 1. Notice also that the language $T(M)$ is not slt since it clearly violates the *noncounting property* [4].

The next known proposition follows by taking as the local-language ranked alphabet Λ the product $\Delta \times Q$.

Proposition 1 (Medvedev's theorem for trees). *A tree language $T \subseteq \mathcal{T}_\Delta$ is regular if, and only if, there exists a ranked alphabet Λ, a local tree language $T' \subseteq \mathcal{T}_\Lambda$ and a projection $\eta : \Lambda \to \Delta$ such that $T = \eta(T')$. Moreover, if there is a tree automaton recognizing T with a state set Q, then the* alphabetic ratio $\frac{|\Lambda|}{|\Delta|}$ *is less or equal to $|Q|$.*

Comma-free codes. Let Λ be an alphabet. A finite set $X \subset \Lambda^+$ is a *code* [2] if every word in Λ^+ has at most one factorization in words (a.k.a *codewords*) of X, more precisely: for any $u_1 u_2 \ldots u_m$ and $v_1 v_2 \ldots v_n$ in X, where the u and v are codewords, the identity $u_1 u_2 \ldots u_m = v_1 v_2 \ldots v_n$ holds only if $m = n$ and $u_i = v_i$ for $1 \leq i \leq n$.

We use a code X to represent a finite alphabet Γ by means of a one-to-one homomorphism, denoted by $[\![\ ]\!]_X : \Gamma^* \to \Lambda^*$, called *encoding*, such that $[\![a]\!]_X \in X$ for every $\alpha \in \Gamma$.

Let $k \geq 1$. A set $X \subset \Lambda^k$ is a *comma-free* code of length k, if, intuitively, no codeword overlaps the concatenation of two codewords: more precisely, for any $t, u, v, w \in \Lambda^*$, if tu, uv, vw are in X, then $u = w = \varepsilon$, or $t = v = \varepsilon$. This condition can equivalently be restated as $X^2 \cap \Lambda^+ X \Lambda^+ = \emptyset$.

We need the following result (see [11] and its references) on the number of codewords in a comma-free code of length k over an alphabet with cardinality $|\Lambda| = n$.

Let $\ell_k(n) = \frac{1}{k} \sum \mu(d) n^{k/d}$, where the summation ranges over all divisors d of k, and μ is the Möbius function: $\mu(d) = \begin{cases} 1 \text{ if } d = 1 \\ 0 \text{ if } d \text{ has any square factor} \\ (-1)^r \text{ if } d = p_1\, p_2 \ldots p_r \\ \quad \text{where } p_1\, p_2 \ldots p_r \text{ are distinct primes.} \end{cases}$

It is well-known that $\ell_k(n)$ counts the number of primitive cyclic words of length k.

Proposition 2. *For every alphabet with n letters and for every odd integer $k > 1$ there is a comma-free code of length k with $\ell_k(n)$ words.*

The definition of the Möbius function is such that, if k is a prime number, the preceding summation is equal to $n^k - n$, and we may use the simpler formula: for all prime numbers k, $\ell_k(n) = \frac{n^k - n}{k}$. We always use comma-free codes with $n = 2$ over the binary alphabet $\{0, 1\}$. An example of codes is in Example 2, below.

3   Main Results

In this section we present a constructive proof that any regular tree language T in $\mathcal{T}_\Delta$, defined by a TA with state set Q, is the projection of an slt tree language in $\mathcal{T}_\Delta \times \{0, 1\}$. Preliminarily, we encode the states of Q with a (binary) comma-free code X of length k. Then, we present a tree transformation τ from the S-trees in $\widehat{T} = \widehat{T(M)}$ to the slt trees, denoted by $\widetilde{T}$, that we call *encoding trees* (*E-trees*) since they encode on their path labels the states traversed by M's computations. The mapping from S-trees to E-trees preserves the structure of trees. E-tree labels are couples of the form $\langle a, b \rangle$ with $a \in \Delta$, $b \in \{0, 1\}$. Clearly, the bits b needed to encode all states traversed by M would exceed the number of E-tree nodes by a factor k, therefore we can only encode a sampled subset of the states. Then, we analyze and prove in a series of propositions the properties of E-trees needed to state in Lemma 1 the slt property for the language of E-trees. From this the main Theorem 1 easily follows. The theorem is accompanied by two corollaries, one stating the minimality of the alphabetic ratio $2 = |\Delta \times \{0, 1\}|/|\Delta|$, the other applying the main result to the context-free syntax trees.

From Trees to Encoding Trees. Let $M = (Q, \Delta, \delta, S, F)$ be a tree automaton as in Definition 1. Let $X \subset \{0,1\}^k$, $k \geq 2$ be a binary code for Q, so that each $q \in Q$ has the encoding $[\![q]\!]_X$. Using Proposition 2, we compute the value of k: let $h = |Q|$ and let k be such that $\ell_k(2) > h$, i.e., $\frac{2^k - 2}{2} = 2^{k-1} - 1 \geq h$. Therefore, the value of k is in $O(\lg|Q|)$.

A tree from $T(M)$ is input to a tree transformation that nondeterministically produces a structurally equivalent E-tree. In the latter, each node label carries a bit, which is used to encode the states traversed by M's computation on the tree, sampled at distances multiple of k from the root (where computation starts).

We need some definitions and conventions to prepare for the main one. Let $t \in T(M) \subseteq \mathcal{T}_\Delta$ be a tree. Consider the accepting computations of the TA M over t. As said, each computation may be described as an S-tree $\hat{t} \in \widehat{T(M)} \subseteq \mathcal{T}_{\Delta \times Q}$. Given an S-tree $\hat{t}$, its *state-root* sr is defined as the state component of its root label. More formally, $sr(\langle a, q \rangle) = q$ for $a \in \Delta_0$, $q \in F_a$, and $sr(\langle a, q \rangle[t_1, \ldots, t_k]) = q$ with $k \geq 1$, $a \in \Delta_k$, $q \in Q$. In the following, we denote by $\widehat{T(M, q)}$ the S-trees of computations that start in a given state $q \in Q$ (i.e., those $\hat{t}$ such that $sr(\hat{t}) = q$), regardless of state q being starting.

Next, for every S-tree $\hat{t}$, we define its *encoding tree* (E-tree), denoted by $\tilde{t}$.

Definition 4 (translation from S- to E-trees). *We define a tree transformation by means of a partial map $\tau : \mathcal{T}_{\Delta \times Q} \rightarrow \mathcal{T}_{\Delta \times \{0,1\}}$ from state-labeled trees to structurally equivalent encoding trees.*
Let $S_X = \{w \in \{0,1\}^+ \mid yw \in X, y \in \{0,1\}^*\}$, i.e., w is in S_X if it is a codeword or a suffix thereof.
To define τ, we introduce a family of mappings $\{\rho^w\}_{w \in S_X}$ with each $\rho^w : \mathcal{T}_{\Delta \times Q} \rightarrow \mathcal{T}_{\Delta \times \{0,1\}}$. For all $\hat{t} \in \mathcal{T}_{\Delta \times Q}$, $\rho^w(\hat{t})$ is recursively defined as follows, for all $b \in \{0,1\}$, $w \in \{0,1\}^*$ such that $bw \in S_X$:

1. *if $\hat{t} = \langle a, q \rangle$, with $a \in \Delta_0, q \in F_a$, then let $\rho^{bw}(\hat{t}) = \langle a, b \rangle$;*
2. *if $\hat{t} = \langle a, q \rangle[t_1, \ldots, t_n], n \geq 1, a \in \Delta_n, q \in Q$, then let*

$$\rho^{bw}(\hat{t}) = \langle a, b \rangle[t_1', \ldots, t_n'] \text{ where } \begin{cases} t_i' = \rho^w(t_i) \text{ if } w \neq \varepsilon \\ t_i' = \rho^{[\![q_i]\!]_X}(t_i) \text{ otherwise,} \\ \qquad \text{where } q_i = sr(t_i), 1 \leq i \leq k. \end{cases}$$

Then τ is defined for every $\hat{t} \in \mathcal{T}_{\Delta \times Q}$ of height at least k as $\tau(\hat{t}) = \rho^{[\![sr(\hat{t})]\!]_X}(\hat{t})$. The E-tree language $\tilde{T}$ is defined as $\{\tau(\hat{t}) \mid \hat{t} \in \widehat{T(M)}\}$. $\square$

Properties of E-trees. Consider a path of $\tilde{t} \in \mathcal{T}_{\Delta \times \{0,1\}}$; its *binary label* is the projection on $\{0,1\}$ of the path label.
Given a node x of a tree t and a value $j > 0$, define $dist(x, j)$ as the set of nodes of $t_{|x}$ at distance j from x.

Definition 5 (profile). *Let $\tilde{t} \in \mathcal{T}_{\Delta \times \{0,1\}}$ be a tree of height at least k, such that the binary label of any path of length $h \leq k$ from the root agrees with the* *binary label of every other path of length h from the root. The unique binary label of every path of length k (from the root) is called the* profile of length k of *$\tilde{t}$, denoted $pr_k(\tilde{t})$.*

For instance, in Fig. 2, the subtree whose root is labeled with $\langle$ code-gram $+, \textcircled{1}q_2\rangle$ has profile 1110.

The next two propositions assert that when $\tilde{t}$ is $\tau(\hat{t})$ for some $\hat{t} \in \mathcal{T}_{\Delta \times Q}$, the profile of length k of $\tilde{t}$ is always defined and it is a codeword.

Proposition 3 (identity of binary path labels). *Given a state $q \in Q$ and a tree $\hat{t} \in \widehat{T(M, q)} \subseteq \mathcal{T}_{\Delta \times Q}$, for every $i \geq 0$, for all $0 < j < k$, if $\hat{t}$ has height at least $k * i + j$ and x is a node in $dist(root(t), k * i)$, then for all nodes of $\tau(\hat{t})$ in $dist(x, j)$ the projections of their labels to $\{0, 1\}$ are identical.*

In any E-tree $\tau(\hat{t})$, the codewords are regularly spaced from the root, as stated next.

Proposition 4 (position of codewords). *For $i \geq 0$, let $\hat{t}$ be an S-tree of height at least $k * (i + 1)$, and let the E-tree $\tilde{t}$ be $\tau(\hat{t})$. For all nodes x at distance $k * i$ from the root of $\hat{t}$, if x has label $\langle a, p\rangle$ for some $a \in \Delta, p \in Q$, then $pr_k(\tilde{t}_{|x})$ is defined and it is equal to the codeword $[\![p]\!]_X$.*

This proposition allows us to single out the notion of *code-gram*: a k-gram whose profile of length k is a codeword in X.

We define the test sets to be used later to prove that the language of E-trees is slt.

Definition 6 (test sets). *Given the S-tree language $\widehat{T(M)} \subseteq \mathcal{T}_{\Delta \times Q}$, let k be the length of the binary code X that encodes the states Q of the TA M. We associate with $\widehat{T(M)}$ the following set of 2k-grams (computed on the encoding trees using Definition 2):*

$$\Gamma_{2k} = \bigcup_{q \in Q} \bigcup_{\hat{t} \in \widehat{T(M, q)}} \langle\!\langle \tau(\hat{t})\rangle\!\rangle_{2k}$$

and, for all $q \in Q$, the following sets of root 2k-grams:

$$\Theta_{2k}(q) = \bigcup_{\hat{t} \in \widehat{T(M, q)}} \{2k\text{-grams at the root of } \tau(\hat{t})\}.$$

The set of all root 2k-grams is defined as $\Theta_{2k} = \bigcup_{q \in Q} \Theta_{2k}(q)$. Clearly, $\Theta_{2k} \subseteq \Gamma_{2k}$.
The above sets define by Definition 3 the 2k-slt language $T(\Gamma_{2k}, \Theta_{2k})$ and, for every $q \in Q$, the languages $T(\Gamma_{2k}, \Theta_{2k}(q))$.

For convenience, in the following proposition we gather some properties of root 2k-grams, following immediately from the definitions of τ and $\Theta_{2k}(q)$.

Proposition 5. (properties of Θ_{2k}). *Given the S-tree language $\widehat{T} \subseteq \mathcal{T}_{\Delta \times Q}$, let Γ_{2k} and $\Theta_{2k}(q)$ be as in Definition 6. For all $q \in Q$ and for all 2k-grams $\widetilde{y} \in \Theta_{2k}(q)$ of height at least k:*

1. *$\widetilde{y} \in \Gamma_{2k}$ and its profile $pr_k(\widetilde{y})$ is $[\![q]\!]_X$*
2. *every k-gram of $\widetilde{y}$ of height k with root in $dist(root(\widetilde{y}), k)$ has profile in X*
3. *there exists an S-tree $\widehat{t} \in \widehat{T}$ with state-root q such that $\widetilde{y}$ is the 2k-gram at the root of $\tau(\widehat{t})$.*

Proposition 6 (codeword-state agreement in root). *Given the S-tree language $\widehat{T} \subseteq \mathcal{T}_{\Delta \times Q}$, let $\Gamma_{2k}, \Theta_{2k}(q)$, for all $q \in Q$, be as in Definition 6. Let $\widetilde{y} \in \Gamma_{2k}$ be a 2k-gram of height at least k.*
If the profile $pr_k(\widetilde{y})$ is a codeword $[\![q]\!]_X$ for some $q \in Q$, then $\widetilde{y} \in \Theta_{2k}(q)$.

Proof. If $\widetilde{y}$ has height less than $2k$, then the statement is obvious. If $\widetilde{y}$ has height $2k$, it is a (possibly non-root) 2k-gram of an E-tree $\widetilde{t} = \tau(\widehat{t})$ associated with an S-tree $\widehat{t} \in T(M)$. Let x be one of the nodes in $\widetilde{t}$ (and in $\widehat{t}$) corresponding to the root of $\widetilde{y}$.

By contradiction, suppose that in $\widehat{t}$ the node x does not correspond to a node with a state label $q \in Q$. Then, by Proposition 4, x is not at a distance multiple of k from the root of $\widehat{t}$, i.e., there exist $i \geq 0$ and j, $0 < j \leq k-1$, such that x is at distance $k * i + j$ from the root. Let x_i, x_{i+1} and x_{i+2} be nodes at distance, respectively $k * i$, $k * (i+1)$ and $k * (i+2)$ from the root, such that there is a path from x_i to x to x_{i+1} to x_{i+2}. These nodes must exist since $\widetilde{y}$ has height $2k$, hence the tree $\widetilde{t}$ has height at least $k * i + j + 2k > k * (i+2)$.

By Proposition 4, the profiles of length k of the subtrees $\widetilde{t}_{|x_i}$ and $\widetilde{t}_{|x_{i+1}}$ are codewords, say, $[\![p]\!]_X$ and $[\![r]\!]_X$, respectively, for some $p, r \in Q$. Consider the path from x_i to x_{i+2}, which includes nodes x and x_{i+1}: its binary label is $[\![p]\!]_X$ $[\![r]\!]_X$. However, this label also contains by hypothesis $[\![q]\!]_X$ as a (proper) factor, i.e., it is of the form $u[\![q]\!]_X v$, for some $u, v \in \{0,1\}^+$, a contradiction with the comma-free property of code X.

□

We prove that an E-tree language is an slt language of order $2k$.

Lemma 1 (slt property of encoding trees). *Let $\widehat{T}$ be the S-tree language recognized by TA $M = (Q, \Delta, \delta, q_0, F)$. There is $k > 1$ such that the E-tree language $\tau(\widehat{T})$ is the 2k-slt language defined by the test sets in Definition 6, i.e., $\tau(\widehat{T}) = T(\Gamma_{2k}, \Theta_{2k}(q_0))$.*

Proof. We prove the more general statement $\bigcup_{q \in Q} \tau(\widehat{T(M,q)}) = T(\Gamma_{2k}, \Theta_{2k})$.
The inclusion $\tau(\widehat{T(M,q)}) \subseteq T(\Gamma_{2k}, \Theta_{2k}(q))$ is obvious by Definition 6, since every $\widetilde{t} \in \tau(\widehat{T(M,q)})$ comprises only 2k-grams in Γ_{2k}, with the one at the root being in $\Theta_{2k}(q)$.
The proof of the converse inclusion is by induction on the height $n \geq 1$ of a tree in $T(\Gamma_{2k}, \Theta_{2k})$. More precisely, the induction hypothesis is that for every $n > 1$,

if a tree $\tilde{t}$ of height less than n is in $T(\Gamma_{2k}, \Theta_{2k})$, then there exists $q \in Q$ such that $\tilde{t}$ has the form $\tau(\hat{t})$ for some $\hat{t} \in \widehat{T(M, q)}$ and $\tilde{t} \in T(\Gamma_{2k}, \Theta_{2k}(q))$.

Base case: The base case includes all trees of height up to $2k$. Let $\tilde{t}$ be a tree of height up to $2k$ in $T(\Gamma_{2k}, \Theta_{2k})$. Then, there exists $q \in Q$ such that $\tilde{t} \in \Theta_{2k}(q)$. Therefore, there exists an S-tree $\hat{s} \in \widehat{T(M, q)}$ such that $\tilde{t}$ is the $2k$-gram at the root of $\tau(\hat{s})$. Since the $2k$-gram at the root of $\hat{s}$ is structurally equivalent to $\tilde{t}$ and the yield of $\tilde{t}$ is in $(\Delta_0 \times \{0, 1\})^+$, the S-tree $\hat{s}$ must also be of same height as $\tilde{t}$. Therefore, $\tilde{t} = \tau(\hat{s})$.

Induction step: let $\tilde{t} \in T(\Gamma_{2k}, \Theta_{2k})$ be a tree of height $n > 2k$. Hence, the $2k$-gram $\tilde{y}$ at the root of $\tilde{t}$ is in $\Theta_{2k}(q)$, for some $q \in Q$. By Proposition 5, Part (1), the profile of length k of $\tilde{y}$ is $[\![q]\!]_X$.

By Proposition 4, every node z at distance k from the root of $\tilde{t}$ is such that the profile of length k of the subtree $\tilde{t}_{|z}$ (rooted at z) is a codeword, say, $[\![q_z]\!]_X$, where $q_z \in Q$.

All $2k$-grams of the subtree $\tilde{t}_{|z}$ are obviously in Γ_{2k} and, by Proposition 6, the $2k$-gram at the root z of $\tilde{t}_{|z}$ is in $\Theta_{2k}(q_z)$. Hence, $\tilde{t}_{|z}$ is in $T(\Gamma_{2k}, \Theta_{2k}(q_z)) \subseteq T(\Gamma_{2k}, \Theta_{2k})$.

The induction hypothesis thus applies to $\tilde{t}_{|z}$: there exists a (structurally equivalent) tree $\hat{t}_z$ in $\widehat{T(M, q_z)}$ such that $\tilde{t}_{|z} = \tau(\hat{t}_z)$.

The tree $\tilde{t}$ is completely defined by the root $2k$-gram $\tilde{y}$ and by all those subtrees $\tilde{t}_{|z}$; in fact, let x be a node of $\tilde{t}$; if x is at distance less than $2k + 1$ from the root, then x is in $\tilde{y}$; otherwise, there exists a node z at distance k from the root such that x is a node of $\tilde{t}_{|z}$.

By Proposition 5, Part (3), the $2k$-gram $\tilde{y}$ is the $2k$-gram at the root of an E-tree $\tau(\hat{t}')$, where $\hat{t}'$ is an S-tree in $\widehat{T(M, q)}$; moreover, q_z as above is the state label of the node z of $\hat{t}'$.

For each node z as above, replace the subtree $\hat{t}'_{|z}$ in the S-tree $\hat{t}'$ with the S-tree $\hat{t}_z$ found by the induction hypothesis. The resulting S-tree $\hat{t}$ is thus such that the $2k$-gram at the root of $\tau(\hat{t})$ is still $\tilde{y}$ and every subtree $\hat{t}_{|z}$, with z as above, is structurally equivalent to $\tilde{t}_{|z}$. Therefore, $\tilde{t} = \tau(\hat{t})$. $\square$

The following result descends immediately from Lemma 1, since $T(M)$ is the projection on alphabet Δ of the E-tree language $\tau(\widehat{T(M)})$.

Theorem 1. *For every finite ranked alphabet Δ, there exist $k \geq 2$, a finite ranked alphabet Λ, with alphabetic ratio $\frac{|\Lambda|}{|\Delta|} \leq 2$, and a projection $\eta : T_\Lambda \to T_\Delta$, such that for every regular tree language $T \subseteq T_\Delta$ there exists a $2k$-slt tree language $\tilde{T} \subseteq T_\Lambda$ such that $T = \eta(\tilde{T})$.*

Example 2 (a counting language - Part 2). Consider again the tree language defined by the TA in Example 1, with the tree and the corresponding S-tree in Fig. 1. A code X of length $k = 4$ for the state set Q is the following: $[\![q_0]\!]_X = 0010$, $[\![q_1]\!]_X = 0011$, $[\![q_2]\!]_X = 1110$. Applying the transformation of Definition 4 to the S-tree we obtain the E-tree in Fig. 2. Observe the presence of

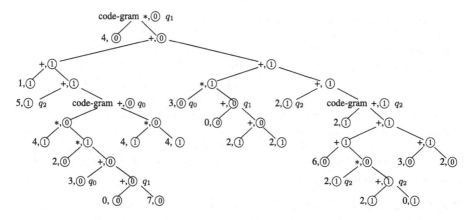

Fig. 2. E-tree with codewords $[\![q_0]\!]_x = 0010$ (blue), $[\![q_1]\!]_x = 0011$ (black), $[\![q_2]\!]_x = 1110$ (red). For readability, we report also the states at distance multiple of k from the root. (Color figure online)

three code-grams having their roots as shown. The E-tree language has the 8- slt property. □

The next statement combines Theorem 1 with a result in [5] for regular word languages.

Corollary 1 (minimality). *For every $n \geq 1$ there exist a finite ranked alphabet Δ, with $|\Delta| = n$, and a regular tree language in $\mathcal{T}_\Delta$ such that the alphabetic ratio of Theorem 1 cannot be less than 2.*

Proof. In Theorem 5 of [5], we proved that for every $n \geq 1$, the word language $L = \{(a_i a_i)^+ \mid 1 \leq i \leq n\}$ over the alphabet $\Delta = \{a_1, \ldots, a_n\}$, can be expressed as the image under a letter-to-letter homomorphism $h : \Lambda \to \Delta$ of a (word) slt language R over an alphabet Λ, only if $|\Lambda| \geq 2|\Delta|$. Consider the TA $M = (Q, \Delta, \delta, S, F)$ where $\Delta_0 = \Delta_1 = \Delta$, $Q = \{q_0, q_1, \ldots, q_n, q_1', \ldots, q_n'\}$, $S = \{q_0\}$ is the set of starting states, and for all $1 \leq i \leq n$ we let $F_{a_i} = \{q_i'\}$. Let the transition relation δ be defined as the family of mappings $\delta_{a_i}^1 = \{(q_i, q_i'), (q_i', q_i), (q_0, q_i)\}$, $1 \leq i \leq n$. The language defined by M is the set of trees of the form $a_i[a_i[a_i[\ldots [a_i] \ldots]]]$, for all $1 \leq i \leq n$, having the (only) path from root to leaf of even length. Therefore, $T(M)$ has a linear structure and can be viewed also as the word language $W(M) = L$.

By contradiction, suppose to define $T(M)$ as $T(M) = \eta(T')$, where T' is an slt tree language over an alphabet Λ such that $|\Lambda| < 2|\Delta|$, and $\eta : \mathcal{T}_\Lambda \to \mathcal{T}_\Delta$ is a projection. Language T' is structurally equivalent to $T(M)$, therefore it can also be regarded as a word language $W' \subseteq \Lambda^+$, hence the projection η can be regarded as a letter-to-letter homomorphism $\eta : \Lambda^* \to \Delta^*$ such that $L = \eta(W')$, violating Theorem 5 of [5]. □

Application to Context-Free Grammars. It is interesting to apply Theorem 1 to the syntax trees of context-free grammars. Let $G = (V_N, \Sigma, P, S)$ be a grammar. It is known [3,6,7] that the syntax trees of $L(G)$ make a local tree language, denoted by $T(G)$, over the ranked alphabet $\Delta = \Sigma \cup V_N$ with the ranks defined as $\Delta_0 = \Sigma$ and $\Delta_i \subseteq V_N$, $i \geq 1$ such that $A \in \Delta_i$ if, and only if, P contains a production $A \to \alpha$ with $\alpha \in (\Sigma \cup V_N)^i$. Since Σ is disjoint from V_N by hypothesis, a terminal symbol may not label an internal node of a syntax tree. By a slight modification of Definition 4 and of the proof of Lemma 1, it is possible to restate Theorem 1 by assigning symbols in Δ_0, rather than in $\Delta_0 \times \{0,1\}$. as labels of the leaves of the E-trees used in the proof. This leads to the following more expressive version of Theorem 1.

Corollary 2 (syntax trees). *Let $L = L(G)$ be a word language over the alphabet Σ generated by a context-free grammar G.*
The tree language $T(G)$ is structurally equivalent to a strictly locally testable tree language $\check{T}$ over the alphabet Δ, with $\Delta_0 = \Sigma$ and, for every rank $i > 0$, $\Delta_i = \{0,1\}$.

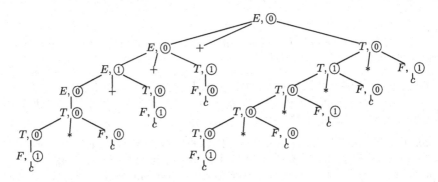

Fig. 3. A representation of both a syntax tree of the grammar G of Example 3 and of its encoding tree. The frontier of both trees are the same; the labels of the syntax tree are the nonterminal symbols of G; the labels of the encoding tree are the encircled symbols.

Example 3 (context-free language defined by code-grams). Given the grammar $G = \{E \to E + T \mid T,\ T \to T * F \mid F,\ F \to (E) \mid c\}$, a code for the nonterminal symbols is: $[\![E]\!]_X = 0010$, $[\![T]\!]_X = 0011$, $[\![F]\!]_X = 1110$. By Corollary 2, the language of syntax trees $T(G)$ is structurally equivalent to an slt language $\check{T}$, that we still call an encoding tree language since its differences from the E-trees in the proof of Lemma 1 are minor: the internal nodes carry bits without TA states as labels and the leaves carry terminal symbols without bits. We show one E-tree in Fig. 3. Notice that the nonterminal names can be viewed as the states of a (top-down) TA.

There is only one 4-code-gram in this tree, comprising the root and the nodes at distance up to 3:

$$0\,[0\,[1[0+0]+1[0]]+0\,[1\,[0*0]*1[c]]]$$

The whole E-tree (disregarding nonterminal labels) in Fig. 3 is one of the many 8-grams (omitted for brevity) that define the 8-slt language $\breve{T}$. The set of such 8-grams, with the ones located at the root marked as such, acts as a structurally equivalent definition of language $L(G)$.   □

4   Conclusion

We see some natural continuations for the present research. Theorem 1 has generalized to regular trees of the ranked type, the extended Medvedev's theorem proved in [5] for regular word languages, i.e., for trees generated by right-linear grammars. It would be of some interest to examine the case of unranked trees.

The new possibility offered by Corollary 2 for defining context-free languages, not by means of grammars or push-down automata, but by means of encoding trees, suggests to reexamine some classical properties, in particular the Chomsky-Schützenberger theorem.

At last, the successful lifting of the Extended Medvedev's Th. from words to the more complex mathematical object, namely trees, invites to investigate whether other generalizations to objects of different nature may be possible. A natural candidate is the family of the two-dimensional picture languages [8] whose definition by tiling systems is essentially a Medvedev's theorem using local tiles; recent work on two-dimensional comma-free codes [1] may bridge the conceptual distance from our work.

Acknowledgment. We thank an anonymous reviewer for pointing out the relevance of reference [13].

References

1. Anselmo, M., Madonia, M.: Two-dimensional comma-free and cylindric codes. Theor. Comput. Sci. **658**, 4–17 (2017)
2. Berstel, J., Perrin, D., Reutenauer, C.: Codes and Automata, volume 129 of Encyclopedia of Mathematics and its Applications. CUP (2009)
3. Comon, H., et al.: Tree automata techniques and applications (2007). http://www.grappa.univ-lille3.fr/tata
4. Crespi-Reghizzi, S., Guida, G., Mandrioli, D.: Noncounting context-free languages. J. ACM **25**(4), 571–580 (1978)
5. Crespi-Reghizzi, S., San Pietro, P.: From regular to strictly locally testable languages. Int. J. Found. Comput. Sci. **23**(8), 1711–1728 (2012)
6. Engelfriet, J.: Tree automata and tree grammars. CoRR, abs/1510.02036 (2015)
7. Gécseg, F., Steinby, M.: Tree languages. In: Rozenberg, G., Salomaa, A. (eds.) Handbook of Formal Languages, pp. 1–68. Springer, Heidelberg (1997). https://doi.org/10.1007/978-3-642-59126-6_1

8. Giammarresi, D., Restivo, A.: Two-dimensional languages. In: Rozenberg, G., Salomaa, A. (eds.) Handbook of Formal Languages, pp. 215–267. Springer, Heidelberg (1997). https://doi.org/10.1007/978-3-642-59126-6_4

9. McNaughton, R., Papert, S.: Counter-Free Automata. MIT Press, Cambridge (1971)

10. Medvedev, Y.T.: On the class of events representable in a finite automaton. In: Moore, E.F. (ed.) Sequential Machines - Selected Papers, pp. 215–227. Addison-Wesley (1964)

11. Perrin, D., Reutenauer, C.: Hall sets, Lazard sets and comma-free codes. Discret. Math. **341**(1), 232–243 (2018)

12. Place, T., Segoufin, L.: A decidable characterization of locally testable tree lan guages. Log. Methods Comput. Sci. **7**(4), 1–25 (2011)

13. Thomas, W.: Classifying regular events in symbolic logic. J. Comput. Syst. Sci. **25**(3), 360–376 (1982)

Approximated Determinisation of Weighted Tree Automata

Frederic Dörband[1]([✉])[iD], Thomas Feller[1][iD], and Kevin Stier[2][iD]

[1] Faculty of Computer Science, Technische Universität Dresden, Dresden, Germany
{frederic.doerband,thomas.feller}@tu-dresden.de
[2] Institute of Computer Science, Universität Leipzig, Leipzig, Germany
stier@informatik.uni-leipzig.de

Abstract. We introduce the notion of *t-approximated determinisation* and the *t-twinning property* of weighted tree automata (WTA) over the tropical semiring. We provide an algorithm that accomplishes *t*-approximated determinisation of an input automaton $\mathscr{A}$, whenever it terminates. Moreover, we prove that the *t*-twinning property of $\mathscr{A}$ is a sufficient condition for the termination of our algorithm. Ultimately, we show decidability of the *t*-twinning property for WTA.

Keywords: Weighted automata · Approximation · Approximated determinisation · Tree automata · Twinning property

1 Introduction

In theoretical computer science, automata theory arose as a very potent field of research. Besides having manifold applications in areas like natural language processing, model checking, and computational biology, automata are studied in a vast number of syntactical variations. The most prominent case of finite string automata has been extended to handle more complex input structures like pictures, trees, and forests (cf. [14,15]). Another direction of generalisation is to allow quantitative calculations rather than simple binary acceptance. Well-studied examples of such automata are weighted string automata and weighted tree automata over some weight structure S (cf. [8] for exhaustive references). Prominent weight structures include commutative semirings [1], and strong bimonoids [9].

One of the major research fields in automata theory is the determinisation of automata. While this problem has a well-known solution for unweighted automata, very little results are known in the weighted setting. In fact, not every weighted automaton can be determinised [5, Example 5.9]. One endeavour

---

Research of the first and third author was supported by the DFG through the Research Training Group QuantLA (GRK 1763). The second author was supported by the European Research Council (ERC) through the ERC Consolidator Grant No. 771779 (DeciGUT).

A. Leporati et al. (Eds.): LATA 2021, LNCS 12638, pp. 255–266, 2021.
https://doi.org/10.1007/978-3-030-68195-1_20

to simplify automata that cannot be determinised is to aim for *approximated determinisation*. Different approaches to this paradigm have been proposed, see e.g. [2,3], and [4]. The main idea of these papers is to take an automaton $\mathscr{A}$ and construct a deterministic automaton that recognizes a "similar" language to the one of $\mathscr{A}$. The notions of similarity differ in the literature. As the present paper aims to generalise [2] from the string case to the tree case, we subsequently focus on [2].

In [2], the weight structure is the tropical semiring $(\mathbb{R}_\infty, \min, +, \infty, 0)$. The notion of approximation is given as follows. Let $t \geq 1$ be a real number, called the *approximation factor*. A weighted automaton $\mathscr{A}'$ *t-approximates* $\mathscr{A}$, if for every input string $w \in \Sigma^*$ it holds that $[\![\mathscr{A}]\!](w) \leq [\![\mathscr{A}']\!](w) \leq t \cdot [\![\mathscr{A}]\!](w)$, where $[\![\mathscr{A}]\!]$ denotes the weighted language recognised by $\mathscr{A}$.

Aminof et al. [2] give an algorithm, called tDet, that takes as input a weighted string automaton $\mathscr{A}$ and an approximation factor t and (if the algorithm terminates) outputs a deterministic weighted string automaton $\mathscr{A}'$ such that $\mathscr{A}'$ t-approximates $\mathscr{A}$. The algorithm tDet executes a weighted powerset construction (with a fixed factorisation) similar to the one given by Kirsten and Mäurer [11]. That is, the states of $\mathscr{A}'$ are essentially subsets of the state set of $\mathscr{A}$, where each state of $\mathscr{A}$ gets assigned a residual weight. These residual weights keep track of the difference between the weights of runs of $\mathscr{A}'$ and runs of $\mathscr{A}$. For approximated determinisation however, tDet keeps track of two *bounds* for every state of $\mathscr{A}$ rather than a single residual weight. Namely, a *lower bound* and an *upper bound*. These bounds describe intervals of residual weights in order to ensure t-approximation.

Next, Aminof et al. [2] prove that tDet terminates if $\mathscr{A}$ satisfies the so-called *t-twinning property*. The *t-twinning property* is a generalisation of the classical twinning property (see [12]). Ultimately, it is proven in [2] that the *t-twinning* property is decidable.

The approach of the present paper closely follows the approach by Aminof et al. [2]. In Sect. 2 we introduce some elementary technical machinery and our automaton model. Next, in Sect. 3 we define t-approximation for weighted tree automata, give an algorithm for t-approximate determinisation, and prove its partial correctness. In Sect. 4 we introduce the *t-twinning property* for weighted tree automata and show that it is a sufficient condition for the termination of our algorithm. In Sect. 5 we prove that our *t-twinning property* is decidable and in Sect. 6 we conclude the paper by posing some open questions.

2   Preliminaries

We denote the set of *integers* by $\mathbb{Z}$, the set of *nonnegative integers* by $\mathbb{N}$, and the set of *positive integers* by $\mathbb{N}_+$. Moreover, we denote the set of *real numbers* by $\mathbb{R}$ and define the set $\mathbb{R}_\infty := \{x \in \mathbb{R} \mid x \geq 0\} \cup \{\infty\}$. Analogously, we denote the set of *rational numbers* by $\mathbb{Q}$ and define the set $\mathbb{Q}_\infty := \{x \in \mathbb{Q} \mid x \geq 0\} \cup \{\infty\}$. For every $x, y \in \mathbb{R}$, we define the *interval* $[x, y] := \{z \in \mathbb{R} \mid x \leq z \leq y\}$ and denote the set $[\infty, \infty] := \{\infty\}$. For every $k \in \mathbb{N}$, we denote the set $\{i \in \mathbb{N} \mid 1 \leq i \leq k\}$

by $[k]$. Note that $[0] = \emptyset$. For a set A we denote the *size* of A by $\#A$ and for every $k \in \mathbb{N}_+$ we denote by A^k the k-fold cartesian power of A.

An *alphabet* is a finite and non-empty set A and $A^* = \bigcup_{k \in \mathbb{N}} A^k$ is the set of all (finite) *words* over A, where $A^0 = \{\varepsilon\}$ contains solely the *empty word* ε. We denote by $|w|$ the *length* of the word $w \in A^*$. Given words $v, w \in A^*$, their concatenation is written $v.w$ or simply vw. We write $v \preceq w$ provided that there exists $u \in A^*$ such that $vu = w$. The relation $\preceq$ is in fact a partial order, called the *prefix order*.

A *ranked alphabet* is a pair (Σ, rk) consisting of an alphabet Σ and a mapping $\text{rk}: \Sigma \to \mathbb{N}$ that assigns a *rank* to each symbol of Σ. We refer to the ranked alphabet (Σ, rk) by the set Σ whenever the map rk is clear from the context. Furthermore, for every $k \in \mathbb{N}$, we let $\Sigma^{(k)} = \{\sigma \in \Sigma \mid \text{rk}(\sigma) = k\}$ and we write $\sigma^{(k)}$ to indicate that $\text{rk}(\sigma) = k$.

Throughout the rest of this paper, we assume Σ to be a ranked alphabet and $\Sigma^{(0)} \neq \emptyset$.

Given a set Z, the set of Σ-*trees* indexed by Z, denoted by $T_\Sigma(Z)$, is the smallest set T such that $Z \subseteq T$ and $\sigma(\xi_1, \ldots, \xi_r) \in T$ for every $r \in \mathbb{N}$, $\sigma \in \Sigma^{(r)}$, and $\xi_1, \ldots, \xi_r \in T$. We abbreviate $T_\Sigma = T_\Sigma(\emptyset)$ and call every subset $L \subseteq T_\Sigma$ a *tree language*.

Next, we recall some common notions and notations for trees. In the following, let $\xi \in T_\Sigma(Z)$. The set $\text{pos}(\xi)$ of *positions* of ξ is defined inductively by $\text{pos}(z) = \{\varepsilon\}$ for all $z \in Z$, and $\text{pos}(\sigma(\xi_1, \ldots, \xi_r)) = \{\varepsilon\} \cup \{i.w \mid i \in [r], w \in \text{pos}(\xi_i)\}$ for every $r \in \mathbb{N}$, $\sigma \in \Sigma^{(r)}$, and $\xi_1, \ldots, \xi_r \in T_\Sigma(Z)$. The *height* of ξ is defined by $\text{height}(\xi) = \max_{w \in \text{pos}(\xi)} |w|$, and the *size* of ξ is defined by $\text{size}(\xi) = \#\text{pos}(\xi)$. A *leaf* is a position $w \in \text{pos}(\xi)$ such that $w.1 \notin \text{pos}(\xi)$. We denote the set of leaves of ξ by $\text{leaf}(\xi)$. Given a position $w \in \text{pos}(\xi)$, the *label* of ξ at w is denoted by $\xi(w)$. The *subtree* of ξ at w, denoted $\xi|_w$, is defined for every $z \in Z$ by $z|_\varepsilon = z$ and for every $r \in \mathbb{N}$, $\sigma \in \Sigma^{(r)}$, and $\xi_1, \ldots, \xi_r \in T_\Sigma(Z)$ by

$$\sigma(\xi_1, \ldots, \xi_r)|_w = \begin{cases} \sigma(\xi_1, \ldots, \xi_r) & \text{if } w = \varepsilon \\ \xi_i|_{w'} & \text{if } w = i.w' \text{ with } i \in \mathbb{N} \text{ and } w' \in \text{pos}(\xi_i). \end{cases}$$

Let Y be a set. The set of *positions of ξ labeled by elements in Y*, denoted by $\text{pos}_Y(\xi)$, is the set $\{w \in \text{pos}(\xi) \mid \xi(w) \in Y\}$. Moreover, the *replacement* of the leaf $w \in \text{leaf}(\xi)$ by the tree $\eta \in T_\Sigma(Z)$, denoted $\xi[\eta]_w$, is given for every $z \in Z$ by $z[\eta]_\varepsilon = \eta$ and for every $r \in \mathbb{N}$, $i \in [r]$, $\sigma \in \Sigma^{(r)}$, $\xi_1, \ldots, \xi_r \in T_\Sigma(Z)$, and $w' \in \text{pos}(\xi_i)$ by $\sigma(\xi_1, \ldots, \xi_r)[\eta]_{i.w'} = \sigma(\xi_1, \ldots, \xi_{i-1}, \xi_i[\eta]_{w'}, \xi_{i+1}, \ldots, \xi_r)$.

The set $\text{path}(\xi) \subseteq (\Sigma \cup Z)^*$ of *paths* of ξ is defined inductively by $\text{path}(z) = \{z\}$ for all $z \in Z$, and $\text{path}(\sigma(\xi_1, \ldots, \xi_r)) = \{\sigma w \mid i \in [r], w \in \text{path}(\xi_i)\}$ for every $r \in \mathbb{N}$, $\sigma \in \Sigma^{(r)}$, and $\xi_1, \ldots, \xi_r \in T_\Sigma(Z)$.

We fix the set $X = \{x_1, x_2, \ldots\}$ of *variables* (which we impose to be disjoint from any other set we consider), and $X_n = \{x_1, \ldots, x_n\}$ for every $n \in \mathbb{N}_+$. A tree $\xi \in T_\Sigma(X_1)$ is a *context*, if $\#\text{pos}_{x_1}(\xi) = 1$. The set of all contexts is denoted by C_Σ.

Given a context $\zeta \in C_\Sigma$ and a tree $\xi \in T_\Sigma(Z)$, the *substitution* of ξ into ζ, denoted by $\zeta[\xi]$, is the tree $\zeta[\xi]_w$, where w is the unique position in $\mathrm{pos}_X(\zeta)$. Note that, given $\zeta, \zeta' \in C_\Sigma$, also $\zeta[\zeta'] \in C_\Sigma$. We write ζ^k for $\zeta[\zeta[\cdots \zeta[\zeta]\cdots]]$ containing the context ζ a total of k times.

We recall the *tropical semiring* $(\mathbb{R}_\infty, \min, +, \infty, 0)$, where $\min$ and $+$ are binary operations on $\mathbb{R}_\infty$ and are the natural extensions of the respective real-valued operations.

Definition 1 (WTA). *A* weighted tree automaton *(short: WTA) is a tuple* $(Q, \Sigma, \mathbb{R}_\infty, \mathrm{final}, T)$, *where Q is an alphabet of* states, $\mathrm{final}: Q \to \mathbb{R}_\infty$ *is a map of* final weights, *and T is a family $(T_\sigma: Q^r \times Q \to \mathbb{R}_\infty \mid r \geq 0, \sigma \in \Sigma^{(r)})$ of maps of* transition weights.

We call a tuple $t = (q_1, \ldots, q_r, \sigma, x, q) \in Q^r \times \Sigma \times \mathbb{R}_\infty \times Q$ a transition *if $\mathrm{rk}(\sigma) = r$ and $T_\sigma(q_1, \ldots, q_r, q) = x$. We sometimes denote t by $\sigma(q_1, \ldots, q_r) \xrightarrow{x} q$.*

Definition 2 (run). *Let $\mathscr{A} = (Q, \Sigma, S, \mathrm{final}, T)$ be a WTA and $\xi \in T_\Sigma \cup C_\Sigma$ be a tree or a context. A* run *of $\mathscr{A}$ on ξ is a map $\rho: \mathrm{pos}(\xi) \to Q$.*

Let $w \in \mathrm{pos}(\xi)$. The weight *of ρ at position w, denoted $\mathrm{wt}(\rho, w)$, is an element of $\mathbb{R}_\infty$ defined inductively as follows. If $\mathrm{label}(\xi, w) \in X$, then we define $\mathrm{wt}(\rho, w) := 0$ and if $\mathrm{label}(\xi, w) = \sigma$ is in $\Sigma^{(r)}$, then we define $\mathrm{wt}(\rho, w) := \mathrm{wt}(\rho, w1) + \cdots + \mathrm{wt}(\rho, wr) + T_\sigma(\rho(w1), \ldots, \rho(wr), \rho(w))$. Furthermore, the* weight *of ρ, denoted $\mathrm{wt}(\rho)$, is defined by $\mathrm{wt}(\rho) := \mathrm{wt}(\rho, \varepsilon)$.*

We say that ρ contains a state $q \in Q$ if there exists $w \in \mathrm{pos}(\xi)$ such that $q = \rho(w)$. We say that ρ is non-vanishing *if $\mathrm{wt}(\rho) \neq \infty$.*

Remark 3. We use the following notation for a run ρ of $\mathscr{A}$ on a tree or context ξ. Let $q := \rho(\varepsilon)$ and $x := \mathrm{wt}(\rho)$. If $\xi \in T_\Sigma$, then we write $\xrightarrow{\xi|\rho|x} q$. If $\xi \in C_\Sigma$, then we write $p \xrightarrow{\xi|\rho|x} q$, where $p := \rho(w)$ for the unique $w \in \mathrm{pos}_X(\xi)$. Whenever we do not care about the name of the run, we simply write $\xrightarrow{\xi|x} q$ and $p \xrightarrow{\xi|x} q$, respectively. Furthermore, if $\xrightarrow{\xi|x} q$ for some tree ξ and $x \neq \infty$, then we call the state q *reachable*.

Definition 4 (sets of runs). *Let $\mathscr{A} = (Q, \Sigma, S, \mathrm{final}, T)$ be a WTA and $\xi \in T_\Sigma \cup C_\Sigma$ be a tree or a context. The* set of runs *of $\mathscr{A}$ on ξ is denoted by $\mathrm{Run}_{\mathscr{A}}(\xi)$. For every $q \in Q$ and $\xi \in T_\Sigma$ we define the set $\mathrm{Run}_{\mathscr{A}}(\xi, q) := \{\rho \in \mathrm{Run}_{\mathscr{A}}(\xi) \mid \xrightarrow{\xi|\rho|\mathrm{wt}(\rho)} q\}$ and the* run weight *of ξ into q as $\theta_{\mathscr{A}}(\xi, q) := \min\{\mathrm{wt}(\rho) \mid \rho \in \mathrm{Run}_{\mathscr{A}}(\xi, q)\}$. Analogously, for every $p, q \in Q$ and $\xi \in C_\Sigma$ we define the set $\mathrm{Run}_{\mathscr{A}}(p, \xi, q) := \{\rho \in \mathrm{Run}_{\mathscr{A}}(\xi) \mid p \xrightarrow{\xi|\rho|\mathrm{wt}(\rho)} q\}$ and the* run weight *of ξ from p into q as $\theta_{\mathscr{A}}(p, \xi, q) := \min\{\mathrm{wt}(\rho) \mid \rho \in \mathrm{Run}_{\mathscr{A}}(p, \xi, q)\}$.*

Definition 5 (semantics of WTA). *Let $\mathscr{A} = (Q, \Sigma, \mathbb{R}_\infty, \mathrm{final}, T)$ be a WTA. The* weighted tree language *accepted by $\mathscr{A}$ is the map $[\![\mathscr{A}]\!]: T_\Sigma \to \mathbb{R}_\infty$, where for every $\xi \in T_\Sigma$ we define*

$$[\![\mathscr{A}]\!](\xi) := \min_{q \in Q} \left(\theta_{\mathscr{A}}(\xi, q) + \mathrm{final}(q)\right).$$

Two WTA $\mathscr{A}$ and $\mathscr{B}$ are called equivalent *if they accept the same weighted tree language, that is, if $[\![\mathscr{A}]\!] = [\![\mathscr{B}]\!]$.*

Note that our weighted tree automata are classical semiring-weighted tree automata (cf. [10]) where we fix the semiring $S = \mathbb{R}_\infty$.

Definition 6 (deterministic). *Let $\mathscr{A} = (Q, \Sigma, \mathbb{R}_\infty, \text{final}, T)$ be a WTA. We call $\mathscr{A}$* deterministic *if for all $r \geq 0, \sigma \in \Sigma^{(r)}$, and $q_1, \ldots, q_r \in Q$ there exist at most one $q \in Q$ such that $T_\sigma(q_1, \ldots, q_r, q) \neq \infty$. Moreover, we call $\mathscr{A}$* unambiguous *if for every $\xi \in T_\Sigma$ there exists at most one non-vanishing run of $\mathscr{A}$ on ξ. If $\mathscr{A}$ is unambiguous, then we define for every $\xi \in T_\Sigma$ the value $\theta_\mathscr{A}(\xi) := \text{wt}(\rho)$ as the weight of the unique non-vanishing run ρ of $\mathscr{A}$ on ξ (if such a run exists, and as ∞ otherwise).*

A map $f: T_\Sigma \to \mathbb{R}_\infty$ is called deterministically recognizable *if there exists a deterministic WTA $\mathscr{A}$ such that $[\![\mathscr{A}]\!] = f$.*

Example 7. Let $\Sigma = \{\alpha^{(0)}, \beta^{(0)}, \sigma^{(2)}\}$ and consider $\mathscr{A} := (Q, \Sigma, \mathbb{R}_\infty, \text{final}, T)$ where $Q := \{q_1, q_2\}$, final $:= 0$, and T is ∞ except in the cases

$$\alpha \xrightarrow{1} q_1, \qquad\qquad \alpha \xrightarrow{2} q_2, \qquad\qquad \sigma(q_1, q_1) \xrightarrow{0} q_1,$$

$$\beta \xrightarrow{0} q_1, \qquad\qquad\qquad\qquad \sigma(q_2, q_2) \xrightarrow{0} q_2.$$

We depict WTA by hypergraphs (see Figs. 1 and 2) which are read in the following way. Each state of the WTA is represented by a circle labeled by the name of the state. A transition of the form $\sigma(q_1, \ldots, q_r) \xrightarrow{x} q$ with $x \neq \infty$ is represented by a box labeled by σ and having r incoming edges and a single outgoing edge. The outgoing edge includes the weight of the transition, x, and is indicated by an arrow. The incoming edges are ordered by counter-clockwise traversal starting to the left of the outgoing edge. A depiction of the automaton $\mathscr{A}$ can be found in Fig. 1.

Let $\xi \in T_\Sigma$. One easily verifies the following statements using the definition of $\mathscr{A}$. If ξ contains at least one β, then there exists a unique non-vanishing run ρ of $\mathscr{A}$ on ξ and it holds that $\text{wt}(\rho) = \#\text{pos}_\alpha(\xi)$. If ξ contains no β, then there exist exactly two non-vanishing runs ρ_1 and ρ_2 of $\mathscr{A}$ on ξ and it holds that $\text{wt}(\rho_1) = \#\text{pos}_\alpha(\xi)$ and $\text{wt}(\rho_2) = 2 \cdot \#\text{pos}_\alpha(\xi)$. In total, we obtain that $[\![\mathscr{A}]\!](\xi) = \#\text{pos}_\alpha(\xi)$.

Clearly, $\mathscr{A}$ is not deterministic, as T contains the transitions $\alpha \xrightarrow{1} q_1$ and $\alpha \xrightarrow{2} q_2$. Furthermore $\mathscr{A}$ is not unambiguous, as there exist two non-vanishing runs of $\mathscr{A}$ on α.

3 Approximated Determinisation

In this section we present an algorithm that takes a weighted tree automaton $\mathscr{A}$ as input and generates a tuple $\mathscr{A}'$. Under certain conditions, this tuple is a deterministic weighted tree automaton that approximates $\mathscr{A}$. After applying

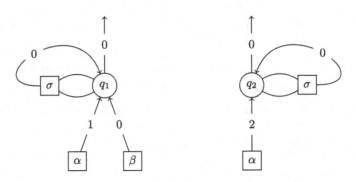

Fig. 1. Non-deterministic WTA $\mathscr{A}$ from Example 7.

the algorithm to the automaton from Example 7, we show the partial correctness of the algorithm. That is, if the algorithm terminates, the tuple $\mathscr{A}'$ is in fact a deterministic weighted tree automaton that approximates $\mathscr{A}$. Our approach closely follows [2] and we start by defining approximation of weighted tree automata.

Throughout the rest of this section, we assume $\mathscr{A} = (Q, \Sigma, \mathbb{R}_\infty, \text{final}, T)$ to be an arbitrary WTA.

Definition 8 (t-approximation [2]). *Let $t \in \mathbb{R}$ be a real number such that $t \geq 1$ and let $\mathscr{B} = (Q', \Sigma, \mathbb{R}_\infty, \text{final}', T')$ be a WTA.*
We say that $\mathscr{B}$ t-approximates $\mathscr{A}$ if for every $\xi \in \mathrm{T}_\Sigma$ it holds that

$$[\![\mathscr{A}]\!](\xi) \leq [\![\mathscr{B}]\!](\xi) \leq t \cdot [\![\mathscr{A}]\!](\xi).$$

Moreover, we call $\mathscr{A}$ t-approximate deterministic (or t-determinisable) if there exists a deterministic WTA $\mathscr{B}$ such that $\mathscr{B}$ t-approximates $\mathscr{A}$.

Remark 9. Note that if $\mathscr{B}$ t-approximates $\mathscr{A}$, then $\text{supp}([\![\mathscr{A}]\!]) = \text{supp}([\![\mathscr{B}]\!])$. Moreover, $\mathscr{B}$ 1-approximates $\mathscr{A}$ if and only if $[\![\mathscr{A}]\!] = [\![\mathscr{B}]\!]$.

Throughout the rest of this section, we assume that $t \in \mathbb{R}$ such that $t \geq 1$.

Remark 10. Note that, in general, $\mathscr{A}$ is not t-determinisable. In fact, for every Σ containing two distinct symbols $\sigma^{(r)}$ and $\tau^{(s)}$ (where $r, s > 0$), there exists a WTA $\mathscr{B}$ such that $\mathscr{B}$ is not t'-determinisable for any $t' \geq 1$.

In [2, Theorem 1], this is proven for strings and the constructions can easily be adapted to the tree case by considering so-called comb trees over σ and τ, which behave similarly to strings.

Next we introduce our approximate determinisation algorithm. For a summary of the conceptional details of our approach and how it fits into the existing literature, we refer to Sect. 1. Recall that our algorithm executes a weighted powerset construction with a fixed factorisation (see [11]). In this intermediate text, we present the intuitive idea of our algorithm and the relevant technicalities.

Given the automaton $\mathscr{A}$ and the approximation factor t, the algorithm builds up a deterministic automaton $\mathscr{A}' = (Q', \Sigma, \mathbb{R}_\infty, \text{final}', T')$ by iteratively adding new states and transitions to $\mathscr{A}'$ (which is initially empty). The states of $\mathscr{A}'$ are subsets of $(\mathbb{R}_\infty \times \mathbb{R}_\infty)^Q$, which we think about as follows. Each state $P \in Q'$ maps every state $q \in Q$ to a *lower bound* l_q^P and an *upper bound* u_q^P. Thus, we denote $(l_q^P, u_q^P) := P(q)$. These bounds represent an interval in $\mathbb{R}_\infty$ and will be determined by the algorithm such that the following holds.

Let ρ be the (unique) non-vanishing run of $\mathscr{A}'$ on a tree ξ and let $\rho(\varepsilon) = P$. For every $q \in Q$ it holds that the interval $[\theta_{\mathscr{A}}(\xi, q), t \cdot \theta_{\mathscr{A}}(\xi, q)]$ contains the interval $[l_q^P + \text{wt}(\rho), u_q^P + \text{wt}(\rho)]$ (see Lemma 14). Note that $[\theta_{\mathscr{A}}(\xi, q), t \cdot \theta_{\mathscr{A}}(\xi, q)]$ is the interval which t-approximates $\theta_{\mathscr{A}}(\xi, q)$. Therefore, $\mathscr{A}'$ t-approximates $\mathscr{A}$ as long as the final weight map of $\mathscr{A}'$ respects the lower and upper bounds.

Moreover, we use of the following concept. Given two states $P, P' : Q \to \mathbb{R}_\infty \times \mathbb{R}_\infty$, we say that P' *refines* P if for every $q \in Q$ it holds that $[l_q^{P'}, u_q^{P'}] \subseteq [l_q^P, u_q^P]$. Refinement plays a major role in ensuring the termination of Algorithm 1.

The overall structure of Algorithm 1 is the following. We initialise $\mathscr{A}'$ as an empty WTA (line 1). Next, we iteratively generate non-vanishing transitions for $\mathscr{A}'$, which in some cases add new states to the state set of $\mathscr{A}'$. The family of sets $(\text{Stack}(\sigma) \mid \sigma \in \Sigma)$ is used to keep track of transitions that have already been processed. Given a left-hand side $\sigma(P_1, \dots, P_r)$ of a transition that has not been processed (lines 4 and 5), we calculate an intermediate successor state P by accumulating the lower bounds and the upper bounds respectively with the transition weights (lines 7–9). Next, we determine the new transition weight c' as the minimal resulting *upper* bound in P (line 8). If c' is not ∞, then we define P' as $P - c'$ (lines 11 and 12). We check if P' is refined by some already existing state P'' (line 13). If this is the case, we add a *red* transition to $\mathscr{A}'$ by letting $T'_\sigma(P_1, \dots, P_r, P'') = c'$ (line 14). Otherwise, we add P' as a new state and add a *green* transition to $\mathscr{A}'$ by letting $T'_\sigma(P_1, \dots, P_r, P') = c'$ (lines 16 and 18). We ultimately define the new final weights (line 17).

We distinguish between red and green transitions for the following reason. A transition $t = (P_1, \dots, P_r, \sigma, c, P)$ is green if and only if it was the first non-vanishing transition with successor state P which was generated by Algorithm 1. Otherwise, t is either vanishing or a red transition. This defines a green subgraph of $\mathscr{A}'$ (viewed as a hypergraph). The proofs of our main theorems rely on the green subgraph of $\mathscr{A}'$ in order to use induction over the set of states of $\mathscr{A}'$.

Note that we define states of $\mathscr{A}'$ using a relational notation (see lines 7 and 12) rather than a functional notation, for better readability. Moreover, note that line 3 is merely a technical requirement that forces the second execution of the outermost while-loop (line 2) to happen immediately after each symbol from $\Sigma^{(0)}$ has been processed.

Algorithm 1: Procedure ttDet with input $\mathscr{A}$ and t

---

1 $Q' := \emptyset$, $\text{final}' := \infty$, $(\text{Stack}(\sigma) := \emptyset \mid \sigma \in \Sigma)$, $T' := (T'_\sigma \mid \sigma \in \Sigma)$ where
   $T'_\sigma := \infty$

2 **while** $\exists \sigma \in \Sigma : (Q')^{\text{rk}(\sigma)} \setminus \text{Stack}(\sigma) \neq \emptyset$ **do**

3      $Q'' := Q'$

4   **foreach** $r \in \mathbb{N}, \sigma \in \Sigma^{(r)}$ **do**

5   **foreach** $((P_1, \ldots, P_r) \in (Q'')^r \setminus \text{Stack}(\sigma))$ **do**

6              $\text{Stack}(\sigma) := \text{Stack}(\sigma) \cup \{(P_1, \ldots, P_r)\}$

7              $P := \{(q, (l_q, u_q)) \mid q \in Q\}$ where

8                  $l_q := \min\{l_{q_1}^{P_1} + \cdots + l_{q_r}^{P_r} + T_\sigma(q_1, \ldots, q_r, q) \mid q_1, \ldots, q_r \in Q\}$

9                  $u_q := \min\{u_{q_1}^{P_1} + \cdots + u_{q_r}^{P_r} + t \cdot T_\sigma(q_1, \ldots, q_r, q) \mid q_1, \ldots, q_r \in Q\}$

10             $c' := \min_{q \in Q} u_q^P$

11   **if** $c' < \infty$ **then**

12                 $P' := \{(q, (l_q^P - c', u_q^P - c')) \mid q \in Q\}$

13   **if** $\exists P'' \in Q'$ such that P'' refines P' **then**

14                     $T'_\sigma(P_1, \ldots, P_r, P'') := c'$ // red transition

15   **else**

16                     $Q' := Q' \cup \{P'\}$

17                     $\text{final}'(P') := \min_{q \in Q}(u_q^{P'} + t \cdot \text{final}(q))$

18                     $T'_\sigma(P_1, \ldots, P_r, P') := c'$ // green transition

19 **return** $(Q', \Sigma, \mathbb{R}_\infty, \text{final}', T')$

Definition 11. *We define $\mathscr{A}'$ as the tuple returned[1] by ttDet applied to $\mathscr{A}$ and t and denote its components by $\mathscr{A}' = (Q', \Sigma, \mathbb{R}_\infty, \text{final}', T')$.*

Example 12. We continue Example 7 by applying ttDet to $\mathscr{A}$ and t for $t \geq 2$. First consider $\alpha \in \Sigma^{(0)}$. Via lines 7–10 we calculate

$$P = \{(q_1, (1, t)), (q_2, (2, 2t))\} \qquad \text{and} \qquad c' = t.$$

By line 12 we obtain $P' = \{(q_1, (1 - t, 0)), (q_2, (2 - t, t))\}$. As Q' is still empty, P' is not refined by some other state and we enter the else-case (lines 16–18). We denote $P'_1 := P'$ and execute lines 16, 17, and 18 to update

$$Q' = \{P'_1\}, \qquad \text{final}'(P'_1) = 0, \qquad \text{and} \qquad T'_\alpha(P'_1) = t.$$

Note that this transition is a green transition.

    Next consider $\beta \in \Sigma^{(0)}$. We calculate $P = \{(q_1, (0, 0)), (q_2, (\infty, \infty))\}$ (lines 7–9) and $c' = 0$ (line 10). By line 12 we obtain $P' = P$. As P'_1 does not refine

---

[1] We denote by $\mathscr{A}'_i$ the tuple $(Q', \Sigma, \mathbb{R}_\infty, \text{final}', T')$ during the i-th execution of line 2. If ttDet does not terminate on the input $\mathscr{A}$ and t, the limit of these tuples for $i \to \infty$ is their componentwise union and we say that ttDet returns this limit.

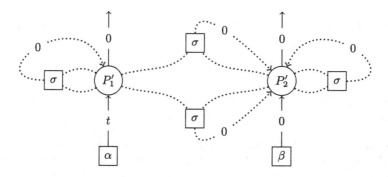

Fig. 2. Deterministic WTA $\mathscr{A}'$ t-approximating the WTA $\mathscr{A}$ from Example 7.

P', we again enter the else-case (lines 16–18). We denote $P_2' := P'$ and execute lines 16, 17, and 18 to update

$$Q' = \{P_1', P_2'\}, \qquad \text{final}'(P_2') = 0, \qquad \text{and} \qquad T_\beta'(P_2') = 0.$$

Note that this transition is a green transition.

Next consider $\sigma \in \Sigma^{(2)}$. As Stack(σ) is still empty, we consider $(P_1', P_1') \in (Q')^2 \setminus \text{Stack}(\sigma)$ (line 5). By lines 7–12 we obtain $P = \{(q_1, (2 - 2t, 0)), (q_2, (4 - 2t, 2t))\}$, $c' = 0$, and $P' = P$. Note that P_2' does not refine P' and that P_1' refines P' if and only if

$$2-2t \leq 1-t, \qquad 0 \leq 0, \qquad 4-2t \leq 2-t, \qquad \text{and} \qquad t \leq 2t.$$

Therefore, P' is refined by P_1' if and only if $t \geq 2$, which is true by assumption. Hence, we enter the if-case (line 14) and add the red transition $T_\sigma'(P_1', P_1', P_1') = 0$ to T'.

By continuing to execute the algorithm, we add more red transitions to T_σ' and arrive at the automaton $\mathscr{A}' = (Q', \Sigma, \mathbb{R}_\infty, \text{final}', T')$, where $Q' = \{P_1', P_2'\}$, $\text{final}' = 0$, and T' is ∞ except in the cases

$$\alpha \xrightarrow{t} P_1', \qquad \sigma(P_1', P_1') \xrightarrow{0} P_1', \qquad \sigma(P_1', P_2') \xrightarrow{0} P_2',$$
$$\beta \xrightarrow{0} P_2', \qquad \sigma(P_2', P_1') \xrightarrow{0} P_2', \qquad \sigma(P_2', P_2') \xrightarrow{0} P_2'.$$

A depiction of $\mathscr{A}'$ can be found in Fig. 2. Note that green transitions are depicted by continuous lines and red transitions are depicted by dotted lines.

Remark 13. Note that ttDet does not preserve the weighted language of $\mathscr{A}$, even if $\mathscr{A}$ is itself deterministic. This is due to the fact that the state normalisation (lines 10 and 12) is done with respect to the *upper* bounds u_q^P. Therefore, $\mathscr{A}'$ realises $t \cdot [\![\mathscr{A}]\!]$ rather than $[\![\mathscr{A}]\!]$ in many basic examples.

A straightforward induction on the depth of states in the green subgraph of $\mathscr{A}'$ shows that if ttDet terminates, then $\mathscr{A}'$ is a deterministic WTA. Similarly, the following lemma can be proven.

Lemma 14. Let $\xi \in T_\Sigma$ and $P \in Q'$ such that $\xrightarrow{\xi \mid \theta_{\mathscr{A}'}(\xi)} P$. For every $q \in Q$ it holds that

$$\theta_{\mathscr{A}}(\xi, q) - \theta_{\mathscr{A}'}(\xi) \leq l_q^P \leq u_q^P \leq t \cdot \theta_{\mathscr{A}}(\xi, q) - \theta_{\mathscr{A}'}(\xi).$$

The following theorem states the partial correctness of Algorithm 1 and follows from Lemma 14 and the definition of final'.

Theorem 15. If ttDet terminates on input $\mathscr{A}$ and t, then $\mathscr{A}'$ is a deterministic WTA that t-approximates $\mathscr{A}$. In this case, $\mathscr{A}$ is in particular t-determinisable.

4   Approximated Twinning Property

In this section, we prove a sufficient condition for the termination of the algorithm, namely the t-twinning property. Our proof closely follows [2]. We start by defining the t-twinning property of weighted tree automata, which is a natural extension of both, the string case [2] and the tree case without approximation (that is, $t = 1$) [6] and [13].

Definition 16 (t-twinning property). Let $\mathscr{A} = (Q, \Sigma, \mathbb{R}_\infty, \text{final}, T)$ be a WTA.

Let $p, q \in Q$. We call p and q siblings if there exists a tree $\xi \in T_\Sigma$ and non-vanishing runs $\rho_1 \in \text{Run}_{\mathscr{A}}(\xi, p)$ and $\rho_2 \in \text{Run}_{\mathscr{A}}(\xi, q)$. Siblings p and q are called t-twins if for every $\zeta \in C_\Sigma$ it holds that either $\theta(p, \zeta, p) = \infty$, $\theta(q, \zeta, q) = \infty$, or $\frac{1}{t} \cdot \theta(q, \zeta, q) \leq \theta(p, \zeta, p) \leq t \cdot \theta(q, \zeta, q)$.

We say that $\mathscr{A}$ has the t-twinning property if for all siblings $p, q \in Q$ it holds that p and q are t-twins.

Throughout the rest of this section, we assume $\mathscr{A} = (Q, \Sigma, \mathbb{Q}_\infty, \text{final}, T)$ to be a WTA with rational weights in $\mathbb{Q}_\infty$ and $t \in \mathbb{R}$ such that $t \geq 1$.

Theorem 17. If $\mathscr{A}$ satisfies the t-twinning property, ttDet terminates on input $\mathscr{A}$ and t.

The proof of Theorem 17 is very similar to the proof of [2, Theorem 8]. The main difference is that, in the tree case, we apply a version of König's Lemma that can handle the hypergraph structure of $\mathscr{A}'$. Note that in [2, Theorem 8], t is a rational number, whereas we allow for t to be a real number. This can be resolved by multiplying t and all weights occurring in $\mathscr{A}$ by $\frac{1}{t}$. Because of the fact that $(\frac{1}{t} \cdot \mathbb{N}, +)$ is isomorphic to $(\mathbb{N}, +)$, we can follow the proof of [2, Theorem 8].

Corollary 18. If $\mathscr{A}$ satisfies the t-twinning property, $\mathscr{A}$ is t-determinisable.

Proof of Corollary 18. This follows immediately from Theorems 15 and 17.   □

Example 19. We continue Example 7 by showing that $\mathscr{A}$ satisfies the 2-twinning property but not the 1-twinning property. First note that q_1 and q_2 are siblings as there are two runs ρ_1 and ρ_2 on $\xi = \alpha$ ending in q_1 and q_2, respectively.

Let $\zeta \in C_\Sigma$ and ρ be a non-vanishing run of $\mathscr{A}$ on ζ. If ζ contains a β, we have that $\theta(q_2, \zeta, q_2) = \infty$ and hence we only have to check the 2-twinning property for the case that ζ does not contain a β. One easily sees that ρ either maps each position to q_1 (in this case $\mathrm{wt}(\rho) = \#\mathrm{pos}_\alpha(\zeta)$) or to q_2 (in this case $\mathrm{wt}(\rho) = 2 \cdot \#\mathrm{pos}_\alpha(\zeta)$). In particular, $\theta(q_2, \zeta, q_2) = 2 \cdot \theta(q_1, \zeta, q_1)$. This proves that $\mathscr{A}$ satisfies the 2-twinning property.

Moreover, $\mathscr{A}$ does not satisfy the 1-twinning property, as q_1 and q_2 are siblings but $\zeta = \sigma(\alpha, x_1)$ does not satisfy $\theta(q_1, \zeta, q_1) = \theta(q_2, \zeta, q_2)$.

Note that ttDet does not terminate on input $\mathscr{A}$ and 1. In Example 12, we generated the state $P' = \{(q_1, (2 - 2t, 0)), (q_2, (4 - 2t, 2t))\}$ by considering the input $\sigma(P_1', P_1')$. If $t = 2$, P' is refined by P_1'. For $t = 1$, however, $P' = \{(q_1, (0, 0)), (q_2, (2, 2))\}$ is not refined and therefore added to the state space. Next, considering $\sigma(P', P')$, we obtain another unrefineable state, namely $P'' = \{(q_1, (0, 0)), (q_2, (4, 4))\}$. One easily sees that the construction continues to generate every state of the form $\{(q_1, (0, 0)), (q_2, (2^k, 2^k))\}$ and hence ttDet does not terminate on input $\mathscr{A}$ and 1.

5   Decidability of the Twinning Property

In the following theorem we prove the decidability of the t–twinning property. This is due to the fact, that if a WTA $\mathscr{A}$ does not satisfy the t-twinning property, then this non-satisfaction is already witnessed by a small context tree.

Theorem 20. *The t-twinning property is decidable for every WTA $\mathscr{A}$ and $t \geq 1$.*

6   Outlook

In this paper we generalised [2] from the string case to the tree case. First we provided an algorithm for t-determinisation and proved its correctness, assuming termination of the algorithm. Next, we introduced the t-twinning property for trees and showed that, for WTA with weights in $\mathbb{Q}_\infty$, the t-twinning property implies the termination of our algorithm. We ultimately showed that our t-twinning property is decidable.

We conclude this paper by listing future research directions. Recent work has shown that the twinning property is equivalent to determinisability in some cases (e.g. [7]). It would be worthwhile to determine whether in our case the t-twinning property is necessary for t-determinisability. Another interesting research direction is to introduce approximated determinisation for general classes of semirings rather than only considering the tropical semiring. Moreover, it seems rather arbitrary to say $x \in \mathbb{R}$ is approximated exactly by the values in the interval $[x, t \cdot x]$. It would be interesting to introduce more general notions of "approximation" and find sufficient conditions for this general approximated determinisability.

References

1. Alexandrakis, A., Bozapalidis, S.: Weighted grammars and Kleene's theorem. Inf. Process. Lett. **24**(1), 1–4 (1987)
2. Aminof, B., Kupferman, O., Lampert, R.: Rigorous approximated determinization of weighted automata. Theor. Comput. Sci. **480**, 104–117 (2013)
3. Boker, U., Henzinger, T.: Exact and approximate determinization of dis-counted-sum automata. Log. Methods Comput. Sci. **10**(1) (2014). https://doi.org/10.2168/LMCS-10(1:10)2014
4. Boker, U., Henzinger, T.A.: Approximate determinization of quantitative automata. In: IARCS Annual Conference on Foundations of Software Technology and Theoretical Computer Science (FSTTCS 2012). Schloss Dagstuhl-Leibniz-Zentrum fuer Informatik (2012)
5. Borchardt, B.: A pumping lemma and decidability problems for recognizable tree series. Acta Cybern. **16**(4), 509–544 (2004)
6. Büchse, M., May, J., Vogler, H.: Determinization of weighted tree automata using factorizations. J. Autom. Lang. Comb. **15**(3/4), 229–254 (2010)
7. Daviaud, L., Jecker, I., Reynier, P.-A., Villevalois, D.: Degree of sequentiality of weighted automata. In: Esparza, J., Murawski, A.S. (eds.) FoSSaCS 2017. LNCS, vol. 10203, pp. 215–230. Springer, Heidelberg (2017). https://doi.org/10.1007/978-3-662-54458-7_13
8. Droste, M., Kuich, W., Vogler, H. (eds.): Handbook of Weighted Automata.EATCS Monographs in Theoretical Computer Science. Springer, Heidelberg (2009)
9. Droste, M., Stüber, T., Vogler, H.: Weighted finite automata over strong bimonoids. Inf. Sci. **180**, 156–166 (2010)
10. Fülöp, Z., Vogler, H.: Weighted tree automata and tree transducers. In: Droste, M., Kuich, W., Vogler, H. (eds.) Handbook of Weighted Automata. Monographs in Theoretical Computer Science. An EATCS Series. Springer, Heidelberg (2009). https://doi.org/10.1007/978-3-642-01492-5_9
11. Kirsten, D., Mäurer, I.: On the determinization of weighted automata. J. Autom. Lang. Comb. **10**, 287–312 (2005)
12. Mohri, M.: Finite-state transducers in language and speech processing. Comput. Linguist. **23**(2), 269–311 (1997)
13. Paul, E.: Finite sequentiality of unambiguous max-plus tree automata. In: 36th International Symposium on Theoretical Aspects of Computer Science (STACS 2019). Schloss Dagstuhl-Leibniz-Zentrum fuer Informatik (2019)
14. Rozenberg, G., Salomaa, A. (eds.): Handbook of Formal Languages, vol. 1 Word, Language, Grammar. Springer, Heidelberg (1997). https://doi.org/10.1007/978-3-642-59136-5
15. Rozenberg, G., Salomaa, A. (eds.): Handbook of Formal Languages, vol. 3 Beyond-Words. Springer, Heidelberg (1997). https://doi.org/10.1007/978-3-642-59126-6

Sequentiality of Group-Weighted Tree Automata

Frederic Dörband[1]([⊠])[iD], Thomas Feller[1][iD], and Kevin Stier[2][iD]

[1] Faculty of Computer Science, Technische Universität Dresden, Dresden, Germany
{frederic.doerband,thomas.feller}@tu-dresden.de
[2] Institute of Computer Science, Universität Leipzig, Leipzig, Germany
stier@informatik.uni-leipzig.de

Abstract. We introduce the notion of group-weighted tree automata over commutative groups and characterise sequentialisability of such automata. In particular, we introduce a fitting notion for tree distance and prove the equivalence between sequentialisability, the so-called Lipschitz property, and the so-called twinning property.

Keywords: Weighted automata · Deterministic automata · Tree automata · Twinning property

1 Introduction

In theoretical computer science, automata theory arose as a very potent field of research. Besides having manifold applications in areas like natural language processing, model checking, and computational biology, automata are studied in a vast number of syntactical variations. The most prominent case of finite string automata has been extended to handle more complex input structures like pictures, trees, and forests (cf. [17,18]). Another direction of generalisation is to allow quantitative calculations rather than simple binary acceptance. Well-studied examples of such automata are weighted string automata and weighted tree automata over some weight structure S (cf. [8] for exhaustive references). Prominent weight structures include commutative semirings [1] and strong bimonoids [9]. In the present paper we consider so-called group-weighted tree automata (short: group-WTA), which are particular semiring-weighted tree automata. We have adapted the notion of group-weighted tree automata from [7], where group-weighted string automata are studied.

One of the major research fields in automata theory is the determinisation of automata. While this problem has a well-known solution for unweighted string automata, very little results are known in the weighted setting. In fact, not

Research of the first and third author was supported by the DFG through the Research Training Group QuantLA (GRK 1763). The second author was supported by the European Research Council (ERC) through the ERC Consolidator Grant No. 771779 (DeciGUT).

A. Leporati et al. (Eds.): LATA 2021, LNCS 12638, pp. 267–278, 2021.
https://doi.org/10.1007/978-3-030-68195-1_21

every weighted automaton can be determinised [3, Example 5.9] and hence, the problem has shifted towards the question of a *characterisation* of determinisable weighted automata. Two recent approaches to this question involve maximal factorisations [5] and automata with set semantics [2,7]. Note however, that [2,7] deal with sequentiality rather than determinism, which makes a subtle difference (Remark 10).

The main goal of the present paper is to characterise *sequential* weighted tree languages (i.e. weighted tree languages accepted by sequential group-WTA) by the so-called Lipschitz property and the so-called twinning property, namely

Theorem 1. For every group-WTA $\mathscr{A}$ it holds that

$$[\![\mathscr{A}]\!] \text{ is sequential } \iff [\![\mathscr{A}]\!] \text{ satisfies the Lipschitz property}$$
$$\iff \mathscr{A} \text{ satisfies the twinning property .}$$

Hereby, our paper generalises [2,6] from the string case to the tree case. Note however, that [2,6] are proven for free monoids rather than (infinitary commutative) groups as in our case. The idea for the proof of Theorem 1 is based on [2] and our proof applies the terminology and proof techniques given in [7]. Note that [2] provides merely an implication of the form "twinning property $\implies$ sequential", whereas [7] provides a full characterisation of sequentiality. In fact, [7] proves a more general theorem for unions of k sequential automata and the present paper only covers the case $k = 1$. Moreover, [7] is based on [6], which first introduced an equivalence similar to Theorem 1.

The present paper executes the proof of Theorem 1 in the following way. In Sect. 2, we introduce some elementary technical machinery and our automaton model.

In Sect. 3, we first introduce the *Lipschitz property* of weighted tree languages, which essentially says that close trees have close values in $\mathbb{G}$ (with respect to a metric on the set of trees and the Cayley distance on $\mathbb{G}$). Second, we introduce the *twinning property* of group-WTA, which states that if the automaton can loop[1] on a context tree in two different states, then the weights of these loops are equal. Next, we prove two implications of Theorem 1, namely "sequential $\implies$ Lipschitz" and "Lipschitz $\implies$ twinning".

In Sect. 4, we prove the implication "twinning $\implies$ sequential" by applying a construction similar to the well-known weighted power set construction.

In Sect. 5, we give a brief presentation of our endeavours to lift the cases $k > 1$ from [7]. Most importantly we show where the approach from [7] fails in the tree case.

We conclude this introductory section by comparing our results to the existing sequentialisation/determinisation results from the literature. The major references for our proofs are [2,6,7]. As stated above, our results generalise [2,6] to the case of group-weighted tree automata. Furthermore, we lift [7, case $k = 1$] from the string case to the tree case. Another major result in the theory of

---

[1] A *loop* is a run on a context tree such that the state at the context variable is the same as the state at the root of the context.

determinisation is given in [5, Theorem 5.2], which subsumes the determinisation results from [3,4,13–15]. Besides the fact that our paper is concerned with sequentiality rather than determinism, our class of weight structures is not subsumed by [5]. In particular, [5] provides a determinisation result only if either $\mathscr{A}$ is nonrecursive, the semiring S is locally finite, or S is extremal, none of which apply to our semirings of the form $\mathcal{P}_{\text{fin}}(\mathbb{G})$. Similarly, the determinisation result given in [16, Section 6] deals only with locally finite strong bimonoids and hence again does not subsume our results.

2 Preliminaries

We denote the set of *nonnegative integers* by $\mathbb{N}$ and the set of *positive integers* by $\mathbb{N}_+$. For every $k \in \mathbb{N}$, we denote the set $\{i \in \mathbb{N} \mid 1 \leq i \leq k\}$ by $[k]$. Note that $[0] = \emptyset$. For a set A we denote the *size* of A by $\#A$ and for every $k \in \mathbb{N}_+$ we denote by A^k the k-fold cartesian power of A.

An *alphabet* is a finite and non-empty set A and $A^* = \bigcup_{k \in \mathbb{N}} A^k$ is the set of all (finite) *words* over A, where $A^0 = \{\varepsilon\}$ contains solely the *empty word* ε. We denote by $|w|$ the *length* of the word $w \in A^*$. Given words $v, w \in A^*$, their concatenation is written $v.w$ or simply vw. We write $v \preceq w$ provided that there exists $u \in A^*$ such that $vu = w$. The relation $\preceq$ is in fact a partial order, called the *prefix order*.

A *ranked alphabet* is a pair (Σ, rk) consisting of an alphabet Σ and a mapping $\text{rk} \colon \Sigma \to \mathbb{N}$ that assigns a *rank* to each symbol of Σ. We refer to the ranked alphabet (Σ, rk) by the set Σ whenever the map rk is clear from the context. Furthermore, for every $k \in \mathbb{N}$, we let $\Sigma^{(k)} = \{\sigma \in \Sigma \mid \text{rk}(\sigma) = k\}$ and we write $\sigma^{(k)}$ to indicate that $\text{rk}(\sigma) = k$. Moreover we define $\text{maxrk}(\Sigma) := \max(\text{rk}(\Sigma))$.

Throughout the rest of this paper, we assume Σ to be an arbitrary ranked alphabet.

Given a set Z, the set of Σ-*trees* indexed by Z, denoted by $\mathrm{T}_\Sigma(Z)$, is the smallest set T such that $Z \subseteq \mathrm{T}$ and $\sigma(\xi_1, \ldots, \xi_s) \in \mathrm{T}$ for every $s \in \mathbb{N}$, $\sigma \in \Sigma^{(s)}$, and $\xi_1, \ldots, \xi_s \in \mathrm{T}$. We abbreviate $\mathrm{T}_\Sigma = \mathrm{T}_\Sigma(\emptyset)$ and call every subset $L \subseteq \mathrm{T}_\Sigma$ a *tree language*.

Next, we recall some common notions and notations for trees. In the following, let $\xi \in \mathrm{T}_\Sigma(Z)$. The set $\text{pos}(\xi)$ of *positions* of ξ is defined inductively by $\text{pos}(z) = \{\varepsilon\}$ for all $z \in Z$, and $\text{pos}(\sigma(\xi_1, \ldots, \xi_s)) = \{\varepsilon\} \cup \{i.w \mid i \in [s],\ w \in \text{pos}(\xi_i)\}$ for every $s \in \mathbb{N}$, $\sigma \in \Sigma^{(s)}$, and $\xi_1, \ldots, \xi_s \in \mathrm{T}_\Sigma(Z)$. The *height* of ξ is defined by $\text{height}(\xi) = \max_{w \in \text{pos}(\xi)} |w|$, and the *size* of ξ is defined by $\text{size}(\xi) = \#\text{pos}(\xi)$. A *leaf* is a position $w \in \text{pos}(\xi)$ such that $w.1 \notin \text{pos}(\xi)$. We denote the set of leaves of ξ by $\text{leaf}(\xi)$. Given a position $w \in \text{pos}(\xi)$, the *label* of ξ at w is denoted by $\xi(w)$. The *subtree* of ξ at w, denoted $\xi|_w$, is defined for every $z \in Z$ by $z|_\varepsilon = z$ and for every $s \in \mathbb{N}$, $\sigma \in \Sigma^{(s)}$, and $\xi_1, \ldots, \xi_s \in \mathrm{T}_\Sigma(Z)$ by

$$\sigma(\xi_1, \ldots, \xi_s)|_w = \begin{cases} \sigma(\xi_1, \ldots, \xi_s) & \text{if } w = \varepsilon \\ \xi_i|_{w'} & \text{if } w = i.w' \text{ with } i \in \mathbb{N} \text{ and } w' \in \text{pos}(\xi_i). \end{cases}$$

Let Y be a set. The set of *positions of ξ labeled by elements in Y*, denoted by $\mathrm{pos}_Y(\xi)$, is the set $\{w \in \mathrm{pos}(\xi) \mid \xi(w) \in Y\}$. Moreover, the *replacement* of the leaf $w \in \mathrm{leaf}(\xi)$ by the tree $\eta \in T_\Sigma(Z)$, denoted $\xi[\eta]_w$, is given for every $z \in Z$ by $z[\eta]_\varepsilon = \eta$ and for every $s \in \mathbb{N}$, $i \in [s]$, $\sigma \in \Sigma^{(s)}$, $\xi_1, \ldots, \xi_s \in T_\Sigma(Z)$, and $w' \in \mathrm{pos}(\xi_i)$ by $\sigma(\xi_1, \ldots, \xi_s)[\eta]_{i.w'} = \sigma(\xi_1, \ldots, \xi_{i-1}, \xi_i[\eta]_{w'}, \xi_{i+1}, \ldots, \xi_s)$.

We fix the set $X = \{x_1, x_2, \ldots\}$ of *variables* (which we impose to be disjoint from any other set we consider), and $X_n = \{x_1, \ldots, x_n\}$ for every $n \in \mathbb{N}_+$. A tree $\xi \in T_\Sigma(X_1)$ is a *context*, if $\#\mathrm{pos}_{x_1}(\xi) = 1$. The set of all contexts is denoted by C_Σ.

Given a context $\zeta \in C_\Sigma$ and a tree $\xi \in T_\Sigma(Z)$, the *substitution* of ξ into ζ, denoted by $\zeta[\xi]$, is the tree $\zeta[\xi]_w$, where w is the unique position in $\mathrm{pos}_X(\zeta)$. Note that, given $\zeta, \zeta' \in C_\Sigma$, also $\zeta[\zeta'] \in C_\Sigma$. We write ζ^k for $\zeta[\zeta[\cdots \zeta[\zeta] \cdots]]$ containing the context ζ a total of k times.

Let $\xi_1, \xi_2 \in T_\Sigma$ be two trees. A *pair-cut* between ξ_1 and ξ_2 is a triple $(\zeta_1, \zeta_2, \eta) \in C_\Sigma \times C_\Sigma \times T_\Sigma$ such that $\zeta_i[\eta] = \xi_i$ for $i \in [2]$. In this case, we call η an *overlap* of ξ_1 and ξ_2. The set of pair-cuts between ξ_1 and ξ_2 is denoted $\mathrm{PairCut}(\xi_1, \xi_2)$. We moreover define the *distance* between ξ_1 and ξ_2 as

$$\mathrm{dist}(\xi_1, \xi_2) := \mathrm{size}(\xi_1) + \mathrm{size}(\xi_2) - 2 \cdot \mathrm{maxoverlap}(\xi_1, \xi_2),$$

where $\mathrm{maxoverlap}(\xi_1, \xi_2)$ is the maximal size of an overlap of ξ_1 and ξ_2.

A *group* $(\mathbb{G}, \otimes, 1)$ is a set $\mathbb{G}$ with an associative operation $\otimes : \mathbb{G}^2 \to \mathbb{G}$, a neutral element $1 \in \mathbb{G}$ such that for all $\alpha \in \mathbb{G}$ there exists $\beta \in \mathbb{G}$ satisfying $\alpha \otimes \beta = \beta \otimes \alpha = 1$. We refer to β as the *inverse element* of α and denote it by α^{-1}. We simply write $\alpha\beta$ for $\alpha \otimes \beta$. A group is *commutative* if $\otimes$ is commutative. We call a group *infinitary* if for every $\alpha, \beta, \gamma \in \mathbb{G}$ with $\alpha\beta\gamma \neq \beta$, the set $\{\alpha^n\beta\gamma^n \mid n \in \mathbb{N}\}$ is infinite (cf. [7,10]). We define the *delay* of $\alpha \in \mathbb{G}$ and $\beta \in \mathbb{G}$, denoted $\mathrm{delay}(\alpha, \beta)$, by $\alpha^{-1}\beta$.

Throughout the rest of this paper, we assume $\mathbb{G}$ to be a finitely generated, infinitary, commutative[2] group, 1 to be its neutral element and Γ to be a finite generating set of $\mathbb{G}$.

The *undirected Cayley graph* for $\mathbb{G}$ and Γ is the graph (V, E), where $V = \mathbb{G}$ is the set of vertices and for every $\alpha \in \mathbb{G}$ and $\beta \in \Gamma$, we have that $(\alpha, \alpha\beta), (\alpha\beta, \alpha) \in E$. The *Cayley distance* $d(\alpha, \beta)$ between $\alpha \in \mathbb{G}$ and $\beta \in \mathbb{G}$ is defined as the length of the shortest path between α and β in the undirected Cayley Graph. For $\alpha \in \mathbb{G}$ we define the *Γ-length* of α as the Cayley distance between 1 and α and denote it by $|\alpha|_\Gamma$.

2.1 Weighted Tree Automata

Definition 2. A *(group-)weighted tree automaton* over Σ and $\mathbb{G}$ (short: group-WTA or simply WTA) is a tuple $(Q, \Sigma, \mathbb{G}, \mathrm{final}, T)$, where Q is a finite set of

---

[2] In fact, we do not require commutativity for the proof our results. However, in order to limit the notational complexity of the present paper, we require $\mathbb{G}$ to be commutative.

states, final $\subseteq Q \times \mathbb{G}$ is the finite *final relation*, and T is a family $(T_\sigma \subseteq Q^s \times \mathbb{G} \times Q \mid s \geq 0, \sigma \in \Sigma^{(s)})$ of finite sets of *transitions*.

We call $q \in Q$ *final* if there exists $\alpha \in \mathbb{G}$ such that $(q, \alpha) \in$ final, which we depict as $q \xrightarrow{\alpha}$. For every $\sigma \in \Sigma^{(s)}$ and $t = (q_1, \ldots, q_s, \alpha, q) \in T_\sigma$, we denote $\mathrm{out}(t) := q$, $\mathrm{in}(t) := (q_1, \ldots, q_s)$, and $\mathrm{wt}(t) := \alpha$. For notational convenience, we use the notation $\mathrm{out}(q) := q$ and $\mathrm{wt}(q) := 1$. To aid readability, we denote the fact that $(q_1, \ldots, q_s, \alpha, q) \in T_\sigma$ by $\sigma(q_1, \ldots, q_s) \xrightarrow{\alpha} q$.

Definition 3. Let $\mathscr{A} = (Q, \Sigma, \mathbb{G}, \text{final}, T)$ be a WTA and let $\xi \in T_\Sigma \cup C_\Sigma$ be a tree or a context. A *run* of $\mathscr{A}$ on ξ is a map $\rho \colon \mathrm{pos}(\xi) \to T \cup Q$ such that

- for every $w \in \mathrm{pos}_\Sigma(\xi)$ we have $\rho(w) \in T_\sigma$ where $\sigma = \xi(w)$ and $\mathrm{in}(\rho(w)) = (\mathrm{out}(\rho(w1)), \ldots, \mathrm{out}(\rho(ws)))$ where $s = \mathrm{rk}(\sigma)$, and
- for every $w \in \mathrm{pos}_X(\xi)$ we have $\rho(w) \in Q$.

We denote by $\mathrm{out}(\rho)$ the state $\mathrm{out}(\rho(\varepsilon))$ and if $\xi \in C_\Sigma$ we denote by $\mathrm{in}(\rho)$ the state $\rho(w)$ where w is the unique position in $\mathrm{pos}_X(\xi)$. The *weight* of such a run ρ is $\mathrm{wt}(\rho) := \prod_{w \in \mathrm{pos}(\xi)} \mathrm{wt}(\rho(w))$. Moreover, we say that ρ *contains* a state $q \in Q$ if there exists $w \in \mathrm{pos}(\xi)$ such that $q = \mathrm{out}(\rho(w))$. A run ρ is called *accepting* if $\mathrm{out}(\rho)$ is final.

Remark 4. We use the following notation for a run ρ of $\mathscr{A}$ on a tree or context ξ. Let $q := \mathrm{out}(\rho)$ and $\alpha := \mathrm{wt}(\rho)$. If $\xi \in T_\Sigma$, then we write $\xrightarrow{\xi|\rho|\alpha} q$. If $\xi \in C_\Sigma$, then we write $p \xrightarrow{\xi|\rho|\alpha} q$, where $p := \mathrm{in}(\rho)$. Whenever we do not care about the name of the run, we simply write $\xrightarrow{\xi|\alpha} q$ and $p \xrightarrow{\xi|\alpha} q$, respectively. Furthermore, if $\xrightarrow{\xi|\alpha} q$ for some tree ξ and some weight α, then we call the state q *reachable*.

Remark 5. Throughout this paper, we assume that all considered WTA are *trim*. For a WTA $\mathscr{A}$, this condition means that every state appears in some accepting run. In particular, for every state p, there exists a context $\xi \in C_\Sigma$, a final state q, a weight $\alpha \in \mathbb{G}$, and a run $p \xrightarrow{\xi|\alpha} q$.

Note moreover that, without loss of generality, the size of ξ is bounded. If the run on ξ contains a single state q' multiple times on a single branch (excluding the root of the tree), then we can replace the subtree at the topmost occurrence of q' with the subtree at the bottommost occurrence of q'. Therefore, we can assume $\mathrm{height}(\xi) \leq \#Q + 1$ and hence $\mathrm{size}(\xi) \leq \mathrm{maxrk}(\Sigma)^{\#Q+2}$.

Definition 6. Let $\mathscr{A} = (Q, \Sigma, \mathbb{G}, \text{final}, T)$ be a WTA. The *weighted tree language accepted by* $\mathscr{A}$ is the relation $[\![\mathscr{A}]\!] \subseteq T_\Sigma \times \mathbb{G}$ containing the pairs $(\xi, \beta\gamma)$ such that $\xrightarrow{\xi|\beta} q \xrightarrow{\gamma}$ for some $q \in Q$.

Two WTA $\mathscr{A}$ and $\mathscr{B}$ are called *equivalent* if they accept the same weighted tree language, that is, $[\![\mathscr{A}]\!] = [\![\mathscr{B}]\!]$.

Moreover, we define the constant

$$M_\mathscr{A} := \max\{|\alpha|_\Gamma \mid (q_1, \ldots, q_k, \alpha, q) \in \bigcup_{\sigma \in \Sigma} T_\sigma \text{ or } (q, \alpha) \in \text{final}\}.$$

That is, $M_\mathscr{A}$ is the maximal Γ-length of weights occurring in T or final.

We will now briefly compare group-WTA to semiring-WTA[3].

Remark 7. Consider the tuple $S = (\mathcal{P}_{\text{fin}}(\mathbb{G}), \cup, \cdot, \emptyset, \{1\})$, where $\mathcal{P}_{\text{fin}}(\mathbb{G})$ is the set of finite subsets of $\mathbb{G}$, $1 \in \mathbb{G}$ is the neutral group element, and $\cdot$ is the group operation lifted to finite sets. It is immediate that S is a semiring.

Let $\mathscr{A} = (Q, \Sigma, \mathbb{G}, \text{final}, T)$ be a group-WTA. In order to syntactically match the definition of group-WTA over $\mathbb{G}$ with the definition of semiring-WTA over S, we replace T_σ by the map $T_\sigma^{\text{sr}} : Q^s \times Q \to \mathcal{P}_{\text{fin}}(\mathbb{G})$ such that $(q_1, \ldots, q_s, q) \mapsto \{\alpha \mid (q_1, \ldots, q_s, \alpha, q) \in T_\sigma\}$. Furthermore we replace the final relation by $\text{final}^{\text{sr}} : Q \to \mathcal{P}_{\text{fin}}(\mathbb{G})$ such that $q \mapsto \{\beta \mid (q, \beta) \in \text{final}\}$. Denote the semiring-WTA $\mathscr{A}^{\text{sr}} = (Q, T^{\text{sr}}, \text{final}^{\text{sr}})$ and note that $[\![\mathscr{A}^{\text{sr}}]\!](\xi) = \{\alpha \mid (\xi, \alpha) \in [\![\mathscr{A}]\!]\}$. Therefore, up to this identification of maps and relations, group-WTA are particular semiring-WTA. However, the important difference is that each run of $\mathscr{A}$ calculates a *single* group element, whereas each run of $\mathscr{A}^{\text{sr}}$ calculates *multiple* aggregated group elements at once.

Definition 8. Let $\mathscr{A}_i = (Q_i, \Sigma, \mathbb{G}, \text{final}_i, T_i)$ be WTA for $i \in [2]$.

The *direct product* of $\mathscr{A}_1$ and $\mathscr{A}_2$, denoted $\mathscr{A}_1 \times \mathscr{A}_2$, is the WTA $(Q_1 \times Q_2, \Sigma, \mathbb{G}, \text{final}, T)$, where $\text{final} := \{((q, p), \alpha\beta) \mid (q, \alpha) \in \text{final}_1 \wedge (p, \beta) \in \text{final}_2\}$ and

$$T_\sigma := \{((q_1, p_1), \ldots, (q_s, p_s), \alpha\beta, (q, p))$$
$$\mid (q_1, \ldots, q_s, \alpha, q) \in (T_1)_\sigma \wedge (p_1, \ldots, p_s, \beta, p) \in (T_2)_\sigma\}.$$

Again, without loss of generality, we assume that $Q_1 \cap Q_2 = \emptyset$. This definition naturally extends to finitely many WTA.

Definition 9. Let $\mathscr{A} = (Q, \Sigma, \mathbb{G}, \text{final}, T)$ be a WTA. We call $\mathscr{A}$ *sequential* if for all $s \geq 0, \sigma \in \Sigma^{(s)}$, and $q_1, \ldots, q_s \in Q$ there exist at most one $\alpha \in \mathbb{G}$ and $q \in Q$ such that $(q_1, \ldots, q_s, \alpha, q) \in T_\sigma$.

A relation $R \subseteq T_\Sigma \times \mathbb{G}$ is called *sequential* if there exists a sequential WTA $\mathscr{A}$ such that $[\![\mathscr{A}]\!] = R$.

Remark 10. Note that Definition 9 is highly similar to the definition of *deterministic* semiring-WTA [5, preceding Example 3.1]. However, sequentiality forces the weight of transitions to be at most one *single* group element, whereas determinism merely forces the weight of transitions to be at most one *set* of group elements. This difference results in sequentiality being a properly more restrictive condition on the automaton than determinism.

Example 11. Let $\Sigma = \{\sigma^{(2)}, \alpha^{(0)}\}$ and $\mathbb{G} = (\mathbb{Z}, +, 0)$. Note that $\mathbb{G}$ is a commutative, finitely generated, infinitary group with finite generating set $\Gamma = \{1\}$.

---

[3] As a reference, we use the definition of semiring-weighted tree automata from [5]. For a more thorough introduction to semirings confer [12] and for semiring-WTA we refer to [11].

Define the WTA $\mathscr{A} := (Q, \Sigma, \mathbb{G}, \text{final}, T)$ where $Q := \{q_\alpha, q_0, q_1\}$, final $:= \{(q_0, 0)\}$ and T is defined by

$$T_\alpha \cup T_\sigma = \{\alpha \xrightarrow{0} q_\alpha, \qquad \sigma(q_\alpha, q_\alpha) \xrightarrow{1} q_1, \qquad \sigma(q_\alpha, q_\alpha) \xrightarrow{3} q_0,$$
$$\sigma(q_\alpha, q_1) \xrightarrow{1} q_0, \qquad \sigma(q_\alpha, q_0) \xrightarrow{1} q_1\}.$$

Consider the context $\eta = \sigma(\alpha, x_1) \in C_\Sigma$. One easily sees that all trees $\xi \in T_\Sigma$ occurring in $[\![\mathscr{A}]\!]$ are of the form $\xi = \eta^\ell[\alpha]$ for some $\ell \geq 1$. In this case, if $\#\text{pos}_{\{\sigma\}}(\xi) = 2n$ for some $n \in \mathbb{N}$ we have $(\xi, 2n) \in [\![\mathscr{A}]\!]$ and if $\#\text{pos}_{\{\sigma\}}(\xi) = 2n + 1$ we have $(\xi, 2n + 3) \in [\![\mathscr{A}]\!]$. Clearly, $\mathscr{A}$ is not sequential.

3   Lipschitz and Twinning Property

In this section we formally introduce the two characterisations of sequentiality from Theorem 1, and prove the implications "sequential $\Longrightarrow$ Lipschitz" in Theorem 13 and "Lipschitz $\Longrightarrow$ twinning" in Theorem 16.

3.1   The Lipschitz Property

Definition 12. A relation $R \subseteq T_\Sigma \times \mathbb{G}$ satisfies the *Lipschitz property* if there exists $L \in \mathbb{N}$ such that for all pairs $(\xi_0, \alpha_0), (\xi_1, \alpha_1) \in R$ it holds that $d(\alpha_0, \alpha_1) \leq L \cdot (\text{dist}(\xi_0, \xi_1) + 1)$.

Theorem 13. Let $R \subseteq T_\Sigma \times \mathbb{G}$ be a sequential relation. Then R satisfies the Lipschitz property.

The proof of Theorem 13 primarily uses the fact that a sequential WTA has a unique non-vanishing run weight on every overlap of two input trees ξ_1 and ξ_2.

3.2   The Twinning Property

Throughout the rest of this paper, we assume $\mathscr{A} = (Q, \Sigma, \mathbb{G}, \text{final}, T)$ to be a WTA.

Definition 14. We say that $\mathscr{A}$ satisfies the *twinning property* if for all runs ρ_0 and ρ_1 of $\mathscr{A}$, states $q_0, q_1 \in Q$, a tree $\xi \in T_\Sigma$, a context $\zeta \in C_\Sigma$, and weights $\alpha_0, \alpha_1, \beta_0, \beta_1 \in \mathbb{G}$, such that ρ_j $(j = 0, 1)$ equals

$$\xrightarrow{\xi | \alpha_j} q_j \xrightarrow{\zeta | \beta_j} q_j,$$

it holds that $\beta_0 = \beta_1$.

Example 15. We continue Example 11 by showing that $\mathscr{A}$ satisfies the twinning property. Let ρ_0 and ρ_1 be runs of $\mathscr{A}$ quantified as in Definition 14. Recall that $\zeta[\xi]$ has the form $\eta^\ell[\alpha]$. Moreover, non-empty runs cannot loop in the state q_α by definition of the transition relation. Therefore, we have that $\xi \neq \alpha$, whence

$\xi = \eta^j[\alpha]$ for some $1 \le j < \ell$. However, in this case the single non-deterministic choice already occurs in ξ and hence $\beta_0 = \beta_1 = \#\mathrm{pos}_{\{\sigma\}}(\zeta)$. In particular, this proves the twinning property.

Next we provide a WTA $\mathscr{B}$ over $\Delta = \{\sigma^{(2)}, \alpha^{(0)}, \beta^{(0)}\}$ and $\mathbb{G} = (\mathbb{Z}, +, 0)$ which does not satisfy the twinning property. Define $\mathscr{B} = (\widetilde{Q}, \Delta, \mathbb{G}, \widetilde{\mathrm{final}}, \widetilde{T})$, where $\widetilde{Q} = \{q_\alpha, q_\beta\}$, $\widetilde{\mathrm{final}} = \{(q_\alpha, 0), (q_\beta, 0)\}$, and

$$\widetilde{T}_\alpha \cup \widetilde{T}_\beta \cup \widetilde{T}_\sigma = \{\alpha \xrightarrow{1} q_\alpha, \quad \beta \xrightarrow{0} q_\alpha, \quad \sigma(q_\alpha, q_\alpha) \xrightarrow{0} q_\alpha, \quad \text{(counting } \alpha\text{)}$$
$$\alpha \xrightarrow{0} q_\beta, \quad \beta \xrightarrow{1} q_\beta, \quad \sigma(q_\beta, q_\beta) \xrightarrow{0} q_\beta\}. \quad \text{(counting } \beta\text{)}$$

One easily verifies the fact that for every $\xi \in T_\Sigma$ there are exactly two runs of $\mathscr{B}$ on ξ and we obtain $(\xi, \#\mathrm{pos}_{\{\alpha\}}(\xi)), (\xi, \#\mathrm{pos}_{\{\beta\}}(\xi)) \in [\![\mathscr{B}]\!]$. Clearly, $\mathscr{B}$ is not sequential. Consider the tree $\xi = \sigma(\alpha, \beta)$, the context $\zeta = \sigma(\alpha, x_1)$, and the two runs of $\mathscr{B}$ on $\zeta[\xi]$

$$\xrightarrow{\xi|\alpha_0} q_\alpha \xrightarrow{\zeta|\beta_0} q_\alpha \qquad \text{and} \qquad \xrightarrow{\xi|\alpha_1} q_\beta \xrightarrow{\zeta|\beta_1} q_\beta.$$

By the definition of $\widetilde{T}$ we calculate the values $\beta_0 = 0 + 1 + 0$ and $\beta_1 = 0 + 0 + 0$. This proves $\beta_0 \ne \beta_1$, whence we obtain that $\mathscr{B}$ does not satisfy the twinning property.

Theorem 16. If $[\![\mathscr{A}]\!]$ satisfies the Lipschitz property, then $\mathscr{A}$ satisfies the twinning property.

The proof of Theorem 16 is done by contradiction. We take a witness of the non-satisfaction of the twinning property and pump the occurring loops. This makes the run weights diverge in $\mathbb{G}$ (using the fact that $\mathbb{G}$ is infinitary), which contradicts the Lipschitz property.

Remark 17. Note that Theorem 16 implies that, whenever $\mathscr{A}$ does not satisfy the twinning property, no equivalent automaton can satisfy the twinning property .

4   Sequentiality of the Twinning Property

This section executes the proof of the implication "twinning $\implies$ sequential" of Theorem 1. Given a WTA $\mathscr{A}$, we apply a construction similar to the well-known power set construction to $\mathscr{A}$. This yields a (not necessarily finite) sequential WTA $D_{\mathscr{A}}$. However, we prove that $D_{\mathscr{A}}$ is indeed finite if $\mathscr{A}$ satisfies the twinning property (see Corollary 25). The proof of Corollary 25 can be outlined as follows. First we show that all runs of $\mathscr{A}$ on a fixed input tree generate close weights with respect to the Cayley-distance (see Lemma 23). Next, the definition of $D_{\mathscr{A}}$ implies that every (reachable) state of $D_{\mathscr{A}}$ contains only weights that are close to the neutral element $1 \in \mathbb{G}$ (see Lemma 24), which implies that the set of reachable states of $D_{\mathscr{A}}$ is finite. We derive the fact that $D_{\mathscr{A}}$ is equivalent to $\mathscr{A}$ from the definition of $D_{\mathscr{A}}$ (see Theorem 26). We conclude this chapter by applying our construction to the automata from Examples 11 and 15.

Theorem 18. If $\mathscr{A}$ satisfies the twinning property, then $[\![\mathscr{A}]\!]$ is sequential.

Throughout this section, we assume $\mathscr{A}$ to satisfy the twinning property.

Definition 19. We define the infinite WTA[4] $D_{\mathscr{A}} = (Q', \Sigma, \mathbb{G}, \text{final}', T')$ as follows. The states of $D_{\mathscr{A}}$ are $Q' := \mathcal{P}(Q \times \mathbb{G})$, the final relation is

$$\text{final}' := \{(S, \alpha\beta) \mid S \in Q', \exists q \in Q \colon (q, \alpha) \in S \text{ and } (q, \beta) \in \text{final}\},$$

and the transitions are constructed as follows. For every $\sigma \in \Sigma^{(s)}$ and every $S_1, \ldots, S_s \in Q'$, consider the set

$$S := \{(q, \alpha_1 \cdots \alpha_s \beta) \mid \exists p_1, \ldots, p_s \in Q \colon \\ (\forall i \in [s] \colon (p_i, \alpha_i) \in S_i) \text{ and } (p_1, \ldots, p_s, \beta, q) \in T_\sigma\}$$

and fix an arbitrary element[5] $(p, \alpha) \in S$. We define the set

$$S' := \{(q, \alpha^{-1}\gamma) \mid (q, \gamma) \in S\}$$

and ultimately add $(S_1, \ldots, S_s, \alpha, S')$ to T'_σ.

Remark 20. Note that $D_{\mathscr{A}}$ is indeed sequential. This follows directly from the construction. Moreover, in Definition 19 we first calculate an intermediate successor state S, which is then shifted by a fixed value α occurring in some pair $(p, \alpha) \in S$. We call this shifting process the *factorisation* of S.

We will show in Corollary 25 that every reachable state S of $D_{\mathscr{A}}$ satisfies $\#S \leq K$ for a global constant K and hence after trimming $D_{\mathscr{A}}$, also the final relation final$'$ is finite.

Lemma 21. Let $\xi \in T_\Sigma$ and consider the (unique) run of $D_{\mathscr{A}}$ on ξ, $\xrightarrow{\xi|\alpha} S'$. It holds that

$$S' = \{(q, \beta) \mid \exists \text{ run} \xrightarrow{\xi|\delta} q \text{ of } \mathscr{A} \colon \alpha\beta = \delta\}.$$

Definition 22. We define the constant $N_{\mathscr{A}} := 2M_{\mathscr{A}} \operatorname{maxrk}(\Sigma)^{(\#Q^2+2)}$.

Lemma 23. For every tree $\xi \in T_\Sigma$ and every two runs $\xrightarrow{\xi|\alpha} q$ and $\xrightarrow{\xi|\beta} p$ of $\mathscr{A}$ on ξ it holds that

$$d(\alpha, \beta) < N_{\mathscr{A}}.$$

Note that the proof of Lemma 23 uses the fact that $\mathscr{A}$ satisfies the twinning property.

Lemma 24. Let S be a reachable state of $D_{\mathscr{A}}$ and let $(q, \alpha) \in S$. It holds that $|\alpha|_\Gamma \leq N_{\mathscr{A}}$.

---

[4] That is, a tuple satisfying the conditions of a WTA, except for finiteness.
[5] Formally, we have a globally fixed choice function $f \colon \mathcal{P}(Q \times \mathbb{G}) \to Q \times \mathbb{G}$ and then simply define $(p, \alpha) := f(S)$.

Proof. The reachability of S implies the existence of a tree $\xi \in T_\Sigma$ such that there exists a run $\xrightarrow{\xi|\delta} S$. If $\alpha = 1$, then we are done. Therefore, we assume that $\alpha \neq 1$ and hence there exists $(p, \beta) \in S$ such that $\beta = 1$. Therefore by Lemma 21, there are two runs of $\mathscr{A}$ on ξ, $\xrightarrow{\xi|\delta\alpha} q$ and $\xrightarrow{\xi|\delta} p$. By Lemma 23 it holds that $d(\delta\alpha, \delta) < N_\mathscr{A}$. The fact that $d(\delta\alpha, \delta) = |\alpha|_\Gamma$ implies that $|\alpha|_\Gamma < N_\mathscr{A}$. □

Corollary 25. *The set of states of $D_\mathscr{A}$ is finite and hence $D_\mathscr{A}$ is a WTA.*

Proof. Denote for every $N \in \mathbb{N}$ the (finite) set $\mathbb{G}_N := \{g \in \mathbb{G} \mid |g|_\Gamma \leq N\}$.

By Lemma 24 every reachable state of $D_\mathscr{A}$ is an element of the finite set $\mathcal{P}(Q \times \mathbb{G}_{N_\mathscr{A}})$, which proves the claim. □

Theorem 26. *$D_\mathscr{A}$ is equivalent to $\mathscr{A}$.*

Proof. We first show that $[\![D_\mathscr{A}]\!] \subseteq [\![\mathscr{A}]\!]$. Let $(\xi, \alpha) \in [\![D_\mathscr{A}]\!]$ and let $\xrightarrow{\xi|\beta} S$ be a run of $D_\mathscr{A}$ on ξ and $(S, \gamma) \in \text{final}'$ such that $\alpha = \beta\gamma$. Note that by the definition of final' there exist $q \in Q$, $(q, \beta') \in S$ and $(q, \gamma') \in \text{final}$ such that $\gamma = \beta'\gamma'$.

By Lemma 21 there exists a run $\xrightarrow{\xi|\delta} q$ of $\mathscr{A}$ such that $\beta\beta' = \delta$. Hence, $\alpha = \beta\gamma = \beta\beta'\gamma' = \delta\gamma'$ and therefore $(\xi, \alpha) = (\xi, \delta\gamma') \in [\![\mathscr{A}]\!]$.

To prove the fact that $[\![\mathscr{A}]\!] \subseteq [\![D_\mathscr{A}]\!]$, we apply a similar argument. Let $(\xi, \alpha) \in [\![\mathscr{A}]\!]$. There is a unique run $\xrightarrow{\xi|\beta} S$ of $D_\mathscr{A}$ on ξ. By definition of $[\![\mathscr{A}]\!]$, there exist a run $\xrightarrow{\xi|\delta} q$ of $\mathscr{A}$ on ξ and a pair $(q, \gamma') \in \text{final}$ such that $\delta\gamma' = \alpha$.

By Lemma 21 there exists an element $(q, \beta') \in S$ such that $\beta\beta' = \delta$ and hence by the definition of final' we obtain $(S, \beta'\gamma') \in \text{final}'$. We obtain $\alpha = \delta\gamma' = \beta\beta'\gamma'$ and hence $(\xi, \alpha) \in [\![D_\mathscr{A}]\!]$. □

Proof of Theorem 18. We have seen that $D_\mathscr{A}$ is a sequential (Remark 20) WTA (Corollary 25) which is equivalent to $\mathscr{A}$ (Theorem 26). This proves the claim. □

Example 27. Recall the WTA $\mathscr{A}$ and $\mathscr{B}$ from Examples 11 and 15. We apply the construction from Definition 19 to both, $\mathscr{A}$ and $\mathscr{B}$, and obtain that $D_\mathscr{A}$ has a finite trim state space, whereas $D_\mathscr{B}$ has an infinite trim state space.

First we consider $D_\mathscr{A}$. Clearly, $T'_\alpha = \{(0, S_0)\}$, where $S_0 = \{(q_\alpha, 0)\}$. By pointwise application of T_σ to (S_0, S_0) we obtain $\{(q_1, 1), (q_0, 3)\}$. We chose $(q_1, 1)$ for the factorisation, which yields the new state $S_1 = \{(q_1, 0), (q_0, 2)\}$ and therefore we have constructed the transition $(S_0, S_0, 1, S_1) \in T'_\sigma$. By continuing this process we arrive at

$$T'_\sigma = \{(S_0, S_0, 1, S_1), (S_0, S_1, 1, S_2), (S_0, S_2, 1, S_1)\},$$

where $S_2 = \{(q_0, 0), (q_1, 2)\}$. The trim state space of $D_\mathscr{A}$ is $Q' = \{S_0, S_1, S_2\}$ and the final relation is $\text{final}' = \{(S_1, 2), (S_2, 0)\}$.

Next we consider $D_\mathscr{B}$. Define $R_1 = \{(q_\alpha, 1), (q_\beta, 0)\}$ and note that $\widetilde{T}'_\alpha = \{(0, R_1)\}$. Pointwise application of $\widetilde{T}_\sigma$ to (R_1, R_1) yields $R_2 = \{(q_\alpha, 2), (q_\beta, 0)\}$, which is already normalised. Another pointwise application of $\widetilde{T}_\sigma$ to (R_1, R_2) results in $R_3 = \{(q_\alpha, 3), (q_\beta, 0)\}$, which is again normalised. One easily sees that repeatedly generating transitions of $D_{\mathscr{B}}$ like this yields an infinite set of reachable states of $D_{\mathscr{B}}$ and hence $D_{\mathscr{B}}$ is not a WTA.

In Sect. 5 we will discuss the approach given in [7, case $k > 1$], which describes how to handle $D_{\mathscr{B}}$ in order to generate a finite union of sequential WTA which is equivalent to $\mathscr{B}$.

5 Outlook

In the present paper, we have successfully lifted the result from [7, case $k = 1$] to weighted tree automata. Recall that [7] characterises unions of k sequential automata. The natural next step is to lift the remaining cases $k > 1$. This section is designed to briefly demonstrate why a straightforward lift of [7, cases $k > 1$] to weighted tree automata fails.

The outline of the proof given in [7] goes as follows. Let $k \in \mathbb{N}$. The notions of k-sequential WTA, the k-Lipschitz property, and the k-branching twinning property are introduced and the directions "sequential $\implies$ Lipschitz" and "Lipschitz $\implies$ twinning" are proven similarly to our Theorems 13 and 16. For the direction "twinning $\implies$ sequential", the automaton $D_{\mathscr{A}}$ is introduced and its properties are studied. As we have seen in the second part of Example 27, $D_{\mathscr{A}}$ is in general infinite. However, if $\mathscr{A}$ satisfies the k-branching twinning property, [7] describes the following construction on $D_{\mathscr{A}}$, yielding a k-sequential automaton which is equivalent to $\mathscr{A}$. First, the set of states Q' of $D_{\mathscr{A}}$ is restricted to a finite set. In fact, the set of reachable states S of $D_{\mathscr{A}}$ containing only "small" weights $|\alpha|_\Gamma < N_{\mathscr{A}}$ is denoted U and the set of states reachable from U in one step is denoted U'. Note that U and U' are finite. $D_{\mathscr{A}}$ is restricted to $U \cup U'$ and each state S in $U' \setminus U$ (i.e. the outer border of U) is replaced by a union of k sequential WTA. These sequential WTA are constructed by induction on k and depend on the state S. The resulting automaton $\bar{D}_{\mathscr{A}}$ can easily be divided into k sequential automata, which concludes the proof.

The tree case differs in the following way. Consider a symbol $\sigma \in \Sigma^{(2)}$ and consider two different states $S, S' \in U' \setminus U$. Surely, in $D_{\mathscr{A}}$ we can find a transition of the form $\sigma(S, S') \to S''$. However, the states S and S' are replaced by different automata in $\bar{D}_{\mathscr{A}}$. Therefore, a run ρ of $D_{\mathscr{A}}$ on a tree ξ ending in S (resp. S') translates into a run of $\bar{D}_{\mathscr{A}}$ on ξ ending in some state $q_S \notin Q'$ (resp. $q_{S'} \notin Q'$). Moreover, q_S and $q_{S'}$ are taken from disjoint sets. We have not been able to find the proper way to construct a transition of the form $\sigma(q_S, q_{S'}) \to q$.

Therefore, we leave the lift of [7, cases $k > 1$] as an open research question.

References

1. Alexandrakis, A., Bozapalidis, S.: Weighted grammars and Kleene's theorem. Inf. Process. Lett. **24**(1), 1–4 (1987)

2. Béal, M.P., Carton, O.: Determinization of transducers over finite and infinite words. Theor. Comput. Sci. **289**(1), 225–251 (2002)
3. Borchardt, B.: A pumping lemma and decidability problems for recognizable tree series. Acta Cybern. **16**(4), 509–544 (2004)
4. Borchardt, B., Vogler, H.: Determinization of finite state weighted tree automata. J. Autom. Lang. Comb. **8**(3), 417–463 (2003)
5. Büchse, M., May, J., Vogler, H.: Determinization of weighted tree automata using factorizations. J. Autom. Lang. Comb. **15**(3/4), 229–254 (2010)
6. Choffrut, C.: Une Caracterisation des Fonctions Sequentielles et des Fonctions Sous-Sequentielles en tant que Relations Rationnelles. Theor. Comput. Sci. **5**(3), 325–337 (1977). https://doi.org/10.1016/0304-3975(77)90049-4
7. Daviaud, L., Jecker, I., Reynier, P.-A., Villevalois, D.: Degree of sequentiality of weighted automata. In: Esparza, J., Murawski, A.S. (eds.) FoSSaCS 2017. LNCS, vol. 10203, pp. 215–230. Springer, Heidelberg (2017). https://doi.org/10.1007/978-3-662-54458-7_13
8. Droste, M., Kuich, W., Vogler, H. (eds.): Handbook of Weighted Automata. EATCS Monographs in Theoretical Computer Science. Springer, Heidelberg (2009). https://doi.org/10.1007/978-3-642-01492-5
9. Droste, M., Stüber, T., Vogler, H.: Weighted finite automata over strong bimonoids. Inf. Sci. **180**, 156–166 (2010)
10. Filiot, E., Gentilini, R., Raskin, J.-F.: Quantitative languages defined by functional automata. In: Koutny, M., Ulidowski, I. (eds.) CONCUR 2012. LNCS, vol. 7454, pp. 132–146. Springer, Heidelberg (2012). https://doi.org/10.1007/978-3-642-32940-1_11
11. Fülöp, Z., Vogler, H.: Weighted tree automata and tree transducers. In: Droste, M., Kuich, W., Vogler, H. (eds.) Handbook of Weighted Automata. Monographs in Theoretical Computer Science. An EATCS Series., pp. 313–403. (2009). https://doi.org/10.1007/978-3-642-01492-5_
12. Golan, J.: Semirings and Their Applications. Kluwer Academic Publishers, Dordrecht (1999)
13. Kirsten, D., Mäurer, I.: On the determinization of weighted automata. J. Autom. Lang. Comb. **10**, 287–312 (2005)
14. May, J., Knight, K.: A Better N-Best List: practical determinization of weighted finite tree automata. In: Proceedings of the Human Language Technology Conference of the NAACL, Main Conference, pp. 351–358. Association for Computational Linguistics, New York City, USA, June 2006
15. Mohri, M.: Finite-state transducers in language and speech processing. Comput. Linguist. **23**(2), 269–311 (1997)
16. Radovanovic, D.: Weighted tree automata over strong bimonoids. Novi Sad J. Math. **40**(3), 89–108 (2010)
17. Rozenberg, G., Salomaa, A. (eds.): Handbook of Formal Languages, vol. 1 Word, Language, Grammar. Springer, Heidelberg (1997). https://doi.org/10.1007/978-3-642-59136-5
18. Rozenberg, G., Salomaa, A. (eds.): Handbook of Formal Languages, vol. 3 Beyond Words. Springer, Heidelberg (1997). https://doi.org/10.1007/978-3-642-59126-6

An Algorithm for Single-Source Shortest Paths Enumeration in Parameterized Weighted Graphs

Bastien Sérée[1,3], Loïg Jezequel[2,3], and Didier Lime[1,3(✉)]

[1] École Centrale de Nantes, Nantes, France
[2] Université de Nantes, Nantes, France
[3] LS2N, UMR CNRS 6004, Nantes, France
{bastien.seree,loig.jezequel,didier.lime}@ls2n.fr

Abstract. We consider weighted graphs with parameterized weights and we propose an algorithm that, given such a graph and a source node, builds a collection of trees, each one describing the shortest paths from the source to all the other nodes of the graph for a particular zone of the parameter space. Moreover, the union of these zones covers the full parameter space: given any valuation of the parameters, one of the trees gives the shortest paths from the source to all the other nodes of the graph when the weights are computed using this valuation.

Keywords: Shortest paths · Weighted graphs · Parameterized graphs

1 Introduction

For many real-world systems and problems there are natural discrete-event abstractions, which can be modelled by a graph or a derived formalism. In many cases, we can also identify resources that need to be optimized (distance, memory, energy, time, etc.) and then (extensions of) weighted graphs are a formalism of choice.

One of the most basic problems is to find optimal paths, for which the accumulated weight (also often called cost) is minimal.

When addressing systems that are not well-known, maybe because we are in the early phases of a design process, one way to cope with this uncertainty is to use parameters for the weights. The interesting problems are parameter synthesis problems, in which one tries to find the values of parameters such that some path is optimal, or such that a target vertex can be reached within a given bound on the accumulated weight, etc.

Surprisingly, those problems have not been studied in detail for the setting for which cost themselves are parameters. Parametric timed automata (PTA) [1] allow clocks to be tested against parameters, which may allow simulation of parametric weights to some extent. The optimal time reachability problem has

© Springer Nature Switzerland AG 2021
A. Leporati et al. (Eds.): LATA 2021, LNCS 12638, pp. 279–290, 2021.
https://doi.org/10.1007/978-3-030-68195-1_22

recently been studied for PTA [2]. Similarly, the bounded-cost reachability problem has been addressed for a formalism related to PTA called parametric time Petri nets in [7]. In [3], the authors do extend PTA with parametric weights on edges, but the topology of the systems considered is always that of a tree.

Much differently, the parametric one-counter machines of [4] should allow one to model a parametric cost, however, the parameter synthesis problems are not addressed in that article.

Finally, in [6] the authors consider graphs in which weights are expressed as a function of a single parameter. They partition the real numbers in a finite way, such that for two parameter values in a given partition, the optimal paths from a given source vertex to all other vertices are the same, and exhibit the corresponding trees those paths form. Such parametric graphs, with a single parameter, were later studied in [8] (directly improving [6]), and in [5] for example. Extending the results of [6] to parameterized graphs with multiple parameters makes the problem more complex since one needs to partition the n-dimensional real space (where n is the number of parameters). This is the subject of our work. To the best of our knowledge, this has not been done prior to this paper. The algorithm that we propose is exponential in the number of vertices and is polynomial in the number of edges, and in the number of parameters of the considered graph.

This article is organised as follows: in Sect. 2 we introduce the basic notations and definitions; in Sect. 3 we informally present our algorithm on a comprehensive example; in Sect. 4, we give the algorithm together with the associated proofs of correctness, completeness, and termination, as well as the complexity; in Sect. 5 we conclude.

2   Definitions

For all $(a, b, i) \in \mathbb{N}$, we denote by $i \in [\![a, b]\!]$ any integer i such that $a \leqslant i \leqslant b$.

2.1   Parametric Graphs

Definition 1 (Parametric graph). *A* parametric graph *with n parameters is a tuple $G = (V, E, f, (\lambda_i)_1^n, (\Lambda_i)_1^n)$ where V is a set of* vertices, *$E \subseteq V \times V$ is a set of* edges, *$f : E \to \mathbb{R}$ is a non-parametric cost function, and for every $i \in [\![1, n]\!]$, λ_i is a parameter, and $\Lambda_i : E \to \{0, 1\}$ is a parametric cost function.*

The parameters can take any value in $\mathbb{R}$. With a slight abuse of notations, we denote by λ_i not only the parameters but also their values. We also write $\overrightarrow{\lambda} = (\lambda_1 \dots \lambda_n)$. All the following definitions are written for a parametric graph $G = (V, E, f, (\lambda_i)_1^n, (\Lambda_i)_1^n)$.

Definition 2 (Edge cost). *The* cost *of an edge $e \in E$ is:*

$$c(e) = f(e) - \sum_{i=1}^{n} \lambda_i \Lambda_i(e).$$

A *Path* p in G is a sequence of edges $p = e_0 e_1 e_2 \ldots e_k$ such that $\forall i \in \llbracket 0, k \rrbracket, e_i = (v_i, v_{i+1})$. In such a path v_0 is the *initial vertex* and v_{k+1} is the *terminal vertex*. The integer k is called the *length* of p.

Definition 3 (Path cost). *Let $p = e_0 e_1 \ldots e_k$ be a path of G, the cost of p is:*

$$c(p, \overrightarrow{\lambda}) = \sum_{j=0}^{k} c(e_j) = \sum_{j=0}^{k} \left(f(e_j) - \sum_{i=1}^{n} \lambda_i \Lambda_i(e_j) \right) = \sum_{j=0}^{k} f(e_j) - \sum_{i=1}^{n} \lambda_i \sum_{j=0}^{k} \Lambda_i(e_j).$$

We call *path of minimal cost* for $\overrightarrow{\lambda} \in \mathbb{R}^n$ a path of G such that there is no other path of G with the same initial and terminal vertices, and with a strictly lower path cost for $\overrightarrow{\lambda}$.

Finally, we call a *cycle* a path where the initial vertex and the terminal vertex are the same. And we call a *negative cycle* for $\overrightarrow{\lambda} \in \mathbb{R}^n$ a cycle with a negative cost for $\overrightarrow{\lambda}$.

2.2 Trees over Parametric Graphs

In the following we suppose that there is a distinguished vertex s in the graph G, from which we will search for paths of minimal cost toward all other vertices. As we are interested in paths from s to each vertex, in the following we assume that the graphs we consider are such that these paths exist.

Definition 4 (Tree). *We define a tree of G rooted at s (or with source s) as any tuple $T = (V_T, E_T, f_T, (\lambda_{T,i})_1^n, (\Lambda_{T,i})_1^n)$ such that $V_T = V$, $E_T \subseteq E$ is such that for all $v \in V \setminus \{s\}$, $|\{(u, v) : (u, v) \in E_T\}| = 1$, $\{(u, s) : (u, s) \in E_T\} = \emptyset$ and there is no cycle in T, $f_T = f_{|E_T}$, $\forall i, \lambda_{T,i} = \lambda_i$, and $\forall i, \Lambda_{T,i} = \Lambda_{T,i|E_T}$.*

Notice that, in such a tree, there are always $|V| - 1$ edges. Moreover, there is exactly one path from s to each vertex.

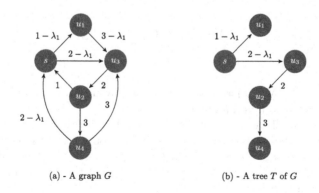

(a) - A graph G (b) - A tree T of G

Fig. 1. A graph G and an example of a tree T of G.

In Fig. 1, we have represented an example of a graph with one parameter, λ_1 (a) and a tree T of G rooted at s (b).

In all the following definitions T is a tree rooted at s.

Definition 5 (Distance). *Let v be a vertex of G, the distance $d(T, v, \overrightarrow{\lambda})$ is the cost of the unique path from s to v in T.*

Definition 6 (Partial distances). *Let v be a vertex of G and let $e_0 e_1 \cdots e_k$ be the unique path from s to v in T. The partial non-parametric distance between s and v in T is*

$$d_f(T, v) = \sum_{j=0}^{k} f(e_j).$$

The partial parametric distances between s and v are the

$$d_{\Lambda_i}(T, v) = \sum_{j=0}^{k} \Lambda_i(e_j),$$

for all $i \in [\![1, n]\!]$.

Notice that $d(T, v, \overrightarrow{\lambda}) = d_f(T, v) - \sum_{i=1}^{n} \lambda_i d_{\Lambda_i}(T, v)$.

Definition 7 (Tree of minimal distances). *We say that T is a tree of minimal distances for $\overrightarrow{\lambda} \in \mathbb{R}^n$ if for all $v \in V$, the unique path from s to v in T is a path of minimal cost $\overrightarrow{\lambda}$ from s to v in G.*

Moreover, if $S \subseteq \mathbb{R}^n$, T is a tree of minimal distances for all $\overrightarrow{\lambda} \in S$, we say that T is a tree of minimal distances on S.

Definition 8 (Neighbour). *Let $e = (u, v)$ be an edge in E, the neighbour of T generated by e is the tuple $N(T, e) = (V_N, E_N, f_N, (\lambda_{N,i})_1^n, (\Lambda_{N,i})_1^n)$ where:*

- $V_N = V$
- $E_N = (E_T \setminus \{(u', v) : (u', v) \in E_T\}) \cup \{e\}$, $f_N = f_{|E_N}$
- $\forall i, \lambda_{N,i} = \lambda_i$
- $\forall i, \Lambda_{N,i} = \Lambda_{i|E_N}$

In other words, $N(T, e)$ is obtained from T by deleting the only edge $e' = (u', v')$ such that $v' = v$ and adding e.

Notice that for all $e \in E_T$, $N(T, e) = T$ and that, for $(u, v) \in E \setminus E_T$, an edge (u', v) does not necessarily exist in T, since T is rooted at s. Indeed an edge e such that $e = (u, s)$ will not be in T (as T is a tree rooted at s) and can generate neighbours as any edge. In particular, this means that the neighbour of a tree is not necessarily a tree.

As an example, consider G from Fig. 1 and let $e = (s, u_3)$, $e_1 = (u_1, u_3)$ and $e_2 = (u_4, u_3)$. Figure 1 (b) represents a tree T rooted at s, Fig. 2 (a) represents

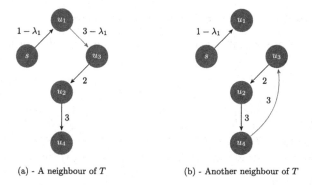

(a) - A neighbour of T (b) - Another neighbour of T

Fig. 2. Some neighbours of the tree T of Fig. 1.

$N(T, e_1)$, which is also a tree, and Fig. 2 (b) represents $N(T, e_2)$, illustrating the fact that a neighbour of a tree is not necessarily a tree itself.

The following proposition specifies in which case a neighbour of a tree is actually a tree. Its proof is omitted due to space constraints

Proposition 1. *Let $e = (u, v) \in E$ be an edge, $N(T, e)$ is a tree if and only if v is not on the unique path from s to u in T.*

Before giving other properties of trees and their neighbours, we have to take a few notations. For an edge $e = (u, v) \in E$ we note:

$$\Delta_f(T, e) = d_f(T, u) + f(e) - d_f(T, v),$$
$$\forall i \in [\![1, n]\!], \Delta_{\Lambda_i}(T, e) = d_{\Lambda_i}(T, u) + \Lambda_i(e) - d_{\Lambda_i}(T, v).$$

These deltas represent the differences in distance from s to v between T and its neighbour generated by e. Δ_f is the difference in the non-parametric part of the distance. Each Δ_{Λ_i} represents the difference of the number of occurrences of the corresponding parameter λ_i. Notice that for any $e \in E_T$, one always has $\Delta_{\Lambda_i}(T, e) = 0$.

For example in Fig. 1, with the same notations as before, $\Delta_f(T, e_1) = d_f(T, u_1) + f(e_1) - d_f(T, u_3) = 1 + 3 - 2 = 2$, and $\Delta_{\Lambda_1}(T, e_1) = d_{\Lambda_1}(T, u_1) + \Lambda_1(e_1) - d_{\Lambda_1}(T, u_3) = 1 + 1 - 1$.

Proposition 2. *Let $e = (u, v)$ be an edge not in T. If $N(T, e)$ is a tree then $\forall w \in V, d(T, w, \overrightarrow{\lambda}) = d(N(T, e), w, \overrightarrow{\lambda})$ if and only if $\Delta_f(T, e) - \sum\limits_{i=1}^{n} \lambda_i \Delta_{\Lambda_i}(T, e) = 0$.*

Proposition 3. *Let $e = (u, v)$ be an edge. If $N(T, e)$ is not a tree then $N(T, e)$ has a cycle of cost 0 if and only if $\overrightarrow{\lambda}$ is such that $\Delta_f(T, e) - \sum\limits_{i=1}^{n} \lambda_i \Delta_{\Lambda_i}(T, e) = 0$.*

The proofs of these propositions are omitted due to space constraints.

2.3   Constraints and Zones Associated to Trees

Definition 9 (Constraint). *Let $e \in E \setminus E_T$ be an edge, the* constraint *associated with e is*

$$C_{T,e} = \Delta_f(T,e) - \sum_{i=1}^{n} \lambda_i \Delta_{\Lambda_i}(T,e).$$

Definition 10 (Zone). *Let $E_c \subseteq E \setminus E_T$ be a set of edges such that $\forall e_c \in E_c, \exists i \in [\![1,n]\!], \Delta_{\Lambda_i}(T,e_c) > 0$. The* zone *defined by the constraints associated to the edges in E_c is the set S such that: $\overrightarrow{\lambda} \in S$ if and only if $\forall e_c \in E_c, \Delta_f(T,e_c) - \sum_{i=1}^{n} \lambda_i \Delta_{\Lambda_i}(T,e_c) \geq 0$.*

We can notice that zones are convex by construction.

Definition 11 (Active constraint). *Let $E_c \subseteq E \setminus E_T$ be a set of edges such that $\forall e_c \in E_c, \exists i \in [\![1,n]\!], \Delta_{\Lambda_i}(T,e_c) > 0$. Let $e_c \in E_c$ be an edge. Let S be the zone defined by the constraints associated to the edges in E_c. Let $S_{/e_c}$ be the zone defined by the constraints associated to the edges in $E_c \setminus \{e_c\}$. The constraint C_{T,e_c} is said to be* active *if and only if $S_{/e_c} \neq S$.*

3   Presentation of Our Algorithm for Minimal Distances

In Sect. 3, we propose an algorithm which, given a graph G returns a list of trees and a list of disjoint zones. Each tree T is associated with one zone S, such that T is a tree of minimal distances on S. Moreover, the union of the returned zones is the zone of $\mathbb{R}^n$ for which there is no negative cycle in G. In this section, we demonstrate how this algorithm works on the example of G_{ex}, a parametric graph with two parameters presented in Fig. 3.

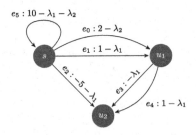

Fig. 3. Graph G_{ex}

The first step of the algorithm consists in finding the zone $Z_{noCycle}$ for which the concept of minimal distances makes sense, that is the values of λ_1 and λ_2 for which there is no negative cycle. Here, the only cycle is $e_5 e_5 e_5 \ldots$. So, the only possible negative cycle is when the cost of e_5 is negative, therefore, when $10 - \lambda_1 - \lambda_2 < 0$. Hence, we have $Z_{noCycle} = \{(\lambda_1, \lambda_2) \in \mathbb{R}^2 : \lambda_1 + \lambda_2 \leqslant 10\}$.

The goal will be to cover $Z_{noCycle}$ with zones associated to trees of minimal distances. For that the algorithm enumerates zones, associated with trees, going from one zone to another by considering the neighbours of the associated tree. The algorithm begins by computing a first tree T_0. This tree is a tree of minimal distances for a particular pair $(\lambda_{1,init}, \lambda_{2,init})$, where $-\lambda_{1,init} = -\lambda_{2,init} = 1 + \sum_{e \in E} |f(e)|$. Here we have $(\lambda_{1,init}, \lambda_{2,init}) = (-20, -20)$ and T_0 is represented in Fig. 4.

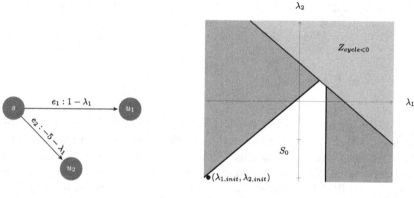

Fig. 4. T_0 **Fig. 5.** $(\lambda_{1,init}, \lambda_{2,init})$ and S_0

Now look at the constraints associated with the neighbours of T_0 to characterize the associated zone S_0. The active constraints also tell which trees will be considered next. Here, the edges that generate neighbours are e_0, e_3, e_4 and e_5. So, one has to look at the constraints C_0, C_3, C_4 and C_5 associated respectively with $N(T_0, e_0)$, $N(T_0, e_3)$, $N(T_0, e_4)$ and $N(T_0, e_5)$.

We have $C_0 : 1 + \lambda_1 - \lambda_2 = 0$, $C_3 : 6 - \lambda_1 = 0$, $C_4 : 7 - \lambda_1 = 0$ and $C_5 : 10 - \lambda_1 - \lambda_2 = 0$. Among these constraints, C_0, C_3 and C_5 are active. Thus, we take $S_0 = \{(\lambda_1, \lambda_2) \in \mathbb{R}^2 : 1 + \lambda_1 - \lambda_2 \geqslant 0, 6 - \lambda_1 \geqslant 0, 10 - \lambda_1 - \lambda_2 \geqslant 0\}$, as represented in Figure 5.

As $N(T_0, e_0)$ and $N(T_0, e_3)$ are associated with active constraints and have not been considered yet, they are added in a list of trees to be considered later, called $listToDo$. $N(T_0, e_5)$ is not added to $listToDo$ because it is not a tree. The pair (T_0, S_0) is added to a list called $listExplored$. This list will be returned by the algorithm at the end of its computation.

From now on, the algorithm iteratively considers the trees of $listToDo$. Assume, for example, that it begins with $N(T_0, e_0) = T_1$, represented in Fig. 6. The edges that generate neighbours are e_1, e_3, e_4 and e_5. The associated constraints are $C_1 : -1 - \lambda_1 + \lambda_2 = 0$, $C_3 : 7 - \lambda_2 = 0$, $C_4 : 8 - \lambda_2 = 0$ and $C_5 : 10 - \lambda_1 - \lambda_2 = 0$. Among them, C_1, C_3, and C_5 are active. From that we can define S_1, represented in Fig. 7. As $N(T_1, e_1) = T_0$ has already been considered, only $N(T_1, e_3)$ is added to $listToDo$ (recall that $N(T_1, e_5)$ is not a tree). (T_1, S_1) is added to $listExplored$.

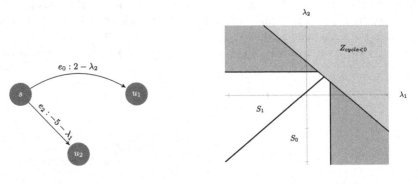

Fig. 6. T_1 **Fig. 7.** S_0 and S_1

At the moment, $listToDo = \{N(T_0, e_3), N(T_1, e_3)\}$ and $listExplored = \{(T_0, S_0), (T_1, S_1)\}$. Assume that the algorithm considers $N(T_0, e_3) = T_2$ next. This tree is represented in Fig. 8. The constraints used to define S_2 are the ones associated with e_0, e_2 and e_5. e_4 is not considered because $\Delta_{\Lambda_1} T_2, e_4 = \Delta_{\Lambda_2} T_2, e_4 = 0$. All constraints are active. The zone S_2 is represented in Fig. 9. No tree is added to $listToDo$ as $N(T_2, e_0) = N(T_1, e_3)$, which is already in $listToDo$ and $N(T_2, e_2) = T_0$, which has already been considered. (T_2, S_2) is added to $listExplored$.

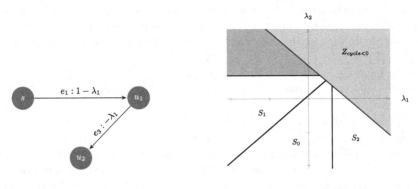

Fig. 8. T_2 **Fig. 9.** S_0, S_1 and S_2

Then, it remains to consider $N(T_1, e_3) = T_3$, represented in Fig. 10. Three constraints are considered, this time associated with e_1, e_2 and e_5. All these constraints are active. The obtained S_3 is represented in Fig. 11. No tree is added to *listToDo* because all neighbours either have already been considered or are not trees. (T_3, S_3) is added to *listExplored*.

At that point, the algorithm terminates because $S_0 \cup S_1 \cup S_2 \cup S_3 = Z_{noCycle}$. It returns *listExplored*. Note that *listToDo* is empty.

4 Formal Presentation of Our Algorithm

In this section, we formalize the algorithm that has been presented on the example in the previous section. We then prove that this algorithm is correct.

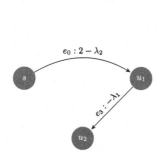

Fig. 10. T_3

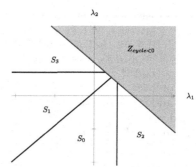

Fig. 11. S_0, S_1, S_2, and S_3

4.1 The Algorithm

Algorithm 1 is an algorithm that, given a parametric graph G and a vertex s, returns a list of pairs (T, S) such that every T is a tree of minimal distances on S and that the union of all S in the list is equal to the zone such that there is no negative cycle in G. In this algorithm, $\overrightarrow{\lambda}_{init}$ is the n-components vector so that each component is equal to $-\left(1 + \sum_{e \in E} |f(e)|\right)$.

In the following, we will refer to the zone where there are no negative cycles as $Z_{noCycle}$ and the union of the zones in *listExplored* as $Z_{explored}$.

Algorithm 1 starts by computing T_0 (the tree from which the space will be explored) as a tree of minimal distances for $\overrightarrow{\lambda}_{init}$. This initial tree is chosen to ensure that the algorithm will cover $Z_{noCycle}$ by only exploring toward greater λ_i, $i \in [\![1, n]\!]$ as proven in the next section. T_0 is added to *listToDo* which is the list of the trees the algorithm needs to consider.

Algorithm 1. Trees of minimal distances of $G = (V, E, f, (\lambda_i)_1^n, (\Lambda_i)_1^n)$

1: **let** $listExplored = \emptyset$.
2: **let** T_0 be a tree of minimal distances for $\overrightarrow{\lambda}_{init}$
3: **let** $listToDo = \{T_0\}$
4: **while** $listToDo \neq \emptyset$ **do**
5:     **choose** a tree T in $listToDo$ (and delete it from $listToDo$)
6:     **let** $Neighbours$ be the set of all possible edges e such that $\exists i \in [\![1, n]\!]$, $\Delta_{\Lambda_i}(T, e) > 0$.
7:     **let** S be the zone defined by the constraints associated to the edges of $Neighbours$
8:     **let** $Active$ be the subset of $Neighbours$ containing the edges giving active constraints
9:     **for** each edge e in $Active$ such that $N(T, e)$ is a tree **do**
10:        **let** $T_N = N(T, e)$
11:        **if** $T_N \notin listToDo$ **and** $T_N \notin listExplored$ **then**
12:            **add** T_N to $listToDo$
13:        **end if**
14:     **end for**
15:     **add** (T, S) to $listExplored$
16: **end while**
17: **return** $listExplored$

For each tree T, the algorithm begins by enumerating all the edges e that can generate a neighbour. From the associated constraints it characterizes S, a zone where T is a tree of minimal distances. Then the algorithm does two things: (1) it adds to $listToDo$ all the neighbour trees that have not been considered yet (those that are not already in $listToDo$ or in $listExplored$) and (2) it adds the new result to the returned list by adding T and S to $listExplored$. We can notice that the zones in $listExplored$ are disjoint by construction.

When $listToDo$ is empty the algorithm returns $listExplored$.

Notice that when there is only one parameter this algorithm is equivalent to the algorithm presented in [6] as constraints have a better form that make computing active constraints equivalent to look for a maximum. So it is possible to do a more efficient exploration because of the fact that there is only one dimension.

4.2   Proof of the Algorithm

A first Lemma expresses that – with the value that has been chosen for $\overrightarrow{\lambda}_{init}$ – it is not necessary to consider all the neighbours of each tree in the main loop of the algorithm. It is sufficient to consider neighbours with increasing number of occurrences of (at least) one parameter, as enforced by line 6 of Algorithm 1.

Lemma 1. *Let S_0 be the first zone computed by the algorithm. $\forall \overrightarrow{\lambda} \in Z_{noCycle}$, $\exists \overrightarrow{\lambda'} \in \mathbb{R}^n$ and $\overrightarrow{\lambda_{S_0}} \in S_0$ such that (1) $\exists i \in [\![1, n]\!], \lambda_i' \geqslant \lambda_{i, S_0}$ and (2) for all trees T of minimal distances for $\overrightarrow{\lambda'}$, T is also a tree of minimal distances for $\overrightarrow{\lambda}$.*

Proof is omitted due to space constraints

A second lemma exhibits a loop invariant that will be instrumental in showing the correctness of the results of the algorithm.

Lemma 2 (loop invariant). *Let $Z_{notExplored} = Z_{noCycle} \setminus Z_{explored}$, at each loop of the while loop we have:*

1: $Z_{explored} \cup Z_{notExplored} = Z_{noCycle}$ *(in particular there is no $\overrightarrow{\lambda} \in Z_{explored}$ such that there is a negative cycle in G for $\overrightarrow{\lambda}$)*

2: *for all $(T, S) \in ListExplored$, T is a tree of minimal distances for all $\overrightarrow{\lambda} \in S$.*

Proof is omitted due to space constraints

Building on the two above lemmas, we give our main theorem, that states that the proposed algorithm terminates and returns correct results.

Theorem 1. *The algorithm terminates and returns ListReturned such that for all $(T, S) \in ListReturned$, T is a tree of minimal length for all $\overrightarrow{\lambda} \in S$, $\bigcup_{(T,S) \in ListReturned} S = Z_{noCycle}$ and for all (T, S) and (T', S') in ListReturned such that $S \neq S'$, $S \cap S' = \emptyset$.*

Proof. If the algorithm does not terminate it means that there is an infinite loop.

As we consider a new tree in each loop and there is a finite number of (possible) trees it is impossible to have an infinite loop so the algorithm does terminate. We also need that $Z_{explored} = Z_{noCycle}$ at the end of the algorithm, i.e. when $listToDo$ is empty. If it is not the case it means that the algorithm has missed one or more zones and this is only possible if there is some zones such that $\overrightarrow{\lambda} \in \mathbb{R}^n$, $\nexists \overrightarrow{\lambda'} \in \mathbb{R}^n$ and $\overrightarrow{\lambda_{S_0}} \in S_0$ such that $\exists i \in [\![1, n]\!], \lambda_i' \geqslant \lambda_{i,S_0}$. Which is impossible by Lemma 1 so the algorithm terminates and we have $\bigcup_{(T,S) \in ListReturned} S = Z_{noCycle}$.

For each $(T, S) \in ListReturned$ we also have that T is a tree of minimal length for all $\overrightarrow{\lambda} \in S$ and $\bigcup_{(T,S) \in ListReturned} S = Z_{noCycle}$ by the loop invariant.

4.3 Complexity

We conclude the presentation of Algorithm 1 by giving its worst-case complexity.

Theorem 2. *The worst case complexity of Algorithm 1 is exponential in the number $|V|$ of vertices and is polynomial in the number $|E|$ of edges and in the number n of parameters. Moreover it is logarithmic in the largest constant value M appearing in the constraints.*

The proof of this theorem is omitted due to space constraints and shows that this complexity is $O(|V|^{|V|-1}(|E| - |V|)^3 n^3 \log(M))$.

5   Conclusion

We have proposed an algorithm to find the optimal paths from a single source to all other vertices in a weighted graph in which weights involve an arbitrary number of real-valued parameters. Since those paths change with the values of the parameters, the result of our algorithm is a finite set of trees, each with a zone of the parameter space on which it is optimal. Those zones cover the parameter space for which there are no negative cost cycles in the graph. This algorithm generalizes a previous work by Karp and Orlin in which only one parameter was considered [6].

Further work includes implementing the algorithm, and evaluating its efficiency on real-world case-studies.

References

1. Alur, R., Henzinger, T.A., Vardi, M.Y.: Parametric real-time reasoning. In: ACM Symposium on Theory of Computing, pp. 592–601 (1993)
2. André, É., Bloemen, V., Petrucci, L., van de Pol, J.: Minimal-time synthesis for parametric timed automata. In: Vojnar, T., Zhang, L. (eds.) TACAS 2019. LNCS, vol. 11428, pp. 211–228. Springer, Cham (2019). https://doi.org/10.1007/978-3-030-17465-1_12
3. André, É., Lime, D., Ramparison, M., Stoelinga, M.: Parametric analyses of attack-fault trees. In: Keller, J., Penczek, W. (eds.) 19th International Conference on Application of Concurrency to System Design (ACSD 2019), Aachen, Germany. IEEE Computer Society, June 2019
4. Bundala, D., Ouaknine, J.: On parametric timed automata and one-counter machines. Inf. Comput. **253**, 272–303 (2017)
5. Chakraborty, S., Fischer, E., Lachish, O., Yuster, R.: Two-phase algorithms for the parametric shortest path problem. In: Marion, J.-Y., Schwentick, T. (eds.) 27th International Symposium on Theoretical Aspects of Computer Science, STACS 2010, Nancy, France, 4–6 March 2010, vol. 5 of LIPIcs, pp. 167–178. Schloss Dagstuhl - Leibniz-Zentrum für Informatik (2010)
6. Karp, R.M., Orlin, J.B.: Parametric shortest path algorithms with an application to cyclic staffing. Discrete Appl. Math. **3**(1), 37–45 (1981)
7. Lime, D., Roux, O.H., Seidner, C.: Parameter synthesis for bounded cost reachability in time petri nets. In: Donatelli, S., Haar, S. (eds.) PETRI NETS 2019. LNCS, vol. 11522, pp. 406–425. Springer, Cham (2019). https://doi.org/10.1007/978-3-030-21571-2_22
8. Young, N., Tarjan, R., Orlin, J.: Faster parametric shortest path and minimum balance algorithms. Networks **21**(2), 205–221 (1991)

Words and Strings

On Balanced Sequences and Their Asymptotic Critical Exponent

Francesco Dolce$^{(\boxtimes)}$, Ľubomíra Dvořáková, and Edita Pelantová

FNSPE, Czech Technical University in Prague, Prague, Czech Republic
{francesco.dolce,lubomira.dvorakova,edita.pelantova}@fjfi.cvut.cz

Abstract. We study aperiodic balanced sequences over finite alphabets. A sequence **v** of this type is fully characterised by a Sturmian sequence **u** and two constant gap sequences **y** and **y′**. We study the language of **v**, with focus on return words to its factors. We provide a uniform lower bound on the asymptotic critical exponent of all sequences **v** arising by **y** and **y′**. It is a counterpart to the upper bound on the least critical exponent of **v** conjectured and partially proved recently in works of Baranwal, Rampersad, Shallit and Vandomme. We deduce a method computing the exact value of the asymptotic critical exponent of **v** provided the associated Sturmian sequence **u** has a quadratic slope. The method is used to compare the critical and the asymptotic critical exponent of balanced sequences over an alphabet of size $d \leq 10$ which are conjectured by Rampersad et al. to have the least critical exponent.

Keywords: Balanced sequence · Critical exponent · Sturmian sequence · Return word · Bispecial factors

1 Introduction

An infinite sequence over a finite alphabet is balanced if, for any two of its factors u and v of the same length, the number of occurrences of each letter in u and v differs by at most 1. Over a binary alphabet aperiodic balanced sequences coincide with Sturmian sequences, as shown by Hedlund and Morse [13]. Hubert [14] provided a construction of balanced sequences. It consists in colouring of entries of a Sturmian sequence **u** by two constant gap sequences **y** and **y′**. In this paper we study combinatorial properties of balanced sequences. We first show that such sequences belong to the class of eventually dendric sequences introduced in [5]. We give formulæ for the factor complexity and the number of return words to each factor. The main goal of this paper is to develop a method computing the asymptotic critical exponent of a given balanced sequence. Our work can be understood as a continuation of research on balanced sequences with the

The research received funding from the project CZ.02.1.01/0.0/0.0/16_019/0000778. We would like to thank Daniela Opočenská for her careful and readily usable implementation of our program computing the asymptotic critical exponent.

A. Leporati et al. (Eds.): LATA 2021, LNCS 12638, pp. 293–304, 2021.
https://doi.org/10.1007/978-3-030-68195-1_23

least critical exponent initiated by Rampersad, Shallit and Vandomme [18]. The relation between factor complexity and critical exponent for binary and ternary sequences was studied as well in [19]

Finding the least critical exponent of sequences is a classical problem. The answer is known as Dejean's conjecture [8], and the proof was provided step by step by many people. The least critical exponent was determined also for some particular classes of sequences: by Carpi and de Luca [6] for Sturmian sequences, and by Currie, Mol and Rampersad [7] for binary rich sequences. Recently, Rampersad, Shallit and Vandomme [18] found balanced sequences with the least critical exponent over alphabets of size 3 and 4 and also conjectured that the least critical exponent of balanced sequences over a d-letter alphabet with $d \geq 5$ is $\frac{d-2}{d-3}$. Their conjecture was confirmed for $d \leq 8$ [3,4].

Here we focus on the asymptotic critical exponent of balanced sequences. We show that the asymptotic critical exponent depends on the slope of the associated Sturmian sequence and, unlike the critical exponent, on the length of the minimal periods of $\mathbf{y}$ and $\mathbf{y}'$, but not on $\mathbf{y}$ and $\mathbf{y}'$ themselves. We also give a lower bound on the asymptotic critical exponent. We provide an algorithm computing the exact value of the asymptotic critical exponent for balanced sequences originated in Sturmian sequences with a quadratic slope (in this case the continued fraction of the slope is eventually periodic).

2 Preliminaries

An *alphabet* $\mathcal{A}$ is a finite set of symbols called *letters*. A *word* over $\mathcal{A}$ of *length* n is a string $u = u_0 u_1 \cdots u_{n-1}$, where $u_i \in \mathcal{A}$ for all $i \in \{0, 1, \ldots, n-1\}$. The length of u is denoted by $|u|$. The set of all finite words over $\mathcal{A}$ together with the operation of concatenation forms a monoid, denoted $\mathcal{A}^*$. Its neutral element is the *empty word* ε and we denote $\mathcal{A}^+ = \mathcal{A}^* \setminus \{\varepsilon\}$. If $u = xyz$ for some $x, y, z \in \mathcal{A}^*$, then x is a *prefix* of u, z is a *suffix* of u and y is a *factor* of u. We sometimes use the notation $yz = x^{-1}u$. To any word u over $\mathcal{A}$ with cardinality $\#\mathcal{A} = d$, we assign its *Parikh vector* $\mathbf{V}(u) \in \mathbb{N}^d$ defined as $(\mathbf{V}(u))_a = |u|_a$ for all $a \in \mathcal{A}$, where $|u|_a$ is the number of letters a occurring in u.

A *sequence* over $\mathcal{A}$ is an infinite string $\mathbf{u} = u_0 u_1 u_2 \cdots$, where $u_i \in \mathcal{A}$ for all $i \in \mathbb{N}$. A sequence $\mathbf{u}$ is *eventually periodic* if $\mathbf{u} = vwww \cdots = v(w)^\omega$ for some $v \in \mathcal{A}^*$ and $w \in \mathcal{A}^+$. It is *periodic* if $\mathbf{u} = w^\omega$. If $\mathbf{u}$ is not eventually periodic, then it is *aperiodic*. A *factor* of $\mathbf{u} = u_0 u_1 u_2 \cdots$ is a word y such that $y = u_i u_{i+1} u_{i+2} \cdots u_{j-1}$ for some $i, j \in \mathbb{N}$, $i \leq j$. We usually denote $y = \mathbf{u}_{[i,j)}$. The number i is called an *occurrence* of the factor y in $\mathbf{u}$. In particular, if $i = j$, the factor y is the empty word ε and any index i is its occurrence. If $i = 0$, the factor y is a *prefix* of $\mathbf{u}$. If each factor of $\mathbf{u}$ has infinitely many occurrences in $\mathbf{u}$, the sequence $\mathbf{u}$ is *recurrent*. Moreover, if for each factor the distances between its consecutive occurrences are bounded, $\mathbf{u}$ is *uniformly recurrent*.

The *language* $\mathcal{L}(\mathbf{u})$ of a sequence $\mathbf{u}$ is the set of all its factors. We also define $\mathcal{L}(\mathbf{u})^+ = \mathcal{L}(\mathbf{u}) \setminus \{\varepsilon\}$. A factor w of $\mathbf{u}$ is *right special* if wa, wb are in $\mathcal{L}(\mathbf{u})$ for at least two distinct letters $a, b \in \mathcal{A}$. Analogously, we define a *left special* factor. A factor is *bispecial* if it is both left and right special. The *factor complexity* of a sequence $\mathbf{u}$ is the mapping $\mathcal{C}_\mathbf{u} : \mathbb{N} \to \mathbb{N}$ defined by $\mathcal{C}_\mathbf{u}(n) = \#\{w \in \mathcal{L}(\mathbf{u}) : |w| = n\}$. The first difference of the factor complexity is $s_\mathbf{u}(n) = \mathcal{C}_\mathbf{u}(n+1) - \mathcal{C}_\mathbf{u}(n)$. Aperiodic sequences with the lowest possible factor complexity, i.e., such that $\mathcal{C}_\mathbf{u}(n) = n + 1$ for all $n \in \mathbb{N}$, are called *Sturmian sequences* (for other equivalent definitions see [2]). Clearly, all Sturmian sequences are defined over a binary alphabet, e.g., $\{\mathsf{a}, \mathsf{b}\}$. If both sequences au and bu are Sturmian, then $\mathbf{u}$ is called a *standard Sturmian sequence*. It is well-known that for any Sturmian sequence there exists a unique standard Sturmian sequence with the same language. For other facts about Sturmian sequences see [17].

A sequence $\mathbf{u}$ over the alphabet $\mathcal{A}$ is *balanced* if for every letter $a \in \mathcal{A}$ and every pair of factors $u, v \in \mathcal{L}(\mathbf{u})$ with $|u| = |v|$, we have $||u|_a - |v|_a| \leq 1$. The class of Sturmian sequences and the class of aperiodic balanced sequences coincide over a binary alphabet (see [13]). Vuillon [21] provides a survey on some previous work on balanced sequences.

A *morphism* over $\mathcal{A}$ is a mapping $\psi : \mathcal{A}^* \to \mathcal{A}^*$ such that $\psi(uv) = \psi(u)\psi(v)$ for all $u, v \in \mathcal{A}^*$. The morphism ψ can be naturally extended to sequences by setting $\psi(u_0 u_1 u_2 \cdots) = \psi(u_0)\psi(u_1)\psi(u_2) \cdots$.

Consider a factor w of a recurrent sequence $\mathbf{u} = u_0 u_1 u_2 \cdots$. Let $i < j$ be two consecutive occurrences of w in $\mathbf{u}$. Then the word $u_i u_{i+1} \cdots u_{j-1}$ is a *return word* to w in $\mathbf{u}$. The set of all return words to w in $\mathbf{u}$ is denoted by $\mathcal{R}_\mathbf{u}(w)$. If $\mathbf{u}$ is uniformly recurrent, the set $\mathcal{R}_\mathbf{u}(w)$ is finite for each prefix w. The opposite is true if $\mathbf{u}$ is recurrent. In this case $\mathbf{u}$ can be written as a concatenation $\mathbf{u} = r_{d_0} r_{d_1} r_{d_2} \cdots$ of return words to w. The *derived sequence* of $\mathbf{u}$ to w is the sequence $\mathbf{d}_\mathbf{u}(w) = d_0 d_1 d_2 \cdots$ over the alphabet of cardinality $\#\mathcal{R}_\mathbf{u}(w)$. The concept of derived sequences was introduced by Durand [11].

Given a sequence $\mathbf{u}$ over an alphabet $\mathcal{A}$ and $w \in \mathcal{L}(\mathbf{u})$, we define the sets of left extensions, right extensions and bi-extensions of w in $\mathcal{L}(\mathbf{u})$ respectively as $L_\mathbf{u}(w) = \{a \in \mathcal{A} : aw \in \mathcal{L}(\mathbf{u})\}$, $R_\mathbf{u}(w) = \{b \in \mathcal{A} : wb \in \mathcal{L}(\mathbf{u})\}$ and $B_\mathbf{u}(w) = \{(a, b) \in \mathcal{A} \times \mathcal{A} : awb \in \mathcal{L}(\mathbf{u})\}$. The *extension graph* of w in $\mathcal{L}(\mathbf{u})$, denoted $\mathcal{E}_\mathbf{u}(w)$, is the undirected bipartite graph whose set of vertices is the disjoint union of $L_\mathbf{u}(w)$ and $R_\mathbf{u}(w)$ and with edges the elements of $B_\mathbf{u}(w)$. A sequence $\mathbf{u}$ (resp. a language $\mathcal{L}(\mathbf{u})$) is said to be *eventually dendric* with *threshold* $m \geq 0$ if $\mathcal{E}_\mathbf{u}(w)$ is a tree for every word $w \in \mathcal{L}(\mathbf{u})$ of length at least m. It is said to be *dendric* if we can choose $m = 0$. Dendric languages were introduced in [5] under the name of tree sets. It is known that Sturmian sequences are dendric.

Example 1. It is known that the sequence $\mathbf{u}_f = \mathsf{abaababaabaababaababaa} \cdots$, obtained as fixed point of the morphism $f : \mathsf{a} \mapsto \mathsf{ab}, \mathsf{b} \mapsto \mathsf{a}$, is Sturmian (see [17]).

3 Languages of Balanced Sequences

In 2000, Hubert [14] characterised balanced sequences over alphabets of higher cardinality. A suitable tool for their description is the notion of constant gap.

Definition 1. *A sequence* **y** *over an alphabet* $\mathcal{A}$ *is a* constant gap sequence *if for each letter* $a \in \mathcal{A}$ *appearing in* **y** *there is a positive integer* d *such that the distance between successive occurrences of* a *in* **y** *is always* d.

Obviously, any constant gap sequence is periodic. Given a constant gap sequence **y**, we denote its minimal period length by $\text{Per}(\mathbf{y})$.

Example 2. The sequence $\mathbf{y} = (0102)^\omega$ is a constant gap sequence because the distance between consecutive 0s is always 2, while the distance between consecutive 1s (resp. 2s) is always 4. Its minimal period is $\text{Per}(\mathbf{y}) = 4$.

The sequence $(011)^\omega$ is periodic but it is not a constant gap sequence.

The i-th *shift* of a constant gap sequence $\mathbf{y} = (y_0 y_1 \cdots y_{k-1})^\omega$ with minimal period $k \geq 1$ (and $0 \leq i < k$) is the sequence $\sigma^i(\mathbf{y}) = (y_i \cdots y_{k-1} y_0 \cdots y_{i-1})^\omega$.

Example 3. Let **y** be the sequence seen in Example 2. Then we have $\sigma^0(\mathbf{y}) = \mathbf{y}$, $\sigma(\mathbf{y}) = (1020)^\omega$, $\sigma^2(\mathbf{y}) = (0201)^\omega$ and $\sigma^3(\mathbf{y}) = (2010)^\omega$.

Theorem 1 ([14]). *A recurrent aperiodic sequence* **v** *is balanced if and only if* **v** *is obtained from a Sturmian sequence* **u** *over* $\{a, b\}$ *by replacing the* a*s in* **u** *by a constant gap sequence* **y** *over some alphabet* $\mathcal{A}$, *and replacing the* b*s in* **u** *by a constant gap sequence* **y**′ *over some alphabet* $\mathcal{B}$ *disjoint from* $\mathcal{A}$.

Definition 2. *Let* **u** *be a Sturmian sequence over the alphabet* $\{a, b\}$, *and* **y**, **y**′ *be two constant gap sequences over two disjoint alphabets* $\mathcal{A}$ *and* $\mathcal{B}$. *The* colouring *of* **u** *by* **y** *and* **y**′, *denoted* $\mathbf{v} = \text{colour}(\mathbf{u}, \mathbf{y}, \mathbf{y}')$, *is the sequence over* $\mathcal{A} \cup \mathcal{B}$ *obtained by the procedure described in Theorem 1.*

For $\mathbf{v} = \text{colour}(\mathbf{u}, \mathbf{y}, \mathbf{y}')$ we use the notation $\pi(\mathbf{v}) = \mathbf{u}$ and $\pi(v) = u$ for any $v \in \mathcal{L}(\mathbf{v})$ and the corresponding $u \in \mathcal{L}(\mathbf{u})$. Symmetrically, given a word $u \in \mathcal{L}(\mathbf{u})$, we denote by $\pi^{-1}(u) = \{v \in \mathcal{L}(\mathbf{v}) : \pi(v) = u\}$. We say that **u** (resp. u) is a *projection* of **v** (resp. v). The map $\pi : \mathcal{L}(\mathbf{v}) \to \mathcal{L}(\mathbf{u})$ is clearly a morphism.

Example 4. Let $\mathbf{u}_f$ be as in Example 1. Let us take the constant gap sequences $\mathbf{y} = (0102)^\omega$ and $\mathbf{y}' = (34)^\omega$ over the alphabets $\mathcal{A} = \{0, 1, 2\}$ and $\mathcal{B} = \{3, 4\}$ respectively. The sequence $\mathbf{v}_f = \text{colour}(\mathbf{u}_f, \mathbf{y}, \mathbf{y}') = 0310423014023041032401 \cdots$ is balanced according to Theorem 1. One has $\pi(\mathbf{v}_f) = \mathbf{u}_f$. Moreover, $\pi(031) = \pi(041) = aba$, and $\pi^{-1}(aba) = \{031, 032, 041, 042, 130, 140, 230, 240\}$.

Definition 3. *An aperiodic sequence* **u** *over* $\{a, b\}$ *has* well distributed occurrences, *or has the* WDO *property, if for every* $m \in \mathbb{N}$ *and for every* $w \in \mathcal{L}(\mathbf{u})$ *one has* $\{\mathbf{V}(p) \bmod m : pw \text{ is a prefix of } \mathbf{u}\} = \mathbb{Z}_m^2$.

It is known that Sturmian sequences have the WDO property (see [1]).

Example 5. Let $\mathbf{u}_f$ be as in Example 1 and let us consider $m = 2$ and $w = ab \in \mathcal{L}(\mathbf{u}_f)$. Then it is easy to check that $\mathbf{V}(\varepsilon) \equiv \binom{0}{0} \bmod 2$, $\mathbf{V}(aba) \equiv \binom{0}{1} \bmod 2$, $\mathbf{V}(abaab) \equiv \binom{1}{1} \bmod 2$ and $\mathbf{V}(abaababa) \equiv \binom{1}{1} \bmod 2$, where $w, abaw, abaabw$ and $abaababaw$ are prefixes of $\mathbf{u}_f$.

Using the WDO property we can prove that, to study the language of aperiodic recurrent balanced sequences, it is enough to study standard Sturmian sequences.

Proposition 1. *Let* $\mathbf{u}, \mathbf{u}'$ *be two Sturmian sequences such that* $\mathcal{L}(\mathbf{u}) = \mathcal{L}(\mathbf{u}')$, $\mathbf{y}$ *and* $\mathbf{y}'$ *two constant gap sequences over disjoint alphabets and* $i, j \in \mathbb{N}$. *Let* $\mathbf{v} = \mathrm{colour}(\mathbf{u}, \mathbf{y}, \mathbf{y}')$, $\mathbf{v}' = \mathrm{colour}(\mathbf{u}', \mathbf{y}, \mathbf{y}')$ *and* $\mathbf{v}'' = \mathrm{colour}(\mathbf{u}, \sigma^i(\mathbf{y}), \sigma^j(\mathbf{y}'))$. *Then* $\mathcal{L}(\mathbf{v}) = \mathcal{L}(\mathbf{v}') = \mathcal{L}(\mathbf{v}'')$.

Proof. Let $v \in \mathcal{L}(\mathbf{v})$ and w such that wv is a prefix of $\mathbf{v}$. Then $\pi(w)\pi(v)$ is a prefix of $\mathbf{u}$ and $|\pi(w)|$ is an occurrence of $\pi(v)$ in $\mathbf{u}$. Since $\pi(v) \in \mathcal{L}(\mathbf{u}')$, using the WDO property, we can find $p \in \mathcal{L}(\mathbf{u}')$ such that $p\pi(v)$ is a prefix of $\mathbf{u}'$ and $V(\pi(w)) = V(p) \bmod \mathrm{Per}(\mathbf{y})\mathrm{Per}(\mathbf{y}')$. Thus v appears both in $\mathbf{v}$ at occurrence $|\pi(w)|$ and in $\mathbf{v}'$ at occurrence $|p|$. Hence $\mathcal{L}(\mathbf{v}) \subset \mathcal{L}(\mathbf{v}')$. Using the same argument we can prove the opposite inclusion.

Let p be a prefix of $\mathbf{u}$ such that $V(p) = \binom{i}{j} \bmod \mathrm{Per}(\mathbf{y})\mathrm{Per}(\mathbf{y}')$. Denote $\mathbf{u}'' = p^{-1}\mathbf{u}$. Then $\mathrm{colour}(\mathbf{u}'', \sigma^i(\mathbf{y}), \sigma^j(\mathbf{y}'))$ gives the same sequence as the one obtained by erasing the prefix of length $|p|$ from $\mathbf{v}$. Since $\mathcal{L}(\mathbf{u}) = \mathcal{L}(\mathbf{u}'')$, using the same argument as before we have $\mathcal{L}(\mathbf{v}'') = \mathcal{L}(\mathbf{v})$.

Corollary 1. *Let* $\mathbf{v} = \mathrm{colour}(\mathbf{u}, \mathbf{y}, \mathbf{y}')$ *and* $v \in \mathcal{L}(\mathbf{v})$. *For any* i, j *such that* $0 \leq i < \mathrm{Per}(\mathbf{y})$ *and* $0 \leq j < \mathrm{Per}(\mathbf{y}')$, *the word* v' *obtained from* $\pi(v)$ *by replacing the* $\mathtt{a}$s *by* $\sigma^i(\mathbf{y})$ *and the* $\mathtt{b}$s *by* $\sigma^j(\mathbf{y}')$ *is in* $\pi^{-1}(\pi(v))$, *and thus in* $\mathcal{L}(\mathbf{v})$.

Example 6. Let $\mathbf{u}_f, \mathbf{v}_f, \mathbf{y}$ and $\mathbf{y}'$ be as in Example 4. Let $v = 03104 \in \mathcal{L}(\mathbf{v}_f)$ and let us denote $u = \pi(v) = \mathtt{abaab}$. One can easily check that the word $v' = 24013$ obtained from u by replacing the $\mathtt{a}$s by $\sigma^3(\mathbf{y})$ and the $\mathtt{b}$s by $\sigma(\mathbf{y}')$ is in $\mathcal{L}(\mathbf{v}_f)$.

Note that, if $\mathbf{v} = \mathrm{colour}(\mathbf{u}, \mathbf{y}, \mathbf{y}')$, there exists an $m \in \mathbb{N}$ such that every factor $v \in \mathcal{L}(\mathbf{v})$ longer than m contains at least $\mathrm{Per}(\mathbf{y})$ letters in $\mathcal{A}$ and at least $\mathrm{Per}(\mathbf{y}')$ letters in $\mathcal{B}$. Indeed, it is enough to find m such that all factors of length m in $\mathcal{L}(\mathbf{u})$ contain at least $\mathrm{Per}(\mathbf{y})$ $\mathtt{a}$s and at least $\mathrm{Per}(\mathbf{y}')$ $\mathtt{b}$s.

Example 7. Let $\mathbf{u}_f, \mathbf{v}_f, \mathbf{y}$ and $\mathbf{y}'$ be as in Example 4. Then, it easy to check that all factors of length 7 in $\mathcal{L}(\mathbf{u}_f)$ contain at least 4 $\mathtt{a}$s and 2 $\mathtt{b}$s. Thus, all factors of length 7 in $\mathcal{L}(\mathbf{v}_f)$ contain at least four letters in $\mathcal{A}$ and at least two letters in $\mathcal{B}$. On the other hand, $\mathtt{babaab} \in \mathcal{L}(\mathbf{u}_f)$ has length 6 and contains only three $\mathtt{a}$s.

As we saw in Example 4, the set $\pi^{-1}(u)$, for a word $u \in \mathcal{L}(\mathbf{u})$, is not in general a singleton. However, it is not difficult to prove that any long enough factor in $\mathbf{v}$ is uniquely determined, between the words having the same projection in $\mathbf{u}$, by the first $\mathrm{Per}(\mathbf{y})$ letters in $\mathcal{A}$ and the first $\mathrm{Per}(\mathbf{y}')$ letters in $\mathcal{B}$ (the number of needed letters can be reduced by studying the bispecial factors in $\mathbf{y}$ and $\mathbf{y}'$).

Lemma 1. *Let* $\mathbf{v} = \mathrm{colour}(\mathbf{u}, \mathbf{y}, \mathbf{y}')$ *and* $u \in \mathcal{L}(\mathbf{u})$ *such that* $|u|_a \geq \mathrm{Per}(\mathbf{y})$ *and* $|u|_b \geq \mathrm{Per}(\mathbf{y}')$. *Let* $a_0, a_1, \ldots, a_{\mathrm{Per}(\mathbf{y})-1} \in \mathcal{A}$, *and* $b_0, b_1, \ldots, b_{\mathrm{Per}(\mathbf{y}')-1} \in \mathcal{B}$. *There exists at most one word in* $\pi^{-1}(u)$ *having* $a_0, a_1, \ldots, a_{\mathrm{Per}(\mathbf{y})} - 1$ *(in this order) as first letters in* $\mathcal{A}$ *and* $b_0, b_1, \ldots, b_{\mathrm{Per}(\mathbf{y}')}$ *(in this order) as first letters in* $\mathcal{B}$.

Example 8. Let $\mathbf{u}_f, \mathbf{v}_f, \mathbf{y}$ and $\mathbf{y}'$ be as in Example 4 and $u =$ abaabaaba $\in$ $\mathcal{L}(\mathbf{u}_f)$. One has $|u|_a = 6 > \mathrm{Per}(\mathbf{y})$ and $|u|_b = 3 > \mathrm{Per}(\mathbf{y}')$. One can check that the only word in $\pi^{-1}(u)$ having $0, 2, 0, 1$ as first letters in $\mathcal{A}$ and $4, 3$ as first letters in $\mathcal{B}$ is 042031042, that is the word obtained from u by $\sigma^2(\mathbf{y})$ and $\sigma(\mathbf{y}')$. On the other hand, no word in $\mathcal{L}(\mathbf{v})$ can have $0, 0, 1, 2$ (in this order) as first letters in $\mathcal{A}$ or $3, 3$ as first letters in $\mathcal{B}$.

Putting together Corollary 1 and Lemma 1, we obtain the following result.

Lemma 2. *Let* $\mathbf{v} = \mathrm{colour}(\mathbf{u}, \mathbf{y}, \mathbf{y}')$ *and* $u \in \mathcal{L}(\mathbf{u})$ *be such that* $|u|_a \geq \mathrm{Per}(\mathbf{y})$ *and* $|u|_b \geq \mathrm{Per}(\mathbf{y}')$. *Then* $\#(\pi^{-1}(u)) = \mathrm{Per}(\mathbf{y})\mathrm{Per}(\mathbf{y}')$.

Example 9. Let $\mathbf{u}_f, \mathbf{v}_f, \mathbf{y}, \mathbf{y}'$ be as in Example 4 and $u =$ abaabaab $\in$ $\mathcal{L}(\mathbf{u}_f)$. The set $\pi^{-1}(u) = \{03104203, 03204103, 04103204, 04203104, 13024013,$ $14023014, 23014023, 24013024\}$ has exactly 8 elements, according to Lemma 2.

The following result easily follows from Lemma 1 and the WDO property.

Lemma 3. *Let* $\mathbf{v} = \mathrm{colour}(\mathbf{u}, \mathbf{y}, \mathbf{y}')$ *and* $u \in \mathcal{L}(\mathbf{u})$ *be such that* $|u|_a \geq \mathrm{Per}(\mathbf{y})$ *and* $|u|_b \geq \mathrm{Per}(\mathbf{y}')$. *Let* $v \in \pi^{-1}(u)$. *Then* v *is right special (resp. left special) if and only if* u *is right special (resp. left special). Moreover, in this case the unique two right (resp. left) extensions of* v *belong to different alphabets* $\mathcal{A}$ *and* $\mathcal{B}$.

Proposition 2. *The language* $\mathcal{L}(\mathbf{v})$ *is eventually dendric.*

Proof. Let m be a positive integer such that for every word $w \in \mathcal{L}(\mathbf{u})$ of length at least m one has $|w|_a \geq \mathrm{Per}(\mathbf{y})$ and $|w|_b \geq \mathrm{Per}(\mathbf{y}')$. Let $v \in \mathcal{L}(\mathbf{v})$ and $u = \pi(v)$, and suppose that $|v| \geq m$. It easily follows from Lemmata 1 and 3 that $\mathcal{E}_\mathbf{v}(u)$ is isomorphic to $\mathcal{E}_\mathbf{u}(u)$ via the projection π. Since $\mathbf{u}$ is Sturmian, then $\mathcal{L}(\mathbf{u})$ is dendric. Thus $\mathcal{E}_\mathbf{v}(v)$ is a tree. Hence $\mathcal{L}(\mathbf{v})$ is eventually dendric of threshold m.

The following result easily follows from Lemma 2.

Proposition 3. *Let* $\mathbf{v} = \mathrm{colour}(\mathbf{u}, \mathbf{y}, \mathbf{y}')$ *and* m *be a positive integer such that every word in* $\mathcal{L}(\mathbf{u})$ *of length* m *has at least* $\mathrm{Per}(\mathbf{y})$ *as and at least* $\mathrm{Per}(\mathbf{y}')$ *bs. Then for any* $n \geq m$ *one has* $\mathcal{C}_\mathbf{v}(n) = \mathrm{Per}(\mathbf{y})\mathrm{Per}(\mathbf{y}')(n+1)$.

Example 10. Let $\mathbf{u}_f, \mathbf{v}_f, \mathbf{y}$ and $\mathbf{y}'$ be as in Example 4. The language $\mathcal{L}(\mathbf{v}_f)$ is eventually dendric with threshold 7. The factor complexity of $\mathbf{v}_f$ is defined by $\mathcal{C}_{\mathbf{v}_f}(n) = 8(n+1)$ for every $n \geq 7$, according to Proposition 3.

Proposition 4. *Let* $\mathbf{v} = \mathrm{colour}(\mathbf{u}, \mathbf{y}, \mathbf{y}')$ *and* $v \in \mathcal{L}(\mathbf{v})$ *such that* $|\pi(v)|_a \geq$ $\mathrm{Per}(\mathbf{y})$ *and* $|\pi(v)|_b \geq \mathrm{Per}(\mathbf{y}')$. *Then* $\#(\mathcal{R}_\mathbf{v}(v)) = 1 + \mathrm{Per}(\mathbf{y})\mathrm{Per}(\mathbf{y}')$.

Proof. From Proposition 3 we have $s_\mathbf{v}(n) = \mathrm{Per}(\mathbf{y})\mathrm{Per}(\mathbf{y}')$ for every n large enough. The result thus follows from Proposition 2 and [10, Theorem 7.3].

Corollary 2. *A recurrent aperiodic balanced sequence is uniformly recurrent.*

Proof. A recurrent language is uniformly recurrent if and only if the number of return words to a given word in the language is finite. The result then just follows from Proposition 4 and [10, Theorem 7.3].

Given a vector $b = \begin{pmatrix} b_1 \\ b_2 \end{pmatrix} \in \mathbb{N}^2$ and two periodic sequences $\mathbf{y}, \mathbf{y}'$, we use the notation $b \bmod \boldsymbol{Per}(\mathbf{y}, \mathbf{y}') := \begin{pmatrix} b_1 \bmod \mathrm{Per}(\mathbf{y}) \\ b_2 \bmod \mathrm{Per}(\mathbf{y}') \end{pmatrix}$.

Lemma 4. *Let* $\mathbf{v} = \mathrm{colour}(\mathbf{u}, \mathbf{y}, \mathbf{y}')$, $u \in \mathcal{L}(\mathbf{u})$ *with* $|u|_a \geq \mathrm{Per}(\mathbf{y})$, $|u|_b \geq \mathrm{Per}(\mathbf{y}')$ *and* $v, w \in \mathcal{L}(\mathbf{v})$ *such that* $\pi(v) = \pi(w) = u$. *Let* i, j *be occurrences of* v *and* w *in* $\mathbf{v}$ *respectively and let us assume that* $i < j$. *Then* $v = w$ *if and only if* $\boldsymbol{V}(\mathbf{u}_{[i,j)}) = \begin{pmatrix} 0 \\ 0 \end{pmatrix} \bmod \boldsymbol{Per}(\mathbf{y}, \mathbf{y}')$.

Proof. By Lemma 1, $v = w$ if and only if there exist $0 \leq s < \mathrm{Per}(\mathbf{y})$ and $0 \leq t < \mathrm{Per}(\mathbf{y}')$ such that both v and w are obtained from u by replacing the as by $\sigma^s(\mathbf{y})$ and the bs by $\sigma^t(\mathbf{y}')$. Furthermore, in this case we have $\boldsymbol{V}(\mathbf{u}_{[0,i)}) = \boldsymbol{V}(\mathbf{u}_{[0,j)})$ $\bmod \boldsymbol{Per}(\mathbf{y}, \mathbf{y}')$, that is $\boldsymbol{V}(\mathbf{u}_{[i,j)}) = \begin{pmatrix} 0 \\ 0 \end{pmatrix} \bmod \boldsymbol{Per}(\mathbf{y}, \mathbf{y}')$. $\square$

Lemma 5. *Let* $\mathbf{v} = \mathrm{colour}(\mathbf{u}, \mathbf{y}, \mathbf{y}')$, $u \in \mathcal{L}(\mathbf{u})$ *with* $|u|_a \geq \mathrm{Per}(\mathbf{y})$, $|u|_b \geq \mathrm{Per}(\mathbf{y}')$ *and* $v, w \in \mathcal{L}(\mathbf{v})$ *with* $\pi(v) = \pi(w) = u$. *Then* $\pi(\mathcal{R}_{\mathbf{v}}(v)) = \pi(\mathcal{R}_{\mathbf{v}}(w))$.

Proof. Let $r \in \mathcal{R}_{\mathbf{v}}(v)$. Then u is both a prefix and a suffix of $\pi(rv)$. By Lemma 1 there exist a unique $0 \leq s < \mathrm{Per}(\mathbf{y})$ and a unique $0 \leq t < \mathrm{Per}(\mathbf{y}')$ such that w is obtained from u by replacing the as by $\sigma^s(\mathbf{y})$ and the bs by $\sigma^t(\mathbf{y}')$. By Corollary 1 the word obtained from $\pi(rv)$ by replacing the as by $\sigma^s(\mathbf{y})$ and the bs by $\sigma^t(\mathbf{y}')$ is in $\mathcal{L}(\mathbf{v})$ and has w as a prefix. This factor is equal to $r'w$ and it contains only two occurrences of w. Indeed, it follows from Lemma 4 that $\pi(r') = \pi(r)$ is the unique non-empty prefix of $\pi(rv)$ satisfying $\boldsymbol{V}(\pi(r)) = \begin{pmatrix} 0 \\ 0 \end{pmatrix} \bmod \boldsymbol{Per}(\mathbf{y}, \mathbf{y}')$. Thus, r' is a return word to w with $\pi(r') = \pi(r)$, which implies $\pi(\mathcal{R}_{\mathbf{v}}(v)) \subset \pi(\mathcal{R}_{\mathbf{v}}(w))$. The opposite inclusion can be proved symmetrically. $\square$

4   Critical Exponent and Its Relation to Return Words

Let $z \in \mathcal{A}^+$ be a prefix of a periodic sequence u^ω with $u \in \mathcal{A}^+$. We say that z has *fractional root* u and the *exponent* $e = |z|/|u|$. We usually write $z = u^e$. Let us emphasise that a word z can have multiple exponents and fractional roots.

Definition 4. *Given a sequence* $\mathbf{u}$ *and* $u \in \mathcal{L}(\mathbf{u})^+$, *we define the* index *of* u *in* $\mathbf{u}$ *as* $\mathrm{ind}_{\mathbf{u}}(u) = \sup\{e \in \mathbb{Q} : u^e \in \mathcal{L}(\mathbf{u})\}$. *The* critical exponent *of a sequence* $\mathbf{u}$ *is defined as* $E(\mathbf{u}) = \sup\{\mathrm{ind}_{\mathbf{u}}(u) : u \in \mathcal{L}(\mathbf{u})^+\}$. *Its* asymptotic critical exponent *is defined as* $E^*(\mathbf{u}) = \lim_{n \to \infty} (\sup\{\mathrm{ind}_{\mathbf{u}}(u) : u \in \mathcal{L}(\mathbf{u}), |u| \geq n\})$.

Clearly, $E(\mathbf{u}) \geq E^*(\mathbf{u})$. If $\mathbf{u}$ is eventually periodic, then both $E(\mathbf{u})$ and $E^*(\mathbf{u})$ are infinite. If $\mathbf{u}$ is aperiodic and uniformly recurrent, then each factor of $\mathbf{u}$ has finite index. Nevertheless, $E^*(\mathbf{u})$ may be infinite. An example of such a sequence is given by Sturmian sequences whose continued fraction expansions of their slope have unbounded partial quotients (see [9]).

Lemma 6. *Let u, w be non-empty factors of a recurrent sequence $\mathbf{u}$. If $u \in \mathcal{R}_{\mathbf{u}}(w)$, then $w = u^e$ for some $e \in \mathbb{Q}$. Moreover, if $\mathbf{u}$ is aperiodic and uniformly recurrent, then u is a return word to a finite number of factors in $\mathbf{u}$.*

Proof. Since $u \in \mathcal{R}_{\mathbf{u}}(w)$, w is a prefix of uw. Hence there exists $z \in \mathcal{L}(u)$ such that $uw = wz$. A known result from equations on words implies that there exist $x, y \in \mathcal{L}(u)$ and a non-negative integer i such that $u = xy$, $z = yx$ and $w = (xy)^i x$. Thus, w is a prefix of $u^\omega = (xy)^\omega$.

Let us now suppose that u is a return word to infinitely many factors. By the previous argument, u is a fractional root of all those factors. This implies that $u^n \in \mathcal{L}(u)$ for all $n \in \mathbb{N}$. Thus, $\mathbf{u}$ is either periodic or not uniformly recurrent. $\square$

Lemma 7 ([12]). *Let $\mathbf{u}$ be a uniformly recurrent aperiodic sequence and $f \in \mathcal{L}(\mathbf{u})^+$ such that $ind_{\mathbf{u}}(f) > 1$. Then there exist a factor $u \in \mathcal{L}(\mathbf{u})$ and a bispecial factor w in $\mathbf{u}$ such that $|f| = |u|$, $ind_{\mathbf{u}}(f) \leq ind_{\mathbf{u}}(u) = 1 + \frac{|w|}{|u|}$ and $u \in \mathcal{R}_{\mathbf{u}}(w)^+$.*

Proposition 5. *Let $\mathbf{u}$ be a uniformly recurrent aperiodic sequence. Let $(w_n)_{n \in \mathbb{N}}$ be a sequence of all bispecial factors ordered by their length. For every $n \in \mathbb{N}$, let v_n be a shortest return word to w_n in $\mathbf{u}$. Then*

$$E(\mathbf{u}) = 1 + \sup_{n \in \mathbb{N}} \left\{ \frac{|w_n|}{|v_n|} \right\} \qquad and \qquad E^*(\mathbf{u}) = 1 + \limsup_{n \to \infty} \frac{|w_n|}{|v_n|}.$$

Proof. By Lemma 6, $v_n w_n = v_n^{e_n}$ for some exponent $e_n \in \mathbb{Q}$ and thus $ind_{\mathbf{u}}(v_n) \geq e_n = \frac{|v_n w_n|}{|v_n|} = 1 + \frac{|w_n|}{|v_n|}$. Hence $E(\mathbf{u}) \geq 1 + \sup\{\frac{|w_n|}{|v_n|}\} > 1$. By the second statement of the same lemma, $\lim\limits_{n \to \infty} |v_n| = \infty$. Therefore, $E^*(\mathbf{u}) \geq 1 + \limsup \frac{|w_n|}{|v_n|} \geq 1$.

To show the opposite inequality, let $\delta > 0$ be such that $E(\mathbf{u}) - \delta > 1$. Thus there exists $f \in \mathcal{L}(\mathbf{u})$ satisfying $E(\mathbf{u}) - \delta < ind_{\mathbf{u}}(f)$. Using Lemma 7, we find $u \in \mathcal{L}(\mathbf{u})$ and a bispecial factor w such that $ind_{\mathbf{u}}(f) \leq ind_{\mathbf{u}}(u) = 1 + \frac{|w|}{|u|}$, where $u \in \mathcal{R}_{\mathbf{u}}(w)^+$. Therefore, for some index $m \in \mathbb{N}$, one has $w = w_m$ and $|u| \geq |v_m|$. Altogether, for arbitrarily positive δ we have

$$E(\mathbf{u}) - \delta < ind_{\mathbf{u}}(f) \leq ind_{\mathbf{u}}(u) = 1 + \frac{|w|}{|u|} \leq 1 + \frac{|w_m|}{|v_m|} \leq 1 + \sup\left\{ \frac{|w_n|}{|v_n|} \right\}.$$

Consequently, $E(\mathbf{u}) \leq 1 + \sup\left\{\frac{|w_n|}{|v_n|}\right\}$.

If $E^*(\mathbf{u}) = 1$, then the above proven inequality $E^*(\mathbf{u}) \geq 1 + \limsup \frac{|w_n|}{|v_n|} \geq 1$ implies the second statement of the proposition. If $E^*(\mathbf{u}) > 1$, then there exists a sequence of factors $f^{(n)} \in \mathcal{L}(\mathbf{u})$ with $ind_{\mathbf{u}}(f^{(n)}) > 1$ such that $|f^{(n)}| \to {}'\infty$ and $ind_{\mathbf{u}}(f^{(n)}) \to E^*(\mathbf{u})$. For each n, we find the factor $u^{(n)}$ and the bispecial factor $w^{(n)}$ with the properties given in Lemma 7 and we proceed analogously as before. $\square$

5   Asymptotic Critical Exponent of Balanced Sequences

To describe the asymptotic critical exponent of a balanced sequence, we first list important facts on Sturmian sequences. They are partially taken from [12], where they are used to compute the critical exponent of complementary symmetric Rote sequences.

In the sequel, we use the characterisation of standard Sturmian sequences by their directive sequences. To introduce them, we define two morphisms $G : \mathtt{a} \to \mathtt{ba}, \mathtt{b} \to \mathtt{b}$ and $D : \mathtt{a} \to \mathtt{a}, \mathtt{b} \to \mathtt{ab}$.

Proposition 6 ([15]). *For every standard Sturmian sequence $\mathbf{u}$ there exists a unique sequence $\mathbf{\Delta} = \Delta_0 \Delta_1 \Delta_2 \cdots \in \{G, D\}^{\mathbb{N}}$ of morphisms and a sequence $(\mathbf{u}^{(n)})_{n \geq 0}$ of standard Sturmian sequences such that $\mathbf{u} = \Delta_0 \Delta_1 \cdots \Delta_{n-1}(\mathbf{u}^{(n)})$ for every $n \in \mathbb{N}$. Moreover, the sequence $\mathbf{\Delta}$ contains infinitely many letters G and infinitely many letters D, i.e., for some sequence $(a_i)_{i \geq 1}$ of positive integers we can write $\mathbf{\Delta} = G^{a_1} D^{a_2} G^{a_3} D^{a_4} \cdots$ or $\mathbf{\Delta} = D^{a_1} G^{a_2} D^{a_3} G^{a_4} \cdots$.*

We call the sequence $\mathbf{\Delta}$ in Proposition 6 the *directive sequence* of $\mathbf{u}$.

Let us fix the notation by adopting the following **convention**: To a standard Sturmian sequence $\mathbf{u}$ with directive sequence $\mathbf{\Delta} = G^{a_1} D^{a_2} G^{a_3} D^{a_4} \cdots$, we assign an irrational number $\theta \in (0, 1)$ having the continued fraction expansion $\theta = [a_0, a_1, a_2, a_3, \ldots]$ with $a_0 = 0$. The frequencies of the letters in the Sturmian sequence $\mathbf{u}$ are $\frac{\theta}{1+\theta}$ (for the least frequent letter) and $\frac{1}{1+\theta}$ (for the most frequent letter). For every $N \in \mathbb{N}$, we define the Nth *convergent* to θ as $\frac{p_N}{q_N}$ and the Nth convergent to $\frac{\theta}{1+\theta}$ as $\frac{P_N}{Q_N}$, where p_N, q_N, Q_N satisfy the following recurrence relation for all $N \geq 1$: $X_N = a_N X_{N-1} + X_{N-2}$, but they differ in their initial values: $p_{-1} = 1, p_0 = 0; q_{-1} = 0, q_0 = 1; Q_{-1} = Q_0 = 1$. This implies $p_N + q_N = Q_N$ for all $N \in \mathbb{N}$. Note that $\mathbf{u}$ has directive sequence $G^{a_1} D^{a_2} G^{a_3} D^{a_4} \cdots$ if and only if $\mathbf{u}$ after exchange of letters $\mathtt{a} \leftrightarrow \mathtt{b}$ has directive sequence $D^{a_1} G^{a_2} D^{a_3} G^{a_4} \cdots$.

By Vuillon's result [20], every factor of any Sturmian sequence has exactly two return words and its derived sequence is Sturmian as well. The Parikh vectors of the bispecial factors in $\mathbf{u}$ and the corresponding return words can be easily expressed using the convergents $\frac{p_N}{q_N}$ to θ. In the following proposition we order the bispecial factors in the Sturmian sequence by their length.

Proposition 7 ([12]). *Let $\theta = [a_0, a_1, a_2, a_3, \ldots]$ be the irrational number associated with a Sturmian sequence $\mathbf{u}$ and let us suppose that $\mathtt{b}$ is the most frequent letter. Let b be the nth bispecial factor of $\mathbf{u}$. Then there exists a unique pair $(N, m) \in \mathbb{N}^2$ with $0 \leq m < a_{N+1}$ such that $n = m + a_0 + a_1 + a_2 + \cdots + a_N$. The Parikh vectors of the most frequent return word r to b, of the least frequent return word s to b and of b itself are $V(r) = \binom{p_N}{q_N}$, $V(s) = \binom{m p_N + p_{N-1}}{m q_N + q_{N-1}}$ and $V(b) = V(r) + V(s) - \binom{1}{1}$. The irrational number associated with the derived sequence $\mathbf{d_u}(b)$ to b in $\mathbf{u}$ is $\theta' = [0, a_{N+1} - m, a_{N+2}, a_{N+3}, \ldots]$.*

We will describe how to compute the asymptotic critical exponent of the balanced sequence $\mathbf{v} = \text{colour}(\mathbf{u}, \mathbf{y}, \mathbf{y}')$ associated with a standard Sturmian sequence $\mathbf{u}$ with θ having an eventually periodic continued fraction expansion. Our main tool for computing $E^*(\mathbf{v})$ is Proposition 5. Thus we need to find for any bispecial factor of length $|b|$ in $\mathbf{v}$ the length $|v|$ of its shortest return word.

As stated in Lemma 2, if w is a bispecial factor of $\mathbf{v}$ and w is long enough, then there exist $\text{Per}(\mathbf{y})\text{Per}(\mathbf{y}')$ bispecial factors of the same length in $\mathbf{v}$, all of them having the same projection in $\mathbf{u}$, and this projection is bispecial in $\mathbf{u}$. By Lemma 5, the shortest return words to these bispecial factors of $\mathbf{v}$ have the same length. Therefore, we can consider only one representative. In the sequel, we denote by w_n a bispecial factor of $\mathbf{v}$ such that $b_n = \pi(w_n)$ is the nth bispecial factor in the Sturmian sequence $\mathbf{u}$, when these are ordered by length. We want to compute $\mathrm{E}^*(\mathbf{v}) = 1 + \limsup \frac{|w_n|}{|v_n|}$, where v_n is a shortest return word to w_n in $\mathbf{v}$. The fact that the continued fraction expansion of θ is eventually periodic enables us to split the sequence $(|w_n|/|v_n|)$ into a finite number of subsequences such that each of them has a finite limit. The largest limit of these subsequences is the searched $\mathrm{E}^*(\mathbf{v})$. To find a suitable partition of the index set $\mathbb{N}$ into a finite number of subsets of indices describing subsequences, we define an equivalence on $\mathbb{N}$. First we fix our notation: $\theta = [a_0, a_1, a_2, a_3, \dots] = [0, a_1 a_2 \dots a_h (z_0 z_1 \dots z_{M-1})^\omega]$. In particular, $a_i = z_j$, if $i > h$ and $i - 1 - h = j \mod M$.

Definition 5. *To any $n \in \mathbb{N}$ we assign a unique pair $(N, m) \in \mathbb{N}^2$ as in Proposition 7. Let (N_1, m_1) and (N_2, m_2) be assigned to the integers n_1 and n_2 respectively. We say that n_1 is equivalent to n_2 and write $n_1 \sim n_2$ if*

$$m_1 = m_2, \qquad\qquad N_1 = N_2 \mod M,$$
$$\binom{p_{N_1-1}}{q_{N_1-1}} = \binom{p_{N_2-1}}{q_{N_2-1}} \mod \boldsymbol{Per}(\mathbf{y}, \mathbf{y}'), \quad \binom{p_{N_1}}{q_{N_1}} = \binom{p_{N_2}}{q_{N_2}} \mod \boldsymbol{Per}(\mathbf{y}, \mathbf{y}').$$

Obviously, the above defined relation on $\mathbb{N}$ is an equivalence and there are only finitely many equivalence classes, say $C_1, C_2, \dots, C_T$. Now we can define subsequences of the sequence $(|w_n|/|v_n|)$: if $\#C_t = \infty$, then we insert $|w_n|/|v_n|$ into the tth subsequence for each $n \in C_t$. For each $n \in \mathbb{N}$, up to a finite number of exceptions, $|w_n|/|v_n|$ belongs to a subsequence. The number of subsequences is at most $Z \operatorname{Per}(\mathbf{y})^2 \operatorname{Per}(\mathbf{y}')^2$, where $Z = z_0 + z_1 + \dots + z_{M-1}$. We obtain thus the following algorithm computing the asymptotic critical exponent.

Algorithm for determining $\mathrm{E}^*(\mathbf{v})$, where $\mathbf{v} = \operatorname{colour}(\mathbf{u}, \mathbf{y}, \mathbf{y}')$:
Input: $\theta = [0, a_1 a_2 \cdots a_h (z_0 z_1 \cdots z_{M-1})^\omega]$, $\operatorname{Per}(\mathbf{y})$ and $\operatorname{Per}(\mathbf{y}')$.

Step 1. Find all infinite equivalence classes C_t introduced in Definition 5.
Step 2. For each class C_t
  - insert $|w_n|/|v_n|$ into the tth subsequence for each $n \in C_t$;
  - find the limit e_t of the tth subsequence.

Output: $\mathrm{E}^*(\mathbf{v}) = 1 +$ the maximum value among all limits e_t.

Proposition 5 and a thorough study of short return words provide a lower bound on the asymptotic critical exponent.

Theorem 2. *Let $\mathbf{u}$ be a Sturmian sequence, $\mathbf{y}, \mathbf{y}'$ two constant gap sequences and $\mathbf{v} = \operatorname{colour}(\mathbf{u}, \mathbf{y}, \mathbf{y}')$. Then $E(\mathbf{v}) \geq E^*(\mathbf{v}) \geq 1 + \frac{1}{\operatorname{Per}(\mathbf{y})\operatorname{Per}(\mathbf{y}')}$. Moreover, $E^*(\mathbf{v})$ depends only on $\operatorname{Per}(\mathbf{y})$ and $\operatorname{Per}(\mathbf{y}')$ (not on the structure of $\mathbf{y}$ and $\mathbf{y}'$).*

On one hand, the asymptotic critical exponent depends only on the length of the periods of $\mathbf{y}$ and $\mathbf{y}'$ and it does not depend on their structure, in contrast to the critical exponent. On the other hand, the asymptotic critical exponent depends on the preperiod of the continued fraction of θ, in contrast to the asymptotic critical exponent of the associated Sturmian sequence (see [16]).

Example 11. Let $\mathbf{v}$ be the balanced sequence given by the parameters $\theta = [0, \overline{2}]$, $\mathrm{Per}(\mathbf{y}) = 1$ and $\mathrm{Per}(\mathbf{y}') = 2$. One can check that $\mathrm{E}^*(\mathbf{v}) = 3 + \sqrt{2} \doteq 4.41$.

For the balanced sequence $\mathbf{v}'$ given by the parameters $\theta = [0, 1, \overline{2}]$, $\mathrm{Per}(\mathbf{y}) = 1$ and $\mathrm{Per}(\mathbf{y}') = 2$, one has $\mathrm{E}^*(\mathbf{v}') = 2 + \frac{\sqrt{2}}{2} \doteq 2.7$.

Table 1. The balanced sequences with the least critical exponent over alphabets of size d. We denote by question mark the conjectures that are not yet proved.

d	θ	$\mathbf{y}$	$\mathbf{y}'$	$\mathrm{E}(\mathbf{v})$	$\mathrm{E}^*(\mathbf{v})$
3	$[0, 1, \overline{2}]$	0^ω	$(12)^\omega$	$2 + \frac{1}{\sqrt{2}}$	$2 + \frac{1}{\sqrt{2}}$
4	$[0, \overline{1}]$	$(01)^\omega$	$(23)^\omega$	$1 + \frac{1+\sqrt{5}}{4}$	$1 + \frac{1+\sqrt{5}}{4}$
5	$[0, 1, \overline{2}]$	$(01)^\omega$	$(2324)^\omega$	$\frac{3}{2}$	$\frac{3}{2}$
6	$[0, 2, 1, 1, \overline{1, 1, 1, 2}]$	0^ω	$(123415321435)^\omega$	$\frac{4}{3}$	$\frac{4}{3}$
7	$[0, 1, 3, \overline{1, 2, 1}]$	$(01)^\omega$	$(234526432546)^\omega$	$\frac{5}{4}$	$\frac{5}{4}$
8	$[0, 3, 1, \overline{2}]$	$(01)^\omega$	$(234526732546237526432576)^\omega$	$\frac{6}{5} = 1.2$	$\frac{12+3\sqrt{2}}{14} \doteq 1.16$
9	$[0, 2, 3, \overline{2}]$	$(01)^\omega$	$(234567284365274863254768)^\omega$	$?\frac{7}{6} \doteq 1.167$	$1 + \frac{2\sqrt{2}-1}{14} \doteq 1.13$
10	$[0, 4, 2, \overline{3}]$	$(01)^\omega$	$(3456728496325476829436527 4869)^\omega$	$?\frac{8}{7} \doteq 1.14$	$1 + \frac{\sqrt{13}}{26} \doteq 1.139$

We used a program implemented by our student Daniela Opočenská computing the asymptotic critical exponent of balanced sequences $\mathbf{x}_d$ defined in [18] for $d \in \{3, 4, \ldots, 10\}$. The authors of [18] conjectured that the least critical exponent over an alphabet of cardinality d equals $\frac{d-2}{d-3}$ and this is achieved on the sequences $\mathbf{x}_d$. This conjecture was proved for $d = 3$ and $d = 4$ in [18]. [1] Later, in [3, 4] it is shown that $\mathbf{x}_d$ are indeed the sequences with the least critical exponent over alphabets of size 5 to 8. The balanced sequences $\mathbf{x}_d$ and their critical exponent are listed in Table 1.

The table is taken from [18] (instead of the slope α of a Sturmian sequence, used in the original table, we use the parameter θ corresponding to the directive sequence). We also added to the table a column containing the asymptotic critical exponent. We see that $\mathrm{E}^*(\mathbf{x}_d) = \mathrm{E}(\mathbf{x}_d)$ for $d = 3, 4, 5, 6, 7$. However $\mathrm{E}^*(\mathbf{x}_8) < \mathrm{E}(\mathbf{x}_8)$. Moreover, using the table we can deduce that there exists a balanced sequence $\mathbf{x}$ over an 8-letter alphabet with $\mathrm{E}^*(\mathbf{x}) < \mathrm{E}^*(\mathbf{x}_8)$. The sequence $\mathbf{x}$ uses the same pair $\mathbf{y}$ and $\mathbf{y}'$ as $\mathbf{x}_8$. The parameter θ corresponding to $\mathbf{x}$ is $\theta = [0, 2, 3, \overline{2}]$. Since $\mathbf{x}_8$ and $\mathbf{x}_9$ have same θ and same lengths of constant gap sequences, we have $\mathrm{E}^*(\mathbf{x}) = \mathrm{E}^*(\mathbf{x}_9) < \mathrm{E}^*(\mathbf{x}_8)$. The method used for finding the candidates with the least critical exponent cannot be applied to find a suitable

---

[1] More precisely, the minimality in the case $d = 4$ was proved by Peltomäki in a private communication to Rampersad.

$\mathbf{x}_d$ for a general d. The same is true for the least asymptotic critical exponent. Indeed, even a proof that the candidates should be given by θ with an eventually periodic continued fraction expansion is still missing.

References

1. Balková, L., Bucci, M., De Luca, A., Hladký, J., Puzynina, S.: Aperiodic pseudorandom number generators based on infinite words. Theoret. Comput. Sci. **647**, 85–100 (2016)
2. Balková, L., Pelantová, E., Starosta, Š.: Sturmian jungle (or garden?) on multiliteral alphabets. RAIRO-Theoret. Inf. Appl. **44**, 443–470 (2010)
3. Baranwal, A.R.: Decision algorithms for Ostrowski-automatic sequences, the master thesis, University of Waterloo (2020). http://hdl.handle.net/10012/15845
4. Baranwal, A.R., Shallit, J.: Critical exponent of infinite balanced words via the Pell number system. In: Mercaş, R., Reidenbach, D. (eds.) WORDS 2019. LNCS, vol. 11682, pp. 80–92. Springer, Cham (2019). https://doi.org/10.1007/978-3-030-28796-2_6
5. Berthé, V., et al.: Acyclic, connected and tree sets. Monatshefte für Mathematik **176**(4), 521–550 (2014). https://doi.org/10.1007/s00605-014-0721-4
6. Carpi, A., de Luca, A.: Special factors, periodicity, and an application to Sturmian words. Acta Inform. **36**, 983–1006 (2000). https://doi.org/10.1007/PL00013299
7. Currie, J.D., Mol, L., Rampersad, N.: The repetition threshold for binary rich words. Discrete Math. Theoret. Comput. Sci. **22**(1) (2020). https://doi.org/10.23638/DMTCS-22-1-6. no. 6
8. Dejean, F.: Sur un théorème de Thue. J. Combin. Theory Ser. A **13**, 90–99 (1972)
9. Damanik, D., Lenz, D.: The index of Sturmian sequences. Eur. J. Comb. **23**, 23–29 (2002)
10. Dolce, F., Perrin, D.: Eventually dendric shift spaces. In: Ergodic Theory and Dynamical Systems, pp. 1–26 (2020)
11. Durand, F.: A characterization of substitutive sequences using return words. Discrete Math. **179**, 89–101 (1998)
12. Dvořáková, L., Medková, K., Pelantová, E.: Complementary symmetric Rote sequences: the critical exponent and the recurrence function. Discrete Math. Theoret. Comput. Sci. **20**(1) (2020). https://doi.org/10.23638/DMTCS-22-1-20. #20.
13. Hedlund, G.A., Morse, M.: Symbolic dynamics II - Sturmian trajectories. Am. J. Math. **62**, 1–42 (1940)
14. Hubert, P.: Suites équilibrées. Theoret. Comput. Sci. **242**, 91–108 (2000)
15. Justin, J., Pirillo, G.: Episturmian words and episturmian morphisms. Theoret. Comput. Sci. **276**, 281–313 (2002)
16. Justin, J., Pirillo, G.: Fractional powers in Sturmian words. Theoret. Comput. Sci. **223**, 363–376 (2001)
17. Pytheas Fogg, N.: Substitutions in Dynamics, Arithmetics and Combinatorics. Lecture Notes in Mathematics, vol. 313. Springer, Heidelberg (2002). Edited by Berthé, V., Ferenczi, S., Mauduit, C., Siegel, A
18. Rampersad, N., Shallit, J., Vandomme, É.: Critical exponents of infinite balanced words. Theoret. Comput. Sci. **777**, 454–463 (2019)
19. Shallit, J., Shur, A.: Subword complexity and power avoidance. Theoret. Comput. Sci. **792**, 96–116 (2019)
20. Vuillon, L.: A characterization of Sturmian words by return words. Eur. J. Comb. **22**, 263–275 (2001)
21. Vuillon, L.: Balanced words. Bull. Belgian Math. Soc. **10**, 787–805 (2003)

Completely Reachable Automata, Primitive Groups and the State Complexity of the Set of Synchronizing Words

Stefan Hoffmann[(✉)]

Informatikwissenschaften, FB IV, Universität Trier, Universitätsring 15,
54296 Trier, Germany
hoffmanns@informatik.uni-trier.de

Abstract. We give a new characterization of primitive permutation groups tied to the notion of completely reachable automata. Also, we introduce sync-maximal permutation groups tied to the state complexity of the set of synchronizing words of certain associated automata and show that they are contained between the 2-homogeneous and the primitive groups. Lastly, we define k-reachable groups in analogy with synchronizing groups and motivated by our characterization of primitive permutation groups. But the results show that a k-reachable permutation group of degree n with $6 \leqslant k \leqslant n - 6$ is either the alternating or the symmetric group.

Keywords: Finite automata · Synchronization · Completely reachable automata · Primitive permutation groups · State complexity

1   Introduction

A deterministic semi-automaton is synchronizing if it admits a reset word, i.e., a word which leads to some definite state, regardless of the starting state. This notion has a wide range of applications, from software testing, circuit synthesis, communication engineering and the like, see [20,23]. The famous Černý conjecture [8] states that a minimal length synchronizing word has length at most $(n-1)^2$ for an n state automaton. We refer to the mentioned survey articles for details [20,23]. An automaton is completely reachable, if for each subset of states we can find a word which maps the whole state set onto this subset. This is a generalization of synchronizability, as a synchronizing word maps the whole state set to a singleton set. The class of completely reachable automata was formally introduced in [5], but already in [9,15] such automata appear in the results. The time complexity of deciding if a given automaton is completely reachable is unknown. A sufficient and necessary criterion for complete reachability of a given automaton in terms of graphs and their connectivity is known [6], but it is not known if these graphs could be constructed in polynomial time. A special case

© Springer Nature Switzerland AG 2021
A. Leporati et al. (Eds.): LATA 2021, LNCS 12638, pp. 305–317, 2021.
https://doi.org/10.1007/978-3-030-68195-1_24

of the general graph construction, which gives a sufficient criterion for complete reachability [5], is known to be constructible in polynomial time [10].

The size of a minimal automaton accepting a given regular language is called the state complexity of that language. The set of synchronizing words of a given automaton is a regular ideal language whose state complexity is at most exponential in the size of the original automaton [15,16]. The Černý family of automata [8,23], the first given family yielding the lower bound $(n-1)^2$ for the length of synchronizing words, is completely reachable and the corresponding sets of synchronizing words have maximal state complexity [15,16].

The notion of primitive permutation groups could be traced back to work by Galois [17] on the solubility of equations by radicals. Nowadays, it is a core notion of the theory of permutation groups [7].

Outline and Contribution: In Sect. 2, we give definitions and state known results. Then, in Sect. 3, for completely reachable automata, we state a sufficient and necessary condition for the set of synchronizing words to have maximal state complexity, which yields a polynomial time decision procedure.

In Sect. 4, we introduce a new characterization of primitive permutation groups, motivated by work on synchronizing and completely reachable automata and on detecting properties of permutation groups by functions [3–5,9]. We relate this to the notion of the state complexity of the set of synchronizing words. Beside the Černý family [8,23], the properties that a minimal length synchronizing word has quadratic length and that the set of synchronizing words has maximal state complexity are shared by a wealth of different slowly synchronizing automata [1, 2,15,16]. Motivated by this, we introduce the class of sync-maximal permutation groups and show that they fit between the 2-homogeneous and the primitive groups.

Lastly, in Sect. 5 we introduce k-reachable groups motivated by our investigations and the definition of synchronizing groups [4]. We show that for almost all k, only the symmetric and alternating groups are k-reachable.

2   Preliminaries and Definitions

General Notions: Let $\Sigma = \{a_1, \ldots, a_k\}$ be a finite set of symbols, called an *alphabet*. By Σ^*, we denote the *set of all finite sequences*, i.e., of all words or strings. The *empty word*, i.e., the finite sequence of length zero, is denoted by ε. We set $\Sigma^+ = \Sigma^* \setminus \{\varepsilon\}$. For a given word $w \in \Sigma^*$, we denote by $|w|$ its *length*. The subsets of Σ^* are called *languages*.

For $n > 0$, we set $[n] = \{0, \ldots, n-1\}$ and $[0] = \varnothing$. For a set X, we denote the *power set* of X by $\mathcal{P}(X)$, i.e, the set of all subsets of X.

Any function $f : X \to Y$ induces a function $\hat{f} : \mathcal{P}(X) \to \mathcal{P}(Y)$ by setting $\hat{f}(Z) := \{f(z) \mid z \in Z\}$. Here, we will denote this extension also by f.

Let $k \geqslant 1$. A k-*subset* $Y \subseteq X$ is a finite set of cardinality k. A 1-set is also called a *singleton set*. For functions $f : A \to B$ and $g : B \to C$, the *functional composition* $gf : A \to C$ is the function $(gf)(x) = g(f(x))$, i.e., the function on the right is applied first[1].

Automata-Theoretic Notions: A *finite, deterministic* and *complete* automaton will be denoted by $\mathscr{A} = (\Sigma, Q, \delta, s_0, F)$ with $\delta : Q \times \Sigma \to Q$ the state transition function, Q a finite set of states, $s_0 \in Q$ the start state and $F \subseteq Q$ the set of final states. The properties of being deterministic and complete are implied by the definition of δ as a total function.

The transition function $\delta : Q \times \Sigma \to Q$ could be extended to a transition function on words $\delta^* : Q \times \Sigma^* \to Q$ by setting $\delta^*(s, \varepsilon) := s$ and $\delta^*(s, wa) := \delta(\delta^*(s, w), a)$ for $s \in Q$, $a \in \Sigma$ and $w \in \Sigma^*$. In the remainder we drop the distinction between both functions and will also denote this extension by δ. For $S \subseteq Q$ and $w \in \Sigma^*$, we write $\delta(S, w) = \{\delta(s, w) \mid s \in S\}$ and $\delta^{-1}(S, w) = \{q \in Q \mid \delta(q, w) \in S\}$.

The *language accepted*, or *recognized*, by $\mathscr{A} = (\Sigma, S, \delta, s_0, F)$ is $L(\mathscr{A}) = \{w \in \Sigma^* \mid \delta(s_0, w) \in F\}$. A language $L \subseteq \Sigma^*$ is called *regular* if $L = L(\mathscr{A})$ for some finite automaton $\mathscr{A}$.

For a language $L \subseteq \Sigma^*$ and $u, v \in \Sigma^*$ we define the *Nerode right-congruence* with respect to L by $u \equiv_L v$ if and only if $\forall x \in \Sigma : ux \in L \leftrightarrow vx \in L$. The equivalence class for some $w \in \Sigma^*$ is denoted by $[w]_{\equiv_L} := \{x \in \Sigma^* \mid x \equiv_L w\}$. A language is regular if and only if the above right-congruence has finite index, and it could be used to define the minimal deterministic automaton $\mathscr{A}_L = (\Sigma, Q, \delta, [\varepsilon]_{\equiv_L}, F)$ with $Q := \{[w]_{\equiv_L} \mid w \in \Sigma^*\}$, $\delta([w]_{\equiv_L}, a) := [wa]_{\equiv_L}$ for $a \in \Sigma$, $w \in \Sigma^*$ and $F := \{[w]_{\equiv_L} \mid w \in L\}$. It is indeed the smallest automaton accepting L in terms of states, and we will refer to this construction as the minimal automaton [11] of L. The *state complexity* of a regular language is defined as the number of Nerode right-congruence classes. We will denote this number by $sc(L)$.

Let $\mathscr{A} = (\Sigma, Q, \delta, s_0, F)$ be an automaton. A state $q \in Q$ is *reachable*, if $q = \delta(s_0, u)$ for some $u \in \Sigma^*$. We also say that a state q is reachable from a state q' if $q = \delta(q', u)$ for some $u \in \Sigma^*$. Two states q, q' are *distinguishable*, if there exists $u \in \Sigma^*$ such that either $\delta(q, u) \in F$ and $\delta(q', u) \notin F$ or $\delta(q, u) \notin F$ and $\delta(q', u) \in F$. An automaton for a regular language is isomorphic to the minimal automaton if and only if all states are reachable and distinguishable [11].

A *semi-automaton* $\mathscr{A} = (\Sigma, Q, \delta)$ is like an ordinary automaton, but without a designated start state and without a set of final states. Sometimes, we will also call a semi-automaton simply an automaton if the context makes it clear what is meant. Also, definitions without explicit reference to a start state and a set of final states are also valid for semi-automata.

Let $\mathscr{A} = (\Sigma, Q, \delta)$ be a finite semi-automaton. A word $w \in \Sigma^*$ is called *synchronizing*, if $\delta(q, w) = \delta(q', w)$ for all $q, q' \in Q$, or equivalently $|\delta(Q, w)| = 1$. Set $\mathrm{Syn}(\mathscr{A}) = \{w \in \Sigma^* \mid |\delta(Q, w)| = 1\}$.

---

[1] In group theory, usually the other convention is adopted, but we stick to the convention most often seen in formal language theory.

The *power automaton (for synchronizing words)* associated to $\mathscr{A}$ is $\mathcal{P}_{\mathscr{A}} = (\Sigma, \mathcal{P}(Q), \delta, Q, F)$ with start state Q, final states $F = \{\{q\} \mid q \in Q\}$ and the transition function of $\mathcal{P}_{\mathscr{A}}$ is the transition function of $\mathscr{A}$, but applied to subsets of states. Then, as observed in [22], the automaton $\mathcal{P}_{\mathscr{A}}$ accepts the set of synchronizing words, i.e., $L(\mathcal{P}_{\mathscr{A}}) = \mathrm{Syn}(\mathscr{A})$. As for $\{q\} \in F$, we also have $\delta(\{q\}, x) \in F$ for each $x \in \Sigma^*$, the states in F could all be merged to a single state to get an accepting automaton for $\mathrm{Syn}(\mathscr{A})$. Also, the empty set is not reachable from Q. Hence $\mathrm{sc}(\mathrm{Syn}(\mathscr{A})) \leqslant 2^{|Q|} - |Q|$ and this bound is sharp [15,16].

We call $\mathscr{A}$ *completely reachable*, if for any non-empty $S \subseteq Q$, there exists a word $w \in \Sigma^*$ with $\delta(Q, w) = S$, i.e., in the power automaton, every state is reachable from the start state. When we say a *subset of states* in $\mathscr{A}$ is *reachable*, we mean reachability in $\mathcal{P}_{\mathscr{A}}$. The state complexity of $\mathrm{Syn}(\mathscr{A})$ is maximal, i.e., $\mathrm{sc}(\mathrm{Syn}(\mathscr{A})) = 2^{|Q|} - |Q|$, if and only if all subsets $S \subseteq Q$ with $|S| \geqslant 2$ are reachable and at least one singleton subset of Q, and all these states are distinguishable in $\mathcal{P}_{\mathscr{A}}$. For *strongly connected automata*, i.e., those for which all states are reachable from each other, the state complexity of $\mathrm{Syn}(\mathscr{A})$ is maximal iff $\mathscr{A}$ is completely reachable and all $S \subseteq Q$ with $|S| \geqslant 2$ are distinguishable in $\mathcal{P}_{\mathscr{A}}$.

Transformations and Permutation Groups: Let $n \geqslant 0$. Denote by $\mathcal{S}_n$ the *symmetric group on* $[n]$, i.e., the group of all permutations of $[n]$. A *permutation group (of degree n)* is a subgroup of $\mathcal{S}_n$. For $n > 1$, the *alternating group* is the unique subgroup of size $n!/2$ in $\mathcal{S}_n$, see [7]. The *orbit* of an element $i \in [n]$ for a permutation group G is the set $\{g(i) \mid g \in G\}$. A permutation group G over $[n]$ is *primitive*, if it preserves no non-trivial equivalence relation[2] on $[n]$, i.e., for no non-trivial equivalence relation $\sim \subseteq [n] \times [n]$ we have $p \sim q$ if and only if $g(p) \sim g(q)$ for all $g \in G$ and $p, q \in [n]$. A permutation group G over $[n]$ is called *k-homogeneous* for some $k \geqslant 1$, if for any two k-subsets S, T of Q, there exists $g \in G$ such that $g(S) = T$. A *transitive* permutation group is the same as a 1-homogeneous permutation group. Note that here, all permutation groups with $n \leqslant 2$ are primitive, and for $n > 2$ every primitive group is transitive. Because of this, some authors exclude the trivial group for $n = 2$ from being primitive. A permutation group G over $[n]$ is called *k-transitive* for some $k \geqslant 1$, if for any two tuples $(p_1, \ldots, p_k), (q_1, \ldots, q_k) \in [n]^k$, there exists $g \in G$ such that $(g(p_1), \ldots, g(p_k)) = (q_1, \ldots, q_k)$.

By $\mathcal{T}_n$, we denote the set of all maps on $[n]$. A submonoid of $\mathcal{T}_n$ for some n is called a *transformation monoid*. If the set U is a submonoid (or a subgroup) of $\mathcal{T}_n$ (or $\mathcal{S}_n$) we denote this by $U \leqslant \mathcal{T}_n$ (or $U \leqslant \mathcal{S}_n$). For a set $A \subseteq \mathcal{T}_n$ (or $A \subseteq \mathcal{S}_n$), we denote by $\langle A \rangle$ the submonoid (or the subgroup) generated by A. Let $\mathscr{A} = (\Sigma, Q, \delta)$ be an semi-automaton and for $w \in \Sigma^*$ define $\delta_w : Q \to Q$ by $\delta_w(q) = \delta(q, w)$ for all $q \in Q$. Then, we can associate with $\mathscr{A}$ the transformation monoid of the automaton $\mathcal{T}_{\mathscr{A}} = \{\delta_w \mid w \in \Sigma^*\}$, where we can identify Q with $[n]$ for $n = |Q|$. We have $\mathcal{T}_{\mathscr{A}} = \langle \{\delta_x \mid x \in \Sigma\} \rangle$. The *rank* of a map $f : [n] \to [n]$ is the cardinality of its image. For a given semi-automaton $\mathscr{A} = (\Sigma, Q, \delta)$, the *rank of a word* $w \in \Sigma^*$ is the rank of δ_w.

---

[2] The trivial equivalence relations on $[n]$ are $[n] \times [n]$ and $\{(x, x) \mid x \in [n]\}$.

Known Results: The next result appears in [3] and despite it was never clearly spelled out by Rystsov himself, it is implicitly present in arguments used in [19].

Theorem 2.1 (Rystsov [3,19]). *A permutation group G on $[n]$ is primitive if and only if, for any map $f : [n] \to [n]$ of rank $n - 1$, the transformation monoid $\langle G \cup \{f\} \rangle$ contains a constant map.*

In [5] a sufficient criterion for complete reachability was given. It is based on the following graph construction associated to a semi-automaton.

Definition 2.2 (Bondar & Volkov [5]). *Let $\mathscr{A} = (\Sigma, Q, \delta)$. Then, we define the graph $\Gamma_1(\mathscr{A}) = (Q, E)$ with vertex set Q and edge set $E = \{(p, q) \mid \exists w \in \Sigma^* : p \notin \delta(Q, w), |\delta^{-1}(q, w)| = 2, w \text{ has rank } |Q| - 1\}$. For transformation monoids $M \leqslant \mathcal{T}_n$, a similar definition $\Gamma_1(M)$ applies.*

The construction was extended in [6] to give a sufficient and necessary criterion. The graph $\Gamma_1(\mathscr{A})$ could be computed in polynomial time [10].

Theorem 2.3 (Bondar & Volkov [5]). *Let $\mathscr{A} = (\Sigma, Q, \delta)$. If $\Gamma_1(\mathscr{A}) = (Q, E)$ is strongly connected, then $\mathscr{A}$ is completely reachable.*

3 General Results on the State Complexity of Syn($\mathscr{A}$)

The first result of this section will be needed later when we investigate the properties of the sync-maximal groups introduced in Sect. 4.2. But it could also be used for a polynomial time decision procedure to decide if, for a given completely reachable automaton, the set of synchronizing words has maximal state complexity.

Lemma 3.1. *Let $\mathscr{A}$ be completely reachable with n states. Then, $\mathrm{sc}(Syn(\mathscr{A})) = 2^n - n$ if and only if all 2-sets of states are pairwise distinguishable in $\mathcal{P}_{\mathscr{A}}$.*

Hence, we only need to check for all pair states $\{p, q\}$ with $p \neq q$ in the power automaton if they are all distinguishable to each other. This could be done in polynomial time.

Corollary 3.2. *Let $\mathscr{A} = (\Sigma, Q, \delta)$ be a completely reachable semi-automaton with n states. Then, we can decide in polynomial time if $\mathrm{sc}(Syn(\mathscr{A})) = 2^n - n$.*

Next, we state a simple observation.

Lemma 3.3. *Let $\mathscr{A} = (\Sigma, Q, \delta)$ be a strongly connected semi-automaton. If $Syn(\mathscr{A})$ has maximal state complexity, then $\mathscr{A}$ is completely reachable.*

4   Permutation Groups and State Complexity of Syn($\mathscr{A}$)

In Sect. 4.1, we state characterizations of the k-homogeneous and of primitive permutation groups by inspection of the resulting transformation monoid when non-permutations are added. This is in the spirit of "detecting properties of (permutation groups) with functions" as presented in [3]. In Sect. 4.2, we introduce the sync-max permutation groups, defined by stipulating that the resulting semi-automaton, given by the generators of the permutation group and a non-permutation, is synchronizing and its set of synchronizing words has maximal state complexity. This definition is independent of the choice of generators for the group and we will show that the resulting class of groups is contained between the 2-homogeneous and the primitive groups.

4.1   Primitive and k-Homogeneous Permutation Groups

First, as a warm-up, we state two equivalent conditions to k-homogeneity. They are not hard to see, but seem to be unnoticed or at least never stated in the literature before.

Lemma 4.1. *Let $G \leqslant S_n$. The following conditions are equivalent:*

1. *G is k-homogeneous,*
2. *for any map of rank $n - k$, in $\langle G \cup \{f\} \rangle$ every subset of size $n - k$ is reachable,*
3. *there exists a map of rank $n - k$ such that in $\langle G \cup \{f\} \rangle$ every subset of size $n - k$ is reachable.*

Remark 1. As a group is k-homogeneous if and only if it is $(n-k)$-homogeneous, we could also add that k-homogeneity is equivalent to the property that adding any (or some) function whose image contains precisely k elements gives a transformation monoid in which every k-subset is reachable.

An automaton is synchronizing precisely if its transformation monoid contains a constant map. Hence, the next result is a strengthening of Theorem 2.1.

Theorem 4.2. *A finite permutation group $G \leqslant S_n$ with $n \geqslant 3$ is primitive if and only if, for any transformation $f \colon [n] \to [n]$ of rank $n - 1$, in the transformation semigroup $\langle G \cup \{f\} \rangle$ we find, for each non-empty $S \subseteq [n]$, an element $g \in \langle G \cup \{f\} \rangle$ such that $g([n]) = S$.*

Proof. Suppose G is a permutation group on $[n]$ that is not primitive. Then, $n \geqslant 3$ and by Theorem 2.1, for some $f \colon [n] \to [n]$ of rank $n - 1$ the group $\langle G \cup \{f\} \rangle$ does not contain a constant map. So, no element in $\langle G \cup \{f\} \rangle$ can map $[n]$ to a singleton subset of $[n]$.

Conversely, let G be a primitive permutation group on Ω. Take any transformation monoid M generated by G and a transformation of rank $n-1$. If (p, q) is an edge of the graph $\Gamma_1(M)$, then for each $g \in G$, the pair $(g(p), g(q))$ also constitutes an edge of $\Gamma_1(M)$. Since G is transitive, we see that every element in $[n]$

has an outgoing edge in $\Gamma_1(M)$. This clearly implies that $\Gamma_1(M)$ has a directed cycle, and by the definition of $\Gamma_1(M)$ this cycle is not a loop. Assume $\Gamma_1(M)$ is not strongly connected. The partition of $\Gamma_1(M)$ into strongly connected components induces a partition π of $[n]$ which is nontrivial, since $\Gamma_1(M)$ has a directed cycle which is not a loop. As G preserves the edges of $\Gamma_1(M)$, if we have a path between any two vertices, we also have a path between their images. Hence G respects π, which is not possible as G is primitive by assumption. So $\Gamma_1(M)$ must be strongly connected. Then M is completely reachable by Theorem 2.3. $\square$

Remark 2. For finite permutation groups, Theorem 2.1, Theorem 2.3, Higman's orbital graph characterization of primitivity and the fact that, for finite orbital graphs, connectedness implies strongly connectedness (please see [7] for these notions) could be used to give another proof of Theorem 4.2.

Remark 3. Theorem 4.2 could actually be strengthened[3] by assuming $f : [n] \to [n]$ to be idempotent. For let $G \leqslant S_n$ be primitive and suppose the statement is valid for idempotent transformations only. Then, let $f : [n] \to [n]$ be any transformation. Assume $f(a) = f(b)$ with $a, b \in [n]$. By Lemma 4.1, G is transitive. Hence, there exists $g \in G$ such that $a \notin g(f([n]))$ and gf permutes $[n] \setminus \{a\}$. Then, some power of gf acts as the identity on $[n] \setminus \{a\}$, i.e., is idempotent.

Because our main motivation comes from the theory of automata, let us state a variant of Theorem 4.2 formulated in terms of automata.

Corollary 4.3. *Let $n \geqslant 0$. Suppose $G = \langle g_1, \ldots, g_k \rangle \leqslant S_n$. Then G is primitive if and only if for every transformation $f : [n] \to [n]$ of rank $n - 1$, the semiautomaton $\mathscr{A} = (\Sigma, Q, \delta)$ with[4] $\Sigma = \{g_1, \ldots, g_k, f\}$, $Q = [n]$ and $\delta(i, g) = g(i)$ for $i \in Q$ and $g \in \Sigma$ is completely reachable.*

As a last consideration in this subsection, and in view of Theorem 4.2, let us derive a sufficient condition for "almost complete" reachability of all subsets of size strictly smaller than $n - 1$ when adding any function of rank $n - 2$. By Lemma 4.1, any such condition must imply 2-homogeneity.

Proposition 4.4. *Let $G \leqslant S_n$ and $n \geqslant 3$. Suppose the following holds true:*

For any 2-subset $\{a, b\} \subseteq [n]$ and any $A \subseteq [n]$ with $1 \leqslant |A| \leqslant n - 2$ and $c \in [n] \setminus \{a, b\}$ we find $g \in G$ such that $\{c\} \subseteq g(A) \subseteq [n] \setminus \{a, b\}$.

Then, for any function $f : [n] \to [n]$ of rank $n - 2$, in $\langle G \cup \{f\} \rangle$ every non-empty subset of size at most $n - 2$ is reachable. In particular, by Lemma 4.1, G is 2-homogenous.

---

[3] I am thankful to an anonymous referee for this observation.
[4] The elements of Σ are meant to be abstract symbols.

4.2   Sync-Maximal Permutation Groups

Here, we introduce the sync-maximal permutation groups. For their definition, we associate to a given group and a non-permutation a semi-automaton whose letters are generators of the group and the non-permutation. This definition is actually independent of the choice of generators of the group. But before giving the definition of sync-maximal groups, let us first state a result linking the notion of complete reachability to the state complexity of the set of synchronizing words.

Proposition 4.5. *Let* $G = \langle g_1, \ldots, g_k \rangle \leqslant S_n$ *be a permutation group and* $f \colon [n] \to [n]$ *be a non-permutation. Set* $\Sigma = \{g_1, \ldots, g_k, f\}$ *and* $\mathscr{A} = (\Sigma, [n], \delta)$ *with* $\delta(m, g) = g(m)$ *for* $m \in [n]$ *and* $g \in \Sigma$. *If* $n > 2$ *and* $\mathrm{sc}(\mathrm{Syn}(\mathscr{A})) = 2^n - n$, *then* G *is transitive and* $\mathscr{A}$ *completely reachable.*

Proof. Suppose $n > 2$. As $\mathrm{sc}(\mathrm{Syn}(\mathscr{A})) = 2^n - n$, in $\mathcal{P}_{\mathscr{A}}$ sets of size $n - 1$ are reachable[5]. Hence, f must have rank $n - 1$. Then, by Lemma 4.1, the group G is transitive, which implies the semi-automaton $\mathscr{A}$ is strongly connected. So, by Lemma 3.3, the semi-automaton is completely reachable.   □

Remark 4. For $n = 2$, if G only contains the identity transformation, then adding any non-permutation gives an automaton such that the set of synchronizing words has state complexity two, but it is not completely reachable nor is G transitive. Also, note that the assumption $\mathrm{sc}(\mathrm{Syn}(\mathscr{A})) = 2^n - n$ implies that f must have rank $n - 1$, for otherwise sets of size $n - 1$ are not reachable.

Definition 4.6. *A permutation group* $G = \langle g_1, \ldots, g_k \rangle \leqslant S_n$ *is called sync-maximal, if for any map* $f \colon [n] \to [n]$ *of rank* $n - 1$, *for the automaton* $\mathscr{A} = (\Sigma, [n], \delta)$ *with* $\Sigma = \{g_1, \ldots, g_k, f\}$ *and* $\delta(m, g) = g(m)$ *for* $m \in [n]$ *and* $g \in \Sigma$, *we have* $\mathrm{sc}(\mathrm{Syn}(\mathscr{A})) = 2^n - n$,

As written, the definition involves a specific set of generators for G. But the resulting transformation monoids are equal for different generators and we can write one set of generators in terms of another. So reachability of subsets and distinguishability of subsets is preserved by a change of generators. Hence, the definition is actually independent of the specific choice of generators for G. This might be different if we are concerned with the length of shortest words to reach certain subsets, but that is not part of the definition.

Proposition 4.7. *Every sync-maximal permutation group is primitive.*

Remark 5. By case analysis, note that for $n \leqslant 2$ every group is sync-maximal. But also for $n \leqslant 2$, by our definition of primitivity, every permutation group is primitive. See the explanations in Sect. 2.

---

[5] This argument only works for $n > 2$. If $n = 2$, as singletons sets are not distinguishable, if the state complexity is maximal, not all singleton sets need to be reachable from Q. Also, see Remark 4.

Next, we relate the condition that the set of synchronizing words has maximal state complexity to the notion of 2-homogeneity. However, as shown in Example 1, we do not get a characterization of 2-homogeneity similar to Theorem 4.2 for primitivity.

Proposition 4.8. *If $G \leqslant \mathcal{S}_n$ is 2-homogeneous, then G is sync-maximal.*

The next example shows that the converse of Proposition 4.8 does not hold.

Example 1. Let $g: [5] \to [5]$ be given by $g(i) = i + 1$ for $i \in \{0, \ldots, 3\}$ and $g(4) = 0$. Set $G = \langle \{g\} \rangle$. Then, as it is a cycle of prime length, G is primitive. So, by Theorem 4.2, for any $f: [n] \to [n]$ of rank $n - 1$ the transformation semigroup $\langle G \cup \{f\} \rangle$ is completely reachable. Also, we show $\mathrm{sc}(\mathrm{Syn}(\mathscr{A})) = 2^n - n$. But G is not 2-homogeneous. We have the following two orbits on the 2-sets: $A = \{\{1,2\}, \{2,3\}, \{3,4\}, \{4,0\}, \{0,1\}\}$, $B = \{\{0,2\}, \{1,3\}, \{2,4\}, \{3,0\}, \{4,1\}\}$. Let $f: [n] \to [n]$ be any map of rank $n - 1$. Without loss of generality, we can assume $f(0) = f(1)$. Then, two distinct 2-sets are distinguishable, if one could be mapped to $\{0,1\}$, but not the other, as a final application of f gives that one is mapped to a singleton, but not the other. First, note that all 2-sets in A are distinguishable, as for each $\{x,y\} \in A$ we find a unique $0 \leqslant k < |A|$ such that $g^k(\{z,v\}) = \{0,1\}$ if and only if $\{z,v\} = \{x,y\}$ for each $\{z,v\} \in A$, as g permutes A. If we have any $\{x,y\} \in B$ such that $f(\{x,y\}) \in A$, then all sets in B are distinguishable. For if $\{z,u\} \in B$ there exists a unique g^k with $0 \leqslant k < |B|$, $g^k(\{z,u\}) = \{x,y\}$ and $g^k(\{z',u'\}) \neq \{x,y\}$ for each other $\{z',u'\} \in B \setminus \{\{z,u\}\}$. As $\{0,1\} \notin B$ and f is injective on $\{1,2,3,4\}$ and on $\{0,2,3,4\}$, f is injective on B. Hence $f(g^k(\{z',u'\})) \neq f(g^k(\{z,u\}))$ for the previously chosen $\{z,u\} \in B$ and $\{z',u'\} \in B \setminus \{\{z,u\}\}$. Now, choose g^l such that $g^l(f(g^k(\{z,u\}))) = \{0,1\}$. In any case, i.e., whether $f(g^k(\{z',u'\}))$ is in B or in A, we have $g^l(f(g^k(\{z',u'\}))) \neq \{0,1\}$. So, all 2-sets in A are distinguishable and all 2-sets in B. That a 2-set from A is distinguishable from any 2-set in B is clear, as we can map the 2-set from A to $\{0,1\}$ by a power of g, and the one from B would not be mapped to $\{0,1\}$. Lastly, we show that we must have some $\{x,y\} \in B$ with $f(\{x,y\}) \in A$, which gives the claim. Consider the sets $\{0,2\}, \{0,3\}, \{1,4\}$ and $\{1,3\}$ from B and suppose their images are all contained in B, i.e., $\{\{f(0), f(2)\}, \{f(0), f(3)\}, \{f(0), f(4)\}\} \subseteq B$. But in B at most two sets share an element, hence $|\{f(2), f(3), f(4)\}| \leqslant 2$, which is not possible as f has rank $n - 1$ and we already have $f(0) = f(1)$. So, some image of these three sets must be in A. $\qquad\square$

5   *k*-Reachable Permutation Groups

A permutation group $G \leqslant \mathcal{S}_n$ is called *synchronizing*, if for any non-permutation the transformation monoid $\langle G \cup \{f\} \rangle$ contains a constant map. This notion was introduced in [4]. For further information on synchronizing groups and its relation to the Černý conjecture, see the survey [3]. In [3], the question was asked to detect properties of permutation groups by functions. Theorem 4.2 and Theorem 2.1 are in this vain. Note that every synchronizing group is primitive [4], but not conversely [18]. By Theorem 4.2, primitive groups have the property that if we add any function of rank $n - 1$ every non-empty subset is reachable. Motivated by this, we introduce k-reachable groups, which generalize the condition of complete reachability mentioned in Theorem 4.2.

Definition 5.1. *A permutation group over the finite set $[n]$ with $n > 1$ is called k-reachable, if for any map $f : [n] \to [n]$ of rank $n - k$ all subsets of cardinality*

$$n - k, n - 2k, \ldots, n - (\lceil n/k \rceil - 1) \cdot k$$

are reachable, i.e., we have some transformation in the transformation monoid generated by G and f which maps $[n]$ to any such set.

By Theorem 4.2, the 1-reachable group are precisely the primitive groups. Also note that $(n - 1)$-reachable is the same as transitivity.

Proposition 5.2. *A k-reachable permutation group is k-homogeneous.*

Proof. This is implied by Lemma 4.1 and the definition of k-reachability.   □

The reverse implication does not hold in the previous proposition. For example, we find 1-homogeneous, i.e., transitive groups, which are not 1-reachable, i.e., primitive by Theorem 4.2. By a result of Livingstone and Wagner [12,14], for $5 \leqslant k \leqslant n/2$ a k-homogeneous permutation group of degree n is k-transitive, and for $k \leqslant n/2$ a k-homogeneous permutation group is also $(k - 1)$-homogeneous. As 2-homogeneity implies synchronizability [3], combined with the fact that k-homogeneity is equivalent with $(n - k)$-homogeneity, a k-reachable group for any $1 < k < n - 1$ is synchronizable and we get the next statement together with our previous results.

Proposition 5.3. *A k-reachable permutation group of degree n for $1 < k < n-1$ is synchronizable. For $k = 1$ we have precisely the primitive permutation groups, and for $k = n - 1$ precisely the permutation groups which are transitive in their action.*

For $k \in \{2, 3, 4\}$ the non-k-transitive but k-homogeneous groups where determined by Kantor [13]. A list of all possible k-transitive groups of finite degree for $k \geqslant 2$ could be found in [7]. For $k \geqslant 6$ the only cases are the symmetric group or the alternating group. So, if we want to formulate a stronger version of Theorem 4.2 by assuming k-reachability for all $1 \leqslant k \leqslant n$, for $n \geqslant 6$ only the symmetric and the alternating group fulfill this condition.

Proposition 5.4. *If a permutation group of degree $n \geqslant 6$ is k-reachable for all $1 \leqslant k \leqslant n - 1$, then it is either the symmetric group or the alternating group.*

Or to be more specific.

Proposition 5.5. *If a permutation group of degree n is k-reachable for $6 \leqslant k \leqslant n - 6$, then it is either the symmetric or the alternating group.*

Note that the classification of the k-transitive groups of finite degree for $k > 1$ cited above relies on the classification of the finite simple groups, see [21] for an account of this important and highly non-trivial result.

6   Conclusion

We have given a new characterization for primitive permutation groups. In an analogous way, with the property that the set of synchronizing words has maximal state complexity, we have introduced the class of sync-maximal permutation groups. We have shown that the sync-maximal permutation groups are primitive and that 2-homogeneous groups are sync-maximal. Example 1 shows that not every sync-maximal permutation group is 2-homogeneous. However, we do not know if the converse of Proposition 4.7 is true, i.e., does there exist a primitive permutation group that is not sync-maximal? More results on the structure of the sync-maximal permutation groups would be highly interesting and might be the goal of future investigations. Also, for future investigations, their relation to the synchronizing groups, as introduced in [4], is of interest, as synchronizing groups lie strictly between the 2-homogenous and primitive groups [3,4,18]. More specifically, let us mention that in [3] a whole hierarchy of permutation groups was surveyed. The hierarchy is the following, stated without formally introducing every term:

2-transitive $\subsetneq$ 2-homogeneous $\subsetneq$ $\mathbb{Q}I$ $\subseteq$ spreading $\subsetneq$ separating

$\subsetneq$ synchronizing $\subsetneq$ primitive $\subsetneq$ transitive.

Note that it is unknown if the inclusion between the $\mathbb{Q}I$-groups and the spreading groups is proper. The question of fitting the sync-maximal groups more precisely into this hierarchy arises naturally. Lastly, we introduced k-reachable permutation groups. But for most k these do not give any groups beside the symmetric and alternating groups. A more close investigation and characterization of these groups for $k \in \{2, 3, 4\}$ is still open.

Acknowledgement. I thank my supervisor, Prof. Dr. Henning Fernau, for giving valuable feedback, discussions and research suggestions concerning the content of this article. I also thank Prof. Dr. Mikhail V. Volkov for introducing our working group to the idea of completely reachable automata at a joint workshop in Trier in the spring of 2019, from which the present work draws inspiration. Lastly, the argument in the proof of Theorem 4.2, which closely resembles a proof from [4], was communicated to me by an anonymous referee of a considerable premature version of this work. I thereby sincerely thank the referee for this and other remarks. I also thank another anonymous referee for Remark 3 and other suggestions related to the content of this work. In case a referee is wondering why an example from the submitted version is missing, it contained a subtle error I was unable to fix in due time for the final version.

References

1. Ananichev, D., Gusev, V., Volkov, M.: Slowly synchronizing automata and digraphs. In: Hliněný, P., Kučera, A. (eds.) MFCS 2010. LNCS, vol. 6281, pp. 55–65. Springer, Heidelberg (2010). https://doi.org/10.1007/978-3-642-15155-2_7
2. Ananichev, D.S., Volkov, M.V., Gusev, V.V.: Primitive digraphs with large exponents and slowly synchronizing automata. J. Math. Sci. **192**(3), 263–278 (2013). https://doi.org/10.1007/s10958-013-1392-8

3. Araújo, J., Cameron, P.J., Steinberg, B.: Between primitive and 2-transitive: synchronization and its friends. EMS Surv. Math. Sci. **4**(2), 101–184 (2017). http://www.ems-ph.org/doi/10.4171/EMSS/4-2-1

4. Arnold, F., Steinberg, B.: Synchronizing groups and automata. Theor. Comput. Sci. **359**(1–3), 101–110 (2006). https://doi.org/10.1016/j.tcs.2006.02.003

5. Bondar, E.A., Volkov, M.V.: Completely reachable automata. In: Câmpeanu, C., Manea, F., Shallit, J. (eds.) DCFS 2016. LNCS, vol. 9777, pp. 1–17. Springer, Cham (2016). https://doi.org/10.1007/978-3-319-41114-9_1

6. Bondar, E.A., Volkov, M.V.: A characterization of completely reachable automata. In: Hoshi, M., Seki, S. (eds.) DLT 2018. LNCS, vol. 11088, pp. 145–155. Springer, Cham (2018). https://doi.org/10.1007/978-3-319-98654-8_12

7. Cameron, P.J.: Permutation Groups. London Mathematical Society Student Texts. Cambridge University Press, Cambridge (1999). https://doi.org/10.1017/CBO9780511623677

8. Černý, J.: Poznámka k homogénnym experimentom s konečnými automatmi. Matematicko-fyzikálny časopis **14**(3), 208–216 (1964)

9. Don, H.: The Černý conjecture and 1-contracting automata. Electron. J. Comb. **23**(3), P3.12 (2016). http://www.combinatorics.org/ojs/index.php/eljc/article/view/v23i3p12

10. Gonze, F., Jungers, R.M.: Hardly reachable subsets and completely reachable automata with 1-deficient words. J. Automata Lang. Comb. **24**(2–4), 321–342 (2019). https://doi.org/10.25596/jalc-2019-321

11. Hopcroft, J.E., Ullman, J.D.: Introduction to Automata Theory, Languages, and Computation. Addison-Wesley Publishing Company, Boston (1979)

12. Huppert, B., Blackburn, S.: Finite Groups III. Grundlehren der mathematischen Wissenschaften, vol. 243. Springer, Heidelberg (1982). https://doi.org/10.1007/978-3-642-67997-1

13. Kantor, W.M.: k-homogeneous groups. Math. Z. **124**(4), 261–265 (1972). https://doi.org/10.1007/BF01113919

14. Livingstone, D., Wagner, A.: Transitivity of finite permutation groups on unordered sets. Math. Z. **90**(5), 393–403 (1965). https://doi.org/10.1007/BF01112361

15. Maslennikova, M.I.: Reset complexity of ideal languages. CoRR abs/1404.2816 (2014). http://arxiv.org/abs/1404.2816

16. Maslennikova, M.I.: Reset complexity of ideal languages over a binary alphabet. Int. J. Found. Comput. Sci. **30**(6–7), 1177–1196 (2019). https://doi.org/10.1142/S0129054119400343

17. Neumann, P.M.: The Mathematical Writings of Évariste Galois. European Mathematical Society, Heritage of European Mathematics (2011). Doi: https://doi.org/10.4171/104

18. Neumann, P.M.: Primitive permutation groups and their section-regular partitions. Michigan Math. J. **58**, 309–322 (2009)

19. Rystsov, I.K.: Estimation of the length of reset words for automata with simple idempotents. Cybern. Syst. Anal. **36**(3), 339–344 (2000). https://doi.org/10.1007/BF02732984

20. Sandberg, S.: 1 Homing and synchronizing sequences. In: Broy, M., Jonsson, B., Katoen, J.-P., Leucker, M., Pretschner, A. (eds.) Model-Based Testing of Reactive Systems. LNCS, vol. 3472, pp. 5–33. Springer, Heidelberg (2005). https://doi.org/10.1007/11498490_2

21. Solomon, R.: A brief history of the classification of the finite simple groups. Bull. Am. Math. Soc. **38**(3), 315–352 (2001)

22. Starke, P.H.: Eine Bemerkung über homogene Experimente. Elektronische Informationsverarbeitung und Kybernetik (later J. Inf. Process. Cybern.) **2**(4), 257–259 (1966)
23. Volkov, M.V.: Synchronizing automata and the Černý conjecture. In: Martín-Vide, C., Otto, F., Fernau, H. (eds.) LATA 2008. LNCS, vol. 5196, pp. 11–27. Springer, Heidelberg (2008). https://doi.org/10.1007/978-3-540-88282-4_4

State Complexity of the Set
of Synchronizing Words for Circular
Automata and Automata over Binary
Alphabets

Stefan Hoffmann$^{(\boxtimes)}$ (ID)

Informatikwissenschaften, FB IV, Universität Trier, Universitätsring 15,
54296 Trier, Germany
hoffmanns@informatik.uni-trier.de

Abstract. Most slowly synchronizing automata over binary alphabets
are circular, i.e., containing a letter permuting the states in a single cycle,
and their set of synchronizing words has maximal state complexity, which
also implies complete reachability. Here, we take a closer look at generalized circular and completely reachable automata. We derive that over
a binary alphabet every completely reachable automaton must be circular, a consequence of a structural result stating that completely reachable
automata over strictly less letters than states always contain permutational letters. We state sufficient conditions for the state complexity of
the set of synchronizing words of a generalized circular automaton to be
maximal. We apply our main criteria to the family $\mathscr{K}_n$ of automata that
was previously only conjectured to have this property.

Keywords: Finite automata · Synchronization · Completely reachable
automata · State complexity · Set of synchronizing words

1 Introduction

A deterministic semi-automaton is synchronizing if it admits a reset word, i.e.,
a word which leads to some definite state, regardless of the starting state. This
notion has a wide range of applications, from software testing, circuit synthesis,
communication engineering and the like, see [15,17]. The famous Černý conjecture [5] states that a minimal synchronizing word has length at most $(n-1)^2$
for an n state automaton. We refer to the mentioned survey articles [15,17] for
details. An automaton is completely reachable, if for each subset of states we
can find a word which maps the whole state set onto this subset. This is a generalization of synchronizability, as a synchronizing word maps the whole state
set to a singleton set. The class of completely reachable automata was formally
introduced in [3], but already in [6,12] such automata appear in the results.
The time complexity of deciding if a given automaton is completely reachable is
unknown. A sufficient and necessary criterion for complete reachability of a given

© Springer Nature Switzerland AG 2021
A. Leporati et al. (Eds.): LATA 2021, LNCS 12638, pp. 318–330, 2021.
https://doi.org/10.1007/978-3-030-68195-1_25

automaton in terms of graphs and their connectivity is known [4], but it is not known if these graphs could be constructed in polynomial time. A special case of the general graph construction, which gives a sufficient criterion for complete reachability [3], is known to be constructible in polynomial time [9]. The size of a minimal automaton accepting a given regular language is called the state complexity of that language. The set of synchronizing words of a given automaton is a regular ideal language whose state complexity is at most exponential in the size of the original automaton [12,13]. The Černý family of automata [5], a family of automata yielding the lower bound $(n-1)^2$ for the length of shortest synchronizing words, is completely reachable, but also the set of synchronizing words has maximal state complexity [12,13]. These properties are shared by many families of automata that are also slowly synchronizing [1,2,12,13], i.e., those for which a shortest reset word is close to the Černý bound.

Outline and Contribution: In Sect. 2 we give definitions and state known results. Then, in Sect. 3, we give a general criterion for completely reachable automata to deduce that the set of synchronizing words has maximal state complexity. We also state a structural result by which we can deduce that in completely reachable automata, where the number of letters is strictly less than the number of states, we must have permutational letters generating a non-trivial permutation group. In Sect. 4, we state sufficient conditions for generalized circular automata to deduce that their set of synchronizing words has maximal state complexity. In Sect. 5, we apply the results from Sect. 3 to deduce that completely reachable automata over binary alphabets must be circular and must have a letter mapping precisely two states to a single state. Also, with the results from Sect. 4, we show that the family $\mathscr{K}_n$, $n > 5$ odd, from [13] gives completely reachable automata such that the set of synchronizing words has maximal state complexity. This solves an open problem from [13], where this was only conjectured. The Černý family of automata [5,17], the first given family yielding the lower bound $(n-1)^2$ for the length of synchronizing words, is completely reachable and its set of synchronizing words has maximal state complexity [12,13]. These properties are also shared by a wealth of different slowly synchronizing automata [1,2,12,13]. Our criteria apply to all the automata mentioned in this previous work. However, we give an example showing that our stated conditions are only sufficient, but not necessary.

2 Preliminaries and Definitions

General Notions: Let $\Sigma = \{a_1, \ldots, a_k\}$ be a finite set of *symbols* (also called *letters*), called an *alphabet*. The set Σ^* denotes the set of all finite sequences, i.e., of all words or strings. The finite sequence of length zero, or the *empty word*, is denoted by ε. We set $\Sigma^+ = \Sigma^* \setminus \{\varepsilon\}$. For a given word $w \in \Sigma^*$, we denote by $|w|$ its *length*. The subsets of Σ^* are called *languages*. For $n > 0$ we set $[n] = \{0, \ldots, n-1\}$ and $[0] = \varnothing$. If $a, b \in \mathbb{Z}$ and $b \neq 0$, by $a \bmod b$ we denote the unique number $0 \leqslant r < |b|$ with $a = qb + r$ for some $q \in \mathbb{Z}$. For some set X by $\mathcal{P}(X)$ we denote the *power set* of X, i.e, the set of all subsets of X. Every function $f : X \to Y$ induces a function $\hat{f} : \mathcal{P}(X) \to \mathcal{P}(Y)$ by setting $\hat{f}(Z) := \{f(z) \mid z \in Z\}$. Here, we will denote this extension also by f. Let $k \geqslant 1$. A *k-subset* $Y \subseteq X$ is a finite set of cardinality k.

Automata-Theoretic Notions: A *finite, deterministic* and *complete* automaton will be denoted by $\mathscr{A} = (\Sigma, Q, \delta, s_0, F)$ with $\delta : Q \times \Sigma \to Q$ the state transition function, Q a finite set of states, $s_0 \in Q$ the start state and $F \subseteq Q$ the set of final states. The properties of being deterministic and complete are implied by the definition of δ as a total function. The transition function $\delta : Q \times \Sigma \to Q$ could be extended to a transition function on words $\delta^* : Q \times \Sigma^* \to Q$ by setting $\delta^*(s, \varepsilon) := s$ and $\delta^*(s, wa) := \delta(\delta^*(s, w), a)$ for $s \in Q$, $a \in \Sigma$ and $w \in \Sigma^*$. In the remainder we drop the distinction between both functions and will also denote this extension by δ. For $S \subseteq Q$ and $w \in \Sigma^*$, we write $\delta(S, w) = \{\delta(s, w) \mid s \in S\}$ and $\delta^{-1}(S, w) = \{q \in Q \mid \delta(q, w) \in S\}$. The *language accepted* by $\mathscr{A} = (\Sigma, S, \delta, s_0, F)$ is $L(\mathscr{A}) = \{w \in \Sigma^* \mid \delta(s_0, w) \in F\}$. A language $L \subseteq \Sigma^*$ is called *regular* if $L = L(\mathscr{A})$ for some finite automaton $\mathscr{A}$. For a language $L \subseteq \Sigma^*$ and $u, v \in \Sigma^*$ we define the *Nerode right-congruence* with respect to L by $u \equiv_L v$ if and only if $\forall x \in \Sigma : ux \in L \leftrightarrow vx \in L$. The equivalence class for some $w \in \Sigma^*$ is denoted by $[w]_{\equiv_L} := \{x \in \Sigma^* \mid x \equiv_L w\}$. A language is regular if and only if the above right-congruence has finite index, and it could be used to define the minimal deterministic automaton $\mathscr{A}_L = (\Sigma, Q, \delta, [\varepsilon]_{\equiv_L}, F)$ with $Q := \{[w]_{\equiv_L} \mid w \in \Sigma^*\}$, $\delta([w]_{\equiv_L}, a) := [wa]_{\equiv_L}$ for $a \in \Sigma$, $w \in \Sigma^*$ and $F := \{[w]_{\equiv_L} \mid w \in L\}$. It is indeed the smallest automaton accepting L in terms of the number of states, and we will refer to this construction as the minimal automaton [11] of L. The *state complexity* of a regular language is defined as the number of Nerode right-congruence classes. We will denote this number by $\mathrm{sc}(L)$. Let $\mathscr{A} = (\Sigma, Q, \delta, s_0, F)$ be an automaton. A state $q \in Q$ is *reachable*, if $q = \delta(s_0, u)$ for some $u \in \Sigma^*$. We also say that a state q is reachable from a state q' if $q = \delta(q', u)$ for some $u \in \Sigma^*$. Two states q, q' are *distinguishable*, if there exists $u \in \Sigma^*$ such that either $\delta(q, u) \in F$ and $\delta(q', u) \notin F$ or $\delta(q, u) \notin F$ and $\delta(q', u) \in F$. An automaton for a regular language is isomorphic to the minimal automaton if and only if all states are reachable and distinguishable [11]. A *semi-automaton* $\mathscr{A} = (\Sigma, Q, \delta)$ is like an ordinary automaton, but without a distinguished start state and without a set of final states. Sometimes we will also call a semi-automaton simply an automaton if the context makes it clear what is meant. Also, definitions without explicit reference to a start state and a set of final states are also valid for semi-automata. Let $\mathscr{A} = (\Sigma, Q, \delta)$ be a finite semi-automaton. A word $w \in \Sigma^*$ is called *synchronizing* if $\delta(q, w) = \delta(q', w)$ for all $q, q' \in Q$, or equivalently $|\delta(Q, w)| = 1$. Set $\mathrm{Syn}(\mathscr{A}) = \{w \in \Sigma^* \mid |\delta(Q, w)| = 1\}$. The *power automaton (for synchronizing words)* associated to $\mathscr{A}$ is $\mathcal{P}_{\mathscr{A}} = (\Sigma, \mathcal{P}(Q), \delta, Q, F)$ with start state Q, final states $F = \{\{q\} \mid q \in Q\}$ and the transition function of $\mathcal{P}_{\mathscr{A}}$ is the transition function of $\mathscr{A}$, but applied to subsets of states. Then, as observed in [16], the automaton $\mathcal{P}_{\mathscr{A}}$ accepts the set of synchronizing words, i.e., $L(\mathcal{P}_{\mathscr{A}}) = \mathrm{Syn}(\mathscr{A})$. As for $\{q\} \in F$, we also have $\delta(\{q\}, x) \in F$ for each $x \in \Sigma^*$, the states in F could all be merged to a single state to get an accepting automaton for $\mathrm{Syn}(\mathscr{A})$. Also, the empty set is not reachable from Q. Hence $\mathrm{sc}(\mathrm{Syn}(\mathscr{A})) \leqslant 2^{|Q|} - |Q|$ and this bound is sharp [12,13]. We call $\mathscr{A}$ *completely reachable* if for any non-empty $S \subseteq Q$ there exists a word $w \in \Sigma^*$ with $\delta(Q, w) = S$, i.e., in the power automaton, every state is reachable from the start state. When we say a *subset of states* in $\mathscr{A}$ is *reachable*, we mean reachability in $\mathcal{P}_{\mathscr{A}}$. The state complexity of $\mathrm{Syn}(\mathscr{A})$ is maximal, i.e., $\mathrm{sc}(\mathrm{Syn}(\mathscr{A})) = 2^{|Q|} - |Q|$, if and only if at least one singleton subset of Q and all subsets $S \subseteq Q$ with $|S| \geqslant 2$ are reachable, and all non-singleton subsets are distinguishable in $\mathcal{P}_{\mathscr{A}}$. For *strongly connected automata*, i.e., those for which every state is reachable from every other state, is maximal iff $\mathscr{A}$ is completely reachable and all $S \subseteq Q$ with $|S| \geqslant 2$ are distinguishable in $\mathcal{P}_{\mathscr{A}}$. A *permutation* on a finite set Q (which here will always be the set of states of some automaton) is a bijective function, a subset of permutations closed under concatenation (and function inversion, but this is implied in the finite case) is called a *permutation group*. The *orbit* of an element from Q under a given permutation group on Q is the sets of all elements to which this element could be mapped by elements from the permutation group. A permutation group with a single orbit, i.e., every element could be mapped to any other, is called *transitive*. A semi-automaton $\mathscr{A} = (\Sigma, Q, \delta)$ is called *circular*, if some letter acts as a cyclic permutation on all states. This family of automata was one of the first inspected with respect to the Černý-conjecture [14], and the conjecture was finally confirmed for this family [7,8]. A semi-automaton $\mathscr{A} = (\Sigma, Q, \delta)$ is called *generalized circular*, if some word acts as a cyclic permutation on all states[1]. Let $\mathscr{A} = (\Sigma, Q, \delta)$ be an automaton and for $w \in \Sigma^*$ define $\delta_w : Q \to Q$ by $\delta_w(q) = \delta(q, w)$ for all $q \in Q$. Then, we can associate with $\mathscr{A}$ the *transformation monoid* of the automaton $\mathcal{T}_{\mathscr{A}} = \{\delta_w \mid w \in \Sigma^*\}$. The *rank* of a map $f : Q \to Q$ on a finite set Q is the cardinality of its image. For a given automaton, seeing a word as a transformation of its state set, the rank of the word is the rank of this transformation. A *permutational letter* is a letter of full rank, i.e., a letter inducing a permutation on the states.

Known Results: We will need the following result from [6] to deduce complete reachability of some automata families we consider.

Proposition 1 (Don [6]). *Let $\mathscr{A} = (\Sigma, Q, \delta)$ be a finite circular automaton with n states, where b induces a cyclic permutation of the states. Suppose we have another letter $a \in \Sigma$ of rank $n - 1$ and choose $s, t \in Q$ and $0 < d < |Q|$ such that $\delta(Q, a) = Q \setminus \{s\}$, $|\delta^{-1}(t, a)| = 2$ and $\delta(s, b^d) = t$. If d and n are coprime, then for every non-empty set $S \subseteq Q$ of size k, there exists a word w_S of length at most $n(n - k)$ such that $\delta(Q, w_S) = S$.*

---

[1] The circular automata are a proper subfamily of the generalized circular automata, as shown by $\mathscr{A} = (\{a, b\}, [3], \delta)$ with $\delta(0, a) = 1, \delta(1, a) = 0, \delta(2, a) = 2$ and $\delta(0, b) = 0, \delta(1, b) = 2, \delta(2, b) = 1$. The word ba cyclically permutes the states.

3   General Results on the State Complexity of Syn($\mathscr{A}$)

Our first result states that for completely reachable automata, to deduce that the set of synchronizing words has maximal state complexity, we only need to show distinguishability for those subsets of states with precisely two elements.

Lemma 2 (Hoffmann [10]). *Let $\mathscr{A} = (\Sigma, Q, \delta)$ be a completely reachable semi-automaton with n states. Then, $sc(Syn(\mathscr{A})) = 2^n - n$ if and only if all 2-sets of states are pairwise distinguishable in $\mathcal{P}_{\mathscr{A}}$.*

With the next result we can deduce information about the structure of completely reachable automata when the alphabet, or more precisely only the number of letters of rank $n - 1$, is strictly smaller than the number of states. Later, for completely reachable automata over binary alphabet, we can deduce that they must be circular and have to contain a letter of rank $n - 1$. Note that we formulate it with a weaker condition than full complete reachability, merely only with reachability of subsets of size $n - 1$.

Proposition 3. *Let $\mathscr{A} = (\Sigma, Q, \delta)$ be a semi-automaton with n states, m letters of rank $n - 1$ and $n > m$. Then, the following conditions are equivalent:*

1. *every subset of size $n - 1$ is reachable,*
2. *there exists at least one letter of rank $n - 1$ and a subset of letters generating a non-trivial permutation group such that every state is in the same orbit as some state not in the image of a rank $n - 1$ letter. In particular, we have at most m orbits.*

Remark 1. The condition $m < n$ cannot be omitted in Proposition 3. For example, let $\mathscr{A} = (\Sigma, Q, \delta)$ with $Q = [n]$ and $\Sigma = \{a_1, \ldots, a_n, b_1, \ldots, b_n\}$ be such that for $i \in \{1, \ldots, n\}$ we have $Q \setminus \{i\} = \delta(Q, a_i)$ and a_i cyclically permutes $Q \setminus \{i\}$. Furthermore, let b_i map some fixed state $q_i \in Q \setminus \{i\}$ to $\delta(q_i, a_i)$ and act as the identity transformation on the rest. Then, in $\mathscr{A}$, even when only using the alphabet $a_1, \ldots, a_n$ we reach every subset of size $n - 1$. But with the additional letters, $\mathscr{A}$ is also completely reachable, as the subautomaton given by $Q \setminus \{i\}$ and only the letters a_i and b_i equals the Černý-automaton, which is completely reachable [13]. Hence, combining these facts gives complete reachability of $\mathscr{A}$, but we have no permutational letters at all.

With this result, we can derive that a completely reachable automaton whose alphabet is small enough has to contain a non-trivial permutation group as part of its transformation monoid. Or more specifically, if we only have a single letter of rank $n - 1$, this permutation group must be transitive.

Corollary 4. *If $\mathscr{A} = (\Sigma, Q, \delta)$ is completely reachable with only a single non-permutational letter and $|Q| > 2$, then $\mathcal{T}_{\mathscr{A}}$ contains a transitive permutation group as a submonoid.*

If we find a transitive permutation group in the transformation monoid of some given automaton, then this automaton is strongly connected. Hence, let us state the following observation concerning strongly connected automata.

Lemma 5 (Hoffmann [10]). *Let $\mathscr{A} = (\Sigma, Q, \delta)$ be strongly connected. If* $\mathrm{Syn}(\mathscr{A})$ *has maximal state complexity, then $\mathscr{A}$ is completely reachable.*

Combining Corollary 4 and Lemma 5 gives the next lemma.

Lemma 6. *Let $\mathscr{A} = (\Sigma, Q, \delta)$ be completely reachable with only a single non-permutational letter and $|Q| > 2$. Then, if* $\mathrm{Syn}(\mathscr{A})$ *has maximal state complexity, the semi-automaton is completely reachable.*

As circular automata are strongly connected, the next follows by Lemma 5.

Corollary 7. *Let $\mathscr{A} = (\Sigma, Q, \delta)$ be a circular semi-automaton. If* $\mathrm{sc}(\mathrm{Syn}(\mathscr{A})) = 2^n - n$*, then $\mathscr{A}$ is completely reachable.*

4   Generalized Circular Automata

Here, Theorem 8 and Proposition 10 give sufficient conditions to deduce, for completely reachable circular automata, that the set of synchronizing words has maximal state complexity. Both conditions entail all known cases of automata over a binary alphabets for which the set of synchronizing words has maximal state complexity [12,13]. However, at the end of this section, we will show that the stated conditions are not necessary. In Theorem 8, we do not assume the automaton to be completely reachable, but only to be circular and to have a letter of rank $n - 1$ fulfilling a certain condition. If we also suppose complete reachability, then the theorem gives that the set of synchronizing words has maximal state complexity. Also, note that instead of a letter, any word fulfilling the mentioned condition in Theorem 8 will work to give the conclusion. Most of the time, we formulate our results for letters, but in all statements the assumptions could be formulated with words instead, as the notions of distinguishablity do not depend on the length, but only on the existence of certain words[2].

However, we have a slight focus on automata over binary alphabets later on, and the results of Sect. 5 will show that completely reachable automata over binary alphabets with at least three states are always circular and every word that cyclically permutes the states is a power of the cyclic permutation. So, we formulate our result with letters instead of words for simplicity. Intuitively, in Theorem 8, Eq. (1) says that we can apply the letter a to reduce the distance modulo n on the cycle given by b, or, by Eq. (2), that we can map to a state having a specific distance, from which we can then reduce it. Please see Fig. 1 for a graphical depiction.

---

[2] For, if we choose a finite number of words and build the automaton by identifying these words with new letters, distinguishability or reachability of states (or subsets of states) of this new automaton is inherited to the original automaton. Hence, all results are also valid when stated with words instead of letters, but otherwise the same conditions.

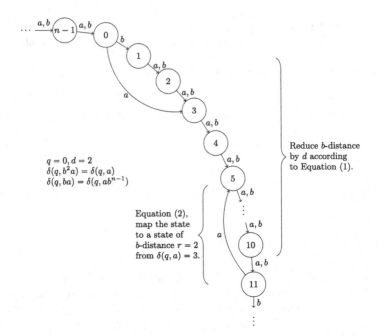

Fig. 1. Illustration of the conditions stated in Theorem 8 for an instance with $d = 2$. Shown are the first twelve states and the state $n - 1$ for a circular automaton with n states. Note that in Theorem 8, we suppose $0 < m < n$, and indeed, for $m \in \{0, n\}$ Eq. (1) does not apply in general.

Theorem 8. *Suppose* $\mathscr{A} = (\Sigma, Q, \delta)$ *has* n *states. Let* $\{a, b\} \subseteq \Sigma$ *(or any two words in* Σ^**). Assume the letter* b *cyclically permutes the states and the letter* a *has rank* $n - 1$*. Then all 2-sets are distinguishable in* $\mathcal{P}_{\mathscr{A}}$*, if we can find a state* $q \in Q$ *and a number* $d > 0$ *coprime to* n *such that for each* $0 < m < n$ *we*[3] *either have*

$$\delta(q, b^m a) = \delta(q, ab^{n+m-d}) \tag{1}$$

or, but only in case m *is not divisible by* d*,*

$$\delta(q, ab^r) = \delta(q, b^m a) \ or \ \delta(q, b^m ab^r) = \delta(q, a) \tag{2}$$

for some number $0 \leqslant r < n$ *divisible by* d*.*

In the formulation of Theorem 8, we have $r > 0$, as in this case $m \neq 0$. Also, note that $\delta(q, b^m ab^r) = \delta(q, a)$ is equivalent with $\delta(q, ab^{n-r}) = \delta(q, b^m a)$, as for any states $s, t \in Q$ and $0 \leqslant k < n$ we have $\delta(s, b^k) = t$ if any only if $\delta(t, b^{n-k}) = s$, as $\delta(s, b^n) = s$. The conditions mentioned in Theorem 8 are the most general ones

---

[3] Note that $0 < m < n$ implies $\delta(q, b^m) \neq q$. Also note that we added n on the right hand side to account for values $d > 1$. In Proposition 10 we only subtract one from the exponent of b, which is always non-zero and strictly smaller than n, and so we do not needed this "correction for the b-cycle" in case of resulting negative exponents.

in this paper, but let us state next, as a corollary, a more relaxed formulation, stating that we can reduce the distance on the cycle by one for each application of some word of rank $n - 1$.

Corollary 9. *Let $\mathscr{A} = (\Sigma, Q, \delta)$ be a circular automaton with n states where the letter b permutes the states with a single orbit. Suppose we find a word $w \in \Sigma^*$ and state $q \in Q$ such that, for*[4] $0 \leqslant m < n$,

$$\delta(q, b^{m+1}w) = \delta(q, wb^m) \tag{3}$$

Then all 2-sets are distinguishable in $\mathcal{P}_{\mathscr{A}}$. In particular, if $\mathscr{A}$ is completely reachable, then $\mathrm{sc}(\mathrm{Syn}(\mathscr{A})) = 2^n - n$.

Proof. Set $s = \delta(q, b)$ and $t = \delta(q, w)$. Then, $\delta(s, w) = t = \delta(q, w)$ and $s \neq q$. For $m \in \{1, \ldots, n-1\}$ we have $\delta(q, b^m w) = \delta(t, b^{m-1})$ and

$$\{\delta(q, b), \delta(q, b^2), \ldots, \delta(q, b^{n-1})\} = Q \setminus \{q\}.$$

So, as b is a permutation, w acts injective on $Q \setminus \{q\}$ and has rank $n - 1$. Now, apply Theorem 8, interpreting w as the letter a of rank $n - 1$.   □

Actually, for the relaxed condition mentioned in Corollary 9 we can give a small strengthening by only requiring that we can reduce the "cyclic distance" for all states which are no more than $\lfloor n/2 \rfloor + 1$ steps, or applications of b, away from some specific state.

Proposition 10. *Let $\Sigma = \{a, b\}$ and suppose $\mathscr{A} = (\Sigma, Q, \delta)$ has n states and is completely reachable with the letter a having rank $n-1$ and the letter b permuting the states with a single orbit. Then $\mathrm{sc}(\mathrm{Syn}(\mathscr{A})) = 2^n - n$ if we can find a state $q \in Q$ such that for all $0 \leqslant m \leqslant \lfloor n/2 \rfloor - 1$ we have*

$$\delta(q, b^{m+1}a) = \delta(q, ab^m). \tag{4}$$

Finally, we show that the mentioned sufficient conditions are not necessary. In Example 1 we will give a circular automaton whose set of synchronizing words has maximal state complexity but for which this could not be derived with any of the results stated here.

Example 1. Let $\mathscr{A} = (\Sigma, [4], \delta)$ with $\Sigma = \{a, b\}$, $\delta(i, b) = (i + 1) \bmod 4$ and $\delta(0, a) = 1, \delta(1, a) = 2, \delta(2, a) = 1, \delta(3, a) = 3$. Please see Fig. 2 for a graphical depiction of $\mathscr{A}$ and $\mathcal{P}_{\mathscr{A}}$. Then, all words of rank 3 are listed in Table 1.

In each word w of rank 3 the distance of the two distinct states mapped to one state is 2. So, in Eq. (1), for each such word of rank 3 (in place of a), we would have $d = 2$. But 2 is not coprime to 4, hence Theorem 8 does not apply here. However, we have $\mathrm{sc}(\mathrm{Syn}(\mathscr{A})) = 2^4 - 4 = 12$. We see in Fig. 2 that every subset is reachable. We also see that a distinguishes $\{0, 2\}$ from every other 2-set of states, ba distinguishes $\{1, 3\}$ from every other, $baba$ distinguishes $\{2, 3\}$ from $\{0, 3\}$, $\{1, 2\}$ and $\{0, 1\}$ and these latter three 2-sets are easily seen to be distinguishable by words in b^*aba. So, by Lemma 2, all non-empty subsets of states are distinguishable.

---

[4] Note that here, even if the bounds for m from Theorem 8 do not include this case, $\delta(q, w) = \delta(q, b^n w) = \delta(q, wb^{n-1})$, which is equivalent with $\delta(q, w) = \delta(q, b)$.

5   Automata over Binary Alphabets

Here, we take a closer look at automata over a binary alphabet. We apply our results and solve an open problem posed in [13]. In general, if a letter has rank k and some subset is mapped to a subset of size k, we must hit the full image of this letter. This gives, if we only have two letters but more than two states and no letter has full rank, that we can only reach at most two subsets of size $n-1$. So, if more $(n-1)$-sets are reachable, we must have precisely one letter of rank $n-1$ and Corollary 4 gives the next result.

Lemma 11. *Let $\Sigma = \{a, b\}$ be a binary alphabet and $\mathscr{A} = (\Sigma, Q, \delta)$ a finite semi-automaton with $n > 2$ states. Then, the following conditions are equivalent:*

1. *every subset of size $n-1$ is reachable,*
2. *exactly one letter acts as a cyclic permutation with a single orbit and the other letter has rank $n-1$.*

In particular, over a binary alphabets, completely reachable automata and those whose set of synchronizing words has maximal state complexity are circular.

Table 1. All rank 3 words for the automaton from Example 1. To the right of each word the induced transformation on the states is written, where $j \in [4]$ written at position $i \in [4]$ means the word maps the state i to state j. The entries are ordered such that for two words u, v in the same row we have $\delta(i, u) = \delta(j, u)$ iff $\delta(i, v) = \delta(j, v)$ for $i \in [4]$ and the images of words in the same column are equal.

Word	Mapping	Word	Mapping	Word	Mapping	Word	Mapping
$b^2a^2b^2$	[0, 1, 0, 3]	a	[1,2,1,3]	ab^3	[0,1,0,2]	ab^2ab	[0,2,0,3]
a^2b^2	[0, 3, 0, 1]	b^2a	[1,3,1,2]	b^2ab^3	[0,2,0,1]	ab^2a^2b	[0,3,0,2]
$ab^2a^2b^2$	[1, 0, 1, 3]	a^2	[2,1,2,3]	a^2b^3	[1,0,1,2]	b^2ab	[2,0,2,3]
ab^2ab^2	[1, 3, 1, 0]	b^2a^2	[2,3,2,1]	$b^2a^2b^3$	[1,2,1,0]	ab	[2,3,2,0]
ab^2	[3, 0, 3, 1]	ab^2a	[3,1,3,2]	ab^2ab^2	[2,0,2,1]	b^2a^2b	[3,0,3,2]
b^2ab^2	[3, 1, 3, 0]	ab^2a^2	[2,2,3,1]	$ab^2a^2b^3$	[2,1,2,0]	a^2b	[3,2,3,0]
$bab^2a^2b^2$	[0,1,3,1]	ba^2	[1,2,3,2]	ba^2b^3	[0,1,2,1]	b^3ab	[0,2,3,2]
bab^2	[0,3,1,3]	bab^2a	[1,2,3,2]	bab^2ab^3	[0,2,1,2]	b^3a^2b	[0,3,2,3]
$b^3a^2b^2$	[1,0,3,0]	ba	[2,1,3,1]	bab^3	[1,0,2,0]	bab^2ab	[2,0,3,1]
b^3ab^2	[1,3,0,3]	bab^2a^2	[2,3,1,3]	$bab^2a^2b^3$	[1,2,0,2]	ba^2b	[2,3,0,3]
ba^2b^2	[3,0,1,0]	b^3a	[3,1,2,1]	b^3ab^3	[2,0,1,0]	bab^2a^2b	[3,0,2,1]
bab^2ab^2	[3,1,0,1]	b^3a^2	[3,2,1,3]	$b^3a^2b^3$	[2,1,0,2]	bab	[3,2,0,2]

Remark 2. In Lemma 11, we need $n > 2$. For let $\mathscr{A} = (\{a, b\}, \{p, q\}, \delta)$ with $\delta(p, a) = \delta(q, a) = q$ and $\delta(p, b) = \delta(q, b) = p$. Then $\mathscr{A}$ is completely reachable, but no letter acts as a non-trivial permutation.

With Theorem 8, we can solve an open problem from [13]. For $n > 5$, define the automata[5] $\mathscr{K}_n = (\Sigma, [n], \delta)$, introduced in [13], with

$\delta(i, b) = i + 1$ for $i \in \{0, \ldots, n-2\}$, and $\delta(n-1, b) = 0$;

$\delta(i, a) = i + 1$ for $i \in \{1, \ldots, n-3\}, \delta(n-1, a) = 0, \delta(n-2, a) = 1, \delta(0, a) = 3$.

Please see Example 2 for an illustration of this automata family. In [13], it was conjectured that $\mathrm{sc}(\mathrm{Syn}(\mathscr{K}_n)) = 2^n - n$ for every odd $n > 5$. With Theorem 8, together with Proposition 1 and Lemma 2, we can confirm this.

Proposition 12. *Let $n > 5$ be odd. Then we have $\mathrm{sc}(\mathrm{Syn}(\mathscr{K}_n)) = 2^n - n$.*

Proof. First, we will show, using Proposition 1, that the automata $\mathscr{K}_n$, for odd $n > 5$, are completely reachable. Then, we will show, using Proposition 8, that all 2-subsets of states are distinguishable in the power automaton $\mathcal{P}_{\mathscr{K}_n}$. With Lemma 2, this would then give $\mathrm{sc}(\mathrm{Syn}(\mathscr{K}_n)) = 2^n - n$.

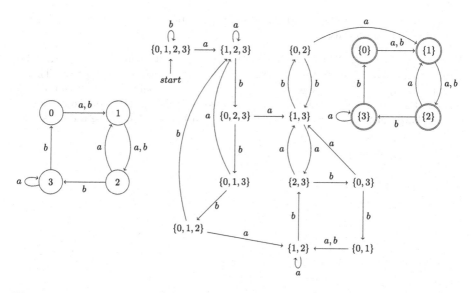

Fig. 2. The automaton from Example 1 and its power automaton. An example of an automaton whose set of synchronizing words has maximal state complexity but for which Theorem 8 or Proposition 10 do not apply, not for a and not for any word of rank 3. The final states in the power automaton are marked with double circles.

1. For $n > 5$ odd, the automata $\mathscr{K}_n$ are completely reachable: We have two letters, the letter a has rank $n-1$ and the letter b is a cyclic permutation of all the states. Also $\delta(Q, a) = Q \setminus \{n-1\}$, $\delta^{-1}(3, a) = \{0, 2\}$ and $\delta(n-1, b^4) = 3$. If n is odd, then n and 4 are coprime. We have listed the prerequisites of Proposition 1, hence applying it gives that $\mathscr{K}_n$ is completely reachable.

---

[5] I slightly changed the numbering of the states with respect to the action of the letter a compared to [13].

2. For $n > 5$ odd, in $\mathcal{K}_n$ all 2-sets are distinguishable in $\mathcal{P}_{\mathcal{K}_n}$: Let $q = 0$. Then

$$\delta(q, ba) = 2 = \delta(q, ab^{n-1}) = \delta(q, ab^{n+1-2}).$$

For $m \in \{2, \ldots, n-3\}$, we have $\delta(0, b^m) = m$ and

$$\delta(q, b^m a) = m + 1 = \delta(3, b^{m-2}) = \delta(q, ab^{m-2}) = \delta(q, ab^{n+m-2}).$$

The value $m = n - 2$ does not follow the above pattern, but we have $\delta(q, b^{n-2}ab^2) = \delta(1, b^2) = 3 = \delta(q, a)$. And lastly, for $m = n - 1$, we have

$$\delta(q, b^{n-1}a) = q = \delta(q, ab^{n-1-2}).$$

So, with $d = 2$ and $r = 2$, for odd n, as then $n - 2$ is not divisible by d, and with $q = 0$, the prerequisites of Theorem 8 are fulfilled and give the claim.

So, both statements together with Lemma 2 yield $\mathrm{sc}(\mathrm{Syn}(\mathcal{K}_n)) = 2^n - n$.   □

Lastly, let us give some additional examples from the literature [1,2,12,13] for which our results apply.

Example 2. Please see Fig. 3 for the automata families. The automata $\mathcal{C}_n$ gives the Černý family, the automata $\mathcal{L}_n$, $\mathcal{V}_n$, $\mathcal{F}_n$ and $\mathcal{K}_n$ were introduced in [1,12,13]. There, except for $\mathcal{K}_n$, it was established that in each case (for $\mathcal{F}_n$ only if n is odd and $n > 3$) the set of synchronizing words has maximal state complexity. Note that our results, namely Theorem 8, together with Proposition 1 and Lemma 2 also give these results.

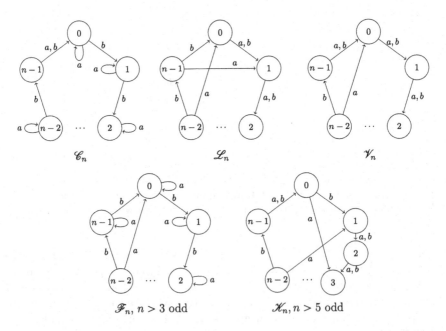

Fig. 3. Families of automata whose sets of synchronizing words have maximal state complexity. Please see Example 2 for explanation.

6   Conclusion

We have stated sufficient criteria for completely reachable generalized circular automata with a letter of rank $n - 1$ to deduce that their set of synchronizing words has maximal state complexity. Note that by our results, every completely reachable automaton over a binary alphabet must have this form. It is natural to ask if we can generalize this to obtain a sufficient and necessary criterion. As a step in this direction, another family for which this might be tackled first is the family of circular automata $\mathscr{A} = (\Sigma, Q, \delta)$ over a binary alphabet with a rank $n - 1$ letter a such that $\delta(Q, aa) = \delta(Q, a)$, i.e., the set $\delta(Q, a)$ is permuted by a. For these automata, we can find a power of a such that a acts as the identity on $\delta(Q, a)$ and the single state in $Q \setminus \delta(Q, a)$ is mapped into $\delta(Q, a)$. Note that these automata closely resemble those of the Černý family. Hence, we can find easy sufficient criteria for these automata by applying our obtained results to the resulting automaton, where the power of a is considered as the new rank $n - 1$ letter. However, as Example 1 shows, such a criterion is also not necessary. But a more finer analysis might give sufficient and necessary criteria.

References

1. Ananichev, D., Gusev, V., Volkov, M.: Slowly synchronizing automata and digraphs. In: Hliněný, P., Kučera, A. (eds.) MFCS 2010. LNCS, vol. 6281, pp. 55–65. Springer, Heidelberg (2010). https://doi.org/10.1007/978-3-642-15155-2_7
2. Ananichev, D.S., Volkov, M.V., Gusev, V.V.: Primitive digraphs with large exponents and slowly synchronizing automata. J. Math. Sci. **192**(3), 263–278 (2013)
3. Bondar, E.A., Volkov, M.V.: Completely reachable automata. In: Câmpeanu, C., Manea, F., Shallit, J. (eds.) DCFS 2016. LNCS, vol. 9777, pp. 1–17. Springer, Cham (2016). https://doi.org/10.1007/978-3-319-41114-9_1
4. Bondar, E.A., Volkov, M.V.: A characterization of completely reachable automata. In: Hoshi, M., Seki, S. (eds.) DLT 2018. LNCS, vol. 11088, pp. 145–155. Springer, Cham (2018). https://doi.org/10.1007/978-3-319-98654-8_12
5. Černý, J.: Poznámka k homogénnym experimentom s konečnými automatmi. Matematicko-fyzikálny časopis **14**(3), 208–216 (1964)
6. Don, H.: The Černý conjecture and 1-contracting automata. Electron. J. Comb. **23**(3), P3.12 (2016)
7. Dubuc, L.: Les automates circulaires biaisés vérifient la conjecture de Cerný. ITA **30**(6), 495–505 (1996)
8. Dubuc, L.: Sur les automates circulaires et la conjecture de Cerný. ITA **32**(1–3), 21–34 (1998)
9. Gonze, F., Jungers, R.M.: Hardly reachable subsets and completely reachable automata with 1-deficient words. J. Automata Lang. Comb. **24**(2–4), 321–342 (2019)
10. Hoffmann, S.: Completely reachable automata, primitive groups and the state complexity of the set of synchronizing words. In: Leporati, A., et al. (eds.) LATA 2021. LNCS, vol. 12638, pp. 305–317. Springer, Cham (2021)
11. Hopcroft, J.E., Ullman, J.D.: Introduction to Automata Theory, Languages, and Computation. Addison-Wesley Publishing Company, Boston (1979)

12. Maslennikova, M.I.: Reset complexity of ideal languages. CoRR abs/1404.2816 (2014)
13. Maslennikova, M.I.: Reset complexity of ideal languages over a binary alphabet. Int. J. Found. Comput. Sci. **30**(6–7), 1177–1196 (2019)
14. Pin, J.E.: Sur un cas particulier de la conjecture de Cerny. In: Ausiello, G., Böhm, C. (eds.) ICALP 1978. LNCS, vol. 62, pp. 345–352. Springer, Heidelberg (1978). https://doi.org/10.1007/3-540-08860-1_25
15. Sandberg, S.: 1 Homing and synchronizing sequences. In: Broy, M., Jonsson, B., Katoen, J.-P., Leucker, M., Pretschner, A. (eds.) Model-Based Testing of Reactive Systems. LNCS, vol. 3472, pp. 5–33. Springer, Heidelberg (2005). https://doi.org/10.1007/11498490_2
16. Starke, P.H.: Eine Bemerkung über homogene Experimente. Elektronische Informationsverarbeitung und Kybernetik (later J. Inf. Process. Cybern.) **2**(4), 257–259 (1966)
17. Volkov, M.V.: Synchronizing automata and the Černý conjecture. In: Martín-Vide, C., Otto, F., Fernau, H. (eds.) LATA 2008. LNCS, vol. 5196, pp. 11–27. Springer, Heidelberg (2008). https://doi.org/10.1007/978-3-540-88282-4_4

Cadences in Grammar-Compressed Strings

Julian Pape-Lange[✉][iD]

Technische Universität Chemnitz, Straße der Nationen 62, 09111 Chemnitz, Germany
julian.pape-lange@informatik.tu-chemnitz.de

Abstract. Cadences are structurally maximal arithmetic progressions of indices corresponding to equal characters in an underlying string.

This paper provides a detection algorithm for 3-cadences in binary strings which runs in linear time on uncompressed strings and in polynomial time on grammar-compressed strings.

Furthermore, this paper proves that several variants of the cadence detection problem are $\mathcal{NP}$-complete on grammar-compressed strings and that the equidistant subsequence matching problem with patterns of length three is $\mathcal{NP}$-complete on grammar-compressed ternary strings.

Keywords: String-cadences · String algorithms · Compressed pattern matching

1 Introduction

A sub-cadence in a string is an arithmetic progression of indices corresponding to equal characters. Van der Waerden shows in [12] that for each k and each alphabet size $|\Sigma|$, there is a natural number $m(k, |\Sigma|)$, such that each sequence of characters in Σ with length greater than or equal to $m(k, |\Sigma|)$ has a sub-cadence consisting of k indices. However, the term cadence in the context of strings was first used by Gardelle in [5] in the year 1964.

In this paper, we use the notation of Amir et al. in [1] and say that a cadence is a sub-cadence which is structurally maximal in the sense that the extension of the arithmetic progression to the left or to the right would not result in a valid index of the string.

For example, in the string $S = 10101$, the three indices $1, 3, 5$ form a cadence, since the indices -1 and 7 are both outside of the string. On the other hand, in the string $S = 01110$, the three indices $2, 3, 4$ do not form a cadence, since, for example, the index 1 is inside the string.

Funakoshi and Pape-Lange prove in [4] that if the underlying alphabet has a constant size, the number of 3-cadences in an uncompressed string of length n can be counted in $\mathcal{O}(n(\log n)^2)$ time using fast Fourier transform.

Furthermore, Funakoshi et al. present in [3] the more general problem of equidistant subsequence matching which extends the sub-cadences to arbitrary arithmetic factors, and showed that techniques for cadence-detection can be adopted to solve equidistant subsequence matching with similar time complexity.

© Springer Nature Switzerland AG 2021
A. Leporati et al. (Eds.): LATA 2021, LNCS 12638, pp. 331–342, 2021.
https://doi.org/10.1007/978-3-030-68195-1_26

Strings can be compressed by straight-line programs, which are context-free grammars whose languages contain exactly one string each. Since this grammar-based compression is able to compress some strings to logarithmic size, we are interested which polynomial time problems on uncompressed strings can also be solved in polynomial time with respect to the compressed size of the string. For example, grammar-based compression allows for fast algorithms as the fully compressed pattern matching by Jeż presented in [6]. Also, the size of the smallest grammar is comparable to other strong string compression algorithms as LZ77 (as proven simultaneously by Rytter in [11] and by Charikar et al. in [2]) and hence as the run-length encoded Burrows-Wheeler transform (as recently proven independently by Kempa and Kociumaka in [7] and by Pape-Lange in [10]).

In this paper, we will consider several variants of the k-cadence detection problem. I.e. the decision problem on whether such a k-cadence occurs in a given string.

We prove that the 3-cadence detection problem can be solved in linear time on an uncompressed binary string and in polynomial time on a grammar-compressed binary string.

Furthermore, on grammar-compressed strings, the cadence detection problem becomes $\mathcal{NP}$-complete for longer cadences or 3-cadences over a ternary alphabet.

In order to obtain these algorithms, this paper introduces two new special cases of the sub-cadence, the L-R-cadence, which starts and ends in given intervals, and the even/odd 3-sub-cadence, which starts at even/odd indices.

2 Preliminaries

A string S of length n is the concatenation $S = S[1]S[2]S[3]\dots S[n]$ of characters from an alphabet Σ. Strings naturally split into runs of equal characters. For example, the string 00010101100 splits into $000 \cdot 1 \cdot 0 \cdot 1 \cdot 0 \cdot 11 \cdot 00$. In this paper, these runs of equal characters are just called runs for the sake of simplicity.

For the sub-cadences and cadences, this paper uses the definitions of Amir et al. in [1]. These definitions are slightly different from the definition by Gardelle in [5] and by Lothaire in [9]. Funakoshi and Pape-Lange present a comparison of these definitions in [4].

Definition 1. *A k-sub-cadence is an arithmetic progression*

$$(i, i + d, \dots, i + (k - 1)d)$$

of indices given by the triple (i, d, k) of integers with $d, k > 0$ such that

$$S[i] = S[i + d] = \dots = S[i + (k - 1)d]$$

holds.

As a special case, cadences additionally have to be *structurally maximal* in the sense that neither of the extensions of the underlying arithmetic progression is contained in the integer interval $\{1, 2, 3, \dots, n\}$ anymore. More formally:

Definition 2. *A k-cadence is a k-sub-cadence (i, d, k) such that the inequalities $i - d \leq 0$ and $n < i + kd$ hold.*

In this paper, we will also consider a new special case of the sub-cadence, in which the first element and the last element of the sub-cadence are contained in given intervals:

Definition 3. *For two disjoint intervals L and R, an L-R-k-cadence is a k-sub-cadence (i, d, k) which starts in the interval L and ends in the interval R. I.e. $i \in L$ and $i + (k - 1)d \in R$ hold.*

Since the first element and the third element of each 3-sub-cadence have the same parity, it is useful to divide the L-R-3-cadences and 3-cadences according to this parity. Without loss of generality, we will only consider the even sub-cadences.

Definition 4. *An L-R-3-cadence/3-cadence is* even *if its first element is an even number and* odd *otherwise.*
    For each set M, we define $M_{even} := M \cap 2\mathbb{Z}$ and $M_{odd} := M \cap (2\mathbb{Z} + 1)$ and for each $M = \{a_1, a_2, \ldots, a_l\} \subset \mathbb{Z}$ with $1 \leq a_1 < a_2 < a_3 < \cdots < a_l \leq n$, we define the string $S[M] = S[a_1]S[a_2] \ldots S[a_l]$ as the subsequence of characters with indices given by M.
    The string S_{even} is defined by $S_{even} := S\left[\left\{2, 4, 6, \ldots, 2\left\lfloor\frac{|S|}{2}\right\rfloor\right\}\right]$.

For the compressed problems, we consider the strings to be given by straight-line grammars.

Definition 5. *A* straight-line grammar *is a context-free grammar (V, Σ, R, S) with variables $V = \{v_1, v_2, \ldots, v_i\}$ such that for each variable v_i there is exactly one rule $v_i \rightarrow u_1 u_2 \ldots u_j$ and each u_k on the right-hand side is either a character in Σ or a variable $v_{k'}$ in V with a smaller index than v_i. I.e. $k' < i$.*
    The size *of a straight-line grammar is given by the total length of the right-hand sides of the rules.*

These straight-line grammars allow on the one hand compression to logarithmic size and on the other hand fully compressed pattern matching in polynomial time with respect to the compressed sizes of the string and the pattern.

3 NP-Complete Cadence Problems

In this section, we will prove the following theorem:

Theorem 1. *If at least one of the following conditions holds, the k-cadence detection problem on compressed strings is $\mathcal{NP}$-complete:*

- $k \geq 3$ *and* $|\Sigma| \geq 2$ *and we only consider k-cadences with a given character,*
- $k \geq 3$ *and* $|\Sigma| \geq 3$ *or*
- $k \geq 4$ *and* $|\Sigma| \geq 2$.

Since we can test for a given candidate (i, d, k) of a k-cadence in polynomial time, whether (i, d, k) forms indeed a k-cadence, all three problems mentioned above belong to $\mathcal{NP}$ and it is left to show that they are $\mathcal{NP}$-hard.

To show the $\mathcal{NP}$-hardness, we will reduce the following problem, which Lohrey proves in Theorem 3.13 of [8] to be $\mathcal{NP}$-complete, to the problems above:

input: Two compressed strings P and P' over the alphabet $\{0, 1\}$.
output: Is there an index l with $P[l] = P'[l] = 1$?

Let P and P' be compressed strings over the alphabet $\{0, 1\}$. Without loss of generality, the inequality $|P'| \leq |P|$ holds. Define $P'' = (P'0^{|P|-|P'|})_{\text{rev}}$.

In this setting, for every index l, the equation $P[l] = P'[l] = 1$ holds if and only if the equation $P[l] = P''[|P| + 1 - l] = 1$ holds as well.

Consider the string

$$S = \left(0^{(k-1)|P|} \cdot P \cdot 0 \cdot 0^{k|P|}\right) \left(0^{k|P|} \cdot 1 \cdot 0^{k|P|}\right) \left(0^{k|P|} \cdot 0 \cdot P'' \cdot 0^{(k-1)|P|}\right) \left(1^{2k|P|+1}\right)^{k-3},$$

A grammar of this string can be built by the grammars of P and P' and $\mathcal{O}\left(\log(k^2|P|)\right)$ additional nonterminals. Since the compression of a string P needs at least $\Omega(\log|P|)$ nonterminals, the compressed size of S is, for fixed k, polynomial in the compressed size of the inputs.

If there is an index l with $P[l] = P''[|P| + 1 - l] = 1$, the corresponding indices are contained in the arithmetic progression starting at $i = (k-1)|P| + l$ with distance $d = 2k|P| + 1 + (|P| + 1 - l)$ and length k.

For each $-1 \leq j \leq k$ the inequality $j(2k|P| + 1) < i + jd \leq (j+1)(2k|P| + 1)$ holds. Therefore, the indices of the arithmetic progression starting at the index $i = (k-1)|P| + l$ with distance $d = 2k|P| + 1 + (|P| + 1 - l)$ and length k correspond to 1s in S and the inequalities $i - d \leq 0$, $i > 0$, $i + (k-1)d \leq n$ and $i + kd > n$ hold. Therefore, this arithmetic progression is a k-cadence.

Conversely, if the triple (i, d, k) defines a k-cadence with character 1 in S, the inequalities $i - d \leq 0 < i$ and $i + (k-1)d \leq n < i + kd$ of the cadence imply

$$\frac{j}{k}n < \frac{k-j}{k}i + \frac{j}{k}(i+kd) = i+jd = \frac{k-j-1}{k}(i-d) + \frac{j+1}{k}(i+(k-1)d) \leq \frac{j+1}{k}n.$$

In particular, the index $i + d$ has to be the single 1 in the second bracket which has the index $(2k|P| + 1) + (k|P|) + 1$. Furthermore, the first element of the k-cadence has to be a 1 in P in the first bracket and the third element of the k-cadence has to be a 1 in P'' in the third bracket.

By construction, the two indices of these characters have the same distance to the index $(2k|P| + 1) + (k|P|) + 1$, and the two strings P and P'' have the same distance to the index $(2k|P| + 1) + (k|P|) + 1$ as well. Therefore, the first element of the k-cadence and the third element of the k-cadence define an index l with $P[l] = P''[|P| + 1 - l] = 1$.

Therefore, the string S has a k-cadence with character 1 if and only if there is an index l such that $P[l] = P'[l] = 1$ holds.

If $k > 3$ holds, the requirement that the underlying character has to be 1 can be dropped since there is at least one bracket in S containing only 1s.

For 3-cadences on a ternary alphabet we consider the string

$$S = \left(0^{(k-1)|P|} \cdot P \cdot 0 \cdot 0^{k|P|}\right)\left(2^{k|P|} \cdot 1 \cdot 2^{k|P|}\right)\left(0^{k|P|} \cdot 0 \cdot P'' \cdot 0^{(k-1)|P|}\right)$$

which similarly has a 3-cadence if and only if there is an index l such that $P[l] = P'[l] = 1$ holds.

This concludes the proof of Theorem 1.

4   L-R-Cadences

The algorithm of Funakoshi and Pape-Lange in [4] counts the 3-cadences of an uncompressed string of length n in $\mathcal{O}\left(n(\log n)^2\right)$ time. This is done by counting the L-R-3-cadences in $\mathcal{O}\left((|L| + |R|)(\log(|L| + |R|))\right)$ time. It therefore seems reasonable to understand the L-R-cadences to be a simplification of cadences.

Nevertheless, in this section, we will show that all detection problems on compressed strings discussed in the last section are also $\mathcal{NP}$-complete for the L-R-cadences, even if $k = 3$ and $|\Sigma| = 2$ hold.

However, for $k = 3$ and $|\Sigma| = 2$, we will provide a detection algorithm for L-R-3-cadences which needs on uncompressed strings only $\mathcal{O}\left(|L| + |R|\right)$ time and on compressed strings polynomial time with respect to the compressed size of the string and the additional variable $\max\left(\frac{|L|}{|R|}, \frac{|R|}{|L|}\right)$.

Theorem 2. *For $k \geq 3$ and $|\Sigma| \geq 2$, the L-R-k-cadence detection problem is $\mathcal{NP}$-complete on compressed strings.*

Proof. If either $k > 3$ or $|\Sigma| > 2$ hold or if we only consider L-R-k-cadences with a given character, the proof is essentially equal to the corresponding proof in the last section, since for $L = \left\{1, 2, \ldots, \frac{1}{k}n\right\}$ and $R = \left\{\frac{k-1}{k}n + 1, \frac{k-1}{k}n + 2, \ldots, n\right\}$, all k-cadences in the discussed string S are L-R-k-cadences and vice versa.

Otherwise, we have $k = 3$, $|\Sigma| = 2$ and we consider L-R-3-cadences with any character. Since we can test for every triple (i, d, k), whether this triple forms an L-R-3-cadence, this problem belongs to $\mathcal{NP}$ and it is left to show that even this special case is $\mathcal{NP}$-hard.

Let P and P' be compressed strings over the alphabet $\{0, 1\}$. Without loss of generality, the inequality $|P'| \leq |P|$ holds. Define P'' to be the string which results from duplicating all characters of P'. For example, for $P' = 011$, we define $P'' = 001111$. This can be done by introducing two additional nonterminals.

Consider the string $S = 1(0^{|P|})(P)(P'')$ as well as the intervals $L = \{1\}$ and $R = \{1 + 2|P| + 1, 1 + 2|P| + 2, \ldots 1 + 2|P| + |P''|\}$. In this setting $S[L] = 1$ and $S[R] = P''$ holds. Furthermore, for each index $1 \leq l \leq |P|$, the equations $P[l] = S[1 + (|P| + l)]$ and $P'[l] = P''[2l] = S[1 + 2|P| + 2l] = S[1 + 2(|P| + l)]$ hold.

Therefore, for each index l, the equation $P[l] = 1 = P'[l]$ holds if and only if the equation $S[1] = S[1 + (|P| + l)] = S[1 + 2(|P| + l)]$ holds. This equation, however, defines an L-R-3-cadence.

This concludes the proof of Theorem 2.

Similarly, for two compressed strings P over $\{0,1\}$ and P' over $\{0,2\}$, we can define P'' as above and the string $S = 2(0^{|P|})(P)(P'')$ has an equidistant occurrence of the pattern 212 if and only if there is an index i with $P[i] = 1$ and $P'[i] = 2$. Therefore, equidistant subsequence matching with patterns of length 3 on compressed ternary strings is also $\mathcal{NP}$-complete.

All reductions above used that we could force all cadences to use a fixed character of the string. However, surprisingly, if L and R have similar length, we can detect in polynomial time, whether a compressed binary string has an L-R-3-cadence. Furthermore, with the same idea we can detect in $\mathcal{O}\left(|L| + |R|\right)$ time, whether an uncompressed binary string has an L-R-3-cadence.

The remainder of this section will prove the following theorem:

Theorem 3. *The problem of L-R-3-cadence detection can be done on uncompressed binary strings in $\mathcal{O}\left(|L| + |R|\right)$ time and on compressed binary strings in polynomial time with respect to the compressed size of the string and the additional variable* $\max\left(\frac{|L|}{|R|}, \frac{|R|}{|L|}\right)$.

The key insight for the detection algorithm for L-R-3-cadences is that if the string does not contain L-R-3-cadences, either $S[L_{\text{even}}]$ or $S[R_{\text{even}}]$ is very structured. The following lemma implies that if $S[L_{\text{even}}]$ has the substring 01 and $S[R_{\text{even}}]$ has the substring 10 or vice versa, then S has an L-R-3-cadence:

Lemma 1. *Let S be a binary string and L and R be two intervals.*
   If there are indices i and j with

  - $S[i] = S[j] \neq S[i+2] = S[j-2]$,
  - $i, i+2 \in L$,
  - $j, j-2 \in R$ and
  - $i \equiv j \pmod 2$,

then S has an L-R-3-cadence.

Proof. Since $i \equiv j \pmod 2$ holds, the number $\frac{i+j}{2}$ is an integer. Furthermore, since S is binary and $S[i] = S[j] \neq S[i+2] = S[j-2]$ holds, we either have $S[i] = S[\frac{i+j}{2}] = S[j]$ or $S[i+2] = S[\frac{i+j}{2}] = S[j-2]$. Therefore, there is at least one L-R-3-cadence.

This implies that if S does not contain L-R-3-cadences, then there are only few possibilities for the subsequences $S[L_{\text{even}}]$ and $S[R_{\text{even}}]$:

Corollary 1. *Let S be a binary string and L and R be two intervals such that S has no L-R-3-cadences.*
   Then, either

  - $S[L_{\text{even}}]$ *or* $S[R_{\text{even}}]$ *is of the form* 0^j *or* 1^j,
  - *both* $S[L_{\text{even}}]$ *or* $S[R_{\text{even}}]$ *are of the form* $0^i 1^j$ *or*
  - *both* $S[L_{\text{even}}]$ *or* $S[R_{\text{even}}]$ *are of the form* $1^i 0^j$.

If both $S[L_{\text{even}}]$ and $S[R_{\text{even}}]$ are of the form 0^i1^j or 1^i0^j, we can divide L and R into intervals L', L'', R' and R'' such that $S[L'_{\text{even}}]$ and $S[R'_{\text{even}}]$ are of the form 0^i and $S[L''_{\text{even}}]$ and $S[R''_{\text{even}}]$ are of the form 1^i. This can be done in $\mathcal{O}\left(|L| + |R|\right)$ time on uncompressed strings and in polynomial time on compressed strings.

By construction, all even L-R-3-cadences are either even L'-R'-3-cadences or even L''-R''-3-cadences. The following lemma, which holds by definition of the even L-R-3-cadence, shows that they can be detected in $\mathcal{O}\left(|L| + |R|\right)$ time in uncompressed strings and in polynomial time in compressed strings.

Lemma 2. *Let S be a binary string and L and R be two intervals such that $S[L_{\text{even}}] = 0^i$ and $S[R_{\text{even}}] = 0^j$ hold. Let further $l_{\min}$, $l_{\max}$, $r_{\min}$ and $r_{\max}$ be the minimal and maximal indices of L_{even} and R_{even}, respectively.*

Then, the string $S\left[\left\{\frac{l_{\min}+r_{\min}}{2}, \frac{l_{\min}+r_{\min}}{2} + 1, \ldots, \frac{l_{\max}+r_{\max}}{2}\right\}\right]$ contains a 0 if and only if S has an even L-R-3-cadence.

$S[\{1, 2, \ldots, 16\}_{\text{even}}]$:

$S[\{17, 18, \ldots, 32\}]$:

$S[\{33, 34, \ldots, 48\}_{\text{even}}]$:

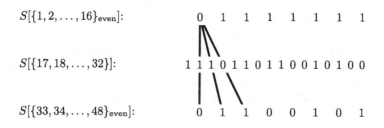

Fig. 1. A string with 48 characters. For $L = \{2\}$ and $R = \{33, 34, \ldots, 48\}$, for each index of R_{even}, there is only one candidate (i, d, k) for forming an L-R-3-cadence.

The more difficult case is that one of the two subsequences contains the substrings 01 and 10 while the other subsequence contains neither of these two substrings. Figure 1 shows that if L is a short interval, we may have to check linearly many pairs with respect to the length of the string in order to find an L-R-3-cadence. This case occurred, for example, in the string $S = 1(0^{|P|})(P)(P'')$ in which the L-R-3-cadence detection was $\mathcal{NP}$-hard. However, it turns out that even in this case, the detection of L-R-3-cadences can be done in $\mathcal{O}\left(|L| + |R|\right)$ time on uncompressed strings and using the additional variable $\max\left(\frac{|L|}{|R|}, \frac{|R|}{|L|}\right)$, the detection can also be done in polynomial time on compressed strings.

By definition of the L-R-3-cadence we get the following lemma:

Lemma 3. *Let S be a string and L and R be two intervals. Let further $S[L_{\text{even}}]$ be of the form 0^i and $l_{\min}$, $l_{\max}$, $r_{\min}$ and $r_{\max}$ be the minimal and maximal indices of L_{even} and R_{even}, respectively.*

Then, for any $r_0 \in R_{\text{even}}$ with $S[r_0] = 0$, there is an even L-R-3-cadence which uses r_0 as last element if and only if $S\left[\left\{\frac{l_{\min}+r_0}{2}, \frac{l_{\min}+r_0}{2} + 1, \ldots, \frac{l_{\max}+r_0}{2}\right\}\right]$ contains a 0.

Also, for any $m_0 \in \left\{ \frac{l_{min}+r_{min}}{2}, \frac{l_{min}+r_{min}}{2} + 1, \ldots, \frac{l_{max}+r_{max}}{2} \right\}$ with $S[m_0] = 0$, define $r'_{min} = \max\left(2m_0 - l_{max}, r_{min}\right)$ and $r'_{max} = \min\left(2m_0 - l_{min}, r_{max}\right)$. There is an even L-R-3-cadence which uses this 0 as middle element if and only if $S\left[\{r'_{min}, r'_{min} + 2, \ldots, r'_{max}\}\right]$ contains a 0.

With Corollary 1, Lemma 2 and Lemma 3, it is possible to efficiently either find an L-R-3-cadence or to shorten R without removing any L-R-3-cadences. The resulting algorithm is also presented in Fig. 2.

Corollary 2. *Let S be a binary string and L and R be two intervals. Let further $S[L_{even}]$ be of the form 0^i and l_{min}, l_{max}, r_{min} and r_{max} be the minimal and maximal indices of L_{even} and R_{even}, respectively.*

If $S[R_{even}]$ is of the form 1^j, there is no even L-R-3-cadence.

Otherwise, define $r_0 = \min\{r \in R_{even} | S[r] = 0\}$ and use Lemma 3 to check whether there is an L-R-3-cadence using an index of L_{even} and r_0.

If such an L-R-3-cadence does not exist, define $m_{min} = \frac{l_{max}+r_0}{2} + 1$ and $m_{max} = \frac{l_{max}+r_{max}}{2}$. If the substring $S\left[\{m_{min}, m_{min} + 1, \ldots, m_{max}\}\right]$ of S is of the form 1^j, then there is no even L-R-3-cadence.

Otherwise, define $m_0 = \min\{m \in \{m_{min}, m_{min} + 1, \ldots, m_{max}\} | S[m] = 0\}$. Then use Lemma 3 to check whether there is an L-R-3-cadence using an index of L_{even} and m_0.

If such an L-R-3-cadence does not exist, define $R' = R \cap \mathbb{Z}_{>2m_0 - l_{min}}$. There is an even L-R-3-cadence if and only if there is an even L-R'-3-cadence.

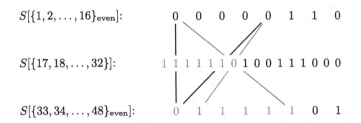

$S[\{1, 2, \ldots, 16\}_{even}]$:   0   0   0   0   0   1   1   0

$S[\{17, 18, \ldots, 32\}]$:   1 1 1 1 1 1 0 1 0 0 1 1 1 0 0 0

$S[\{33, 34, \ldots, 48\}_{even}]$:   0   1   1   1   1   1   0   1

Fig. 2. A string with 48 characters after one application of Corollary 2. Let the intervals $L = \{2, 3, \ldots, 10\}$ and $R = \{33, 34, \ldots, 48\}$ be given. First, the index $r_0 = 34$ is found. The minimal and maximal candidates for 3-cadences with r_0 are given by the black lines. Then, the index $m_0 = 23$ is found. The minimal and maximal candidates for 3-cadences with m_0 are given by the gray lines. Afterwards, the gray characters are guaranteed not to form an L-R-3-cadence with characters from the first run of the string.

In the uncompressed case, iterated application of Corollary 2 reads each character of $S\left[\left\{\frac{l_{min}+r_{min}}{2}, \frac{l_{min}+r_{min}}{2} + 1, \ldots, \frac{l_{max}+r_{max}}{2}\right\}\right]$ and $S[R_{even}]$ at most once. It is therefore easy to see that iterated application of this corollary only uses $\mathcal{O}(|L| + |R|)$ time to decide, whether an L-R-3-cadence exists.

On the other hand, after the application of Corollary 2 either R' is empty or it contains at least $|L|$ elements less than R. Therefore, the algorithm described in Corollary 2 has to be used at most $\mathcal{O}\left(\frac{|R|}{|L|}\right)$ times. Since each application of Corollary 2 only takes polynomial time on compressed strings, the L-R-3-cadence detection on compressed strings can be done in polynomial time with respect to the compressed size of the string and the additional variable $\max\left(\frac{|L|}{|R|}, \frac{|R|}{|L|}\right)$.

Since by symmetry, the detection of odd L-R-3-cadences can also be done as the detection of even L-R-3-cadences, this concludes the proof of Theorem 3.

5   3-Cadences in Binary Strings

In this section, we will show that the results of Theorem 3 also hold for the corresponding 3-cadence problems:

Theorem 4. *The 3-cadence detection problem can be solved in linear time on uncompressed binary strings and in polynomial time on compressed binary strings.*

Let i, d be two integers such that $i - d \leq 0$ and $i + 3d > n$ hold. Let $L = \{1, 2, \ldots, i\}$ and $R = \{i + 2d, i + 2d + 1, \ldots, n\}$ be two intervals. Then each L-R-3-cadence is also a 3-cadence. On the other hand, each 3-cadence defines integers i and d such that $i - d \leq 0$ and $i + 3d > n$ hold. Therefore, Lemma 1 implies that if S has an even 3-cadence, it also has a 3-cadence that either starts in one of the first two runs of S_{even} or ends in one of the last two runs of S_{even}.

The main challenge for the adaption of the detection algorithm for L-R-3-cadences to an detection algorithm for 3-cadences is that while each character in the first/last third of the string can be the first/last index in a 3-cadence, not all 3-sub-cadences which start in the first third and end in the last third are 3-cadences. For example, in the string 001001001 the three 1s form a 3-cadence while the in the string 001010100 the three 1s do not form a 3-cadence. Therefore, Lemma 3 does not quite work on 3-cadences and we have to restrict the strings in Lemma 3 to those indices such that the corresponding 3-sub-cadences are structurally maximal.

The following lemma restricts the bound of Lemma 3 to the allowed indices for 3-cadences with $i - d \leq 0$ and $i + kd > n$ and therefore holds by definition of the 3-cadence:

Lemma 4. *Let S be a string. Let further S_{even} be of the form $0^i S'$ and let $l_{\min} = 2$ and $l_{\max} = \max\left(2i, 2\left\lfloor \frac{1}{6}n\right\rfloor\right)$ be the first and the last indices of the first run of S_{even} which can be the first index of a 3-cadence, respectively.*

Then, for any $r_0 \in S_{\text{even}}$ define $l'_{\max} = \min\left(l_{\max}, 2\left\lfloor \frac{r_0}{6}\right\rfloor, 2\left(\left\lceil \frac{3r_0 - 2n}{2}\right\rceil - 1\right)\right)$. There is an even 3-cadence which uses this r_0 as last element and any of the first i 0s of $S[L_{\text{even}}]$ as first element if and only if $S[r_0] = 0$ holds and the string $S\left[\left\{\frac{l_{\min} + r_0}{2}, \frac{l_{\min} + r_0}{2} + 1, \ldots, \frac{l'_{\max} + r_0}{2}\right\}\right]$ contains a 0.

Conversely, for any $m_0 \in \left\{ \frac{l_{\min}+2\lfloor\frac{n}{3}\rfloor+2}{2}, \frac{l_{\min}+2\lfloor\frac{n}{3}\rfloor+2}{2} + 1, \ldots, \frac{l_{\max}+2\lfloor\frac{n}{2}\rfloor}{2} \right\}$
with $S[m_0] = 0$ *define* $r'_{\min} = 2m_0 - \min\left(l_{\max}, 2\lfloor\frac{m_0}{4}\rfloor, 2\left(\lceil\frac{3m_0-n}{4}\rceil - 1\right)\right)$ *and*
$r'_{\max} = 2m_0 - \max\left(l_{\min}, 2\left(m_0 - \lfloor\frac{n}{2}\rfloor\right)\right)$. *There is an even 3-cadence which uses*
m_0 *as middle element if and only if* $S\left[\{r'_{\min}, r'_{\min}+2, \ldots, r'_{\max}\}\right]$ *contains a* 0.

Similarly to the case of the L-R-3-cadence, we can use this lemma to shrink the interval in which the last element of the arithmetic progression can be.

Corollary 3. *Let S be a binary string. Define $R = \{r_{\min}, r_{\min} + 1, \ldots, n\}$ for an $r_{\min} \geq \lfloor\frac{2}{3}n\rfloor + 1$. Let further S_{even} be of the form $0^i S'$ and let $l_{\min} = 2$ and $l_{\max} = \max\left(2i, 2\lfloor\frac{1}{6}n\rfloor\right)$ be the first and the last indices of the first run of S_{even} which can be the first index of a 3-cadence, respectively. Define the interval $L = \{l_{\min}, l_{\min} + 1, \ldots, l_{\max}\}$.*

If $S[R_{\text{even}}]$ is of the form 1^j, there is no even 3-cadence starting in L.

Otherwise, define $r_0 = \min\{r \in R_{\text{even}} | S[r] = 0\}$ and use Lemma 4 to check whether there is a 3-cadence starting in L and ending with r_0.

If such a 3-cadence does not exist, define $m_{\min} = \frac{l_{\max}+r_0}{2} + 1$ with the variable $l'_{\max} = \min\left(l_{\max}, 2\lfloor\frac{r_0}{6}\rfloor, 2\left(\lceil\frac{3r_0-2n}{2}\rceil - 1\right)\right)$ as defined in Lemma 4 and define $m_{\max} = \frac{l_{\max}+n}{2}$. If $S[\{m_{\min}, m_{\min} + 1, \ldots, m_{\max}\}]$ is of the form 1^j, then there is no even 3-cadence starting in L.

Otherwise, define $m_0 = \min\{m \in \{m_{\min}, m_{\min} + 1, \ldots, m_{\max}\} | S[m] = 0\}$. Then use Lemma 4 to check whether there is a 3-cadence starting in L with m_0 as second element.

If such a 3-cadence does not exist, define $R' = R \cap \mathbb{Z}_{>r'_{\max}}$ with the variable $r'_{\max} = 2m_0 - \max\left(l_{\min}, 2\left(m_0 - \lfloor\frac{n}{2}\rfloor\right)\right)$ as defined in Lemma 4. There is an even 3-cadence starting in L if and only if there is an even 3-cadence starting in L and ending in R'.

An application of this corollary can be seen in Fig. 3.

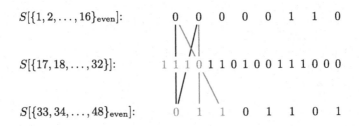

$S[\{1, 2, \ldots, 16\}_{\text{even}}]$:

$S[\{17, 18, \ldots, 32\}]$:

$S[\{33, 34, \ldots, 48\}_{\text{even}}]$:

Fig. 3. A string with 48 characters after one application of Corollary 3. First, the index $r_0 = 34$ is found. The minimal and maximal candidates for 3-cadences with r_0 are given by the black lines. Then, the index $m_0 = 20$ is found. The minimal and maximal candidates for 3-cadences with m_0 are given by the gray lines. Afterwards, the gray characters are guaranteed not to form a 3-cadence with characters from the first run of S_{even}.

In the uncompressed case, each element of the middle third and the last third has to be read at most once in order to decide whether there is a 3-cadence which starts in the first run of S_{even}. Furthermore, we can modify this algorithm to detect the existence of a 3-cadence which start in the second run of S_{even}. By symmetry, we can also decide in linear time, whether there exists a 3-cadence which ends in one of the two last runs of S_{even}. Similarly, we can decide in linear time, whether there is an odd 3-cadence.

In the compressed case, a problem can arise if the first run of S_{even} is short. Let S_{even} be of the form $0^i 1 S'$. Then the 1 has index $2i + 2$. Let r_1 be the smallest even index such that $(2i+2) - \frac{r_1 - (2i+2)}{2} \leq 0$ and $(2i+2) + 3\frac{r_1 - (2i+2)}{2} > n$ hold. Since $S[\{2, 4, 6, \ldots, 2i + 2\}]$ contains 01, we can use Corollary 1 and Lemma 2 to check in polynomial time, whether there is a 3-sub-cadence starting in $\{2, 4, 6, \ldots, 2i + 2\}$ and ending with an index greater than or equal to r_1. By construction, such a 3-sub-cadence would be a 3-cadence.

If such a 3-cadence exists, we are done. Therefore, it is only left to show that even in the compressed case, the application of Corollary 3 is fast enough to find a 3-cadence which start in the first run of S_{even} and end at an index smaller than r_1 in polynomial time if such a cadence exists. Since each application of Corollary 3 can be done in polynomial time, it is left to show that after a polynomial number of applications, the value $r'_{\max}$ is greater than r_1.

In the worst case, each r_0 is $r_{\min}$. Since each 3-sub-cadences with distance of at least $\frac{1}{3}n$ is a 3-cadence, we can assume $r_0 < r_1 \leq 2i + 2 + \frac{2}{3}n$ holds and therefore $l_{\max} \geq r_0 - \frac{2}{3}n$ holds as well. Also, both $2\lfloor \frac{r_0}{6} \rfloor$ and $2(\lceil \frac{3r_0 - 2n}{2} \rceil - 1)$ are greater than or equal to $r_0 - \frac{2}{3}n$. Therefore $l'_{\max} \geq r_0 - \frac{2}{3}n$ holds.

Similarly, in the worst case, each m_0 is directly behind $\frac{l'_{\max} + r_0}{2} \geq r_0 - \frac{1}{3}n$. Hence, we can assume $m_0 = 2r_0 - \frac{1}{3}n + 1$. With $l''_{\min} = \max\left(l_{\min}, 2\left(m_0 - \lfloor \frac{n}{2} \rfloor\right)\right)$, this implies that the inequality $2m_0 - l''_{\min} \geq \min\left(r_0 + (r_0 - \frac{2}{3}n), n - 1\right)$ holds.

Therefore, for $r_0 < r_1$, one application of Corollary 3 checks for an interval of size $r_0 - \frac{2}{3}n$, whether there is 3-cadence which starts in the first run and ends in this interval. Therefore, we only need at most $\log n$ applications of this corollary.

This implies that it can be decided in polynomial time whether a compressed binary string contains any 3-cadences.

6   Conclusion

This paper shows that we can decide in linear time whether an uncompressed binary string contains a 3-cadence. While we should expect that it is more difficult to avoid 3-cadences in binary strings than to include 3-cadences, it is surprising that it is strictly easier to decide whether there is any 3-cadence at all than to decide whether there is a 3-cadence with a given character.

For the compressed case, we have shown that we can decide in polynomial time whether a compressed binary string contains a 3-cadence. However, all even slightly harder problems have been shown to be $\mathcal{NP}$-complete. These hardness-results seem to indicate that cadences may not be very useful in compressed pattern matching.

Regarding k-sub-cadences, there are no known nontrivial bounds on the bit complexity of the detection of k-sub-cadences with a given character. Closely related, it is unknown whether equidistant subsequence matching is $\mathcal{NP}$-hard on compressed binary strings.

Finally, in terms of uncompressed cadence detection, it is still unknown whether we can decide with sub-quadratic bit complexity whether a given string contains a 4-cadence. The currently best result is by Funakoshi et al., who presented in [3] a detection algorithm with sub-quadratic time complexity in the word RAM model.

References

1. Amir, A., Apostolico, A., Gagie, T., Landau, G.M.: String cadences. Theor. Comput. Sci. **698**, 4–8 (2017). https://doi.org/10.1016/j.tcs.2017.04.019
2. Charikar, M., et al.: The smallest grammar problem. IEEE Trans. Inf. Theor. **51**(7), 2554–2576 (2005). https://doi.org/10.1109/TIT.2005.850116
3. Funakoshi, M., Nakashima, Y., Inenaga, S., Bannai, H., Takeda, M., Shinohara, A.: Detecting k-(Sub-)cadences and equidistant subsequence occurrences. In: Gørtz, I.L., Weimann, O. (eds.) 31st Annual Symposium on Combinatorial Pattern Matching (CPM 2020). Leibniz International Proceedings in Informatics (LIPIcs), vol. 161, pp. 12:1–12:11. Schloss Dagstuhl-Leibniz-Zentrum für Informatik, Dagstuhl, Germany (2020). https://doi.org/10.4230/LIPIcs.CPM.2020.12
4. Funakoshi, M., Pape-Lange, J.: Non-rectangular convolutions and (Sub-)cadences with three elements. In: Paul, C., Bläser, M. (eds.) 37th International Symposium on Theoretical Aspects of Computer Science (STACS 2020). Leibniz International Proceedings in Informatics (LIPIcs), vol. 154, pp. 30:1–30:16. Schloss Dagstuhl-Leibniz-Zentrum fuer Informatik, Dagstuhl, Germany (2020). https://doi.org/10.4230/LIPIcs.STACS.2020.30
5. Gardelle, J.: Cadences. Mathématiques et Sci. Humaines **9**, 31–38 (1964). http://www.numdam.org/item/MSH_1964__9__31_0
6. Jeż, A.: Faster fully compressed pattern matching by recompression. ACM Transactions on Algorithms 11(3), Jan 2015. https://doi.org/10.1145/2631920
7. Kempa, D., Kociumaka, T.: Resolution of the burrows-wheeler transform conjecture. CoRR abs/1910.10631. Accepted to the 61st Annual Symposium Foundations of Computer Science (FOCS 2020) (2019). http://arxiv.org/abs/1910.10631
8. Lohrey, M.: Algorithms on compressed words. The Compressed Word Problem for Groups. SM, pp. 43–65. Springer, New York (2014). https://doi.org/10.1007/978-1-4939-0748-9_3
9. Lothaire, M.: Combinatorics on Words. Cambridge Mathematical Library, Cambridge University Press, Cambridge (1997). https://doi.org/10.1017/CBO9780511566097
10. Pape-Lange, J.: On extensions of maximal repeats in compressed strings. In: Gørtz, I.L., Weimann, O. (eds.) 31st Annual Symposium on Combinatorial Pattern Matching, CPM 2020, June 17–19, 2020, Copenhagen, Denmark. LIPIcs, vol. 161, pp. 27:1–27:13. Schloss Dagstuhl - Leibniz-Zentrum für Informatik (2020). https://doi.org/10.4230/LIPIcs.CPM.2020.27
11. Rytter, W.: Application of Lempel-Ziv factorization to the approximation of grammar-based compression. Theor. Comput. Sci. **302**(1–3), 211–222 (2003). https://doi.org/10.1016/S0304-3975(02)00777-6
12. Beweis einer Baudet'schen Vermutung: Waerden, B.L.v.d. Nieuw Archief voor Wiskunde **15**, 212–216 (1927)

Author Index

Printed in the United States
by Baker & Taylor Publisher Services